Communications
in Computer and Information Sci

T0238534

Kang Li   Yusheng Xue
Shumei Cui   Qun Niu (Eds.)

Intelligent Computing in Smart Grid and Electrical Vehicles

International Conference on Life System Modeling
and Simulation, LSMS 2014
and International Conference on Intelligent Computing
for Sustainable Energy and Environment, ICSEE 2014
Shanghai, China, September 20-23, 2014
Proceedings, Part III

 Springer

Volume Editors

Kang Li
Queen's University Belfast, UK
E-mail: k.li@qub.ac.uk

Yusheng Xue
State Grid Electric Power Research Institute, Jiangsu, China
E-mail: xueyusheng@sgepri.sgcc.com.cn

Shumei Cui
Harbin Institute of Technology, China
E-mail: cuism@hit.edu.cn

Qun Niu
Shanghai University, China
E-mail: nq@shu.edu.cn

ISSN 1865-0929   e-ISSN 1865-0937
ISBN 978-3-662-45285-1   e-ISBN 978-3-662-45286-8
DOI 10.1007/978-3-662-45286-8
Springer Heidelberg New York Dordrecht London

Library of Congress Control Number: 2014951655

Typesetting: Camera-ready by author, data conversion by Scientific Publishing Services, Chennai, India

Printed on acid-free paper

Springer is part of Springer Science+Business Media (www.springer.com)

Preface

The 2014 International Conference on Life System Modeling and Simulation (LSMS 2014) and 2014 International Conference on Intelligent Computing for Sustainable Energy and Environment (ICSEE 2014), which were held during September 20–23, in Shanghai, China, aimed to bring together international researchers and practitioners in the field of life system modeling and simulation as well as intelligent computing theory and methodology with applications to sustainable energy and environment. These events built on the success of previous LSMS conferences held in Shanghai and Wuxi in 2004, 2007, and 2010, and ICSEE conferences held in Wuxi and Shanghai in 2010 and 2012, and are based on large-scale RCUK/NSFC jointly funded UK–China collaboration projects on energy.

At LSMS 2014 and ICSEE 2014, technical exchanges within the research community take the form of keynote speeches, panel discussions, as well as oral and poster presentations. In particular, two workshops, namely, the Workshop on Integration of Electric Vehicles with Smart Grid and the Workshop on Communication and Control for Distributed Networked Systems, were held in parallel with LSMS 2014 and ICSEE 2014, focusing on the two recent hot topics on smart grid and electric vehicles and distributed networked systems for the Internet of Things.

The LSMS 2014 and ICSEE 2014 conferences received over 572 submissions from 13 countries and regions. All papers went through a rigorous peer review procedure and each paper received at least three review reports. Based on the review reports, the Program Committee finally selected 159 high-quality papers for presentation at LSMS 2014 and ICSEE 2014. These papers cover 24 topics, and are included into three volumes of CCIS proceedings published by Springer. This volume of CCIS includes 56 papers covering 8 relevant topics.

Shanghai is one of the most populous, vibrant, and dynamic cities in the world, and has contributed significantly toward progress in technology, education, finance, commerce, fashion, and culture. Participants were treated to a series of social functions, receptions, and networking sessions, which served to build new connections, foster friendships, and forge collaborations.

The organizers of LSMS 2014 and ICSEE 2014 would like to acknowledge the enormous contributions made by the following: the Advisory Committee for their guidance and advice, the Program Committee and the numerous referees for their efforts in reviewing and soliciting the papers, and the Publication Committee for their editorial work. We would also like to thank the editorial team from Springer for their support and guidance. Particular thanks are of course due to all the authors, as without their high-quality submissions and presentations the conferences would not have been successful.

Finally, we would like to express our gratitude to:

The Chinese Association for System Simulation (CASS)
IEEE Systems, Man and Cybernetics Society (SMCS) Technical Committee on
    Systems Biology
IET China
IEEE CIS Adaptive Dynamic Programming and Reinforcement Learning
    Technical Committee
IEEE CIS Neural Network Technical Committee
IEEE CC Ireland Chapter and IEEE SMC Ireland Chapter
Shanghai Association for System Simulation
Shanghai Instrument and Control Society
Shanghai Association of Automation
Shanghai University
Queen's University Belfast
Life System Modeling and Simulation Technical Committee of CASS
Embedded Instrument and System Technical Committee of China
    Instrument and Control Society
Central Queensland University
Harbin Institute of Technology
China State Grid Electric Power Research Institute
Cranfield University

September 2014

Bo Hu Li
George W. Irwin
Mitsuo Umezu
Minrui Fei
Kang Li
Qinglong Han
Shiwei Ma
Sean McLoone
Luonan Chen

Organization

Sponsors

Chinese Association for System Simulation (CASS)
IEEE Systems, Man and Cybernetics (SMCS) Technical Committee on Systems Biology
IET China

Organizers

Shanghai University
Queen's University Belfast
Life System Modeling and Simulation Technical Committee of CASS
Embedded Instrument and System Technical Committee of China Instrument and Control Society

Co-Sponsors

IEEE CIS Adaptive Dynamic Programming and Reinforcement Learning Technical Committee
IEEE CIS Neural Network Technical Committee
IEEE CC Ireland Chapter and IEEE SMC Ireland Chapter
Shanghai Association for System Simulation
Shanghai Instrument and Control Society
Shanghai Association of Automation

Co-Organizers

Central Queensland University
Harbin Institute of Technology
China State Grid Electric Power Research Institute
Cranfield University

Honorary Chairs

Li, Bo Hu (China)
Irwin, George W. (UK)
Umezu, Mitsuo (Japan)

Consultant Committee Members

Cheng, Shukang (China)
Hu, Huosheng (UK)
Pardalos, Panos M. (USA)
Pedrycz, Witold (Canada)
Scott, Stan (UK)
Wu, Cheng (China)

Xi, Yugeng (China)
Xiao, Tianyuan (China)
Xue, Yusheng (China)
Yeung, Daniel S. (Hong Kong, China)
Zhao, Guangzhou (China)

General Chairs

Fei, Minrui (China)

Li, Kang (UK)

International Program Committee

Chairs

Han, Qinglong (Australia)
Ma, Shiwei (China)

McLoone, Sean (UK)
Chen, Luonan (Japan)

Local Chairs

Chen, Ming (China)
Chiu, Min-Sen (Singapore)
Ding, Yongsheng (China)
Fan, Huimin (China)
Foley, Aoife (UK)
Gu, Xingsheng (China)
He, Haibo (USA)
Li, Tao (China)

Luk, Patrick (UK)
Park, Poogyeon (Korea)
Peng, Chen (China)
Su, Zhou (Japan)
Yang, Taicheng (UK)
Zhang, Huaguang (China)
Yang, Yue (China)

Members

Albertos, Pedro (Spain)
Cai, Zhongyong (China)
Cao, Jianan (China)
Chang, Xiaoming
  (China)
Chen, Guochu (China)
Chen, Jing (China)
Chen, Qigong (China)
Chen, Weidong (China)

Chen, Rongbao (China)
Cheng, Wushan (China)
Deng, Jing (UK)
Deng, Li (China)
Ding, Zhigang (China)
Du, Dajun (China)
Du, Xiangyang (China)
Emmert-Streib (Frank)
  (UK)

Fu, Jingqi (China)
Gao, Shouwei (China)
Gao, Zhinian (China)
Gu, Juping (China)
Gu, Ren (China)
Han, Liqun (China)
Han, Xuezheng (China)
Harkin-Jones, Eileen
  (UK)

He, Jihuan (China)
Hu, Dake (China)
Hu, Guofen (China)
Hu, Liangjian (China)
Hu, Qingxi (China)
Jiang, Ming (China)
Jiang, Ping (China)
Kambhampati,
    Chandrasekhar (UK)
Keane, Andrew (Ireland)
Konagaya, Akihiko
    (Japan)
Lang, Zi-Qiang (UK)
Li, Donghai (China)
Li, Gang (China)
Li, Guozheng (China)
Li, Tongtao (China)
Li, Wanggen (China)
Li, Xin (China)
Li, Xinsheng (China)
Li, Zhicheng (China)
Lin, Haiou (China)
Lin, Jinguo (China)
Lin, Zhihao (China)
Liu, Guoqiang (China)
Liu, Mandan (China)
Liu, Tingzhang (China)
Liu, Wanquan
    (Australia)
Liu, Wenbo (China)
Liu, Xinazhong (China)
Liu, Zhen (China)
Luo, Wei (China)
Luo, Pi (China)

Luo, Qingming (China)
Maione, Guido (Italy)
Man, Zhihong
    (Australia)
Marion, McAfee
    (Ireland)
Naeem, Wasif (UK)
Ng, Wai Pang (UK)
Nie, Shengdong (China)
Ochoa, Luis (UK)
Ouyang, Mingsan
    (China)
Piao, Xiongzhu (China)
Prasad, Girijesh (UK)
Qian, Hua (China)
Qu, Guoqing (China)
Ren, Wei (China)
Rong, Qiguo (China)
Shao, Chenxi,(China)
Shen, Chunshan (China)
Shen, Jingzi (China)
Song, Zhijian (China)
Song, Yang (China)
Sun, Guangming (China)
Sun, Xin (China)
Teng, Huaqiang (China)
Tu, Xiaowei (China)
Verma, Brijesh,
    (Australia)
Wang, Lei (China)
Wang, Ling (China)
Wang, Mingshun (China)
Wei, Hua-Liang (UK)
Wei, Kaixia (China)

Wen, Guihua (China)
Wen, Tieqiao (China)
Wu, Jianguo (China)
Wu, Lingyun (China)
Wu, Xiaofeng (China)
Wu, Zhongcheng (China)
Xi, Zhiqi (China)
Xu, Daqing (China)
Xu, Sheng (China)
Xu, Zhenyuan (China)
Xue, Dong (China)
Yang, Hua (China)
Yang, Wankou (China)
Yang, Yi (China)
Yao, Xiaodong (China)
Yu, Ansheng (China)
Yu, Jilai (China)
Yuan, Jingqi (China)
Yue, Dong (China)
Zhang, Bingyao (China)
Zhang, Peijian (China)
Zhang, Qianfan (China)
Zhang, Quanxing
    (China)
Zhang, Xiangfeng
    (China)
Zhao, Wanqing (UK)
Zheng, Qingfeng (China)
Zhou, Huiyu (UK)
Zhou, Wenju (China)
Zhu, Qiang (China)
Zhu, Xueli (China)
Zhuo, Jiangang (China)
Zuo, Kaizhong (China)

Organizing Committee

Chairs

Jia, Li (China)
Wu,Yunjie (China)

Cui, Shumei (China)
Laverty, David (UK)

Members

Sun, Xin (China)
Liu, Shixuan (China)   Niu, Qun (China)

Special Session Chairs

Wang, Ling (China)   Zhang, Qianfan (China)
Ng, Wai Pang (UK)   Yu, Jilai (China)

Publication Chairs

Zhou, Huiyu (UK)   Li, Xin (China)

Publicity Chairs

Wasif, Naeem (UK)   Deng, Li (China)
Deng, Jing (UK)

Secretary-General

Sun, Xin (China)   Niu, Qun (China)
Liu, Shixuan (China)

Registration Chairs

Song, Yang (China)   Du, Dajun (China)

Table of Contents

The First Section: Computational Intelligence in Utilization of Clean and Renewable Energy Resources, Including Fuel Cell, Hydrogen, Solar and Winder Power, Marine and Biomass

The Second Section: Intelligent Modeling, Control and Supervision for Energy Saving and Pollution Reduction

The Third Section: Intelligent Methods in Developing Electric Vehicles, Engines and Equipment

The Fourth Section: Intelligent Computing and Control in Distributed Power Generation Systems

The Fifth Section: Intelligent Modeling, Simulation and Control of Power Electronics and Power Networks

The Sixth Section: Intelligent Road Management and Electricity Marketing Strategies

The Seventh Section: Intelligent Water Treatment and Waste Management Technologies

The Eighth Section: Integration of Electric Vehicles with Smart Grid

Wind Power Short-Term Prediction Method Based on Multivariable Mutual Information and Phase Space Reconstruction

Lu-Jie Liu[1,2], Yang Fu[2], and Shi-Wei Ma[1]

[1] School of Mechatronics Engineering and Automation, Shanghai University, Zhabei District, Shanghai 200072, China
[2] Department of Electrical Engineering, Shanghai University of Electric Power, Yangpu District, Shanghai 200090, China

Abstract. Wind power is influenced by multivariable, which usually shows complex nonlinear dynamics. Therefore the wind power is hardly described and traced by single variable prediction model; the precision of which decreases while it contains uncorrelated or redundant variables. The approach is proposed to reconstruct the phase space of multivariable time series and then predicate wind power. First, the delay time of single variable time series is selected by mutual information entropy, and then the embedding dimension of phase space is extended by the false nearest neighborhood method, which can eliminate the redundancy of reconstructed phase space from low space to high space. Then, the vector is utilized as input to predicate the wind power using the radial basis function neural network. Simulation of wind predication of Shanghai wind farms, show that the proposed method can describe the nonlinear system by less variables, and improve the precision and sensitivity of prediction.

Keywords: wind power, mutual information, multivariate chaotic time series, phase space reconstruction, radial basis function neural network.

1   Introduction

The main characteristic of wind power is fluctuation, intermittence and randomness. A large number of wind farms integrate into the power network, which will affect the safety, stability and economic operation of power network. Wind power prediction is one of useful and economical means to promote the ability of rating peak load of power network, the capability of integrated wind power, the availability of wind power and reasonable arrangement of maintenance for wind power farm.

The model for predicting wind power can be classified into three main groups, which is physical, statistical and combined model. In physical model, the numerical meteorological prediction model for the complicated topography of wind farm is established. According to the power curve of wind farm (or wind turbine), wind power is calculated. The point of the model is the precise numerical weather forecast, furthermore, meteorological knowledge, the collection and description of complicated topography is needed as the important part of the model. In the statistical model, linear

K. Li et al. (Eds.): LSMS/ICSEE 2014, Part III, CCIS 463, pp. 1–12, 2014.
© Springer-Verlag Berlin Heidelberg 2014

and nonlinear theory are adopted to forecast wind speed or power in term of historical data which are formulated by the real-time data of the wind speed, direction and output power. The main methods include constant method、 autoregressive moving average model、 artificial neural network, fuzzy, chaos analysis and so on [5-8]. Either model above does not meet the need of precise precision very well; therefore, the combined model is adopted.

The error of wind power prediction is caused by the error of weather prediction and the wind power prediction model. Two methods are usually used, in one method the historical power data is used to forecasts wind power directly, while in the other method, the historical wind speed data is firstly used to forecast wind speed and then wind power is calculated by the power curve of wind farm. The single variable time series prediction of the wind speed or power is researched, however the information of the short single variable time series caused by the limited length is not sufficient for prediction and the precision precision is also decreased while it contains uncorrelated or redundant variables which induce to noise data.

According to the nonlinear characteristic of the wind power, this article provides an approach to predicate short wind power in which the wind power is taken as a multi-dimension nonlinear chaotic system affected by multi-variable, combined the mutual information, phase space reconstruction and radial basis function (RBF) neural network. The theory analysis, method description, simulation result and conclusion are developed in this paper.

2   Basic Theories

2.1   Information Entropy and Mutual Information Method

Suppose two discrete random variables $X \in \{x_1, x_2, \cdots x_n\}$ and $Y \in \{y_1, y_2, \cdots y_n\}$, $p(x_i)$ is the probability of the x_i in X, and $p(y_i)$ is the probability of the y_i in Y, The information entropy is defined as

$$H(X) = -\sum_{i=1}^{n} p(x_i) \log p(x_i) \tag{1}$$

Where, $p(x_i, y_i)$ is the joint probability of x_i and y_i in the two series. The joint entropy of X and Y is defined as.

$$H(X,Y) = -\sum_{i=1}^{n}\sum_{j=1}^{k} p(x_i, y_j) \log p(x_i, y_j) \tag{2}$$

Therefore, the mutual information can be calculated by

$$I(X;Y) = H(X) + H(Y) - H(X,Y) \tag{3}$$

The mutual information reflects the mutual dependence of two variables. The larger value indicates the strong coupling of two variables, which is the X includes more information of the Y, and the correlation between X and Y is large. The mutual information method can measure the nonlinear random correlation of the time series, which is suitable for high dimensional nonlinear system.

2.2   Phase Space Reconstruction of the Multivariate Chaotic Time Series

In the nonlinear system, any evolvement of the component is decided by other components which are interaction, and the relative information of the component is contained in the evolvement of other components. Suppose that the M dimensions chaotic system

$$Y_{n+1} = F(Y_n) \tag{4}$$

Where, $Y_n \in \Re_M$ is the state variable, $F : \Re_M \to \Re_M$ is a Smooth and continuous function. Let the M dimensions multivariate time series are X_1, X_2, ..., X_N, where $X_i = [x_{1,\ i}, x_{2,\ i}, ..., x_{M,\ i}]$, $i = 1$, 2, ..., N, N refers to the number of variables, X_i indicates the time series of the ith variable.

According to the Takens theory , if the delay time τ_j and embedded dimension m_j of each variable are selected appropriately, the deterministic mapping is described as

$$F^m : \Re^m \to \Re^m, \quad X_{n+1} = F^m(X_n) \tag{5}$$

For the formulation (4) and (5) has same dynamic characteristic, the phase space reconstruction of the multivariate time series is expressed

$$X_n = [x_{1,n}, x_{1,n-\tau_1}, \cdots, x_{1,n-(m_1-1)\tau_1}, \cdots, x_{M,n}, x_{M,n-\tau_M}, \cdots, x_{M,n-(m_M-1)\tau_M}, x_{1,j}, x_{1,j-\tau_1}, \cdots, x_{1,j-(m_1-1)\tau_1}, \cdots, x_{M,j}, x_{M,j-\tau_M}, \cdots, x_{M,j-(m_M-1)\tau_M}; \tag{6}$$
$$x_{1,N}, x_{1,N-\tau_1}, \cdots, x_{1,N-(m_1-1)\tau_1}, \cdots, x_{M,N}, x_{M,N-\tau_M}, \cdots, x_{M,N-(m_M-1)\tau_M}]$$

Where phase number is denoted by $n = \max_{1 \le i \le M}((m_j - 1)\tau_j) + 1$, $j = 1$, 2, ..., M, the system embedded dimension is $m = m_1 + m_2 + \cdots + m_M$. When $M = 1$, the series converts into single variable time series, which indicates that is the special case of the multivariate time series.

2.3   Radial Basis Function

Radial Basis Function (RBF) is developed by J.Moody and C.Darken in the late 1980s. it is a type of neural network structure, which has three layers feed forward network with single hidden layer. Suppose $x = [x_1, x_2, \cdots, x_n]^T$ is the input vector, $G = [g_1(x, c_1), g_2(x, c_2), \cdots, g_k(x, c_k)]^T$ is the hidden layer output matrix, $y = [y_1, y_2, \cdots, y_m]$ is output vector, $W = [w_1, w_2, \cdots, w_k]^T$ is weight matrix connecting the hidden layer and output layer, and $w_i = [w_{i1}, w_{i2}, \cdots, w_{in}]^T (i = 1, 2, \cdots, k)$. Therefore, the output of the RBF is described as

$$y_j = \sum_{i=1}^{k} w_{ij} g_i(x), (j = 1, 2, \cdots, n) \tag{7}$$

Where, activation function of hidden layer is radial basis function, in which Gaussian distribution is usually selected. The characteristic of the RBF network structure is adaptive, fast calculating, which can approach any continues function in any precision and suit for the prediction of the nonlinear time series [14-15].

3   Wind Power Prediction model Based on Multivariable Mutual Information and Phase Space Reconstruction

Short-term wind power affected by the wind characteristic such as wind speed and direction presents the randomness and uncertainty. However the fluctuation of wind power overall is presented regularity, which appears the dynamic character of the multivariable. The nonlinear characteristic of the prediction system can't be sufficiently described by single variable; meanwhile excess variable will increase the redundancy of the model.

According to the nonlinear character of the wind power, the prediction is treated as a multi dimension nonlinear chaotic system affected by multivariable in this paper, and wind power prediction model based on multivariable mutual information and phase space reconstruction is proposed. The multivariable influencing the wind power prediction is selected adopting the mutual information method, and then the multivariable time series reduction set is reconstructed using the chaos theory. Combing with the RBF network which has simply structure, strong approaching ability and fast convergent rate, the phase space reconstruction is used as the input of the RBF network, and the wind power prediction model is established. The process of this method is described as below.

(1) The mapping F from the ith state to the $i+1$th state is established based on the historical multivariable state data set. Adopting the mutual information method, which is used to select variables and reduce the weak variables, the feature variable mapping F' is formed. At last, according to the mapping F' and the latest state X_n, the estimate value Y_{n+1} of the wind power state X_{n+1} is deduced.

(2) The following method is used to select the mutual information multivariable characteristic. Suppose all the influencing factors of wind power is treated as a set $X'_{i,n}$, prediction wind power set is Y'_n, therefore the optimal prediction variable subset should merely contain the variables that have correlation with power prediction, the $X'_{i,n}$ in S satisfies the equations.

$$I(X'_{i,n};Y'_n) > \alpha I(Y'_n;Y'_n), \min \frac{I(X'_{i,n};Y'_n)}{I(Y'_n;Y'_n)} \le \alpha \le \max \frac{I(X'_{i,n};Y'_n)}{I(Y'_n;Y'_n)} \tag{8}$$

Where α is the correlation thresh hold, which represents the $X'_{i,n}$ contains the information about Y'_n on multiple of α. The variables unsatisfied with above formula (irrelevant variable and less information contained variable) will be reduced in S, and the relevant variable set F is formed. In theory, the optimal variable subset S should not contain the weak relevant variables, which will lead to the redundancy of the variables, so the set F needs to be reduced further. Reducing $X'_{i,n}$ step by step, the mutual information changing of set F and Y'_n should satisfy the formulation.

$$I(F;Y'_n)-I(S;Y'_n)\geq\beta I(F;Y'_n) \tag{9}$$

Where β is the redundancy thresh hold. The value of α and β is adjustable parameter. The bigger value of α indicates the higher need for the correlation of the variable in the arithmetic, meanwhile, the larger value of α indicates the higher need for the redundancy.

(3) Accomplishing the characteristic selection and reduction, Lyapunov index is calculated, which is used to analysis the chaotic characteristic of multivariable time series. Then adopting the phase space reconstruction of the chaotic time series, the reduced multivariable phase space is reconstructed as

$$X'_n=[x'_{1,n},x'_{1,n-\tau_1},\cdots;x'_{1,n-(m_1-1)\tau_1},\cdots;x'_{i,n},x'_{i,n-\tau_i},\cdots;x'_{i,n-(m_i-1)\tau_i},\cdots] \tag{10}$$

Where, sampling number $n = \max_{1\leq i\leq 2}((m_i-1)\tau_i)+1,\cdots,N$, $x_{i,n}$ is the ith time series after normalization. The point of the phase space reconstruction is the value of the delay time τ_j and embedded dimension m_j, which determine the level of similarity of the phase space reconstruction.

Since the selection of the delay time affects the information contained in the phase space reconstruction, the frequently-used selection method are autocorrelation method and mutual information method. Because of the measurement for nonlinear random correlation of time series and suitability for the high dimension chaotic system, the mutual information method is adopted in this paper. For the certain subseries $X'_{j,i}$, $i=1, 2, \ldots, N$, of multivariable time series, let delay time is τ_i, and the time series turns to $X'_{j,i+\tau_i}$. Suppose the probability of $x'_{j,k}$ appearing in $X'_{j,i}$ is $p(x'_{j,k})$, and the probability of $x'_{j,k+\tau_i}$ appearing in $X'_{j,i+\tau_i}$ is $p(x'_{j,k+\tau_i})$, then the joint probability of $x'_{j,k}$ and $x'_{j,k+\tau_i}$ appearing in two series is $p(x'_{j,k},x'_{j,k+\tau_i})$, therefore the mutual information of the time series and its delay time series is defined as.

$$I(\tau_i)=I(X'_{j,i};X'_{j,i+\tau_i})=H(X'_{j,i})+H(X'_{j,i+\tau_i})-H(X'_{j,i},X'_{j,i+\tau_i}) \tag{11}$$

According to the mutual information method, $I(\tau_i)$ value in different delay time τ_i is computed, the time corresponding to the first partial optima of $I(\tau_i)$ is selected to as the optima delay time τ_i.

The false nearest neighborhood method is used to determine the embedding dimension m [20, 21]. In the phase space reconstruction, the nearest neighborhood $X'_p, p \neq n$ of phase point X'_n satisfies the equation below.

$$\left\| X'_p - X'_n \right\| \leq \left\| X'_i - X'_n \right\|, i = \max_{1 \leq j \leq M}(m_j - 1)\tau_j + 1, \cdots, N, i \neq n \tag{12}$$

When the embedding dimension increases from m_i to m_i+1, the distance of the neighborhood point can be expressed as below, which is called the false nearest neighborhood point.

$$\frac{\left\| X'_p - X'_n \right\|^{(m_1,\cdots,m_i+1,\cdots,m_M)} - \left\| X'_p - X'_n \right\|^{(m_1,\cdots,m_i,\cdots,m_M)}}{\left\| X'_p - X'_n \right\|^{(m_1,\cdots,m_i,\cdots,m_M)}} \geq R_T \tag{13}$$

Where R_T is the thresh hold, the range of value is $15 \leq R_T \leq 50$. The basic idea of this method is that counting the percentage δ of the false nearest neighborhood point, while the δ is lower than certain thresh hold, the geometric construction is totally opened, which the optimal embedding dimension m_i is computed.

(4) Based on the embedding theorem, in the m dimension phase space reconstruction, the smooth function is expressed by $f' : \Re^m \to \Re$, and the prediction model of the multivariable chaotic time series is developed as

$$Y_{i,n+1} = f'_i(X_n) \tag{14}$$

Where $Y_{i, n+1}$ is the prediction value of $X_{i, n+1}$. Considering the random and nonlinear characteristic of wind power, it is difficult to determine the analyzing formulation of the function f'. In this paper, the RBP network is used to approaching the mapping function, and the prediction model is established as below.

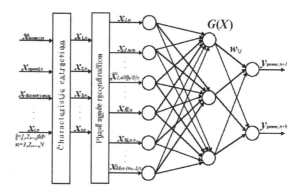

Fig. 1. Model structure of prediction

In this model, using the mutual information and phase space reconstruction, the uncertainty of the single prediction by wind speed or power is reduced, and the main useful information of the short-time wind power prediction is increased, meanwhile, the prediction precision is improved by overcoming the complex model structure, bad generalization and over fitting induced by increasing variable.

(5) The mean absolute error E_{av} and the root mean square error E_{rms} are used to evaluate the precision of the prediction result, which is expressed as

$$E_{av} = \sum_{i=1}^{n} \left| x_{mi} - y_{pi} \right| \Big/ Cn \tag{15}$$

$$E_{rms} = \sqrt{\sum_{i=1}^{n} \left(x_{mi} - y_{pi} \right)^2} \Big/ C\sqrt{n} \tag{16}$$

Where, x_{mi} is the actual power of the ith moment, y_{pi} is the prediction power of the ith moment, C is the operational capacity of wind farm, n is the number of the sample. E_{av} reflects the mean amplitude of the prediction error, and E_{rms} reflects the disperse degree of the prediction error.

4   Case Study

The proposed model has been demonstrated by means of a case study of a wind farm in shanghai consisting of 34 wind turbines. Each turbine has a rated capacity of 3 MW. Considering the influencing factors such as the historic wind speed, direction and power, the wind farm data is sampled every 5 minutes, in 16 days of each season in 2010, the sample space is constructed by the wind power X_1, wind speed and direction in 100 meters (X_2, X_3), in 90 meters (X_4, X_5), speed in 80 meters (X_6), in 40 meters (X_7), in 20 meters (X_8), and the number of each variable are 4608. The model is simulated and analyzed based on the sampling space, and the spring data is chosen to illustrate the presented methodology.

4.1   Data Normalization and Extraction of the Multivariable Characteristic

The sample series $X_i=[x_1, \ _i,\ldots, x_8, \ _i]$, $i=1, 2, \ldots, 4608$. is normalized by the formula.

$$x'_{j,i} = (x_{j,i} - x_{j,\min}) \big/ (x_{j,\max} - x_{j,\min}) \quad j = 1, \cdots, 8 \tag{17}$$

Then the mutual information of variables are calculated by formula (8) and (9), the correlative thresh hold is set to $\alpha=0.7$, therefore the correlative variable set $F=\{X'_{1,n}, X'_{4,n}, X'_{5,n}\}$ of the prediction is obtained. The redundancy thresh hold is set to $\beta=0.3$, the redundancy variable $X'_{5,n}$ is reduced, and the extraction of the multivariable characteristic result is shown in the Tab.1.

Table 1. Mutual information and characteristics extraction of multivariable

Variable	$I(X'_{i,n};Y_n)$	Extraction result
Wind power $\left(X'_{1,n}\right)$	2.4840	
Wind speed in 100m $\left(X'_{2,n}\right)$	2.1941	
Wind direction in 100m $\left(X'_{3,n}\right)$	2.2101	
Wind speed in 90m $\left(X'_{4,n}\right)$	2.2404	
Wind direction in 90m $\left(X'_{5,n}\right)$	2.2379	$X'_{1,n}$, $X'_{4,n}$
Wind speed in 80m $\left(X'_{6,n}\right)$	2.1931	
Wind speed in 40m $\left(X'_{7,n}\right)$	2.1812	
Wind speed in 20m $\left(X'_{8,n}\right)$	2.2056	

4.2   Phase Space Reconstruction of the Multivariable Chaotic Time Series

The time series constructed by the extraction results is analyzed by the chaos theory. In this paper, Wolf method is used to compute the Lyapunov index of the series, which is 0.0614, and the time series appear the characteristic of chaos.

Combining with the extraction results, the power and speed (in 90 meters) time series are reconstructed. Using the formula (10) and (11), the mutual information functions of those two series are calculated respectively, the curve of which are shown in fig.2. The moment corresponding to the first partial optima $I(\tau_i)$ is selected, therefore the optimal delay time for power is $\tau_1 = 14$, and for speed is $\tau_2 = 12$.

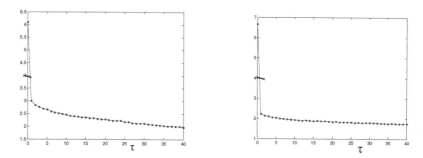

Fig. 2. The mutual information function of wind power and wind speed time series

According to the formula (12) and (13), the false nearest neighborhood function of those two series are obtained respectively, the curve of which are shown in fig.3. Suppose $R_T=15$, $\delta=5\%$, the optimal embedding demission of the power and speed time series are $m_1=7$ and $m_2=4$, then the embedding demission of the whole system is $m=m_1+m_2=11$.

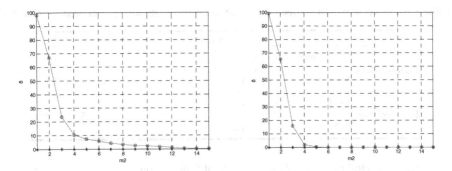

Fig. 3. The false nearest neighborhood rate function of wind power and wind speed time series

4.3   Results of the RBF Network

The historic wind speeds and powers from Mar. 6 to Mar. 20 in 2010 are used to training data, adopting the RBF network model in Fig.1, and the wind power data on Mar. 21 are used to be testing data, step length is 1. The prediction results are shown in Fig.4, the solid line is actual power, while the dash one is prediction value. The mean absolute error is relatively small.

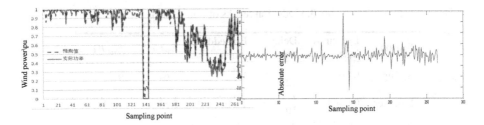

Fig. 4. Prediction results and absolute errors of proposed method

4.4   Analysis of the Results

(1) The comparison analysis of single and multi variable prediction method

In the Tab.2, the mean absolute error and root mean square error of wind power single variable time series are 6.77% and 7.50% respectively and for the wind speed time series are 14.95% and 15.54% respectively; meanwhile the values of the multivariable time series model consisting of speed and power reduce to 5.67% and

9.81%. the results of the multi variable prediction method is superior to the single variable method, which enhances the prediction precision. Based on the prediction results of each point, the maximum values of the mean absolute error for the single variable prediction are 62.12% and 61.1% respectively, while for the multivariable method the error reduces to 27.05%.

Table 2. Results comparison between univariate and multivariate wind power prediction method

Prediction method		Delay time	Embedding dimension	E_{av}/%	E_{rms}/%
Single variable	Wind power $(x'_{1,i})$	14	7	6.77	14.95
	Wind speed $(x'_{2,i})$	12	4	7.50	15.54
Multi variable	Wind power $(x'_{1,i})$	14	11	5.67	9.81
	Wind speed $(x'_{2,i})$	12			

(2)Analysis wind power prediction results of four seasons

The prediction errors are shown in Tab.3. Compared with other multivariable prediction method, the mean absolute error and root mean square error of proposed method are less than 6% and 10%. According to the prediction results of four seasons, when the wind power are continuous series and fluctuates smoothly, the errors of proposed model turn to be reduced, take the autumn and winter samples for example, the mean absolute error are 2.37% and 2.59%, and the root mean square error are 3.52% and 4.01%. While the wind power are intermittent series and fluctuates frequently, the error of proposed model turn to be increased, and the errors mainly focus on the point with abrupt changing, for the spring and summer samples, the mean absolute error are 5.67% and 5.37%, and the root mean square error are 9.81% and 8.64%.

Table 3. Prediction results of four seasons

Season	Lyapunov index	Delay time	Embedding dimension	E_{av}/%	E_{rms}/%
Spring	0.0614	14 12	11	5.67	9.81
Summer	0.0864	7 10	13	5.37	8.64
Autumn	0.0489	9 6	8	2.37	3.52
Winter	0.0210	14 11	10	2.59	4.01

(3)Analysis wind power prediction results with abrupt changing speed

The comparison of prediction results with abrupt changing speed using different methods is shown in the Fig.5. The wind power is mainly affected by the wind characteristic such as random, uncertainty, and constrained by the maximum power and power limit curve of turbine, therefore it is difficult to forecast in real time based on the historical data, the accuracies of the abrupt changing point in showed method are reduced, while the proposed method shows high sensitive to the wind speed changing, which improves the accuracy of short term wind power prediction.

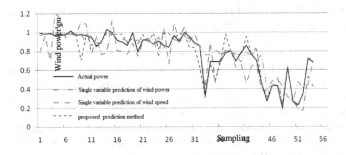

Fig. 5. Comparison of prediction results of different methods

5   Conclusion

The wind power prediction model based on multivariable mutual Information and chaotic phase space reconstruction is proposed in this paper. In the proposed method, the wind power series is treated as a multi dimension nonlinear chaotic system affected by multivariable, the multivariable set is reduced by mutual information method, and the dimension of phase space is reconstructed using delay time and false nearest neighborhood, which remains the integrated dynamic information of the the wind power prediction system during the expanding dissension, can be inputted to the RBP network, and forecasts power. The demonstration shows the benefit of proposed method in improving prediction accuracy compared with single variable method. The proposed prediction model based on the historical data is suit for the short term wind power prediction; the considering with the wind resource distribution and the numerical weather prediction, the time span and accuracy of the prediction will be increased. Furthermore, the fitness of model will be enhanced by adopting the abrupt changing factors into the model, such as abandon the wind and control strategies after integrated power system and O&M.

Acknowledgments. This work was financially supported by Shanghai Green Energy Grid Connected Technology Engineering Research Center（13DZ2251900），the National Natural Science Foundation of China (51177098) and National High Technology Research and Development Program of China（2012AA051703、2012AA051707）and Shanghai Science and Technology Fundation (13510500400).

References

1. Tande, J.: Grid integration of wind farms. Wind Energy 6, 281–295 (2003)
2. Yang, X.Y., Yang, X., Chen, S.Y.: Wind speed and generated power prediction in wind farm. Proceedings of the CSEE 25(11), 1–5 (2005)

3. Kariniotakis, G., et al.: The state of the art in short-term prediction of wind power from an offshore perspective. In: Proceedings of 2004 SeaTechWeek, Brest, France, pp. 20–21 (October 2004)
4. Feng, S.L., Wang, W.S., Liu, C., Dain, H.Z.: Short term wind speed prediction based on the physical principle. Acta Energiae Solaris Sinica 32(5), 611–615 (2011) (in Chinese)
5. Riahy, G.H., Abedi, M.: Short term wind speed prediction for wind turbine applications using linear prediction method. Renew Energy 33(1), 35–41 (2008)
6. Ding, M., Zhang, L.J., Wu, Y.W.: Wind speed forecast model for wind farms based on time series analysis. Electric Power Automation Equipment 25(8), 32–34 (2005) (in Chinese)
7. Fan, G.F., Wang, W.S., Liu, C., et al.: Wind power predication based on artificial neural network. Proceedings of the CSEE 28(34), 118–123 (2008)
8. Damousis, I.G., Alexiadis, M.C., John, B.: A fuzzy model for wind speed prediction and power generation in wind parks using spatial correlation. IEEE Transactions on Energy Conversion 19(2), 352–361 (2004)
9. Zhang, Y.Y., Lu, J.P., Meng, Y.Y., Yan, H., Li, H.: Wind power short-term prediction based on empirical mode decomposition and chaotic phase space reconstruction[J]. Automation of Electric Power Systems 36(5), 24–28 (2012)
10. James, W.T., Patrick, E., Sharry, M.: Wind power density prediction using ensemble predictions and times series models. IEEE Transactions on Energy Conversion 24(3), 775–778 (2009)
11. Zhang, G.Q., Zhang, B.M.: Wind speed and wind turbine output forecast based on combination method. Automation of Electric Power Systems 33(18), 92–95 (2009)

Analysis of the Fault Diagnosis Method for Wind Turbine Generator Bearing Based on Improved Wavelet Packet-BP Neural Network

Quanxian Chen and Mingxing Ye

Shanghai Dianji University, 200240 Shanghai, China
chenqx@sdju.edu.cn, 13774486497@163.com

Abstract. In order to achieve the detection for the fault diagnosis of the wind turbine generator bearing, firstly, the transformation of the wavelet packet is adopted to decompose the vibration signal into several layers, and denoise and reconstruct it. Secondly, this paper takes the combination of the wavelet node energy and the characteristic parameters of the denoised signal both in the time and frequency domain as the input feature vector to BP neural network with the function of self- determining hidden layer neurons. Finally, the results of the fault diagnosis are regarded as the output. The experimental data demonstrate that this method can effectively diagnose the fault types of the wind turbine generator bearing.

Keywords: wind turbine bearings, fault diagnosis, wavelet analysis, BP neural network, Matlab.

1   Introduction

With the development of the wind power technology, the installed capacity of the wind turbines has been increasing greatly in the recent years. Besides, the proportion of wind power as a new kind source of energy is also growing in the global energy structure. However, the rapid growth of the wind power industry has also made the costs of wind turbine operation and maintenance continue to raise. The generator which is the core component of the wind turbine needs the highest maintenance costs. Apart from that, the working status of the generator will also directly affect the machine's performance stability and the power quality.[1-2] Bearing damage is one of the common faults of the generator, and its failure rate accounts for 30 % to 40% of the total generator failure. The large and medium-sized generator bearing failure mechanism, however, is rather complex, and its fault signals display as follows: [3] the weak fault signal submerges in the background of the strong noise signal; the fluctuation range of the characteristic frequency of the signal is slightly larger and even hopping; sometime the signal is transient, discrete and non-stationary. Therefore, the single adoption of wavelet analysis, correlation analysis or spectral analysis is difficult to extract the accurate fault characteristic parameters, neither can it intelligently identify the fault types. To solve the above problems, this paper proposed

K. Li et al. (Eds.): LSMS/ICSEE 2014, Part III, CCIS 463, pp. 13–20, 2014.

a new method, the wavelet packet transform technique combined with the improved BP neural network algorithm, to quickly and intelligently diagnose the generator bearing fault types.

2   Wavelet Transform

Wavelet analysis is a new mathematical theory and method which is developed in the mid-1980s. The wavelet transform decomposes the research object into different scales of space to analyze and process the signal and reconstructs it as needed. [4] Compared with the traditional Fourier analysis, the wavelet transform has better time and frequency localization characteristics.

2.1   Selection of Basic Wavelet Function

Wavelet transform includes continuous wavelet transform and discrete wavelet transform. The continuous wavelet sequence can be described as

$$\psi_{a,b}(t) = \frac{1}{\sqrt{|a|}} \int_{-\infty}^{+\infty} x(t)\psi^*(\frac{t-b}{a})dt \ . \tag{1}$$

Here, $\psi(t)$ is basic wavelet function; $a,b \in R$, and $a \neq 0$, a is stretching factor and b is mobile factor.

The discrete wavelet sequence can be written as

$$\psi_{m,n}(t) = |a_0|^{-\frac{m}{2}} \int_{-\infty}^{+\infty} x(t)\psi^*(a_0^{-m}t - nb_0)dt \ . \tag{2}$$

Here, $a = a_0^m$, $b = nb_0 a_0^m$, $m,n \in R$.

The wavelet function is not unique, when stretching factor and translation factor are changed in equation (1), (2). To judge the quality of the basic wavelet function, and to choose the optimal basic wavelet function, the method of comparing the errors between the theoretical result from the wavelet analysis and the expected result can be adopted.

2.2   Wavelet Packet De-Noising

Wavelet packet theory is another great development in signal processing field of the wavelet theory, which makes it possible to break down the high frequency part left unfinished after the wavelet analysis for further decomposition, and which can select the appropriate frequency bands which match with the signal spectrum based on the characteristics of the signal analyzed.

The wavelet packet recursive formula that decomposes time-domain signal to i layer can be expressed as follows: [5]

$$\begin{cases} P^{(0,0)}(t) = W(t) \\ P^{(i+1,2j+1)}(t) = \sum H(k-2t)P^{(i,j)}(t) \\ P^{(i+1,2j)}(t) = \sum G(k-2t)P^{(i,j)}(t) \end{cases} \tag{3}$$

Here, $P^{(i,j)}$ stands for the coefficient of node j in i-layer, W(t) stands for the original signal of vibration. What's more, $i=0,1,2,\ldots$; $j=0,1,\ldots,2^i-1$; $t=0,1,\ldots2^{n-1}$; $n=\log_2 N$; N means the number of t; H is low-frequency decomposition filter and G is high-frequency decomposition filter which match with the scaling function $\Phi(t)$ and the wavelet function $\psi(t)$.

According to equation (3), the de-noising process of the wind turbine bearing vibration signal can be divided into the following three steps: 1) select db1 wavelet packet to decompose the vibration signal into 3 layers; 2) for the high frequency coefficients under various decomposition scales, choose suitable soft thresholds to quantify them based on the soft threshold function; 3) reconstruct the one-dimensional wavelet based on the low-frequency and high-frequency coefficients of last layer.

2.3   Wavelet Node Energy Extraction

The impact force and the resonance accompanied by the operation of the faulty bearing will change the distribution of signal energy in each band when the bearing fault occurs, what's more, the influence on the distribution of the signal energy may differ because of different types of fault. Therefore, it is advisable to calculate the energy of the wavelet packet coefficient of each node in the wavelet packet decomposition, and to extract the energy value of each frequency band to reflect the changes in the bearing operating.

The following formula (4) is used for calculating the energy of the eight sub-bands in the third layer:

$$E_3^j = \int \left| S_3^j(t) \right|^2 dt = \sum_{k=1}^{N} \left| x_{jk} \right|^2. \tag{4}$$

Here, x_{jk} $(j=0,1,\ldots,7, k=0,1,\ldots,n)$ are the amplitude of all the discrete points of the reconstruction signal S_3^j.

The signal energy within each frequency band will change greatly when the bearing runs in fault, therefore, select several signals which are prominent in energy change as fault characteristic vector, that is:

$$T = [E_3^0, E_3^1, \ldots E_3^j]. \tag{5}$$

3   Selecting Bearing Fault Characteristic Parameters in Time and Frequency Domain

Because the wind turbine bearing is a very complex rotating structure, and there exists a complicated non-linear relationship between the failure modes and its characteristic parameters, it is necessary to simultaneously extract a plurality of characteristic parameters both in time and frequency domain for comprehensive analysis. [6]

3.1   Time-Domain Characteristic Parameters

The frequently used peak index and kurtosis index are chosen as the fault characteristic parameters in time domain. Peak index (P) is an effective characteristic parameter for the detection of the fault bearing impact signal, and its mathematical expression is as follows:

$$P = E\{\max|x(t)|\}. \tag{6}$$

Kurtosis index (K) is often used to detect the deviation degree of the signal from the normal distribution, and the greater the absolute value is, the farther the deviation is from the normal state of the measured object. The expression is as follows:

$$K = \frac{\int_{-\infty}^{+\infty}(x-\mu)^4 P(x)dx}{S^4}. \tag{7}$$

Here, μ stands for signal expectation, $P(x)$ stands for probability density function of the signal, and S stands for standard deviation of the signal.

3.2   Frequency-Domain Characteristic Parameters

When a fault occurs, the vibration amplitude at some certain frequencies will change, which will effect the position of the centre of gravity of the power spectrum to some extent. The power spectrum centre index (PSC) reflects the deviation degree of the gravity centre position, and the harmonic factor index (HF) reflects the distribution and the width of the spectral, as the following expressions indicate:

$$PSC = \frac{\sum_{i=1}^{N} f_i p_i}{\sum_{i=1}^{N} p_i}. \tag{8}$$

$$HF = \frac{\sqrt{\sum_{i=1}^{N} f_i^2 p_i \sum_{i=1}^{N} f_i^{-2} p_i}}{\sum_{i=1}^{N} p_i}. \tag{9}$$

Here f_i is the frequency corresponding to the time i, P_i is the amplitude of the power spectrum at the time i, i=1,2, ..., n.

4   Improved BP Neural Network Algorithm

BP neural network is also known as error back propagation neural network, which is a multilayer forward neural network and consists of input layer, hidden layer and output layer. [7] So far, how to select the optimal number of hidden layer neurons has not had a sound theoretical guidance. Choosing a suitable number of hidden layer nodes may have a great influence on the network performance.[8] If the number of nodes are less than enough, the network fault tolerance is poor, on the contrary, if more than enough, it will not only increase the training time, but also reduces the generalization capability of the network.

In order to find the optimal number of the hidden units quickly, this section proposes an improved method, as follows:

1) Determine the range of the hidden units as $n_1 \leq N_{hid} \leq n_2$ based on the empirical formulas like $N_{hid} = \log_2 N_{in}$, $N_{hid} \leq p/[R+(N_{in}+N_{out})]$, $N_{hid} = 2N_{in}+1$. Among them, p is the total number of the training samples, and $5 \leq R \leq 10$ n_1 is the minimum value of the number of hidden units, while n_2 is the maximum.

2) Take $n=n_1$, train the BP neural network when the hidden node is n_1, and get the mean square error Mse_1. Mse is the performance evaluation parameter to select node number. In the formula $Mse = \dfrac{1}{mp} \sum\limits_{p=1}^{p} \sum\limits_{j=1}^{m} (\hat{y}_{pj} - y_{pj})^2$, m is the number of output nodes, p is the total number of the training samples, $\hat{y}_{pj}$ is the desired output of the network, y_{pj} is the actual output.

3) Take $n=(n_1+n_2)/2$, here, n is an integer. Train the BP neural network when the hidden node is n, and get the mean square error Mse_2.

4) If $Mse_1 < Mse_2$, then take $n=n_1$, $Mse_1 = Mse_2$. Otherwise, take $n_2=n$.

5) If $n_1 < n_2$, then return to step 2), and cycle like this. On the contrary, quit the program.

Finally, the obtained number n is the optimum number of hidden units when the algorithm ends.

5   Simulation Experiment

5.1   Fault Characteristics Extraction

The data used in the experiment was from the Electrical Engineering Laboratory of the U.S. Case Western Reserve University, and the experiment object is the deep groove ball bearing (6205-2RS JEM SKF), whose rotation frequency is 28.82Hz, and sampling frequency is 12kHz.[9]The following example is a set of analysis given to extract the fault characteristics.

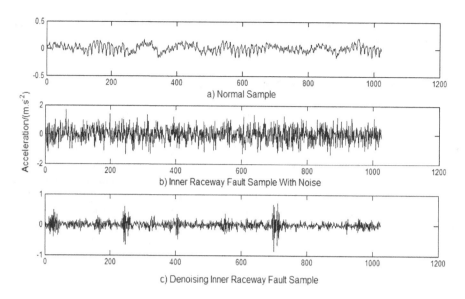

a) Normal Sample

b) Inner Raceway Fault Sample With Noise

c) Denoising Inner Raceway Fault Sample

Fig. 1. Comparison of bearing vibration signal

In figure 5.1, a) is the signal of the bearing in normal operation, b) is the vibration signal of the bearing with inner raceway fault. There is no doubt that the fault vibration signal submerges in the strong noise. In order to extract the fault characteristics, apply equation (3) to decompose the signal into three layers, choose the right soft threshold function to eliminate the noise, and finally reconstruct the signal as c) demonstrates. Furthermore, the signal c) is decomposed again as shown in Figure 5.2, and in Figure 5.2 S stands for original vibration signal, A stands for low-frequency band, D stands for high-frequency band, the number behind means the sequence number of the layers. Finally, obtain the energy of nodes DAA3, DDA3, DAD3, DDD3 which are high frequency nodes of the third layer as the characteristic parameters. Combining with the characteristic parameters of time and frequency domain in section II, it is possible to construct the fault features vector as R = [DAA3 DDA3 DAD3 DDD3 P K PSC HF].

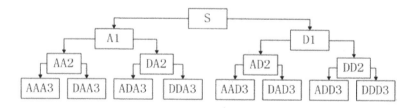

Fig. 2. 3-layer wavelet decomposition tree

5.2   Construction of BP Neural Network

The input neurons are determined as 8 according to the input feature vector R. There are three kinds of output bearing failure, namely, '001'which means normal, '010' which represents the inner raceway fault, and '100 ' which indicates the unknown fault, therefore, it is possible to determine the number of the output neurons are 3. Since 12 is selected as the optimal number of hidden layer nodes through the improved BP neural network algorithm in the third section, the structure of BP neural network is 8-12-3 as shown in figure 3.

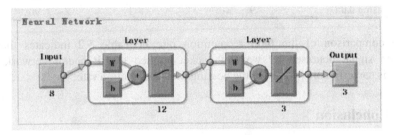

Fig. 3. Structure of BP neural network

5.3   Fault Recognition

According to the input feature vectors, construct two groups of testing sample data, one group of sample data underwent no wavelet packet de-noising and the other experienced de-noising, then compare the fault recognition rate of the two data sets by BP neural network. Parts of the training and testing samples of the two groups are shown in Table 1.

Table 1. Samples of bearing for diagnosis

Data types	Fault types	DAA3	DDA3	DAD3	DDD3	Peak Index (P)	Kurtosis Index (K)	Power Spectral Centre Index (PSC)	Harmonic Factor (HF)	Expected Output
Raw data	Normal	0.03	0.02	0.01	0.01	0.10	0.09	0.01	0.95	001
	Bearing inner raceway damage	0.08	0.09	0.17	0.20	0.40	0.93	0.12	0.46	010
	Test sample	0.10	0.10	0.12	0.16	0.39	0.97	0.11	0.45	010
Wavelet packet denoising data	Normal	0.02	0.02	0.01	0.01	0.12	0.10	0.01	0.93	001
	Bearing inner raceway damage	0.04	0.05	0.10	0.15	0.35	0.87	0.23	0.58	010
	Test sample	0.03	0.04	0.09	0.13	0.30	0.89	0.31	0.62	010

Among the 60 sets of raw data, use 50 sets to train the BP neural network, and the other 10 sets are used to test the network. The same structure of BP neural network is trained by 50 sets of de-noising data, and tested by the other 10 sets of de-noising data. The statistical outputs in both cases are shown in Table 2.

Table 2. Fault recognition rate

Data type	Correct recognition	Inaccurate recognition	Fault recognition rate
Raw data	6	4	60%
De-noising data	9	1	90%

The comparison of the fault recognition rate in Table 5.2 indicates that the vibration signal denoised by wavelet packet applied to BP neural network fault diagnosis can efficiently improve the bearing fault recognition rate.

6   Conclusion

In this paper a new method for the fault diagnosis of wind turbine bearing based on the combination of wavelet packet de-noising and improved BP neural network was proposed. Using wavelet packet to eliminate the strong noise among vibration signal, and taking the combination of the fault characteristic parameters exacted in the time and frequency domain and the energy in the high frequency band after the decomposition of wavelet package as the input feature vector to BP neural network with the function of self- determining hidden layer neurons, this method not only saves the training time for BP neural network, but also improves the accuracy of the bearing fault diagnosis.

References

1. Kennedy, J., Fox, B., Morrow, J.: Working with Wind Power. Engineering &Technology 3(3), 52–55 (2008)
2. Gagari, D.: Nature of Wind Turbine Power. International Journal of Computer and Electrical Engineering 4(2), 145–147 (2012)
3. Antoniadis, I., Glossiotis, G.: Cyclostationary Analysis of Rolling-Element Bearing Vibration Signals. Journal of Sound and Vibration 248(5), 829–845 (2001)
4. Zhang, D.: MATLAB Wavelet Analysis. China Machine Press, Beijing (2011)
5. Wang, D.: Study on Fault Diagnosis Method for Rotor-Bearing. Yanshan University, Qin Huangdao (2012)
6. Meng, H.: Research on the Neural Network and the Grey Theory in the Fault Diagnosis of the Transfer-box. North University of China, Taiyuan (2005)
7. Mohseni, S., Tan, A.: Optimization of Neural Networks Using Variable Structure Systems. Systems 42(6), 1645–1653 (2012)
8. Beigy, H., Meybodi, M.R.: A Learning Automata-based Algorithm for Determination of the Number of Hidden Units for Three-layer Neural Networks. International Journal of Systems Science 40(1), 101–118 (2009)
9. Case Western Reserve University. Bearing Data Center Website: Bearing Data Center Seeded Fault Test Data, http://csegroups.case.edu/bearingdatacenter/pages/download-data-file

An Improved Multi-objective Differential Evolution Algorithm for Allocating Wind Generation and Photovoltaic

Xiao-lin Ge[1] and Shu Xia[2]

[1] College of Electrical Engineering, Shanghai University of Electric Power,
200090 Shanghai, China
[2] State Grid Shibei Power Supply Company, SMEPC, Shanghai 200072, China

Abstract. This paper proposes a multi-objective chance-constrained optimization approach for allocating wind generation and photovoltaic. For reflecting the uncertainty and relevance among wind speed, solar radiation and load, and accelerating the computation, a kind of state selection strategy based on statistical methods is proposed. To evaluate profits from distributed generation more comprehensively, three indices are introduced into the objective function, namely cost index, power loss index and voltage deviation index. To balance the relationship between risk and return, the chance-constrained optimization is adopted. In the process of solution, firstly the novel method integrating the differential evolution for multi-objective optimization and dynamic non-dominated sorting is proposed to get a set of the Pareto-optimal solutions, after that fuzzy multi-attribute decision making method based on information entropy is adopted to select the best compromise solution from the Pareto-optimal solutions. The case studies show that the proposed optimal model is rational, and the algorithm is effective.

Keywords: chance-constrain, multi-objective, optimization, photovoltaic, states selection strategy, wind.

1   Introduction

The installation of distributed generation (DG) in distribution network has impact on distribution system operation, and this influence is associated with the access point and capacity of DG [1-2]. Therefore the optimally allocating DG problem has been widespread concern. Many approaches have been proposed to solve the problem. In [3], the influence of DG on distribution network power losses were analyzed, and an optimization algorithm was presented based on comprehensive use of genetic algorithms and Tabu search to optimize the allocation of DG. In [4], environmental factors were added into the objective function and the optimal location of DG was chosen by calculating equivalent loss factor. In [5], a multi-objective formulation was proposed for the planning of DG with different load models, and the best compromise among power losses, voltage profile, and MVA capacity was decided by weighting method and genetic algorithm. The above works can be used to deal with DG which is

K. Li et al. (Eds.): LSMS/ICSEE 2014, Part III, CCIS 463, pp. 21–31, 2014.
© Springer-Verlag Berlin Heidelberg 2014

dispatchable and controllable, such as, diesel generator and microturbine. However, these methods are difficult to model distributed renewable resources (DRR) which is random and intermittent, for example, wind turbine generators (WTGs) and photovoltaic (PV) panel. In [6], the chance-constrained optimization was employed to deal with the randomness of wind power, and probabilistic power flow was applied to judge whether the planning scheme was satisfying the constraints of both the node voltage and the branch flows, which is provided a good idea to solve the planning problem including uncertain DRRs. Literatures [7] and [8] indicated that wind and solar sources exist complementarities in space and time, for example, the sun radiation is strong at daytime, while wind speed is strong at night. Hence the joint optimization of WTGs and PVs may bring bigger benefits. However, the features that wind and solar are random and correlated each other to a certain extend make the joint optimization of WTGs and PVs become more difficult. At present, most scholars have mainly focused on optimal capacity of hybrid systems without considering the network constraints [7-10], and few articles optimize the access point and capacity of hybrid systems simultaneously.

In this paper, a methodology is proposed to optimize the access point and capacity of WTGs and PVs together. The decision variables include not only the access point and capacity of WTGs and PVs, but also the tilt angle of PVs. To deal with the uncertainty and relevance of wind speed, solar radiation and load, and reduce the number of sampling, wind/solar time-series data is adopted and a kind of state selection strategy are proposed. In the objective function, cost, power losses and voltage deviation are optimized simultaneously. In the constraints, the bus voltage constraints and branch current constraints are described in the form of probability to balance the relationship between risk and return. To solve this problem, differential evolution for multi-objective optimization (DEMO) is employed. Based on the DEMO, dynamic non-dominated sorting (DNS) is applied to provide better distribution of the Pareto-optimal solutions. Then the Pareto-optimal solution set is sorted, and the best compromise solution is obtained by fuzzy multi-attribute decision making method based on information entropy. The feasibility of the proposed approach is verified on a 33-node distribution network test system.

2   Relevance of Wind Speed, Solar Radiation and Load

Using the method proposed in [11], the power output of wind turbine generators (WTGs) is modeled by wind speed and power characteristic.

The power output of PVs is not only affected by local solar radiation and selected photovoltaic cells, but also by PVs' tilt angle. Hence, to describe the performance of PVs more accurately, the tilt angle is taken as a decision variable in the optimization process. The power output of PVs can be calculated as [12]:

$$P_{PV}(t) = N_{PV} \times \psi[H(\theta)] \tag{1}$$

where $P_{PV}(t)$ represents the power output of PVs at hour t, N_{PV} is the number of PV in the panel, θ is the tilt angle, $H(\bullet)$ is the total radiation on the PV surface, $\psi(\bullet)$ is the power output of a photovoltaic panel.

To reflect the relevance of wind speed, solar radiation and load, and describe the system operation more accurately, a states selection strategy is carried out as follows:

Step 1. An annual load profile is divided into three parts according to seasonal feature that is summer, winter, spring/autumn. The daily load profile is also divided into three parts, i.e. peak (7:00-11:00 and 17:00-21:00), shoulder (11:00-17:00 and 21:00-23:00) and off peak (0:00-7:00 and 23:00-24:00). Thus, these nine kinds of load combinations are selected to represent yearly load. Each load combination responds to a time interval, described by l_n, n=1, 2,...,N (N=9), and the probability of each time interval is represented by τ_n. The load value of l_n can be obtained from the forecasting data of planning year.

Step 2. The total radiation on the tilted surface of the PVs can be gotten by the total radiation and diffuse radiation to the horizontal surface and tilt angles according to Eq. (3). Then, according to the total amount of hourly radiation, the range of total radiation, [0, H_{max}], is divided into M radiation states evenly, where H_{max} is the maximum hourly total radiation. Each radiation state is a set, described by h_m, m=1, 2, ..., M.

Step 3. According to the hourly wind speed, the range of wind speed [v_{ci}, v_{co}] can be divided into K-1 states, plus one state which is [0, v_{ci}] or [v_{co}, +∞] due to their same outputs. Thus there are K wind speed states. Each wind speed state is a set, which can be described by v_k, k=1,2,...,K.

Step 4. According to the above division, the number of system operation states is G (G=N×M×K). Then the statistical work of probability of each state should be carried on. Each group of the historical data includes three kinds of information, such as time, solar radiation and wind speed. So these groups of historical data are classified to G states according to the included information. The probability of each state is described as follows:

$$F(l \in l_n, h \in h_m, v \in v_k) = \eta_g, g = 1, 2, ..., G \tag{2}$$

$$\sum_{g=1}^{G} \eta_g = 1 \tag{3}$$

where η_g is the probability when the time drops in the state l_n, the solar radiation in the state h_m, and the wind speed in the state v_k.

3   Multiobjective Chance-Constrained Model of the Optimally Allocating Problem

3.1   Objective Function

For the allocating problem of diesel generator and microturbine, three indices are usually introduced to evaluate distributed generation profits, namely cost index [13], power loss index [5] and voltage deviation index [2], which are deterministic expression. When allocating WTGs and PVs, the three indices have random features because of the randomness of wind speed, solar radiation and load. So, in this paper, the expectations of three indices are taken as the objective function.

(1)Cost index

$$I_C = \frac{C_W + C_{PV} + \sum_{g=1}^{G}(8760 \times \eta_g \times u_g \times P_{DGg})}{\sum_{n=1}^{N}(8760 \times \tau_n \times u_n \times P_n)} \qquad (4)$$

where P_{DGg} is the power supply of distribution network at the system operating state g after planning, P_n is the power supply of distribution network at the time state n before planning, u_g is the electricity price at the state g, u_n is the electricity price at the state n. C_W, C_{PV} is the total cost of WTGs and PVs [13], respectively.

(2)Power loss index

$$I_P = \frac{\sum_{g=1}^{G}(8760 \times \eta_g \times P_{LDGg})}{\sum_{n=1}^{N}(8760 \times \tau_n \times P_{Ln})} \qquad (5)$$

where P_{LDGg} is the power losses at the system operating state g after planning, P_{Ln} is the power loss at the time state n before planning. This index can effectively reflect the change of power losses before and after planning.

(3) Voltage deviation index

$$I_V = \sum_{g=1}^{G}(\eta_g \frac{1}{N_N - 1} \sum_{i=1}^{N_N - 1} w_i \frac{|V_{ig} - V_{0g}|}{V_{0g}}) \qquad (6)$$

where N_N is the number of node, V_{ig}, V_{0g} is the voltage magnitude of the i-th and the first node at the system operating state g, respectively. w_i is the weight factor of the ith node, reflecting the important level of node.

3.2 Constraints

The optimally allocating for DG installations should satisfy some constraints, including power flow constraints, total number constraint of buses with DG installation, total capacity of DG constraint, bus voltage constraints and branch current constraints. Due to the uncertainties of wind speed, solar radiation and load, profits from distributed generation may be poor to guarantee all lines will comply with the thermal capacity and voltages will maintain within prescribed limits at any time. To balance the relationship between risk and return, the chance-constrained optimization (CCO) [14] is adopted in this paper. Therefore, the bus voltage constraints and branch current constraints are described in the form of probability.

(1)Power flow constraints

$$P_{si} = V_i \sum_{j \in i} V_j (G_{ij} \cos \delta_{ij} + B_{ij} \sin \delta_{ij}) \qquad (7)$$

$$Q_{si} = V_i \sum_{j \in i} V_j (G_{ij} \sin \delta_{ij} - B_{ij} \cos \delta_{ij}) \qquad (8)$$

where P_{si}, Q_{si} is the active and reactive power injection at bus i, respectively, $j \in i$ represents those buses linked to the bus i, G_{ij}, B_{ij} is the real and imaginary parts of bus admittance matrix, respectively, δ_{ij} is the phase angle difference between bus i and bus j.

(2)Total number constraint of buses with DG installation

$$\sum_{i=1}^{N_N-1} U_i \le D \qquad (9)$$

where U_i is a 0-1 variable, if bus i is allocated DG, then U_i equals to one, else equals to zero. D is the maximum allowable number of buses with DG installation.

(3)Total capacity of DG constraint

$$\sum_{i=1}^{N_N-1} (P_{Wi} + P_{PVi}) \le P_{DRR\max} \qquad (10)$$

where P_{Wi}, P_{PVi} is the capacity of WTGs and PVs allocated to bus i, respectively, P_{DRRmax} is the maximum allowable capacity on the system.

(4)Bus voltage constraints

$$P_r \{V_{i\min} \le V_i \le V_{i\max}\} \ge \beta_V \qquad (11)$$

where $\Pr\{\cdot\}$ is the probability of the occurrence of the event in the brackets, V_{imin}, V_{imax} is the minimum and the maximum voltage limit of bus i, respectively, β_V is the confidence level.

(5)Branch current constraints

$$P_r \{L_{i\min} \le L_i \le L_{i\max}\} \ge \beta_L \qquad (12)$$

where L_i is current of branch i, L_{imin}, L_{imax} is the minimum and the maximum current limit of branch i, respectively, β_L is the confidence level.

4   Optimization Algorithm for the Problem

4.1   Multiobjective Differential Evolution Based on Dynamic Non-dominated Sorting

To deal with multiobjective problems, traditionally, weighting factors are introduced to convert multiply objectives into single objective [4-5]. However, it is often difficult to determine weighting factors in reality. In recent years, with the development of multiobjective optimization algorithm (MOEA), such as, improved strength Pareto evolutionary algorithm (SPEA2) [15] and improved non-dominated sorting genetic algorithm (NSGA-II) [16], which provides a new approach for solving multiobjective problems. The MOEA does not need to set the weighting factors, and get a uniform distribution of the Pareto-optimal set. Decision-makers can select one or more optimal solutions according to demand. In [17], the DEMO was proposed, which was a new approach to multiobjective optimization based on differential evolution (DE). The DEMO combines the advantages of DE and fast non-dominated sorting (FNS) of NSGA-II, and the performance of the DEMO is better than the NSGA-II in some aspects. As a result, the DEMO is applied to solve the problem in this paper.

The fast non-dominated sorting of DEMO only needs one sorting operation for all individuals, and takes a short time. However, this method does not provide good result. To get better distribution of solutions, dynamic non-dominated sorting (DNS) [18] is adopted instead.

The algorithm combining the DEMO and the DNS (DEMO-DNS) is as follows:
Step 1. Set the algorithm parameters and initialize individuals in the population.
Step 2. Perform mutation and crossover operation.
Step 3. Perform selection operation according to the DNS approach.
Step 4. Return to Step 2 until the given maximum number of generations.

4.2   Fuzzy Multi-attribute Decision Making Method Based on Information Entropy

The output of the DEMO-DNS is a Pareto-optimal set. We need to select one solution, called the best compromise solution, which satisfies different objectives to some extent for practical purpose. Fuzzy multi-attribute decision making method has been successfully implemented in this field [19]. This method requires predetermined information on the relative importance of the attributes, which is usually given by a set of weighting factors. However, it is difficult to set these weighting factors in some problems. Hence, information entropy is employed to derive attribute weighting factors, and if the attribute has similar values across alternatives, it has relatively big entropy, then it is assigned a smaller weight, for which such attribute does not help in differentiating alternatives [20].

5   Case Studies

The proposed approach is tested on a 33 node distribution network, which is shown in Fig. 1. The voltage magnitude of the first node is set 1.05 p.u., the weight factors of all nodes are equal. Bus voltage magnitude should be kept in the range of 0.95 to 1.05pu, and the maximum current of branch is 0.3kA. The electricity price is 0.055EUR/kWh. Data of wind speed and solar radiation are obtained from certain an area in China. The number of the candidate buses for connecting the DGs is within [1, 32]. The maximum allowable number of buses with DG installation is 2, and the allowable maximum capacity on the whole system is 2000kW. By calculating, the total cost, the expected power loss and the expected voltage deviation during a year before installing DG is 1222.07 thousand-EUR, 83.07kW and 0.031pu, respectively.

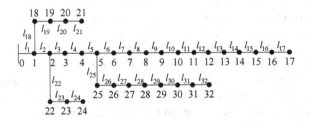

Fig. 1. 33 node distribution network

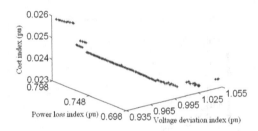

Fig. 2. Pareto-optimal frontiers for 3 objectives

In the simulation, the parameters of the proposed DEMO-DNS are set as: the population size N_p is 100, the largest number of iteration D_{max} is 600, the mutation factor F is 0.5, and the crossover factor C_R is 0.4. The confidence level β_V and β_L both are 0.99. Fig.2 shows the relationship of the three objectives of the Pareto-optimal solutions obtained by the DEMO-DNS, which is called the Pareto frontiers. It is quite clear that these candidate solutions are well distributed.

Table 1. Optimum solutions of different objectives

Objective	Access point number	Capacity of WTGs(kW)	Capacity of PVs(kW)	Tilt angle of PVs(°)	I_C(pu)	I_P(pu)	I_V(pu)
Optimal cost index	5	750	0	--			
	28	1200	0	--	0.9397	0.7813	0.0259
Optimal power loss index	14	450	478	22.53			
	30	750	322	23.02	1.0517	0.6993	0.0234
Optimal voltage deviation index	13	600	272	27.31			
	32	750	378	27.56	1.0306	0.7055	0.0232
Best comparison solution	14	450	396	23.45			
	28	1050	104	25.51	0.9984	0.7088	0.0233

Four solutions are selected from the Pareto-optimal solutions, they are the solutions corresponding to optimal cost index, optimal power loss index, optimal voltage deviation index, and best compromise solution, which are listed in Table 1. From Table 1, it is observed that when the, cost index is optimal, there is only WTGs, no PVs selected. It is because that the cost of PVs is still high at present. By comparing the planning results from optimal cost decision, optimal power loss decision and optimal voltage deviation decision, it can be seen that there are obvious conflicts among the three objectives. When the cost index is optimal, both the power loss and voltage deviation index are poor. Based on the information entropy, the weighting factors are obtained as: $\omega_C=0.402$, $\omega_P=0.316$, $\omega_V=0.282$. Comparing the planning result from the best compromise solution to that situation before planning, it is observed that the annual cost is dropped by 0.16%, the annual power loss expectation is decreased by 29.12%, and the annual expected value of voltage deviation is dropped by 0.0077pu, which proved that fuzzy multi-attribute decision making method based on information entropy can better coordinate the various objectives.

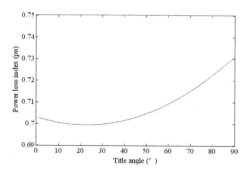

Fig. 3. Relationship between tilt angle of PVs and power loss index

For the planning result from optimal power loss decision, the relationship between tilt angle and power loss is got in the Fig.3 according to the change of the tilt angle of PVs linked to bus 14. From Fig.3, it is seen that the tilt angle has an impact on the power losses for the fixed PV array. Therefore, in the actual optimization process, the tilt angle should be treated as a decision variable in order to get more accurate results.

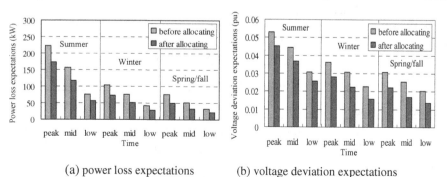

(a) power loss expectations   (b) voltage deviation expectations

Fig. 4. Comparison of the results before and after planning

For the planning result from the best compromise solution, Fig.4(a) and (b) give the comparison of power loss expectations and voltage deviation expectations at each time before and after planning. As shown in Fig.4(a) and (b), the expected values of power loss and voltage deviation are significantly decreased at each time after planning. For the planning result, bus 28 is the most likely violating voltage limit in all buses.

In order to study the relationship between confidence level and planning results, β_V and β_L is set to 1, 0.98, 0.96, respectively, and the corresponding results are listed in Table 2. From the Table 2, it is observed that the three optimization indices are improved to some extent with the decline in confidence level. Therefore, it is needed to balance the relationship between risk and return in the actual planning process.

Table 2. Optimization results of different confidence levels

Confidence levels	Objective	I_C/pu	I_P/pu	I_V/pu
$\beta_V=1$ $\beta_L=1$	Optimal cost index	0.9407	0.8113	0.0267
	Optimal power loss index	1.0702	0.7058	0.0241
	Optimal voltage deviation index	1.0640	0.7122	0.0239
$\beta_V=0.98$ $\beta_L=0.98$	Optimal cost index	0.9386	0.7482	0.0242
	Optimal power loss index	1.0506	0.6987	0.0232
	Optimal voltage deviation index	1.0294	0.7041	0.0229
$\beta_V=0.96$ $\beta_L=0.96$	Optimal cost index	0.9378	0.7308	0.0232
	Optimal power loss index	1.0298	0.6982	0.0230
	Optimal voltage deviation index	1.0077	0.7034	0.0225

6   Conclusions

In this paper, a methodology is proposed to deal with allocating wind generations and photovoltaic arrays in the distribution network, and the conclusions are as follows:

(1) Three indices are adopted to make decision, namely cost index, power loss index and voltage deviation index, in order to evaluate profits from distributed generation more comprehensively.

(2) The tilt angle of PVs has an impact on the planning result for the fixed PV array. Therefore, in the joint optimization of WTGs and PVs, the tilt angle should be treated as a decision variable.

(3) The proposed state selection strategy based on statistical work of wind/solar time-series data has merits in reflecting the relevance of wind speed, solar radiation and load, and in reducing the number of sampling, which gives the planning result more accurately and quickly.

(4) Chance-constrained optimization provides decision-makers an opportunity to balance the relationship between risk and return in the actual planning process.

(5) The proposed DEMO-DNS has the ability of providing some excellent candidates for the planning operators. Fuzzy multi-attribute decision making method based on information entropy can coordinate the various objectives and provide the comprehensive decision-making scheme.

Acknowledgments. This work was supported by Shanghai Green Energy Grid Connected Technology Engineering Research Center, Project Number: 13DZ2251900.

References

1. Luis, F.P., Antonio, G.P.: Evaluating distributed generation impacts with a multiobjective index. IEEE Transactions on Power Systems 21, 1452–1459 (2006)
2. Sujatha, K., Sarika, K., Noel, S.: Impact of distributed generation on distribution contingency analysis. Electric Power Systems Research 78, 1537–1545 (2008)
3. Gandomkar, M., Vakilian, M., Ehsan, M.: A genetic-based tabu search algorithm for optimal DG allocation in distribution networks. Electric Power Components and Systems 33, 1351–1362 (2005)
4. Zhang, Z.H., Qian, A., Gu, C.H., et al: Multi-objective allocation of distributed generation considering environmental factor. Proceeding of the CSEE 29, 23–28 (2009) (in Chinese)
5. Singh, D.K., Verma, S.: Multiobjective optimization for DG planning with load models. IEEE Transactions on Power Systems 24, 427–436 (2009)
6. Zhang, J.T., Cheng, H.Z., Yao, L.Z., et al.: Study on sitting and sizing of distributed wind generation. Proceeding of the CSEE 29, 1–7 (2009)
7. Yang, H.X., Zhou, W., Lou, C.Z.: Optimal design and techno-economic analysis of a hybrid solar-wind power generation system. Applied Energy 86, 163–169 (2009)
8. Chen, H.H., H.Y., Kang, H.I., Amy, L.: Strategic selection of suitable projects for hybrid solar-wind power generation systems. Renewable and Sustainable Energy Reviews 14, 413–421 (2010)
9. Linfeng, W., Singh, C.: Multicriteria design of hybrid power generation systems based on a modified particle swarm optimization algorithm. IEEE Transactions on Energy Conversion 24, 163–172 (2009)
10. Gilles, N., Said, D., Ludmil, S.: Hybrid photovoltaic/wind energy systems for remote locations. Energy Procedia 6, 666–677 (2011)
11. Li, Q., Yuan, Y., Li, Z.J., et al.: Research on energy shifting benefit of hybrid wind power and pumped hydro storage system considering peak-valley electricity. Power System Technology 33, 13–18 (2009)
12. Karira, M., Simsek, M., Babur, Y., et al.: Determining optimum tilt angles and orientations of photovoltaic panels in Sanliurfa. Renewable Energy 29, 1265–1275 (2004)
13. Chen, L., Zhong, J., Ni, Y.X., et al.: A study on grid-connected distributed generation system planning and its operation performance. Automation of Electric Power Systems 31, 26–31 (2007)
14. Mazadi, M., Rosehart, W.D., Malik, O.P., et al.: Modified Chance-Constrained Optimization Applied to the Generation Expansion Problem. IEEE Transactions on Power Systems 24, 1635–1636 (2009)
15. Souza, D., Multiobjective, B.A.: Optimization and Fuzzy Logic Applied to Planning of the Volt/Var Problem in Distributions Systems. IEEE Transactions on Power Systems 25, 1274–1281 (2010)
16. Deb, K., Pratap, A., Agareal, S.: etal: A fast and elitist multiobjective genetic algorithm: NSGA-II. IEEE Transactions on Evolutionary Compution 6, 182–197 (2002)

17. Preetha Roselyn, J., Devaraj, D., Dash, S.S.: Multi Objective Differential Evolution approach for voltage stability constrained reactive power planning problem. International Journal of Electrical Power & Energy Systems 59, 155–165 (2014)
18. Tian, H., Yuan, X., Ji, B.: etal: Multi-objective optimization of short-term hydrothermal scheduling using non-dominated sorting gravitational search algorithm with chaotic mutation. Energy Conversion and Management 81, 504–519 (2014)
19. Wu, L.H., Wang, Y.N., Zhou, S.W., et al.: Environmental/economic power dispatch problem using multi-objective differential evolution algorithm. Electric Power System Reasearch 80, 1171–1181 (2010)
20. Feng, B., Lai, F.: Multi-attribute group decision making with aspirations: A case study. Omega 44, 136–147 (2014)

Li-Ion Battery Management System for Electric Vehicles - A Practical Guide

Jing Deng[1], Kang Li[1], David Laverty[1], Weihua Deng[2], and Yusheng Xue[3]

[1] School of Electronics, Electrical Engineering and Computer Science,
Queen's University Belfast, Belfast, BT9 5AH, UK
[2] Electric Power Engineering, Shanghai University of electric power, Shanghai,
200090
[3] State Grid Electric Power Research Institute, 210003, Jiangsu, China

Abstract. Electric vehicles (EVs) are becoming more popular and have gained better customer acceptance in the past few years due to the improved performances, such as high acceleration rate and long driving distance from a single charging. Recent research also shows some promising benefits from integrating EVs with power grid. One of these is to use EV batteries as distributed energy storage. As a result, the excessive electricity generated from renewable resources can be stored in EVs and release to the power grid when needed. However, compared to traditional Nickel-cadmium and lead-acid batteries, Li-ion battery only can be operated in a narrow window, and needs to be properly monitored, managed and protected. This issue becomes severe when it is deployed for large applications, such as EVs and centralised electricity storage, where a large number of Li-Ion cells are interconnected to provide sufficient voltage and current. The solution mainly relies on a robust and efficient battery management system (BMS). This paper presents a brief review on the features of BMS, followed by a practical guide on selecting a commercial BMS from the market and designing a custom BMS for better control of functionalities. A Lithimate Pro BMS from Elithion is used to demonstrate the effectiveness of BMS in managing and protecting Li-Ion cells during the charge and discharge phases.

Keywords: Lithium-ion Electric vehicle, battery management system, intelligent charging, battery modelling.

1 Introduction

Reducing greenhouse gas emission and fossil fuel consumption are becoming the top issues for future energy systems. UK has set the target of cutting 80% (against the 1990 baseline) greenhouse gas emission by 2050. To achieve this target, deploying renewable electricity generation and promoting electric vehicles are important because the conventional electricity generation contributes 35% of the total UK emissions, while the transportation system accounts for another 21% [1]. Although the use of renewable resources, such as wind and solar, is

K. Li et al. (Eds.): LSMS/ICSEE 2014, Part III, CCIS 463, pp. 32–44, 2014.

increasing in electricity generation, their intermittent, variable, and uncertain nature significantly limit their acceptance in the power system.

Over the past decades, Hybrid electric vehicle (HEV) and battery electric vehicle (BEV) are widely manufactured. A few examples of commercially available BEVs are Tesla Model S, Nissan Leaf, Chevrolet volt, BMW i3, Renault Fluence Z.E, Honda Fit EV, and BYD E6. While EV batteries need to be charged from a power grid, they also can be potentially used as energy storage devices in the power system when EVs are not being used [2]. This provides an opportunity for better acceptance of intermittent renewable power. In off-peak hours, excess electricity can be stored in EV batteries through charging. When the load of the power system is heavy at peak hours, the stored electricity can be released to the grid through discharging. Obviously, this process relies on a bi-directional on-board charger which has attracted lots of R&D worldwide [3]. However, frequent charging and discharging may cause damages to the batteries, thus a battery management system becomes vital in protecting and prolonging the durability of EV batteries.

A standard Li-Ion cell can be in the shape of cylindrical, prismatic, or pouch. The operating voltage is usually between 2.5v and 4.2v, and the energy capacity ranges from 50mAH to 400AH (currently available in the market). This is far from enough to power an electric vehicle. A pure battery powered EV usually needs high voltage (150V-300V) to drive the motor and large capacity to grantee a satisfactory driving distance. Therefore, a EV battery pack often consists of a large number of Li-Ion cells connected in series and parallel. For example, Tesla Model S is a well-known electric car with some distinctive features, such as its specified 350 miles driving distance from a single charge. Undoubtedly, this is supported by a large battery pack which consists of 6,831 Panasonic NCR 18650 cells to provide 375 volts electricity and 56kWH capacity. Obviously, small size Li-Ion cells are easier and safer to be manufactured and used. They also provide great flexibility in customizing large battery packs. However, the risks in using large number of small cells in a battery pack are higher than using a single large cells. Specifically, Li-Ion cell must be operated in a narrow voltage and temperature window. Any situation beyond this range may either cause damage to the battery or lead to safety issues such as fire and explosion. Further, the cell status information cannot be simply obtained through the battery pack terminal voltage. This will lead to two dangerous situations. During charging process, a few cells might reach their top voltage limit quicker than others, and further change will push these cells into an unstable status even though the whole pack is not fully charged. Similarly, during discharging process, some cells might reach their low voltage limits quicker than others, and further discharging beyond this point may cause damage to these cells even though other cells still hold considerable charge. These potential risks are mainly caused by non-uniformity of cells interconnected in a battery pack. Although the quality of Li-Ion cells has been improved over the past decades, the variations in chemical and physical components inside a cell still cause non-uniformity. Thus over charging and deep discharging of battery cells can occur if the status of each cell is not monitored.

Due to the aforementioned reasons, a battery management system (BMS) becomes a vital part in Li-Ion battery applications. The functions of BMS include measurements of cell voltage, current, and temperature, protecting cells from unstable conditions (e.g., over charging, deep discharging, high current flow, either too high or too low operation temperature), balancing cells to maintain high performance, and indicating batter state of charge (SOC), usable capacity, internal resistance, and state of health (SOH). While measurements, protection, and balancing can be easily achieved through different sensors and actuators, the estimation of SOC, usable capacity, internal resistance, and SOH strongly depends on some modelling algorithms embedded in BMSs [4].

In Li-Ion battery applications, such as electric vehicles, the BMS can either be purchased from the DIY market or designed in-house. Most car manufactures choose to design their own BMS for better control of the quality and functionality. By contrast, lots of third-party BMSs are available in the market, such as Lithiumate, Orion BMS, and MiniBMS. More options will inevitably lead to difficulties in choosing the most suitable BMS for a specific application. This paper will discuss some common criteria in comparing and selecting a BMS for electric vehicles. When designing a custom BMS, it is recommended to use special battery management ICs, such as BQ76PL536A-Q1 from Texas Instruments, LTC6804-2 from Linear Technology, ATA6870 from Atmel, AD7280A from Analog Devices, and MAX14921 from Maxim integrated. Obviously, there are some limitations in using application specific ICs, but it requires less designing effort compared with those based on general purpose ICs.

The rest of the paper is organized as follows: Section 2 provides a comprehensive review of BMS and its functionalities. The criteria in selecting market available BMS and the approach in building a custom BMS will be introduced in section 3. Integrating EVs with smart grid requires advanced control and optimization which will be touched in section 4. Finally, a short summary will be provided in section 5.

2   Features of BMS

Similar to a feedback control system, a BMS monitors the status of Li-Ion cells, takes actions when necessary to protect the battery, and optimizes the battery performance to prolong its service life. A typical set up of BMS in electric vehicle is illustrated in Fig 1.

2.1   Li-Ion Cell Characteristics

Li-Ion cell represents a class of rechargeable batteries that involve the movement of Lithium ions. During the charging process, Lithium ions move from anode to cathode, and move back to anode when discharging. The cell capacity is usually determined by the size of electrode which is used to hold lithium ions. Depending on the cathode material, different li-ion cells can be made, such as $LiCoO_2$ (LCO - Lithium Cobalt Oxide),$LiMn_2O_4$ (LMO - Lithium Manganese), $LiFePO_4$ (LFP - Lithium Iron Phosphate), and $LiNiMnCoO_2$ (NMC

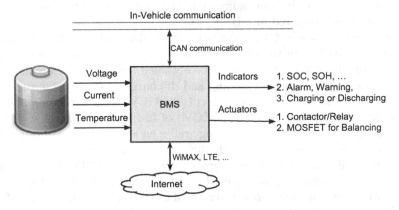

Fig. 1. System configuration of EV battery management system

- Lithium Nickel Manganese Cobalt Oxide). The standard voltage, energy and power density, and safe operation window vary with different chemistry. For example, LCO has higher energy density, but lower safety level. Thus it is widely used in portable electronics. EV manufacturers also have their own preference in choosing Li-Ion cells. For instance, Nissian, Chevrolet, and Renault use LMO, Tesla choose NCA (Lithium Nickel Cobalt Aluminum Oxide, $LiNiCoAlO_2$), and BYD prefers LFP, while Honda select NMC [5].

The capacity of Li-Ion cell is measured by Amp-Hour. For instance, an 40AH CALB Li-Ion cell can be discharged under the current of 40A (1C) for one hour (or 80A, 2C for half hour, or 20A, 0.5C for two hours). However, by considering battery internal resistance, the actual discharging time is shorter, especially at high current rate as more energy is wasted internally in form of heat. The charging of Li-Ion battery usually follows two stages, constant current and constant voltage. The first stage takes about one hour at 1C charging rate (or two hours at 0.5C rate) to reach 85% of its rated capacity, and the second saturation stage takes another 3 hours for the battery to reach its full capacity. However, practically, it is not recommended to charge Li-Ion batteries full as continuously suffering from high voltage stress reduces battery life.

2.2 Cell and Battery Pack Monitoring

The main physical parameters to be measured by a BMS are cell voltage, cell temperature, and battery pack current. Sampling rate can be 1Hz or higher, and recommended accuracy are 10mV for cell voltage, 0.5-1% for current, and 0.1 °C for temperature [6]. As the measured values are analogue signals, an analogue to digital converter (ADC) is usually placed before they can be processed digitally. The resolution for such ADC can be 12bit or higher, and one or more multiplexer may be needed depending on the number of ADC channels available. The most popular current sensors are based on hall effect, which can be embedded in a BMS or mounted on the terminal cable. Both charging and discharging current have to be measured, thus either one bidirectional current sensor of two unidirectional current sensors are required in a BMS.

Open circuit voltage (OCV) is the battery terminal voltage when its internal equilibrium is reached in the absence of load. It provides a good indication of battery status. However, OCV cannot be obtained through the aforementioned voltage measurements during charging and discharging process. Thus some researches have been carried out to estimate the OCV, which are often based on an equivalent electric circuit model (EECM) of Li-Ion cell [7]

For SOC indication, current technology relies on a combination of cell voltage and current integration [4]. Although such method lacks accuracy, it is simple and can be easily embedded in a commercially available BMS. Advanced algorithms have also been proposed over the past few years. A OCV-SOC look-up table may be used if an accurate estimation of OCV can be obtained [8, 9]. Experienced people may be able to estimate battery SOC based on prior knowledge. In this case, a fuzzy logic model can be developed to predict SOC [10]. However, the embedded knowledge will be strongly related to specific applications. This limitation also applies to other model based approaches, such as neural network [11, 12] and Kalman filter [13, 14].

SOH mainly represents the current condition of a battery/cell compared to a freshly manufactured one. However, as the definition of SOH has not yet been clearly defined, the SOH value from one BMS may differ from another. Generally, SOH provides real-time information of battery capacity, age, and the internal resistance (IR). Battery capacity can be measured during a full charge and discharge cycle while the age and resistance strongly relies on a complex model [15]. The internal resistance may be meaningless to the user, but it is useful in IR compensation, OCV and SOC estimation [16]. By contrast, the age is a very useful indicator of the remaining battery life. It is known to be related to the capacity, and a battery is considered to be depleted if the actual capacity is below 80% the the rated value. It is worth mentioning that, in most cases, an unhealthy battery pack is usually caused by one or two individual cells, and replacing those cells can prolong the life of whole battery pack.

2.3   Protection and Control

It is clear that Li-Ion battery must be operated within a small window, and any value (e.g. cell voltage, current, temperature) outside this area may cause damage to the cells and safety issues (e.g. fire). The high voltage and high current from EV battery also introduce potential risk to the driver and passenger. Thus a BMS is a vital component in electric vehicles. The basic protections from BMS are through disconnecting the charge or load to prevent over charging or deep discharging, while advanced control algorithm can be implemented at higher level to regulate energy flow into and out of the battery pack.

The protection to each cell is as important as to the whole battery pack. This includes the cell charging and discharging voltage, maximum continuous and peak current, and upper and lower limit of operating temperature. A BMS should shut down the charger or load for too high or too low of cell voltage.

To achieve this, one or two contactors may be required if the charger and/or motor controller cannot be switched through an external signal. Depending on the weather condition, an air fan or liquid cooling and/or heating may be required to protect the battery pack. In such case, the BMS should be able to regulate the heating/cooling through either on/off control or PWM (Pulse-width modulation). The former can be simply implemented through a contactor or relay while the latter relies on a solid state switcher.

Advanced control algorithms, such as scheduling charge and discharge from and to the power grid, switching between standard and fast charging, and regulating the current flow to suit different driving modes (e.g. Eco, standard, and sport), may not be achievable through a BMS, and higher level of energy management system (EMS) or electric vehicle control system (EVCS) can be adopted. Section 4 will introduce the integration of electric vehicles with smart grid to maximize the acceptance of renewable energy resources.

2.4 Performance Optimization

The aim of optimization from a BMS is to prolong the battery life and maximize energy capacity. These two factors are coupled as reduced capacity indicates shorter battery life. A trade-off needs to be considered when pursuing optimal values for both factors. In practice, Li-Ion battery has limited charge/discharge cycles, which are mainly counted during a full charge and discharge cycle. If the battery is used between 40% - 80% of its rated capacity, its service life will be significantly extended. Most BMSs have the flexibility in configuring battery operation, such as thresholds for charge and discharge voltage, limits on charger and load current, and balancing. The user's behaviour also affects the performance of Li-ion battery. In order to maintain healthy battery conditions, some advices are given below for using Li-Ion batteries in electric vehicles.

1) Leave plenty of space between cells and build an air way to prevent temperature accumulation.
2) Do not charge the battery over 85% of its rated capacity (or set the upper threshold 0.1-0.2v lower than the standard level).
3) Do not discharge the battery lower than 25% of its rated capacity (refer to data sheet for optimal lower threshold).
4) Limit the charge current to 0.5C - 0.7C. The so-called "fast charge" by car manufacturers should be avoided.
5) Avoid long-time acceleration while driving an EV, as high discharge current will cause increased temperature.
6) Leave 40% charge in the battery if the EV is not being used for a long period.
7) Park an EV in a cool and dry place when possible. Expose it to the sun for long period will introduce high stress on the Li-Ion battery.

Besides the above general advices, cell consistency also affects battery performance. Ideally, all the cells connected in parallel and in series have the same specifications. Their SOC changes at the same rate during charging and discharging process. However, non-uniformity may occur after the battery has being used for a period. This is not an issue for traditional lead-acid battery. But for Li-Ion battery, the performance can be significantly affected. For example, as shown in Figure 2a, the variation is small as would be expected in most cases. The battery can be charged and discharged normally without additional care. However, in an extreme case as shown in 2b, the variation is large enough so that the battery can not be charged and discharged, leading to a useless battery until being re-balanced. The reason is that the third cell is already at its upper 85% SOC limit so additional charged is not recommended to prolong the battery life. Similarly, the 7th cell is at the lower discharge limit, and BMS is set to avoid discharge lower than this level. In order to restore the battery to a usable condition, balancing is required, which is a necessary functionality by default in BMS.

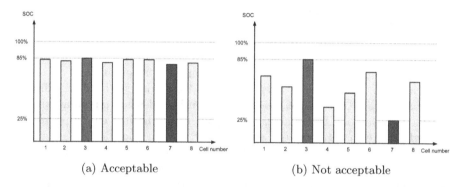

(a) Acceptable   (b) Not acceptable

Fig. 2. Cell non-uniformity in a 8 cells battery module

Basically, balancing it to bring each cell in a battery pack to the same SOC level. This can be achieved by ether removing extra charge from the most charged cells or adding charge to the less charged cells. Depending on the complexity of balancing circuit and algorithm, it can be simply based on cell terminal voltage, or the history of SOC from each cell [15]. Moreover, active balancing is known to be complicated but more efficient than passive balancing. The former moves energy from the most charged cell to less charged ones, while the latter removes extra charge as heat [17].

2.5 Communication

Conventional BMSs are usually built as a standalone device which does not need external communications. Both the sensors (e.g. voltage, current, temperature) and actuators (e.g. contactor, relay) are directly connected to the BMS. Additional analog and digital IOs are available to connect with charger and/or load when necessary. In the case of two of more BMSs are needed in an EV, interconnection between each BMS should be available. Due to the fast developed information technologies, new functionalities are continuously added to the

BMS. For example, lots of BMS has the ability to communicate with other systems in an EV through CAN (Control Area Network) bus as this is a standard protocol in automotive industry. Popular communication methods, such as USB, RS232, Ethernet, WIFI, and WiMax may also be available for the purpose of upgrading firmware, recording data, and advanced control. Bus communication significantly reduced the number of wires needed in a BMS, and increased the flexibility in system configuration.

3 Selecting and Designing BMS

As the BMS is an essential and important part of an electric vehicle, popular car manufacturers prefer to design their own BMS. In the DIY and services market, a lots of third party BMSs are available to choose. Researchers may buy or design a BMS based on their objectives.

3.1 Commercially Available BMSs

There is no standard of what functionalities a BMS should have. This causes difficulty in choosing the right BMS for a specific application. Simple BMS may only be able to monitor the voltage and temperature of Li-Ion cells, while sophisticated BMS can protect the battery and optimize its performance. Some are designed for specific applications, while others are designed for general purpose.

The topology of BMS can be divided into three categories, centralized, modularized, and distributed monitoring. A centralized BMS has direct connections with each cell, thus a lot of cables are required for large battery packs. Centralized BMSs are usually cheaper due to compact design. Some manufacturers also claim it is easier to install than distributed BMS. Modular BMS usually has a slave module for data acquisition and a master module for data processing and control. Each slave module can monitor $6 - 60$ cells, and information is exchanged through CAN bus. In distributed BMS, a monitoring board is attached to each cell, and single or two cable bus communication is adopted to connect all cell boards with the main controller. Table 1 shows some popular BMSs with different topologies.

When selecting a BMS for battery applications, simple design might be preferred as it is more reliable. However, for electric vehicles, better protection and advanced optimization usually require complicated design. Specifically, the following criteria can be considered when selecting a BMS for EV applications.

- **Accuracy**. Recommended accuracy for voltage, current, temperature, and SOC are $\pm 10mV$, $0.5 - 1\%$, $\pm 0.1\,^{\circ}C$, and $3 - 5\%$.
- **Range**. Recommendations are $1 - 5V$ for single cell voltage monitoring, $1 - 500A$ for measuring battery pack current, and $-40\,^{\circ}C$ to $+85\,^{\circ}C$ for temperature monitoring.
- **Number of cells**. EV application may require more than 100 series connection. This is usually an issue when considering a centralized BMS.

Table 1. Popular BMS in the market

Model	Company	Country	Topology	No. Cells	Comm
BMS21	REAP System	UK	Centralized	$4-21$	CAN
BMS 9R	REC	Slovenia	Modularized	$4-225$	CAN
BS313	Guantuo Power	China	Modularized	>1024	CAN
CM series	EV-Power	Australia	Distributed	1-Any	
EK-FT-12	Ligco	China	Distributed	$2-400$	CAN
EMUS BMS	Elektromotus	Lithuania	Distributed	$2-255$	CAN
Lithiumate	Elithion	USA	Distributed	$2-255$	CAN
Orion BMS	Ewert Energy	USA	Centralized	$1-180$	CAN
YN.EV-03	Huizhou Power	China	Modularized	>1024	CAN

- **Balancing**. Active balancing is preferable but complicated. Most BMSs in the market use passive balancing. The balancing current is also important as it determines the time needed for such process.
- **Communication**. CAN bus is recommended, while other techniques, such as USB, ethernet, RS232/485 add more flexibilities.
- **Power consumption**. BMSs are usually powered by the battery pack, thus lower standby and operation current are desirable.
- **Data logging**. Some BMS, such as BS313 from Guantuo power electronics, can record the data to a SD card for further processing and analysis.
- **Battery compatibility**. Some BMS are designed for specific type of Li-Ion batteries while others support wide range of choices.
- **Robust and flexible**. Solid case protection, high voltage isolation, cooling and heating, and circuit breaker can guarantee all weather protection; and GPIOs (general purpose Inputs and Outputs) can provide better flexibilities in system configuration.
- **Upgradable firmware**. Improvements may be released in new version of embedded software, such as more accurate SOC estimation. It should be easy to upgrade to the latest version of firmware. Some BMS can be automatically upgraded through an internet connection while others need to manually download file and upload to BMS through a data cable (USB or RS232).

3.2   BMS Setup

In this paper, Lithiumate Pro battery management system from Elithion is used for research purpose. As a distributed BMS, Lithiumate provides small PCB boards to be attached to each cell and a main controller. The cell boards are interconnected through a single wire, and the connection between terminal cell boards (positive and negative side) and main controller relies on two wires. The system configuration is illustrated in Fig. 3.

In this system, the battery consists of five WINA LiFePO4 10Ah cells connected in series. A XANTREX XFK-300-9 programmable power supply is used

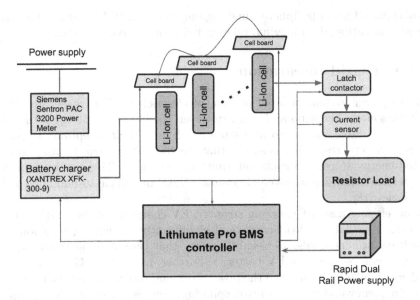

Fig. 3. Block diagram of BMS setup in this study (Red line carries power to and from the battery, and blue line carries control signal)

to charge the battery. The charging voltage and current are set through a customized DAQ box connected to a PC. The charger is also switched by the BMS to prevent over charge. The discharge is through a large variable resister, and the process is monitored and protected through a current sensor and latch contactor connected to BMS. After a few charging and discharging cycles, this system has successfully identified imbalance where the fourth cell has much lower SOC level than others.

3.3   Designing Custom BMS

In some cases, the market available BMS does not meet the requirement for a specific application. For example, more accurate voltage monitoring and SOC estimation are pursuit. A customized BMS can be designed either based on general purpose ICs (integrated circuits) or application specific ICs. The former requires more development effort but easy to customize while the latter relies on a specific battery management ICs which may limits the design.

Due to the increased Li-Ion battery applications, many IC manufactures developed battery management solutions to ease the design effort. For example, LTC6802, LTC6803, and LTC6804 from Linear Technology, AD7280A from Analog Devices, ATA6870 from Atmel, Max 14921 from Maxim integrated, and bq76PL536 from Texas Instruments. The ICs (EL01 and EL01) used in Lithiumate are also available for building custom BMS. These battery managements ICs only provide measurement of cell voltage and temperature, and connections for cell balancing. A micro-controller is still needed to process data and generate control signals. Others peripherals in designing a custom BMS may include CAN bus communication, data logging, relay switch, current sensing, etc..

4   Smart Grid Integration

In future power system, a large portion of electricity will be generated from renewable sources like wind and solar. However, the biggest challenge is their intermittent, variable and uncertain nature which limit their acceptance to power grid. Fortunately, research has shown that electric vehicles can be used as distributed energy storage to cache intermittent electricity. This can be achieved through smart charging and Vehicle-to-grid (V2G) or vehicle-to-building (V2B) technologies [18].

By developing a smart charging strategy, EV charging can be shifted to off-peak period to flat the load curve and significantly reducing generation and network investment needs. Renewable energy that cannot be accepted by the power grid can be stored in EV batteries. Further, with the V2G and V2B, the stored energy can be released to the power grid, and provides regulation services such as frequency and voltage control, spinning reserves and peak-shaving capacity, thus reducing both the operational costs for existing plants and investment in building new power stations.

In reality, the above strategy may be difficult to implement as frequent charging and discharging cycles will reduce the battery life. Besides some compensation scheme that may help to build such structure, a more robust and reliable BMS is vital to keep the battery cells in satisfactory health condition. When necessary, an energy management system (EMS) may be required align with a BMS [19]. EMS takes the control of energy flow at higher level by utilising information from other EVs, power system, and user behaviours. Advanced machine learning, optimal control and telecommunication can be build in an EMS to ease the load of BMS.

5   Summary

Electric vehicles are becoming popular either as transportation or distributed energy storage. One of the most important part of EV is the battery management system (BMS) which is used to protect battery cells, optimise their performance, and prolong the battery life. This paper first explained the importance of BMS in a Li-Ion battery applications. The features include cell voltage, temperature and current monitoring and protection, SOC, OCV, and SOH estimation, and balancing. CAN bus is the most popular communication method used in automotive industry, thus it is also becoming a standard in BMS design.

Depending on available resources and budget, a BMS can either be purchased from the market or designed in-house for better control of the features and functionalities. Some criteria in selecting a commercial BMS are introduced. For example, the accuracy and range of measurements, number of cells it can manage, the balancing technology, power consumption, communication, data logging, etc.. As one of popular choices, Lithiumate Pro BMS from Elithion electronics was used in this paper to illustrate the system setup and Li-Ion cell monitoring. Although this is a small battery charging and discharging system consisting of five Li-Ion cells, it can be easily scaled up to hundreds of cells with larger capacity and higher current. Designing a custom BMS can be based on general ICs or application specific ICs. The former requires more effort in design while the latter has some limitations in BMS features.

Finally, Electric Vehicle can be used as distributed storage in smart grid. This is achieved through smart charging and vehicle-to-grid (V2G) or vehicle-to-building (V2B) technologies. Both techniques strongly rely on a robust and effective Battery management system in protecting Li-Ion cells and maintaining their healthy conditions.

Acknowledgments. This work was supported by the European Social Development Fund and InvestNI under the Proof of Concept Scheme and UK EPSRC under grant EP/L001063/1. It is also partially supported by the Chinese Scholarship Council, National Natural Science Foundation of China under grants 51361130153, 61271347 and 61273040, Science and Technology Commission of Shanghai Municipality under Grant 11ZR1413100, and International Corporation Project of Ningbo under Grant 2013D10009.

Reference

[1] Department of Energy & Climate Change. 2013 UK greenhouse gas emissions, provisional figures and 2012 UK greenhouse gas emissions, final figures by fuel type and end-user. Statistical release (March 2014)

[2] Lund, H., Kempton, W.: Integration of renewable energy into the transport and electricity sectors through v2g. Energy Policy 36(9), 3578–3587 (2008)

[3] Tomić, J., Kempton, W.: Using fleets of electric-drive vehicles for grid support. Journal of Power Sources 168(2), 459–468 (2007)

[4] Ng, K.S., Moo, C.-S., Chen, Y.-P., Hsieh, Y.-C.: Enhanced coulomb counting method for estimating state-of-charge and state-of-health of lithium-ion batteries. Applied Energy 86(9), 1506–1511 (2009)

[5] Lu, L., Han, X., Li, J., Hua, J., Ouyang, M.: A review on the key issues for lithium-ion battery management in electric vehicles. Journal of Power Sources 226, 272–288 (2013)

[6] Rahimi-Eichi, H., Ojha, U., Baronti, F., Chow, M.: Battery management system: An overview of its application in the smart grid and electric vehicles. IEEE Industrial Electronics Magazine 7(2), 4–16 (2013)

[7] Chiang, Y.-H., Sean, W.-Y., Ke, J.-C.: Online estimation of internal resistance and open-circuit voltage of lithium-ion batteries in electric vehicles. Journal of Power Sources 196(8), 3921–3932 (2011)

[8] Plett, G.L.: Extended kalman filtering for battery management systems of lipb-based hev battery packs: Part 2. modeling and identification. Journal of Power Sources 134(2), 262–276 (2004)

[9] Lee, S., Kim, J., Lee, J., Cho, B.: State-of-charge and capacity estimation of lithium-ion battery using a new open-circuit voltage versus state-of-charge. Journal of Power Sources 185(2), 1367–1373 (2008)

[10] Salkind, A.J., Fennie, C., Singh, P., Atwater, T., Reisner, D.E.: Determination of state-of-charge and state-of-health of batteries by fuzzy logic methodology. Journal of Power Sources 80(1), 293–300 (1999)

[11] Cai, C.H., Du, D., Liu, Z.Y.: Battery state-of-charge (soc) estimation using adaptive neuro-fuzzy inference system (anfis). In: The 12th IEEE International Conference on Fuzzy Systems, FUZZ 2003, vol. 2, pp. 1068–1073. IEEE (2003)

[12] Charkhgard, M., Farrokhi, M.: State-of-charge estimation for lithium-ion batteries using neural networks and ekf. IEEE Transactions on Industrial Electronics 57(12), 4178–4187 (2010)

[13] Plett, G.L.: Extended kalman filtering for battery management systems of lipb-based hev battery packs: Part 3. state and parameter estimation. Journal of Power Sources 134(2), 277–292 (2004)

[14] He, H., Xiong, R., Zhang, X., Sun, F., Fan, J.: State-of- charge estimation of the lithium-ion battery using an adaptive extended kalman filter based on an improved thevenin model. IEEE Transactions on Vehicular Technology 60(4), 1461–1469 (2011)

[15] Andrea, D.: Battery Management Systems for Large Lithium Ion Battery Packs. Artech House (2010)

[16] Saint-Pierre, R.: A dynamic voltage-compensation technique for reducing charge time in lithium-ion batteries. In: The Fifteenth Annual Battery Conference on Applications and Advances, pp. 179–184. IEEE (2000)

[17] Moore, S.W., Schneider, P.J.: A review of cell equalization methods for lithium ion and lithium polymer battery systems, pp. 01–0959. SAE Publication (2001)

[18] Kempton, W., Letendre, S.E.: Electric vehicles as a new power source for electric utilities. Transportation Research Part D: Transport and Environment 2(3), 157–175 (1997)

[19] Tie, S.F., Tan, C.W.: A review of energy sources and energy management system in electric vehicles. Renewable and Sustainable Energy Reviews 20, 82–102 (2013)

Optimization and Simulation of Dynamic Stability for Liquid Cargo Ships

Yi-Huai Hu[1], Juan-Juan Tang[1], Yan-Yan Li[1], and She-Wen Liu[2]

[1] Shanghai Maritime University, 1550 Haigang Avenue, Pudong New Area, Shanghai, P.R. China
[2] China Offshore Technology Center, ABS Greater China Division, 5th Floor, Silver Tower, 85 Taoyuan RD, Shanghai, P.R. China
yhhu@shmtu.edu.cn

Abstract. Because of the particularity of liquid cargo transported by liquid cargo ships, traditional ship stability calculation method is not satisfied enough, especially in the respects of matching degree and calculation precision. In order to solve these problems, this paper proposes an optimal method for stability calculation of sea-going liquid cargo ship. Corrections of trimming and free surface are considered, besides, new data preprocessing is adopted. The simulation is demonstrated that this method could satisfy the stability calculation requirements of these liquid cargo ships with good expandability.

Keywords: Ship stability calculation, modeling and simulation, liquid cargo carrier, ship loading system, ship trim.

1   Introduction

Appropriate stability, floating condition and strength should be ensured in the process of loading, transporting and unloading of ships. All ship conditions should be checked by reference of loading manual, both in the process of ship stowage and determining the sequence of loading and unloading in order to determine whether stowage plan and cargo handling sequence meet the requirements of vessel particulars. Otherwise, the original plan should be revised[1-3].

The precision of stability calculation has a decisive effect on the stowage system reliability of liquid cargo ships. Compared with dry cargo vessels, liquid cargo ship stowage has the following characteristics:

(a)   Stability of the ship is more related to its free surface.
(b)   Loading calculation depends more on information directly transmitted from the sensor in ship's hold.
(c)   The stability criterion has higher precision requirements.

By now, the respects mentioned above have not been taken into account in the ship stability calculation. Besides, traditional stowage system requires heavy work and its real-time calculation is not always satisfied.

This paper proposes an optimal method based on the stability requirements of liquid cargo ships. With formula optimization, which improves the arrangements of sensors

K. Li et al. (Eds.): LSMS/ICSEE 2014, Part III, CCIS 463, pp. 45–54, 2014.

and the way of data processing, the precision and matching of traditional stowage system are improved. Thus, the simulation of real-time stowage process could be more realistic and could be used for practice training for senior seafarer crews.

2   Ship Stability

2.1   Basic Stability Calculation

The ability of a ship to return to its equilibrium position after it is displaced there from because of the external forces and external torques M_H is called ship stability. Ship heels under the action of static stability, which supposes the angular velocity is zero. When heeling moment is equal to righting moment, ship will reach an equilibrium state and ship stability could be measured by M_R. On the contrary, if ship is heeled by external forces, only when the work done by external moment is equal to that done by righting moment, can the ship heeling be stopped. Then the ability of a ship to resist external forces is measured by the work done by stability moment, instead of static moment.

$$M_R = \Delta \cdot GZ \tag{1}$$

Where:

Δ——current displacement volume;

GZ——vertical distance from gravity center to action line of buoyancy, in other words, righting arm, which could be classified as static stability lever and dynamic stability lever.

Under certain loading conditions, righting arm changes with heeling angle and this relation could be described as static stability curve. While, the dynamic stability lever is numerically equal to the area enclosed by curve of static stability lever. Thus it could be seen that GZ curve can be used in checking process no matter it is static stability or dynamic stability (see Fig.5).

Observation data is what we usually get from data acquisition. In order to simplify numerical calculation and procedures, cubic spline method is applied to fit GZ curve.

No matter it is static stability or dynamic stability, stability is always classified as initial stability or stability at large angle according to ship heeling angle. If ship heeling angle is less than 15°, initial metacentric height is a fundamental symbol of ship stability.

$$GM = KM - KG \tag{2}$$

Where:

KM——height between transverse metacenter and baseline and it could be found in hydrostatic data;

KG——height between gravity center and baseline and it could be described by the following equation,

$$KG = \sum_i P_i \cdot Z_i / \Delta \tag{3}$$

When stability is calculated at large angle, static stability lever should be taken into account.

$$GZ = KN - KH \tag{4}$$

Where:

KH——lever of weight stability, which is determined by the height of ship gravity center and ship heeling angle.

KN——lever of form stability, which can be looked up in the stability cross curve according to ship displacement and heeling angle.

2.2   Criteria of Stability and Floating Condition

IMO and ship classification societies of different countries all specify minimum requirements for ship stability. Requirements for stability criteria of seagoing vessel traveling on international routes are shown as following[5]:

(1) The initial metacentric height GM shall not be less than 0.15 m;

(2) The area under the curve of static stability lever should not be less than 0.055 m·rad up to 30° angle of heel and not less than 0.090 m·rad up to 40° or the angle of flooding Φ_f if this angle is less than 40°;

(3) The static stability lever should be at least 0.2 m at an angle of heel equal to 30°;

(4) The maximum static stability lever should occur at an angle of heel preferably exceeding 30° ;

(5) For ships whose length are more than 24m, weather criteria K is equal to M_q/M_f and should not be less than 1.

Requirements for floating condition are as following:

(1) Minimum after draft: $d_F \geqslant 0.012L + 2$ (m)

(2) Mean draft: $d_M \geqslant 0.02L + 2$ (m)

(3) When ship is fully loaded(80%~100%), trimming by the stern is preferably between 0.3 and 0.6m; when ship is half loaded, trimming by the stern is preferably between0.6~0.8m; when ship is under-load, trimming by the stern is preferably between 0.9~1.9m. If floating condition does not meet a criterion, trimming should be corrected[8].

3   Optimization of Stability Calculation

3.1   Correction of Stability Parameters of Liquid Cargo Ships

Stability calculation of traditional stowage systems is mostly based on the assumption that flotation center remains steady when the ship heels or trims. However, when a ship heels or trims to a large angle, there are always errors in the stability calculation. In addition, when ship trims to a large angle, the shape of the underwater part will be different from that when ship is floating on even keel, which leads to changes in initial stability and stability at large angle.

Due to the requirements of IMO, all ship information contains values of stability parameters under different trimming conditions, but they are scattered in loading manual. The integrated information are shown in table.1.

Table 1. Hydrostatic data (different trimming conditions)

Trim=0m					
d_M	Δ	KB	LCB	KMT	MTC
11.1	96674	0.991	1.091	19.643	1543.5
11.2	97658	0.923	1.023	19.584	1553.4
11.3	98643	0.857	0.957	19.526	1563.3
11.4	99627	0.791	0.891	19.469	1573.2
...	...	...	...	...	...
Trim=-1m					
d_M	Δ	KB	LCB	KMT	MTC
11.1	96658	0.994	1.094	18.863	1523.8
11.2	97635	0.926	1.026	18.804	1535.6
11.3	98632	0.860	0.960	18.748	1543.2
11.4	99584	0.793	0.893	18.675	1556.4
...	...	...	...	...	...

The reduction of initial metacentric height (δGM) caused by the effect of free surface could be described by the following equation.

$$\delta GM = \sum \rho \cdot i_x / \Delta \tag{5}$$

Where,

ρ——liquid density;

i_x——the moment of inertia of liquid level ,which is defined as $i_x = l \cdot b^2 / 12$,where b is width of cabin, l is length of compartment and the unit of length is meter.

After correction of free surface, righting arm can be described as:

$$GZ = GZ - \delta GZ_\theta = GZ - \sum M_{i\theta} / \Delta \tag{6}$$

Generally, stability calculation are based on the assumption that free surface are independent when ship inclines. However, restrained free surface, which means the upper liquid level in the ship's hold could reach the top of hold ,in other words, it will not cover the whole top of hold. This situation is common in the transportation process of liquid cargo ships. In order to precisely estimate the effect of free surface on stability, the correction under the condition of fully loaded rectangular liquid is studied and the reduction of initial metacentric height (δGM) caused by the effect of free surface is defined as:

$$\delta GM = M_H / (\Delta \cdot \sin \theta) \tag{7}$$

Computational formula of M_H could be found in the reference[7], and the constraint conditions could be obtained from sensors inside a hold. Sensors are generally classified as pressure type and radar type, and the layout of pressure sensors is shown in Fig.1. Pressure sensors are used to measure the pressure of inactive gas. For a liquid cargo ship, in order to precisely measure liquid level and neutral height, there should be a set of sensors on both the left and right sides of central cabin section.

3.2   Data Preprocessing

The utilization of stowage system is based on plenty of ship information, including main characteristics of vessel, cabin arrangement plan, capacity tables, loading lists, hydrostatic data, stability cross-curve, the weight distribution of the whole ship, buoyancy distribution plan, table of extreme height of gravity center, Bonjean's curves and so on. The data above-mentioned is preferably presented in electronic data, and data processing is simplified. In this way errors caused by pattern digitization could be avoided[8].

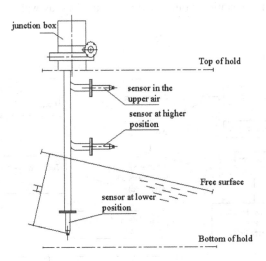

Fig. 1. Pressure sensor layout

These data are usually in different forms. In order to make data more convenient and intuitive, this paper adopts a storage method combining structured table and discrete points fitting. One method is database technology of ADO(Active Data Object),which takes access as a database engine and saves ship information in the form of data sheet. These data in database are relatively independent enhancing the universal property of the system, enabling information management and easy access. For example, the hydrostatic data is saved in database in the form of data sheet, taking ship displacement or draft as abscissas and taking gravity center coordinate or buoyant center coordinate or other stability parameters as longitudinal coordinates. Another method is to utilize a cubic spline fitting method to convert the curve into mathematical polymerization[9].For instance, capacity tables could not only reflect load capacity of all holds, but also contribute to effective data reading.

4   Simulation Example

In order to verify the optimization effect of the modified stability calculation method of liquid cargo ships, Visual Basic was used as programming language, and an LNG carrier "DaPengHao" was taken as an example to simulate the process of loading. The length of the ship is 291.5 m and the width is 43.35m. The ship's summer laden draft is 13.330m and the load displacement is 135000t.

4.1   Constitute of Simulation System

Righting moment M_R is an important symbol of restoration ability[4]. Stability calculation includes calculation and check of floating condition, initial stability, stability at large angle, static stability, dynamic stability, as well as check of weather criterion.

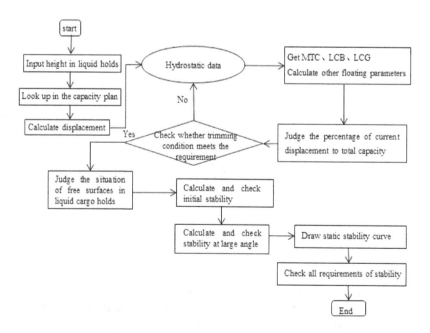

Fig. 2. Flowchart of stability calculation of liquid cargo ships

The basic flowchart of ship stability calculation is shown in Fig.2. Stability calculation could be started after hull form parameters, interior arrangements, etc. are set down. First, input the height of liquid level in all holds, liquid density and other parameters, also calculate the displacement according to capacity tables. Second, look up necessary parameters in hydrostatic data (even keel) and then calculate ship floating conditions (including draft difference, fore draft, after draft and so on) and judge the percentage of current displacement to total capacity. Third, check whether trimming condition meets the requirement of floating conditions .If it does, then judge the situation of free surfaces in liquid cargo holds one by one and then calculate, correct, check other stability parameters. Otherwise, select another value from hydrostatic data again according to trimming condition as shown in Fig. 2.

4.2   Design of Interface

As shown in Fig.3, the stowage system of liquid cargo ships consists of 8 system modules, two of which, digital sensor and database, are connected with computer. In order to realize dynamic simulation, digital sensors are replaced by sensor simulation module and the value of liquid level is read at intervals.

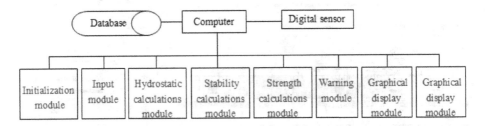

Fig. 3. Constitute of stowage simulation system

As shown in Fig. 4, the main interface of the system includes 7 parts: menu bar, calculation button box, display frame for parameters of floating condition, sensor frame, margin display frame, flow diagram window and stowage data input box.

The change of loading condition is represented by system-defined identifier for change. If there is a change in loading condition, the identifier will become true. However, modifying data under loading condition, pressing the READ button in Sensor Frame, adjusting parameters of software settings could all make identifier true. With the sensors in liquid hold, users could update stowage data either by manual operation or automatic operation, visual warning message could be sent, and check report could be generated in a preset form.

4.3   Process of Trim Modification

After loading condition judgment, situation of trim by the stern should be checked and adjustment calculation should be carried out. Besides, a calculation example is used to explain the process of adjustment calculation of this system.

Assuming that a ship is in port under a certain loading condition and, with the height got from sounding sensors, the calculated displacement is 98211.38t. First, interpolate the hydrostatic data (even keel) and get the value of MTC (Moment to change trim) and LCB (Longitudinal center of buoyancy), which is 1558.9 m and 0.9858 m respectively. In addition, the value of LCG, which is 0.512 m, could be calculated according to the loading condition. Second, use the formula to calculate the draft difference and the value is -0.2986m. Thus, it could be determined that the ship is trimming by the stern. Third, back to the hydrostatic data, interpolate data in table (trim of -1m) to get necessary hydrostatic parameters. Finally, interpolate data in tables again (trim of 0m and trim of -1m) according to the previously calculated draft difference and recall the results to finish the following calculation. All the calculated parameters of floating condition will display in the window. When the criteria of afloat condition does not meet the requirements, there will be a visual warning, and the numbers are red.

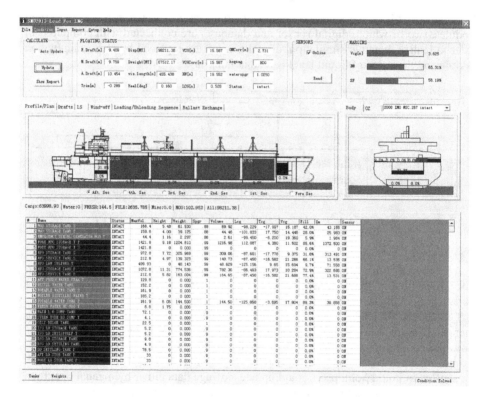

Fig. 4. The main interface of the simulation software

4.4   Precision of Calculated Results

After stability calculation, static stability curve is drawn out under the current loading condition as shown in Fig.5.

Performance parameter deviations between simulated results and data in ship loading manual are shown in Table 2. For the same loading condition, simulated results are almost in accord with the values given by loading manual, and the relative error is less than 5%. Besides, the strength curve and stability curve are both within the rage of vessel particulars.

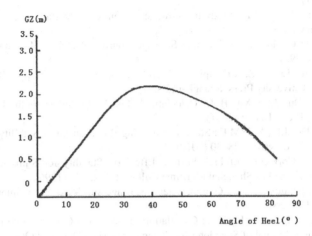

Fig. 5. Static stability curve given by software

Table 2. Contrast of performance parameters

particulars	Result given by software	Value in loading manual	relative deviation
Trim(m)	-0.289	-0.281	2.77%
Disp.(t)	98211.38	98240.2	0.03%
VCG(m)	15.587	15.879	1.87%
KM(m)	19.552	19.557	0.03%
Min.GZ(m)	2.731	2.791	2.15%

5   Conclusion

This paper proposes an optimization method for stability calculation of sea-going liquid cargo ships and applies it to the simulation of a liquid cargo ship. Compared to traditional stowage systems, the most significant characteristic it has is its higher precision. On one hand, it modifies the formulas of stability parameters to improve the matching degree between formulas and trimming, free surfaces and external sensors. On the other hand, it starts with the storage mode of original information to reduce iterated error. As now, there is compulsory requirements to take into account the effect generated from ship's trimming when making stability calculation, therefore, the reference value of this software to existing stowage systems is magnificent. Moreover, the favorable human-machine interaction, high-speed calculation, high accuracy and good expandability of this simulation software make it a good choice for simulation and teaching of stowage system for liquid cargo ships.

References

1. Gao, J.Y.: The Status and Prospect of Study On Ship Stability. Shanghai Shipbuilding (1), 15–24 (2003)
2. Yang, L.J., Wang, D.L.: Study on Stowage System of Bulk Barrier. Traffic Science and Technology (2003)
3. Yang, J.Q.: Computer-aided Tanker Stowage. Journal of Dalian Maritime College 18(4), 401–402 (1992)
4. Sheng, Z.B., Liu, Y.Z.: Principles of Naval Architecture, vol. (9), pp. 20–107. Shanghai Jiao Tong University Press, Shanghai (2003)
5. Shen, H., Du, J.L., Xu, B.Z.: Calculation of stability and strength. Dalian Maritime University Press, Dalian (2001)
6. Li, R.H., Du, J.L., Wu, M.F.: Stability Checking Based on Intact Stability Rules of 2008. Marine Technology (6), 28–30 (2010)
7. Qiu, W.C.: Correction of Free Surface Effect on Stability for Ships under Full Load Condition. Journal of Shanghai Maritime College (4), 23–24 (1996)
8. Song, L.X.: Study on Bulk Grain Carrier's Stowage and Making It Computerized. Dalian Maritime University (2002)
9. Wang, J.X., Hu, Y.H., Li, X.L.: Calculation of Bonjean's Curves Based on Cubic Spline Interpolation. Journal of Shandong Jiao Tong University (1), 64–67 (2011)

Numerical Study of Magneto Thermal Free Convection of Air Under Non-uniform Permanent Magnetic Field

Kewei Song, Yang Zhou, Wenkai Li, and Yuanru Lu

Department of Mechanical Engineering, Lanzhou Jiaotong University, Lanzhou,
Gansu, 730070, China
songkw@mail.lzjtu.cn

Abstract. Magneto thermal free convection of air in a square enclosure under a non-uniform magnetic field provided by a permanent magnet is numerically studied. The results show that the natural convection of air in the square enclosure under magnetic field is quite different from that under the gravity field. The local value of Nusselt number under the magnetic field can reach to a much higher value than the maximum local value under the gravity field. Relatively uniform distribution of Nusselt number can be obtained along the cold wall of the enclosure under the magnetic field. A permanent magnet with high magnetic energy product with B_r reaches to 3.5 Tesla can plays a comparative role on the average Nusselt number compared with that under the gravity environment.

Keywords: magneto thermal convection, permanent magnet, magnetic field.

1   Introduction

The heat transfer and flow characteristics of natural convection in enclosures has attracted much research over the years due to their many practical engineering applications, such as in building insulation, growing crystals, solar energy collection, cooling of electrical industries, and flows in rooms due to thermal energy sources. Natural convection flows in a vertical cavity with two vertical walls at different temperatures and with adiabatic horizontal surfaces are the most considered configuration in the studies of natural convection because of their relative simplicity and practical importance [1-3].

Fast development of super-conducting materials at high temperature has enabled us to utilize commercial super-conducting magnets that produce the magnetic flux density up to 10 Tesla or more. By using such strong magnets, various new findings have reported for the last decades [4-8]. Wakayama [5-8] has been active in finding new and notable effects of a strong magnetic field in fluid convection. In the fields of fluid mechanics and heat transfer, the application of such high magnetic field on the control of natural convection heat transfer has been received considerable attention [9-16]. The control of natural convection heat transfer is based on the principle that the magnetizing force acts at a molecular level depending on the values

K. Li et al. (Eds.): LSMS/ICSEE 2014, Part III, CCIS 463, pp. 55–65, 2014.
© Springer-Verlag Berlin Heidelberg 2014

of magnetic susceptibility of fluid. Therefore, the utilization of the magnetizing force is equivalent to the local gravity control within the region that the high gradient magnetic field exists.

Most of the previous studies about the magneto thermal natural convection in a vertical cavity focus on the effect of magnetic field supplied by super-conducting magnet together with gravity field on natural convection [9-14]. There has been little study on the magneto thermal natural convection under magnetic field alone and seldom magnetic field is supplied by permanent magnet [15-16]. Song [16] recently studied the natural convection of air in a two-dimensional square enclosure under a non-uniform permanent magnetic field and two kinds of the expressions for the magnetizing force are considered. As the magnetic field supplied by permanent magnet is non-uniform and the gravity field is uniform, the natural convection characteristics under such two fields must be quite different. Recently, along with the developing of the technology and improvement of manufacturing process, the residual magnetic flux density of a permanent magnet increases quickly. The maximum value of the residual magnetic flux density of a Neodymium-Iron-Boron magnet can reaches to 1.5 Tesla or more. It is believed that the residual magnetic flux density of a permanent magnet will reach a larger value relying on the fast development of technology. The permanent magnet with large residual magnetic flux density can supply a non-uniform magnetic field and can drive the gas free convection easily under a gradient temperature field. Thus, the permanent magnet which has high magnetic energy product can be widely applied in future for control of natural convection without any power supply from outside like a super-conducting magnet.

The present study focuses on the heat transfer and flowing characteristic of the magneto thermal convection of air in a square enclosure driven by a permanent magnet field alone. Comparison is also carried out for the convection driven by magnetic field and gravity field respectively.

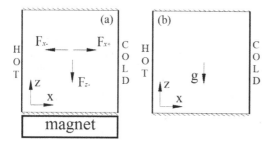
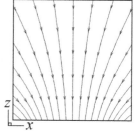

Fig. 1. Physical models. (a) under magnetic field; (b) under gravity field.

Fig. 2. Magnetic field lines in the enclosure

2 Physical Model

The physical model, two-dimensional square enclosure, considered in this paper is shown in Fig. 1. A cubic permanent magnet Neodymium-Iron-Boron (Nd-Fe-B) magnet which has a high magnetic energy product and could drive the gas free convection easily under a gradient temperature field is put on the bottom of the square enclosure with the poles of the magnet parallel to the Z-direction. The length of both the magnet and square enclosure is 0.05 m. The square enclosure is placed on top of the cubic magnet in the middle of the y-side and parallel to the X-Z plane. There is a gap of 1 mm between the magnet and the square enclosure. Figs. 1 shows the physical model in which the left wall is heated and right wall is cooled isothermally and the other two walls are thermally insulated. Fig. 1(a) shows the geometry and direction of magnetizing force of the physical model under magnetic field, Fig. 1(b) shows the model under gravity field. The magnetic field of the permanent magnet is calculated using the numerical method as described in [16]. The computed magnetic field lines supplied by the cubic permanent magnet in the enclosure are shown in Fig. 2.

3   Mathematical Equations

In a single phase with isothermal state, both the magnetizing force and gravitational force are the conservative force and hence convection does not occur by themselves no matter how strong they are. In order to arouse the convection, both the magnetic field gradient and the temperature gradient are necessary. In a normal expression, the magnetizing force is written by the equation as follows [17]:

$$f_{m1} = \frac{\mu_m}{2} \chi \rho \nabla H^2 \cong \frac{1}{2\mu_m} \rho \chi \nabla B^2 \qquad (1)$$

The momentum equation including the magnetizing force alone is as follows:

$$\rho \left[\frac{\partial u}{\partial t} + (u \cdot \nabla) u \right] = -\nabla p + \mu \nabla^2 u + \frac{1}{2\mu_m} \rho \chi \nabla B^2 \qquad (2)$$

And the momentum equation including the gravity buoyancy force alone is:

$$\rho \left[\frac{\partial u}{\partial t} + (u \cdot \nabla) u \right] = -\nabla p + \mu \nabla^2 u - \rho g \qquad (3)$$

We assume that the air is an incompressible Newtonian fluid and the Boussinesq approximation is employed. When we consider the isothermal state, the fluid density ρ and susceptibility χ would be constant, and convection does not arise even in both the gravity and magnetic fields. Parameters under this state are ρ_0, p_0, χ_0 at reference temperature T_0. Hence we get

$$0 = -\nabla p_0 + \frac{1}{2\mu_m} \rho_0 \chi_0 \nabla B^2 . \qquad (4)$$

$$0 = -\nabla p_0 - \rho_0 g . \qquad (5)$$

When there is a temperature difference, the magnetic susceptibility and density also change with temperature. Subtracting Eq. (4) and (5) from Eq. (2) and (3) gives

$$\rho \left[\frac{\partial u}{\partial t} + (u \cdot \nabla) u \right] = -\nabla (p - p_0) + \mu \nabla^2 u + \frac{\rho \chi - \rho_0 \chi_0}{2\mu_m} \nabla B^2 \qquad (6)$$

$$\rho\left[\frac{\partial \boldsymbol{u}}{\partial t}+(\boldsymbol{u}\cdot\nabla)\boldsymbol{u}\right]=-\nabla(p-p_0)+\mu\nabla^2\boldsymbol{u}-(\rho-\rho_0)g \qquad (7)$$

The difference of $\rho\chi$ and ρ can be expressed by a Taylor expansion (keeping two terms only) around a reference state T_0 as follows:

$$\rho-\rho_0=(\frac{\partial\rho}{\partial T})_0(T-T_0)+\cdots \qquad (8)$$

$$\rho\chi-(\rho\chi)_0=(\frac{\partial\rho\chi}{\partial T})_0(T-T_0)+\cdots \qquad (9)$$

For ideal gas, $p=\rho R_g T$, the thermal volume coefficient of expansion β is defined as:

$$\beta=-\frac{1}{\rho}(\frac{\partial\rho}{\partial T})=-\frac{1}{\rho}\frac{\partial}{\partial T}\left(\frac{p}{R_g T}\right)=\frac{p}{\rho R_g T^2}=\frac{1}{T} \qquad (10)$$

Then the difference of ρ can be written as:

$$\rho-\rho_0=(\frac{\partial\rho}{\partial T})_0(T-T_0)=-\rho_0\beta_0(T-T_0) \qquad (11)$$

According to the Curie law, magnetic susceptibility of air is a function of temperature:

$$\chi=\frac{C}{T} \qquad (12)$$

Here, C is a constant, then

$$\frac{\partial\chi}{\partial T}=\frac{\partial}{\partial T}(\frac{C}{T})=-\frac{\chi}{T} \qquad (13)$$

And from Eq. (10)

$$\frac{\partial\rho}{\partial T}=-\rho\beta \qquad (14)$$

So that, Eq. (9) can be written as:

$$\rho\chi-(\rho\chi)_0=(\chi\frac{\partial\rho}{\partial T}+\rho\frac{\partial\chi}{\partial T})_0(T-T_0)=-2\rho_0\beta_0\chi_0(T-T_0) \qquad (15)$$

Given $p=p_0+p'$, the momentum equation Eq. (6) and (7) can be written as follows:

$$\left[\frac{\partial \boldsymbol{u}}{\partial t}+(\boldsymbol{u}\cdot\nabla)\boldsymbol{u}\right]=-\frac{1}{\rho_0}\nabla p'+\frac{\mu}{\rho_0}\nabla^2\boldsymbol{u}-\frac{\beta_0\chi_0(T-T_0)}{\mu_m}\nabla B^2 \qquad (16)$$

$$\left[\frac{\partial \boldsymbol{u}}{\partial t}+(\boldsymbol{u}\cdot\nabla)\boldsymbol{u}\right]=-\frac{1}{\rho_0}\nabla p'+\frac{\mu}{\rho_0}\nabla^2\boldsymbol{u}+g\beta_0(T-T_0). \qquad (17)$$

Continuity equation

$$\nabla\cdot\boldsymbol{u}=0 \qquad (18)$$

Energy equation

$$\rho c_p \left[\frac{\partial T}{\partial t} + (\boldsymbol{u} \cdot \nabla) T \right] = \lambda \nabla^2 T \tag{19}$$

The initial condition (t<0) :

$$u = w = 0, \; T_0 = T_c \tag{20}$$

Velocity on the wall:

$$u = w = 0 \tag{21}$$

Temperature on the wall:

$$T_{x=0} = T_h, \; T_{x=L} = T_c, \; \partial T / \partial z = 0, \; \text{at } z = 0, L \tag{22}$$

The Rayleigh number is:

$$Ra = \frac{Pr \beta g (T_h - T_c) L^3}{v^2} \tag{23}$$

The local Nusselt number Nu_{local} represents the ratio of heat transfer rate by convection to that by conduction in the fluid. Nu_{local} is given by:

$$Nu_{local} = \frac{h_{local} L}{\lambda}, \quad h_{local} = \frac{\lambda}{T_h - T_c} \left| \frac{\partial T}{\partial n} \right| \tag{24}$$

The overall averaged Nusselt number is obtained by averaging Nu_{local} on the walls involved in heat transfer:

$$Nu_m = \frac{1}{L} \int_L Nu_{local} dz \tag{25}$$

The SIMPLE algorithm developed by Patankar [18] is used to solve the coupled heat transfer and fluid flow problem. The power-law scheme is used in the finite difference formulation of convection terms and a fine grid system is selected to raise the simulation precision. The grid-independence test for the solutions is carried out with three grid systems under the gravity field only. The averaged Nusselt numbers at different grid systems are presented in Table 1. The difference of Nusselt number between three grid systems is less than 0.5%, thus the results are grid independent. The validity of the code used in this paper is carried out by comparing the result under gravity field with the reported result by Khanafer [19], as shown in Table 1. When using the grid system 82×82, the difference of Nu is about 0.022%. Thus, grid system 82×82 is selected for all computations.

Table 1. Grid-independence test ($Pr=0.7$, $Ra=10^5$, g=9.8 m/s^2)

Grid system	Nu
62×62×62	4.540
82×82×82	4.523
102×102×102	4.520
K. Khanafer [19]	4.522

4   Results and Discussion

All the computations are carried out from $Ra = 10^4$ to $Ra = 10^6$, and the residual flux densities of permanent magnet changes from 0.5 T to 4.5 T. The flow field, temperature field and the distribution of local Nusselt number under the magnetic field are analyzed and compared with that under gravity field.

4.1   Magnetizing Force Fields

Fig. 3 shows the gravity buoyancy force and magnetizing buoyancy force fields in the enclosure with $Ra=10^5$. From these figures we can find that the magnetizing buoyancy forces are different from the gravity buoyancy force. The gravity buoyancy force parallel along y-direction, but the magnetizing buoyancy force direction changes in the enclosure. The magnetizing buoyancy force increased with increase of magnet strength. When the residual magnetic flux density is 0.5 Tesla, the magnetizing buoyancy force is very weak compared with the gravity buoyancy force. There has large magnetizing force near the bottom wall of the enclosure and the force near the hot wall is larger than that near the cold wall. The magnetizing force decreases quickly along y-direction. But for the gravity field, the distribution of gravity buoyancy force is quite different. The larger gravity buoyancy force locates in the region near the top and hot wall, and the gravity buoyancy force decreases from hot wall to cold wall and increases from bottom wall to top wall. The maximum value of magnetizing buoyancy force is larger than the maximum gravity buoyancy force.

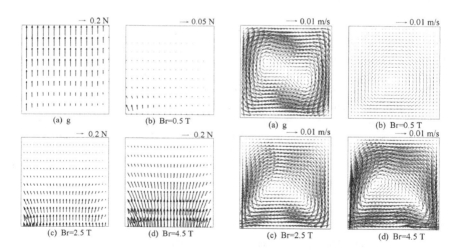

Fig. 3. Gravity and Magnetizing buoyancy force fields at $Ra=10^5$. (a) g; (b) Br=0.5 T; (c) Br=2.5 T;(d) Br=4.5 T.

Fig. 4. Velocity fields at $Ra=10^5$. (a) g; (b) Br=0.5 T; (c) Br=2.5 T; (d) Br=4.5 T.

4.2   Velocity Fields

Fig. 4 shows the velocity fields under gravity field and magnetic field at $Ra=10^5$. When there is gravity field only, as shown in Fig. 4(a), there has two vortices in the center of the enclosure, one is located near the top and hot walls, and another one locates close to the cold and bottom walls. The large value of velocity locates near the walls and the value of velocity decreases from the walls to the center of the enclosure. When there is magnetic field only, the velocity field is different from that under the gravity field. When the magnetic field is weak, as shown in Fig. 4(b), a weak vortex is formed in the enclosure, and the core of the vortex locates close to the bottom wall. The natural convection in the enclosure is enhanced with increasing the magnetic field, the core of the vortex moves upward and towards the hot wall, as shown in Figs. 4(c) and (d).

4.3   Temperature Fields

For the physical model under the gravity field, the temperature field looks rotationally symmetric about the center of the enclosure, as shown in Fig. 6(a). The temperature gradient normal to the hot wall decreases and the temperature gradient near the cold wall increases gently from the bottom wall to the top wall. The largest temperature gradient near the hot wall locates at the position close to the bottom wall, but near the cold wall it locates close to the top wall. The temperature field structure under the magnetic field, as shown in Figs. 6(b)-(d), is quite different from the temperature field under the gravity field. The natural convection in the enclosure is weak when the magnetic field is weak, see Fig. 6(b). The distribution of temperature near the hot and

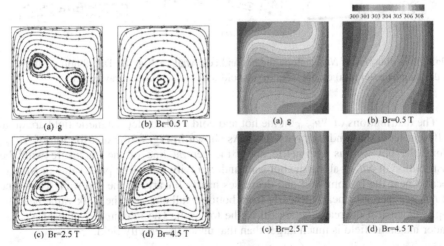

Fig. 5. Streamlines at $Ra=10^5$. (a) g; (b) $Br=0.5$T; (c) $Br=2.5$T; (d) $Br=4.5$T.

Fig. 6. Temperature fields at $Ra=10^5$. (a) g; (b) $Br=0.5$T; (c) $Br=2.5$T; (d) $Br=4.5$T.

walls are quite different. The temperature gradient near the hot wall decreases along z-direction and the maximum gradient of the temperature on the hot wall locates at the bottom of the hot wall. The temperature gradient normal to the cold wall changes a little along the cold wall and the maximum temperature gradient on the cold wall locates at the position close to the bottom wall. The temperature gradient decreases quickly in a small range on the bottom of the cold wall.

4.4   Heat Transfer on the Walls

The distribution curves of the local Nusselt number along the hot and cold walls under magnetic fields are plotted in Fig. 7. The distributions of the local value of Nu_{local} on the hot and cold walls are quite different. $Nu_{local-h}$ gets a high peak value at the bottom of the hot wall and it decreases quickly for a short distance and then continues decreasing at a relatively slow rate till to the top wall. The changing of Nu_{local} along the cold wall is much gentle compared with that along the hot wall. $Nu_{local-c}$ increases slowly from the top wall to the position near the bottom wall and then it decreases quickly till to the bottom wall. The maximum value of $Nu_{local-c}$ is much smaller than the maximum value of $Nu_{local-h}$.

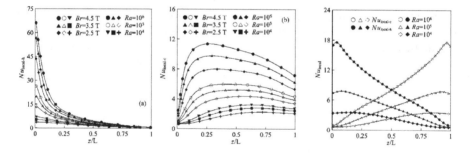

Fig. 7. Distributions of Nu_{local} along the hot and cold walls under magnetic field, (a) on hot wall; (b) on cold wall

Fig. 8. Distributions of Nu_{local} under gravity field

The distributions of Nu_{local} on the hot and cold walls under magnetic field are quite different from that under gravity field. As shown in Fig. 8, the distribution of Nu_{local} along the hot wall is symmetric about the central line of the enclosure with the distribution of Nu_{local} along the hot wall under gravity field. The peak value of $Nu_{local-h}$ locates near the bottom wall and decreases nearly linearly to the top wall. The value of $Nu_{local-c}$ increases nearly linearly from bottom wall and gets the peak value near the top wall, then it decreases again till to the top wall. The maximum value of $Nu_{local-h}$ under magnetic field is much larger than that under the gravity field.

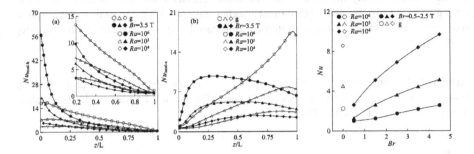

Fig. 9. comparison of Nu_{local} along the hot wall and cold wall. (a) along hot wall; (b) along cold wall

Fig. 10. Comparison of the average Nusselt number

The comparison of the distribution curves of Nu_{local} under gravity field and magnetic field are presented in Fig. 9 for different value of Ra at Br=3.5 T. The value of Nu_{local} along the hot wall is larger than the value under the gravity filed in a short line segment from the bottom wall, the peak value of Nu_{local} relates to magnetic field is about three times larger than the peak value relates to the gravity field. On the cold wall, Nu_{local} relates to magnetic field changes much gentle and has larger value on the lower part of the wall compared with that relates to gravity field.

Fig. 10 shows the distribution of the average Nu with Ra ranging from 10^4 to 10^6 and the residual magnetic flux density of a permanent magnet ranging from 0.5T to 4.5T. The average Nu under the gravity field is also shown in Fig. 10. The average Nu increases with increase of values of Ra and B_r. The value of Nu under magnetic field is smaller than that under gravity field when Br is less than 3.5T. When B_r is about 3.5T, the average Nu is nearly the same with the value of Nu under gravity field. This means that the magnetizing buoyancy force has a comparative effect on Nu in the enclosure compared with the gravity buoyancy force.

5   Conclusion

Natural convection of air in a two-dimensional square enclosure under a non-uniform magnetic field provided by a permanent magnet is carried out. As the magnetizing buoyancy force of air in the magnetic field is quite different with the gravitational buoyancy force, the natural convection in the enclosure under magnetic field has quite different characteristics from that under gravity field. Under the magnetic field, distributions of Nusselt number on the hot and cold walls are different. The local value of Nusselt number has peak value on the bottom of the hot wall and decreases quickly along the hot wall, but the local Nusselt number changes with a relatively gentle rate on the cold wall. The peak value on the hot wall is three times larger than that under the gravity field when B_r=3.5T. When B_r is about 3.5T, the values of average *Nu* are nearly the same with that under gravity field. This means that in the zero gravity environments, a permanent magnet with B_r=3.5T can play a comparative role on the average Nusselt number compared with that under the gravity environment.

Acknowledgment. This work is supported by the National Natural Science Foundation of China (No. 51376086), Gansu Provincial Funds for Distinguished Young Scientists (145RJDA324).

References

1. Vahl Davis, G.: Natural convection of air in a square cavity: a benchmark numerical solution. Int. J. Numer. Methods Fluids 3, 249–264 (1983)
2. Freitas, C.J., Street, R.L., Findikakis, A.N., Kose, J.R.: Numerical simulation of three dimensional flow in a cavity. Int. J. Numer. Meth. Fluids 5, 561–575 (1985)
3. Fusegi, T., Hyun, J.M., Kuwahara, K., Farouk, B.: A numerical study of three dimensional natural convection in a differentially heated cubical enclosure. Int. J. Heat Mass Transfer 34(6), 1543–1557 (1991)
4. Braitewaite, D., Beaughnon, E., Tournier, R.: Magnetically controlled convection in a paramagnetic fluid. Letters to Nature 354(14), 134–136 (1991)
5. Wakayama, N.I.: Behavior of gas flow under gradient magnetic fields. J. Applied Physics 69, 2734–2736 (1991)
6. Wakayama, N.I.: Magnetic promotion of combustion in diffusion flames. Combustion Flame 93, 207–214 (1993)
7. Wakayama, N.I.: Magnetic acceleration and deceleration of O_2 gas stream injected into air. IEEE Trans. Magn. 31(1), 897–901 (1995)
8. Wakayama, N.I., Wakayama, M.: Magnetic acceleration of inhaled and exhaled flows in breathing. Jpn. J. Appl. Phys. 39, 262–264 (2000)
9. Tagawa, T., Shigemitsu, R., Ozoe, H.: Magnetizing force modeled and numerically solved for natural convection of air in a cubic enclosure: Effect of the direction of the magnetic field. Int. J. Heat Mass Transfer 45(2), 267–277 (2002)
10. Bednarz, T., Fornalik, E., Tagawa, T., Ozoe, H., Szmyd, J.S.: Experimental and numerical analyses of magnetic convection of paramagnetic fluid in a cube heated and cooled from opposing verticals walls. International Journal of Thermal Sciences 44, 933–943 (2005)
11. Singh, R.K., Singh, A.K.: Effect of induced magnetic field on natural convection in vertical concentric annuli. Acta Mechanica Sinica 28(2), 315–323 (2012)
12. Sheikholeslami, M., Gorji-Bandpay, M., Ganji, D.D.: Magnetic field effects on natural convection around a horizontal circular cylinder inside a square enclosure filled with nanofluid. International Communications in Heat and Mass Transfer 39, 978–986 (2012)
13. Sheikholeslami, M., Hashim, I., Soleimani, S.: Numerical Investigation of the Effect of Magnetic Field on Natural Convection in a Curved-Shape Enclosure. Mathematical Problems in Engineering. 2013, ID 831725, 10 pages (2013)
14. Mahmoudi, A.H., Pop, I., Shahi, M.: Effect of magnetic field on natural convection in a triangular enclosure filled with nanofluid. International Journal of Thermal Sciences 59, 126–140 (2014)
15. Jalil, J.M., Al-Tae'y, K.A., Ismail, S.J.: Natural Convection in an Enclosure with a Partially Active Magnetic Field. Num. Heat Transfer Part A 64(1), 72–91 (2013)

16. Song, K.W., Tagawa, T., Wang, L.B., Ozoe, H.: Numerical investigation for the modeling of the magnetic buoyancy force during the natural convection of air in a square enclosure. Advances in Mechanical Engineering 2014, ID 873260, 11 pages (2014)
17. Bai, B., Yabe, A., Qi, J., Wakayama, N.I.: Quantitative analysis of air convection caused by magnetic-fluid coupling. AIAA J. 37(12), 1538–1543 (1999)
18. Patankar, S.V.: Numerical heat transfer and fluid flow. Hemisphere Publishing Co., New York (1980)
19. Khanafer, K., Vafai, K., Lightstone, M.: Buoyancy-driven heat transfer enhancement in a two dimensional enclosure utilizing nanofluids. Int. J. Heat Mass Transfer 46(19), 3639–3653 (2003)

Improved LMD and Its Application in Short-Term Wind Power Forecast

Guochu Chen[*], Zhiwei Guan, and Qiaomei Cheng

Electric Engineering School, Shanghai DianJi University, Shanghai 200240, China
chengc@sdju.edu.cn

Abstract. As to the problems of over-moving in local mean decomposition method (LMD), it is proposed to use the cubic Hermite interpolation to get the envelope curve and local mean curve. Instead of moving mean method, it will improve the over moving effect, improving the speed of LMD shifting effectively. Wind power series can be decomposed into different series by improved LMD, and then artificial neural network (ANN) is used to forecast power by each component. The total wind power prediction result is obtained through reconstructing at last. Case study shows that the prediction accuracy has significantly been improved by comparing with other models.

Keywords: Wind power, local mean decomposition method, ANN, prediction, model.

1 Introduction

As the shortage of traditional energy sources such as coal, oil, various countries in the world have put forward urgent request to the renewable energy development and utilization. As a kind of environmental friendly energy, wind power is an important part of renewable energy, most countries in the world take the wind energy as an important energy resource [1, 2]. Development and utilization of new energy sources such as wind power is the important energy strategy in the 21st century in our country.

Due to the randomness and intermittent of wind which bring difficulties to the safe operation of grid connected wind power, however, wind power prediction technology can effectively reduce the impact of wind power on power grid. At present, two methods for the prediction of wind power are mainly used: the indirect method and direct method. The indirect method is to forecast wind speed first and then to predict the power according to the power curve such as the literature [3] using support vector machine (SVM) to predict the wind speed, then get wind power. But because of the wind power can be affected by wind speed, temperature, humidity and other factors, the accuracy of indirect method forecasting is not high and with certain restrictions. The direct method is to forecast wind power directly by using the intelligent algorithm such as support vector machine (SVM), artificial neural network (ANN) and so on,

---

[*] Corresponding author.

K. Li et al. (Eds.): LSMS/ICSEE 2014, Part III, CCIS 463, pp. 66–72, 2014.

combining with historical wind power data, such as the literature [4] using the neural networks to forecast the power. Compared with the indirect method, the forecast precision of direct method is improved markedly. Because of the wind power sequence has the characteristics of nonlinear and non-stationary, single forecasting model usually cann't get very good effect. Literature [5] put forward combined model that combines improved empirical mode decomposition method (EMD) with neural network, compared with single forecast model, the combined model can effectively improve the prediction precision.

This article will combine the improved local mean decomposition method (LMD) with artificial neural network (ANN), and according to the actual data obtained from the wind farm, to forecast wind power. First of all, improve the LMD with Hermite interpolation method and decompose power series with the improved algorithm, then use neural network to forecast each subsequence, last, add up the predicted results, and compared with the actual wind power data.

2   Standard LMD and Its Problems

2.1   Standard LMD Algorithm

LMD is a new method of time domain analysis proposed by Jonathan S.Smith in 2005 [6] which is a kind of self-adaptive analysis method applicable to nonlinear and non-stationary signal and first to be applied in computer electrical signal processing.

LMD method is a kind of adaptive decomposition method which can decompose a complicated multi-component signal into multiple product function (PF) and a margin function according to the order from high frequency to low frequency. PF instantaneous amplitude and instantaneous frequency can be got from itself, the former can be obtained by looking for the envelope signal, while the latter can be found by looking for pure frequency modulation signal.

All the PF component have their physical significance. In this way, the original $x(t)$ can be decomposed into some PF component and one margin $u_n(t)$, namely:

$$x(t) = \sum_i^n PF_i + u_n(t) \qquad (1)$$

From above formula, in the LMD decomposition process, the signal is complete without any missing.

2.2   Problem of LMD Method and Its Improvement

Problem of LMD Method. LMD methods has a lack of unified theoretical basis, at the same time it still has the problem of endpoint effect, over-smoothing and other issues. In LMD decomposition, the quality of obtained initial local mean curve and envelope estimates curve will produce great influence on subsequent decomposition, and to get the curves depends on the realization of the moving average method. There is no clear standard of the moving average method realization yet, and most of the time, for convenience, the choice of smoothing window tend to be consistent in each point. However with the limiting of the smooth termination conditions which may lead smooth phenomenon for many times, this is called over-smoothing effect which will affect the gain of local mean curve and envelope estimate curve.

Restrict and smooth termination conditions may appear, this is the moving average method of smoothing effect. This effect will influence the local mean curve and envelope estimate curve and then affect the correctness and efficiency of the decomposing, meanwhile, during the discrete signal processing, LMD will appear end effect inevitably.

Improvement of LMD Method. In view of the above analysis, in calculating the local mean value curve and envelope estimation, this article put forward that to replace the moving average method with the cubic Hermite interpolation to improve LMD and enhance the decomposing accuracy.

Given the function value and differential coefficient of function $y = f(x)$ at some certain point $a = x_0 < x_1 < \cdots < x_n = b$ are $y_i = f(x_i)$, $y_i' = f'(x_i)$, $(i = 0,1,\cdots n)$, then $H(x)$ needs to meet the follow condition to be cubic Hermite interpolation:

(1) $S(x_i) = y_i$, $S'(x_i) = y_i'$;

(2) In each section $[x_i, x_{i+1}]$, $S(x)$ is a polynomial whose degree is less than three.

For subsection cubic Hermite interpolation, interpolation polynomial $s_3(x)$ and its differential coefficient $s_3'(x)$ are both the continuous function in $[a,b]$, so this is a smooth subsection interpolation, in each $[x_i, x_{i+1}]$, there is :

$$H_i(x) = \frac{y_i}{h_i^3}(h_i + 2\Delta x_i)\Delta x_{i+1}^2 + \frac{y_{i+1}}{h_i^3}(h_i - 2\Delta x_{i+1})\Delta x_i^2$$
$$+ \frac{y_i'}{h_i^2}\Delta x_i \Delta x_{i+1}^2 + \frac{y_{i+1}'}{h_i^2}\Delta x_{i+1}\Delta x_i^2 \quad (2)$$

Among them $h_i = x_{i+1} - x_i$, $\Delta x_i = x - x_i$ .Therefore $S_3(x) = H_i(x)$, $x \in [x_i, x_{i+1}], i = 1, 2 \cdots n$.

For interpolation error, if $f^{(4)}(x)$ exists in $[a,b]$, $R(x)$ is the interpolation margin, for $\forall x \in [a,b]$ there is $|R(x)| = |f(x) - S_3(x)| \le \frac{M_4 h^4}{384}$ (3)

Among them $h = \max\{h_i\}$, $M_4 = \max\{f^{(4)}(x)\}$, when $h \to 0$, $|R(x)| \to 0$, $S_3(x)$ converges to $f(x)$.

For cubic spline interpolation, Hermite interpolation also uses the cubic polynomial approximation which ensure the monotonicity of interpolation curves between the two adjacent maximum or minimum, largely reduce the non-stationary signal decomposition envelope undershoot or overshoot problem. Using the interpolation method to fit the signal, the upper and lower envelope and the mean curve will be more precise, the decomposed PF component will reflect the essential characteristics of signal better.

3 The Improved LMD Application in Short-Term Wind Power Prediction

3.1 Improvement of the Model

Wind power sequence has a strong non-stationary and nonlinear. For nonlinear signal, the neural network fitting effect is very good, but for non-stationary, it will often be ineffective: in most of the research literature, they often use the first separate then close method. In view of the power series is a non-stationary time series, has a strong randomness and periodicity, which contains the random component and periodic components and trend components. So this article base on first separate then close frame, first to decompose the power signal and effectively weaken the non-stationarity, then use the neural network to forecast each components.

3.2 Example and the Result Analysis

This paper adopted the actual wind power data of a wind turbine in a wind farm in June 2013 which is recorded every 10 minutes a group, the total data of 30 days is 30 * 24 * 6 = 4320 groups. This paper studies the short-term wind power prediction, prediction step length is 1 hour ahead of time. For convenient, take the average of 6 groups of data per hour as the measured values for this hour which makes 4320 groups of data be reduced to 720 groups of data.

In order to make the simulation examples persuasive and can show better effect of this prediction model, selecting first 20 days data in June as the modeling data, last 10 days as forecast data.

Due to the poor effect of standard LMD decomposition, here power series are decomposed with LMD based on cubic spline interpolation method and based on the Hermite interpolation, the results are shown in Fig.1 and Fig. 2.

As you can see, using spline interpolation LMD method, one allowance and five PF components can be got, while for the LMD based on Hermite interpolation method, one allowance and six PF components can be got. Component of Hermite interpolation increases compared with the former, the PF1 ~ PF6 is high frequency components, each component amplitude decreases, the change trend allowance $r6$ is obvious, the decomposition have better accuracy than the former.

Because of the poor effect of the standard LMD decomposition, in the MATLAB platform, pure ANN method, LMD - ANN method based on the cubic spline and based on the cubic Hermite are adopted to predict the power series, prediction result of each method are shown in Fig.3. This article adopt normalized MAE (Mean Absolute Error), RMSE (Root Mean Squre Error) and Max-AE (Max-Absolute Error) to evaluate the model performance. If $Y(t)$ is the measured value at t- moment, $y(t)$ is the predicted value, there are:

$$AE = |Y(t) - y(t)| \tag{4}$$

$$MAE = \frac{1}{P_{rated}} \frac{1}{n} \sum_{t=1}^{n} |Y(t) - y(t)| \tag{5}$$

$$RMSE = \frac{1}{P_{rated}} [\frac{1}{n} \sum_{t=1}^{n} (Y(t) - y(t))^2]^{\frac{1}{2}} \tag{6}$$

In the formula, P_{rated} is the rating power of wind turbine, and n is the number of time point.

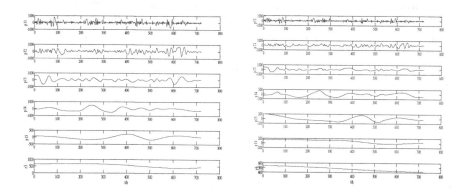

Fig. 1. Power series decomposition using spline LMD

Fig. 2. Power series decomposition using Hermite LMD

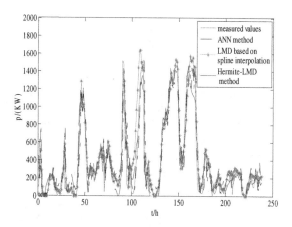

Fig. 3. The prediction results based on different models

Table 1. Error comparison among different wind power prediction methods

Prediction models	MAE/%	RMSE/%	Max-AE/kW
ANN	8.24	13.42	1250.8
Spline LMD-ANN	7.83	11.86	1080.49
Hermite-LMD-ANN	7.51	10.51	725.16

Table 1 shows that prediction precision of LMD-ANN model based on Hermite interpolation is higher. Compared with the ANN model, LMD - ANN model based on the cubic spline interpolation, and its MAE (mean absolute error) dropped to 7.51%, RMSE (root mean square error) dropped to 10.51%, obviously improve the prediction accuracy and also reduce the error of the single point, MAX - AE (single point of maximum error) reduced to 725.16 KW, predict performance is improved significantly. In conclusion, for significant nonlinear non-stationary wind power sequence, LMD - ANN based on the Hermite interpolation forecasting model is superior to ANN prediction model and LMD-ANN based on cubic spline interpolation prediction model.

4   Conclusions

Due to not high prediction accuracy of single model, strong nonlinear and non-stationary wind power sequence, based on the analysis of the excessive sliding in the process of LMD and endpoint drift problems, this paper puts forward the improved methods of these problems, namely the piecewise cubic Hermite interpolation method is adopted to calculate the local mean value curve and estimate envelope curve. By using this improved algorithm for decomposition of wind power, weaken the power of non-stationary signal, and then combining with ANN to predict the result of the each decomposition respectively and has a better effect than ANN prediction model and LMD-ANN based on cubic spline interpolation prediction model.

Acknowledgments. This work was financially supported by scientific research innovation projects of Shanghai municipal education commission (Grant No.13YZ140) and the key disciplines of Shanghai Municipal Education Commission of China (Grant No. J51901).

References

1. Lei, Y.Z., Wang, W.S., Yin, S.H., et al.: Analysys of wind power value to power system operation. Power System Technology 26(5), 10–14 (2002) (in Chinese)
2. Liu, Y.Q., Han, S., Hu, Y.S.: Review on short-term wind power prediction. Modern Electrc Power 24(5), 6–11 (2007) (in Chinese)

3. Qi, S.B., Wang, W.Q., Zhang, X.Y.: Wind speed and wind power prediction based on SVM. East China Electric Power 9(37), 1600–1603 (2009) (in Chinese)
4. Han, S., Qian, L.Y., Yang, Y.P., et al.: Ultra-short term wind power prediction and uncertainty assessment. Acta Energiae Solaris Sinica 8(32), 1251–1256 (2011) (in Chinese)
5. Wang, P., Chen, G.C., Xu, Y.F., et al.: Improved Empirical Mode Decom position and its Application to Wind Power Forecasting. Control Engineering of China 4(18), 588–591 (2011) (in Chinese)
6. Jonathan, S.S.: The local mean decomposition and its application to EEG perception data. Journal of the Royal Society Interface 2(5), 443–454 (2005)

The New Method to Determine the Confidence Probability of Wind Power Prediction Result

Guochu Chen[*] and Weixiang Gong

Electric Engineering School, Shanghai DianJi University, Shanghai, 200240 China
chengc@sdju.edu.cn

Abstract. The uncertainty analysis of power predictive results is very important to the dispatching of wind power. For the shortcomings of traditional methods that determine the confidence probability, this paper proposes a new method to determine the confidence probability based on independent component analysis (ICA) and conditional probability theory. According to the new method, the power independent influence events set can be obtained from ICA, and the problem of determining the confidence probability can be transformed into the problem of determining unconditional probability and conditional probability whose objectives are several independent influence events. The method is clear and easy to be resolved, which fully takes into account occurrence conditions of the objective power and the original content. The simulation results show the confidence probability result obtained by the new method has more realistic sense and scientific guidance value.

Keywords: confidence probability, independent component analysis, wind power, prediction.

1   Introduction

Wind power belongs to the renewable energy, which is pollution-free, has almost no influence on environment and has considerable reserves in various regions of the world. In recent years, China's wind power construction makes a spurt of progress, showing a strong development trend.

The wind power predictive result is an important technical support to spare capacity optimization and economic operation, and is an important basis for power dispatch department to adjust dispatching plan timely, as in[1,2], which makes the confidence probability of predictive results more important than simple predictive result data, the former has the stronger scientific significance.

At present, the confidence interval approach is largely used in analyzing the uncertainty of power prediction results, but the confidence intervals are often asymmetric in prediction point, as in [3], which makes the guidance of confidence interval not strong in the practical application and makes confidence interval difficult to popularize. Reference [3] puts forward sampling method and the weather stability

---

[*] Corresponding author.

K. Li et al. (Eds.): LSMS/ICSEE 2014, Part III, CCIS 463, pp. 73–80, 2014.

indicator, which reduce the confidence interval range to a certain extent, but it needs numerical weather forecast. In references [4, 5], power prediction error is transformed from the power curve. But, in essence, uncertainty is given by the confidence interval and this method is only suitable for physical model. Reference only gives the predictive probability of wind speed, and the probability is determined by sample points between confidence interval. This method has disadvantages. In general, the next moment power is generated by many factors' combined action (such as wind speed, power etc at this moment.). The references above do not consider the conditions of predictive outcomes under objective power and do not deeper mine the existing data and information, which results that the guidance of calculated confidence probability is not strong in the practical application.

Aim at the above problems, this paper puts forward a new method to determine the confidence probability of power predictive results. This new method extends the traditional unconditional probability solution under objective power to conditional probability solving, extracts power independent influence event set by using independent component analysis (ICA). In this way, the problem of power predictive confidence probability can be transformed into the problem of unconditional probability of each independent influence events and condition probability of power independent influence events under the objective power conditions. So, the conditions under the objective power can be considered more fully, and the existing data and information can be utilized deeper.

2   Traditional Methods to Determine the Confidence Probability of Predictive Results

There exist two traditional definitions on confidence probability:

First, the confidence probability of predicted value is the unconditional probability of objective power $P \in (P_a, P_b)$, the calculation is as follows:

$$P \{ P_a < P \le P_b \} = \frac{n}{N} \tag{1}$$

n is the number of samples of power prediction results which fall into the range of (P_a, P_b), and N is the number value of all samples.

Second, the confidence probability of predicted value is defined as the conditional probability of objective power $P \in (P_a, P_b)$ under the influence event $x_i \in (x_{i,a}, x_{i,b}), (i = 1, 2, \cdots, k)$, the calculation is as follows:

$$P \{ P_a < P \le P_b | x_{1,a} < x_1 \le x_{1,b}, \cdots, x_{k,a} < x_k \le x_{k,b} \} = \frac{P \left\{ \begin{matrix} x_{1,a} < x_1 \le x_{1,b}, \\ \cdots, \\ x_{k,a} < x_k \le x_{k,b} \end{matrix} \middle| P_a < P \le P_b \right\} P \{ P_a < P \le P_b \}}{P \{ x_{1,a} < x_1 \le x_{1,b}, \cdots, x_{k,a} < x_k \le x_{k,b} \}} \tag{2}$$

$P\left\{x_{1,a} < x_1 \le x_{1,b}, \cdots, x_{k,a} < x_k \le x_{k,b} \middle| P_a < P \le P_b\right\}$ is the joint probability of objective power which meets $P_a < P < P_b$ and whose influence event is $\left(x_{i,a}, x_{i,b}\right)$. $P\left\{x_{1,a} < x_1 \le x_{1,b}, \cdots, x_{k,a} < x_k \le x_{k,b}\right\}$ is the joint probability of power influence event $x_i \in \left(x_{i,a}, x_{i,b}\right), (i = 1, 2, \cdots, k)$ among the observed range.

The first definition simply analyzes from the predictive results, without considering the occurred conditions of prediction results under objective power, which makes important information lose. The second definition, which analyzes from the occurred conditions and predictive results of prediction results under objective power, is relatively comprehensive. But in the existing methods, power influence events are not independent from each other, nor can simply be assumed as Gauss distribution, which makes solution rather difficult.

3   The New Method to Determine the Confidence Probability of Wind Power Prediction Results

3.1   The Idea of the New Method and the New Model

Independent component analysis (ICA) is a recently developed statistical and computational technology, aiming at revealing the hidden components of the measured data or signal, which has been successfully applied in many fields, among which financial forecast is a rather typical as in. The goal of the ICA is, without knowing the mixing matrix A and source signals s, to find out a unmixed matrix W, which makes the components of the new vector y, transformed from the observation signals x, remain independent, as well as makes it the optimal approximation of the source signals s, i.e.:

$$y = W x \qquad (3)$$

ICA method can extract independent components from current signals, and require each component of the current signal has the non-Gaussian nature. The power influence events don't obey the Gaussian distribution via the statistical analysis, which corresponds to the requirements of ICA methods.

Random variable $y = \left(y_1, y_2, \cdots, y_k\right)^T$ represents the independent influence events set of power estimated by the ICA. Power independent influence event is $y_i \in \left(y_{i,c}, y_{i,d}\right), (i = 1, 2, \cdots, k)$, $y_{i,c}$ and $y_{i,d}$ represents the upper limitation and lower limitation of observations of y_i, $y_{i,c}, y_{i,d}, y_i$ correspond to $x_{i,a}, x_{i,b}, x_i$ respectively.

Let $x_i = \left(x_{i1}, \cdots, x_{in}\right)$, $y_i = \left(y_{i1}, \cdots, y_{in}\right)$, then x and y both are $k \times n$ order matrix, k is the number of the independent component, n is the number of the sample points. Unmixed matrix W is n order square. From (3), the relations between $y_{i,c}, y_{i,d}$ and $x_{i,a}, x_{i,b}$ are:

$$\begin{cases} y_{i,c} = \sum_{m \in A} W_{im} x_{mj,a} + \sum_{m \in B} W_{im} x_{mj,b} \\ y_{i,d} = \sum_{m \in A} W_{im} x_{mj,b} + \sum_{m \in B} W_{im} x_{mj,a} \end{cases} \qquad (4)$$

In which, if $m \notin A$, then $W_{im} > 0$; else if $m \notin B$, then $W_{im} < 0$.

Unmixed matrix W is a linear transformation matrix, and $\det W \neq 0$, so the first order partial derivative of $x = W^{-1} y$ is constant as in, i.e.:

$$q(x_1, x_2, \cdots, x_k) = \frac{p(y_1, y_2, \cdots, y_k)}{|\det W|} \qquad (5)$$

In which, $q(x_1, x_2, \cdots, x_k)$ and $p(y_1, y_2, \cdots, y_k)$ are respectively the joint probability density function of $x_1, x_2, \cdots, x_k$ and $y_1, y_2, \cdots, y_k$. $y_1, \cdots, y_k$ is independent from each other, then:

$$P\{x_{1,a} < x_1 \leq x_{1,b}, \cdots, x_{k,a} < x_k \leq x_{k,b}\} = \frac{1}{|\det W|} \prod_{i=1}^{k} P(y_{i,c} < y_i \leq y_{i,d}) \qquad (6)$$

$$P\{x_{1,a} < x_1 \leq x_{1,b}, \cdots, x_{k,a} < x_k \leq x_{k,b} | P_a < P \leq P_b\} = \frac{1}{|\det W|} \prod_{i=1}^{k} P(y_{i,c} < y_i \leq y_{i,d} | P_a < P \leq P_b) \qquad (7)$$

The conditional probability of objective power under the influence event, i.e. the confidence probability of power prediction results, is :

$$P\{P_a < P \leq P_b | x_{1,a} < x_1 \leq x_{1,b}, \cdots, x_{k,a} < x_k \leq x_{k,b}\} = \frac{\prod_{i=1}^{k} P(y_{i,c} < y_i \leq y_{i,d} | P_a < P \leq P_b)}{\prod_{i=1}^{k} P(y_{i,c} < y_i \leq y_{i,d})} P\{P_a < P \leq P_b\} \qquad (8)$$

In which, $P(y_{i,c} < y_i \leq y_{i,d} | P_a < P \leq P_b)$ is the conditional probability of y_i, among the range $(y_{i,c}, y_{i,d})$, when objective power meets $P_a < P \leq P_b$. $P(y_{i,c} < y_i \leq y_{i,d})$ is the unconditional probability of y_i that among the range $(y_{i,c}, y_{i,d})$.

3.2   Model Solution

1) Get the power independent influence events

Here mainly use the ICA method. The relationship between ICA and the source signals s can be shown in Figure 1, process I is the mixing process, process II is the unmixed process as in.

(Source signal)   (Observation signal)   (Predictive source signal)

Fig. 1. The relationship between ICA and source signals

The relationship between ICA and source signals. In practice application, make the following constraints:

First, Independent components are assumed to be statistical independence; Second, Independent components have non-Gaussian distribution.

At present, the main methods to estimate ICA model are non-Gaussian maximization, mutual information minimization, maximum likelihood function estimation, etc. This paper uses fixed point algorithm (Fast ICA) based on negative entropy to estimate ICA. In addition, "the power independent influence events" via the ICA estimation has no actual physical significance. It just represents the independent quantity, as well as guarantees the independence of events.

2) The unconditional probability of power independent influence events

$$P\left(y_{i,c} < y_i \le y_{i,d}\right) = \int_{y_{i,c}}^{y_{i,d}} p_i(y_i)dy_i \approx \sum_{j=n_i}^{n_i} p_{i,j}(y_{i,j})(y_{i,j+1} - y_{i,j}) \tag{9}$$

where, $i = 1, \cdots, k$, $P_{i,j}(y_i)$ is the unconditional probability density function of power independent influence events y_i, and $p_{i,j}(y_{i,j})$ is the probability density of $y_{i,j}$, which is the sample points of power independent influence events y_i, and $y_{i,c} \le y_{i,n_c} < y_{i,n_d} \le y_{i,d}$.

3) The unconditional probability of power independent influence events under the objective power

$$p\{y_{i,c} < y_{i,} \le y_{i,d} \mid P_a < P \le P_b\} = \int_{y_{i,c}}^{y_{i,d}} P_i(y_i|P_a < P \le P)dy_i \approx \sum_{j=n_i}^{n_i} p_{i,j}(y_i \mid P_a < P \le P_b)(y_{i,j+1} - y_{i,j}) \tag{10}$$

Where, $i = 1, \cdots, k$, $P_i(y_i \mid P_a < P \le P_b)$ is the unconditional probability density function of power independent influence events y_i $P_{i,j}(y_{i,j} \mid P_a < P \le P_b)$ is the probability density of $y_{i,j}$, which is the sample points of power independent influence events y_i when the objective power meets $P_a < P \le P_b$, and $y_{i,c} \le y_{i,n_c} < y_{i,n_d} \le y_{i,d}$.

4) The unconditional probability of the objective power $P \in (P_a, P_b)$

$$P(P_a < P \le P_b) = \frac{n}{N} \tag{11}$$

Where, n is the number value of samples of power predictive results which fall into the range of (P_a, P_b), and N is the number value of all samples.

5) Determine the confidence probability
Use the method of Nonparametric Kernel Density Estimation to get the discrete data of $p_i(y_i)$ and $P_i(y_i \mid P_a < P \le P_b)$ respectively. Use Gaussian Kernel Function $k(x) = \exp(-x^2/2)/\sqrt{2\pi}$ to get the probability nonparametric density estimation .Then $p_i(y_i)$ and $P_i(y_i \mid P_a < P \le P_b)$ can be calculated out, so the confidence probability of power predictive results can be gotten.

4   Simulation Experiment and Results Analysis

Take the No.40 wind turbines unit of a wind farm in China as the research objective, and the measured data of this unit during 5~6 months have been got. This experiment takes the confidence probability of predictive power value on July 3, 2011 8:00 for example, to solve the model, and on July 3, the actual power value is 235.748 KW. Here select power influence events set: x = {the power value at present moment, the power value one hour ago, the wind speed value at present moment, the wind speed value one hour ago} T. The interval of power is 20 KW and the interval of wind speed is 0.5 m/s. According to the model, do ICA estimation to the 4 power influence events, obtaining power independent influence events set y as shown in figure 2.

If use one kind of predictive method, the predictive power of the next one hour is 229.8 KW, then the objective power's range is (220, 240). The power independent influence events set under the condition of objective power is shown in figure 3.

The conditional probability of the objective power 229.8 KW under the condition of power influence events is

$$p\left\{220 < P \le 240 \left| \begin{array}{l} 180 < x_1 \le 200 \\ 360 < x_2 \le 380 \\ 6 < x_3 \le 6.5 \\ 7 < x_4 \le 7.5 \end{array} \right. \right\} = 0.71414 \tag{12}$$

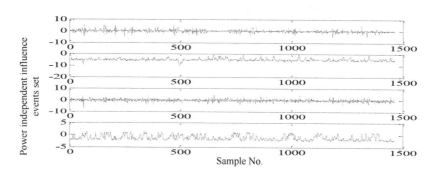

Fig. 2. The sample data of power independent influence events set

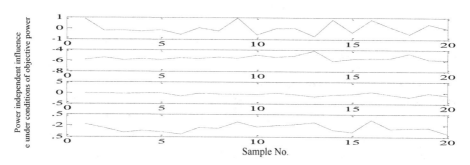

Fig. 3. The sample data of power independent influence events set under conditions of objective power

Table 1. Relative probability of power independent influence events with power forecast value being 229.8KW

$P\{y_{i,c} < y_i \le y_{i,d} \,|\, P_a < P \le P_b\}$	$P\{y_{i,c} < y_i \le y_{i,d}\}$
0.088158	0.023528
0.16714	0.062922
0.7107	0.63755
0.53914	0.11442

Table 2. Relative probability under different power forecast value at next hour (with actual power 235.748KW)

Predictive Value/KW	Scales	Unconditional probability	Conditional probability
205.4	(200, 220)	0.018443	0.22104
229.8	(220, 240)	0.013661	0.71414
255.1	(240, 260)	0.022541	0.25965
274.1	(260, 280)	0.020492	0.15218
288.5	(280, 300)	0.01571	0.29835

According to the traditional confidence probability definition of power prediction, the objective power's unconditional probability is only 0.013661. If defining the confidence probability by using the conditional probability of objective power under influence events, as the traditional methods don't undergo ICA, the probability density of influence events can't obtain, thus the confidence probability will not be able to find out. By using the new model, the calculated confidence probability is 0.71414, indicating that the predictive value closes to the actual value by a large probability.

The Table 1 shows that: First, the conditional probability reflects the forecast accuracy objectively. The closer the predictive value to the actual value, the bigger the condition probability value will, and the more it can reflect the forecast accuracy, which is consistent with the fact. Second, compared with unconditional probability, the sensitivity that conditional probability to power change is significantly enhanced. In practice, the power value of wind farms is changed a lot with time, to which the conditional is of great significance. Last, the conditional probability reflects that power influence events do exist a certain relationship.

5   Conclusion

In view of the defect that exists in the traditional method to determine the confidence probability of power predictive results, this paper puts forward a new model of conditional probability which based on independent component analysis and conditional probability theory. The new model takes the conditions under the objective power into account, and independent component analysis is used for the power influence events set. In this way, conditional probability problems can be transformed into the unconditional probability and conditional probability problems of the power independent influence events, so as to realize objective power's conditional probability solving under influence events successfully. The experiment simulation results show that, this model considers comprehensively and it is practical and effective, and the confidence probability obtained accords with actual significance, fully reflecting its scientific value as a uncertainty indicator.

Acknowledgments. This work was supported by Natural Science Foundation of Shanghai Municipal Education Commission of China (Grant No. 13YZ140) and the key disciplines of Shanghai Municipal Education Commission of China (Grant No. J51901).

References

1. Chen, S., Dai, H., Bai, X., Zhou, X.: Reliablity Model of wind power plants and its application. Proceedings of CSEE 20(3), 26–29 (2000) (Chinese)
2. Liu, Y., Han, S., Hu, Y.: Review on Short-term Wind Power Prediction. Modern Electric Power 24(5), 6–11 (2007) (Chinese)
3. Pinson, P., Kariniotakis, G.N.: Wind Power Forecasting using Fuzzy Neural Networks Enhanced with On-line Prediction Risk Assessment. In: IEEE Bologna Power Tech Conference, Bologna, Italy, June 23-26 (2003)
4. Armin, L., Stefan, B., Hans, G.B.: Analysis of confidence intervals for the prediction of the regional wind power output. In: Proc. of the 2001 European Wind Energy Association Conference, EWEC 2001, Copenhagen, Denmark, July 2-6, pp. 725–728 (2001)
5. Matthias, L., Waldl, H.P.: Assessing the uncertainty of the wind power predictions with regard to specific weather situations. In: Proc. of the 2001 European Wind Energy Association Conference, EWEC 2001, Copenhagen, Denmark, July 2-6, pp. 695–698 (2001)

A High Precision Simulation Model
for Single Phase Heated Tubes of Power Plant Boilers

Ying-wei Kang, Ya-nan Wang, Dao-gang Peng, and Wei Huang

College of Automation Engineering,
Shanghai University of Electric Power, Yangpu Shanghai 200090, China

Abstract. This paper presents a high precision simulation model for single phase heated tubes of power plant boilers. The model takes into account not only the dynamic process of the working fluid velocity, but also the coupling effect of the enthalpy-temperature channel and the pressure-flux channel of working fluid. By simulations of a heated tube of the rear superheater of a 600 MW controlled circulation boiler, the validity and feasibility of the model is verified. The present model can be integrated into virtual power plant simulation platforms for better debugging advanced power plant control systems, e.g. the automatic plant startup and shutdown control system (APS).

Keywords: boiler, single phase heated tube, superheater, simulation model, APS.

1   Introduction

Single phase heated tubes constitute an important part of power plant boiler heated surfaces. The boiler economizers (non-boiling type), superheaters, and reheaters all belong to such type of heated surfaces. Dynamic modeling of single phase heated tubes has always been an important research topic in the power plant simulation field. In the past decades, the domestic and foreign researchers have provided a number of models of single phase heated tubes, relating to many types ranging from linear to nonlinear, from lumped to distributed parameter [1-9].

The work process of single phase heated tubes mainly involves the flow and heat transfer of the working fluid, and the heat transfer inside the metal tube wall, etc. In the past, as the computer performance is not high enough, when modeling a single phase heated tube, its working processes are often strongly simplified, especially for its pressure-flux channel. A common treatment is to ignore the dynamic process of working fluid flux (described by the momentum equation), and the effect of the enthalpy-temperature channel on the pressure-flux channel. This leads to the prominent decrease of the model complexity and solving difficulty, meanwhile the model accuracy is also impaired to some extent, especially in the flux dynamic process, where a distinct error exists between the calculated and the actual results.

In recent years, as the improvement of the automation level of thermal power plants of China, equipping the automatic plant startup and shutdown control system (APS) in some advanced power plants is becoming a trend. In the control system simulation test phase, before the APS is put into operation, to better optimize the control scheme and

K. Li et al. (Eds.): LSMS/ICSEE 2014, Part III, CCIS 463, pp. 81–86, 2014.

logical configurations, a virtual simulation platform with higher accuracy in the full operating range is required. To meet this demand, this paper presents a high precision simulation model of single phase heated tubes. The present model takes into account not only the dynamics of the working fluid velocity, but also the coupling effect between the enthalpy-temperature channel and the pressure-flux channel. Despite the model complexity is improved, but the improvement of the computer performance and the adoption of appropriate numerical solution methods ensure the feasibility of the model. In the following, the model will be described in details, and a simulation case of a heated tube of the rear superheater of a 600 MW controlled circulation boiler superheater is performed to verify the model validity.

2   Model Development

The working principle of a single phase heat tube of power plant boilers is shown in Fig. 1. The single phase working fluid flows within the heat tube, absorbing heat through the metal wall, meanwhile its parameters change accordingly. The present model consists of the mass conservation, momentum conservation and energy conservation equations of the working fluid, the energy conservation equation of metal tube wall, and the state parameter equation of the working fluid. For brevity, the model is given in the form of partial differential equations as follows. However as we know, they need to be transformed to the form of ordinary differential equations by space discretization in the numerical solution.

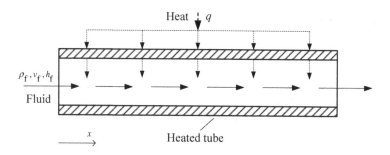

Fig. 1. Working principle of single phase tubes

On the application of heat tube working fluid mass conservation law, the following equation can be obtained

$$\frac{\partial \rho_f}{\partial t} + \frac{\partial (\rho_f v_f)}{\partial x} = 0 \tag{1}$$

where ρ is density, v is velocity, t is time, x is the one-dimensional spatial coordinates; the subscript f represents working fluid.

For heat tube working fluid applied Newton's second law of conservation of momentum equation can be obtained as follows

$$\frac{\partial(\rho_f v_f)}{\partial t} + \frac{\partial p_f}{\partial x} + P = 0 \tag{2}$$

where p is pressure, P is the energy loss. According to the knowledge of fluid mechanics, the energy loss can be calculated using the following Darcy-Weisbach formula

$$P = \zeta \frac{\rho_f v_f^2}{2} \tag{3}$$

where ζ is the friction coefficient of the working fluid flowing along the tube.

Applying energy conservation law to the working fluid, we can obtain the following equation

$$\frac{\partial(\rho_f h_f)}{\partial t} + \frac{\partial(\rho_f v_f h_f)}{\partial x} = S_{h,f} \tag{4}$$

where h is the enthalpy, S_h is the heat power per unit volume. Convective heat transfer heat from the working fluid and the wall between the application of Newton cooling formula, available as $S_{h,f}$ expression

$$S_{h,f} = \frac{4\alpha(T_w - T_f)}{d} \tag{5}$$

where α is the convective heat transfer coefficient between the working fluid and the wall, d is the tube inner diameter; the subscript w represents a metal heat tube wall.

Applying the energy conservation law to the metal heated tube, we can obtain the following equation

$$\frac{\partial(\rho_w c_{p,w} T_w)}{\partial t} = \lambda \frac{\partial^2 T_w}{\partial x^2} + S_{h,w} \tag{6}$$

where c_p is the specific heat capacity per unit mass, λ is the thermal conductivity of the wall. As the heat power conducted in the tube wall is much smaller than that along the radius direction of the tube, the first term on the right hand side of above equation can be ignored. Assuming the absorbed heat of the tube distributes evenly along its wall, the unit volume heat source term for the metal wall can be expressed as

$$S_{h,w} = \frac{q - \alpha(T_w - T_f)}{\delta} \tag{7}$$

where q is a metal tube of heat flow per unit area, δ is the wall thickness.

To make the model closed, the following state parameter equations of working fluids are needed

$$p = f_p(\rho, h) \tag{8}$$

and

$$T = f_T(\rho, h) \tag{9}$$

As thermodynamic properties of single-phase working fluids of power plant boilers, such as water and steam, are very complex, and there is no simple law, the two dimensional Lagrange interpolation method is used to calculate thermodynamic parameters of working fluids based on data of thermodynamic properties of water and steam. The interpolation nodes can be adjusted to obtain a sufficient interpolation precision.

Once the initial conditions and the inlet and outlet boundary conditions are given, equations (1)-(9) constitute a closed single-phase heat tube model. The state variables of the partial differential equations model are: ρ_f, v_f, h_f and T_w.

3   Simulation Case

To verify the effectiveness of the model, a heated tube of the rear superheater of a 600 MW controlled circulation boiler [10] is used to perform simulations. The working fluid in the tube is superheated steam, which absorbs heat to meet the requirements for doing work in turbine. Main parameters used in the simulation are listed in Tab. 1, most of them are based on the design and operation parameters of the boiler; the specific heat capacities of the steam and the metal wall, and other related physical property data come from the physical tables; the convective heat transfer coefficient between the team and the metal tube wall is selected according to engineering experience. Taking both the numerical stability and the solution efficiency into account, the explicit fixed step numerical algorithm, i.e. the 4th order Runge-Kutta method is used to solve the model.

Table 1. Simulation parameters

parameters	units	values
tube length	m	12.73
tube inner diameter	m	4.5×10^{-2}
tube wall thickness	m	7.5×10^{-3}
tube density	$kg \cdot m^{-3}$	7850
tube specific heat capacity	$kJ \cdot kg^{-1} \cdot K^{-1}$	0.46
flowing friction coefficient	m^{-1}	2.72
convective heat transfer coefficient	$kW \cdot m^{-2} \cdot K^{-1}$	2
inlet steam density	$kg \cdot m^{-3}$	58.21
inlet steam velocity	$m \cdot s^{-1}$	21.61
inlet steam enthalpy	$kJ \cdot kg^{-1}$	3300
outlet pressure	Pa	17.4×10^{6}
tube heat power	kW	180

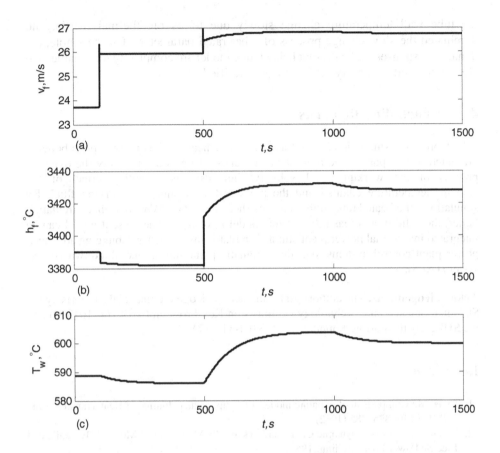

Fig. 2. Dynamic responses of outlet parameters under different disturbances

The following disturbances are imposed on the heated tube: at $t=100$ s, the inlet steam velocity suddenly increases from 21.61 kg·s^{-1} to 23.77 kg·s^{-1}; at $t=500$ s, the inlet steam enthalpy suddenly increases from 3300 kJ·kg^{-1} to 3350 kJ·kg^{-1}; and at $t=1000$ s, the absorbed heat power of the tube suddenly decreases from 180 kJ·s^{-1} to 171 kJ·s^{-1}. Simulation results of the dynamic responses of the outlet steam velocity and enthalpy are given in Fig. 2.

It can be seen from Fig. 2(a), that when the inlet steam velocity increases, the outlet steam velocity increases quickly, and then changes slowly after a dynamic process with overshoot. When the inlet steam enthalpy increases, the dynamic response of the outlet steam velocity has similar characteristics. It's easy to deduce that the initial rapid change is caused by the rapid change of steam pressure, and the subsequent slow change is caused by the corresponding slow temperature change of the heated tube wall. It can be seen from Fig. 2(b), that when the inlet steam velocity increases, the outlet steam enthalpy decreases. This is because that the heat power of the tube is fixed, the steam enthalpy rise will decrease when the steam flux increases. By comparison of the three sub-graphs of Fig. 2, it can be found that, the tube wall temperature change slowly due to its big thermal inertia, and dominated the slow change process of other rated parameters. On the whole, the dynamic responses of the present heated tube model are completely reasonable, and the validity and feasibility of the model is verified.

4   Concluding Remarks

A high precision simulation model for single phase heated tube of power plant boiler is presented in this paper. The present model takes into account not only the dynamic process of the working fluid velocity, but also the coupling effect of the enthalpy-temperature channel and the pressure-flow channel of working fluid. By simulations of a heated tube of the rear superheater of a 600 MW controlled circulation boiler, the validity and feasibility of the model is verified. The present model can be integrated into virtual power plant simulation platforms for better debugging advanced power plant control systems, e.g. the automatic plant startup and shutdown control system (APS).

Acknowledgements. The authors gratefully acknowledge the financial supports by the Shanghai Science and Technology Commission Key Program (No.13111104300) and the SUEP doctor starting Foundation (A-3101-11-017).

References

1. Enns, M.: Comparison of dynamic models of a superheater. Journal of Heat Transfer of the ASME 84(9), 375–385 (1962)
2. Zhang, C.Y.: Boiler Dynamic Characteristics and Its Mathematical Models. Hydraulic and Electric Power Press, Beijing (1987)
3. Li, Y.Z., Yang, X.Y.: Modeling and simulation of the long-term dynamic characteristics of a supercritical once-through boiler. Journal of Engineering for Thermal Energy & Power 18(1), 23–26 (2003)
4. Ren, T.J., Xie, M., Li, Q., Zhang, Z.P.: A new method of chain structure modeling on boiler's single-phase heat-exchanger. Proceedings of the CSEE 23(3), 175–178 (2003)
5. Yang, Y.P., Guo, X.Y., Song, Z.P., Liu, D.H.: A new fast solution method for dynamic process of single-phase medium's temperature. Proceedings of the CSEE 25(3), 158–162 (2005)
6. Liu, S.Q., Yu, S.F., Zhou, L.: Mathematic model and simulation of 900 MW supercritical once-through boiler. Computer Simulation 22(10), 215–218 (2005)
7. Zhuo, X.S., Zhou, Wen, H.C., Chu, Z.Y., Xu, H.: Pressure and temperature dynamics in power plant superheater. Proceedings of the CSEE 27(14), 72–76 (2007)
8. Gao, J.Q., Hao, N., Fan, X.Y., Chen, H.W., Zhao, J.Y., Chen, Z.H.: Dynamic mathematical model and simulation for single-phase heat exchanger of heat recovery steam generator. Journal of North China Electric Power University 36(3), 68–71 (2009)
9. Kang, Y.W., Xue, Y., Huang, W.: Lumped-parameter dynamic modeling and simulation of power plant boiler's superheater. Computer Simulation 29(9), 332–334, 342 (2012)
10. Design Manual of HG-2070/17.5-YM9 Type Boiler for 600 MW Thermal Power Unit. Harbin: Harbin Boiler Company Limited (2005)

Parameter Optimization of Voltage Droop Controller for Voltage Source Converters

Yongling Wu[1,2], Xiaodong Zhao[1], Kang Li[1], and Shaoyuan Li[2]

[1] School of Electronics, Electrical Engineering and Computer Science,
Queen's University Belfast, Belfast, BT9 5AH
[2] Department of Automation,
School of Electronic Information and Electrical Engineering,
Shanghai Jiao Tong University, Shanghai, 200240, China
{ylwu318,syli}@sjtu.edu.cn, {xzhao06,k.li}@qub.ac.uk

Abstract. Mankind is facing the twin challenges of sustainable energy supply and climate change due to the greenhouse gases emissions. Reducing the energy consumption has been given a significant role in addressing these problems. Although modern control technologies have been successfully applied in many engineering systems, PID (Proportional-Integral-Derivative) control and its variants are still the most common control techniques due to their simple control design principle, especially in the power electronics control field. However, little has been done so far to explore how the system performance and control energy are related to each other through the control parameter settings. This paper takes the voltage droop (P) control design in voltage source converters (VSCs) system as the research background to investigate the parameter optimization of droop controllers. The simulation results reveal that there exists a better trade-off between the system performance and control energy consumption through a proper choice of the controller parameters.

Keywords: Parameter Optimization, Droop(P) Controller, Energy Saving, Voltage Source Converters.

1 Introduction

Automatic control is used everywhere in our daily life. The first feedback control device, a water clock of Ktesibios in Alexandria Egypt, can be traced back to around the 3rd century B.C. In the past decades, a number of control technologies have been proposed and successfully applied in many areas, such as model predictive control (MPC) [1,2], linear quadratic regulator (LQR) which is a special case of optimal control [3], and some other intelligent control methods. However, the traditional PID control is still widely used in control systems. It is revealed that more than 90% of feedback control loops are implemented based on PID algorithms [4], especially in the area of electric power systems, power electronic devices, spacecraft attitude control [5], to just name a few. In our societies, the popular adoption of personal cars, air conditioners and large scale industrial production lines imply that more and more energy supplies are needed. The burning

K. Li et al. (Eds.): LSMS/ICSEE 2014, Part III, CCIS 463, pp. 87–96, 2014.
© Springer-Verlag Berlin Heidelberg 2014

of fossil fuels like coal and oil are increasing the atmospheric concentration of carbon dioxide (CO_2) [6]. On one hand, the Mankind is looking for some renewable and clean energy, like wind energy, ocean energy and nuclear power to substitute the traditional energy, on the other hand, we have to find some new ways to reduce the energy consumption and improve the energy efficiency to save energy and protect the environment. Renewables and energy efficiency will have to play a significant role and improved energy efficiency makes an essential contribution to all of the major objectives of EU climate and energy policies [7]. Therefore, energy saving through advanced control can have a profound impact.

Among various renewable energies, wind power is the most mature and technology advanced. The integration of wind farms to the existing alternating current (AC) grid has brought the applications of high voltage direct current (HVDC) transmissions. DC cable has little reactive current and is more suitable for longer distance and submarine cables. Voltage source converters (VSCs) are widely used in many HVDC applications to convert power between AC and DC. Control of VSCs is well developed using voltage oriented vector control (VOC) [8]. Vector control can guarantee high dynamic and static performances with two different control loops [9]. Pulse width modulation (PWM) works as a fundamental method for rectifying an inversion. Many researches have been carried out on the harmonic and loss optimization [10,11], by designing the PWM switching pattern. However, these studies mainly focus on the power loss reduction using PWM or dynamic performance improvement using vector control, the correlation between system performance and power loss has not been properly researched so far. In the context of multi-terminal HVDC (MTDC) system, DC voltage can be maintained using voltage droop control. In this paper, the voltage droop controller design of a VSC used in a MTDC system is investigated.

This paper aims to find a new way to design the droop controller to achieve the desirable requirements and maximize the energy utilization. Unlike early work which are mostly based on simulation [12], the correlations between control energy and tracking performance with the changing of controller parameters will be investigated from the control theory perspective. It shows there exists a nonlinear relationship between the control energy and the tracking performance. With an in-depth study, it is demonstrated that a good choice of controller parameters could target a better balance between the tracking performance and control energy consumption.

2   System Model

The system studied in this paper is a VSC working as DC voltage droop controller, and Fig. 1 shows a typical VSC model represented by the average value. In order to control the DC side voltage of the converter, both voltage drops on the AC side inductor and current in the DC side capacitor need to be compensated. This can be achieved by using two separate control loops, current control loop and voltage control loop. Fig. 2 shows the control structure of the VSC controller. DC voltage is maintained by using a voltage feedback from the capacitor. Output from the voltage droop controller is the reference current for the

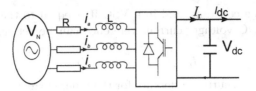

Fig. 1. VSC connection with AC network

current loop. Then current loop is used to control AC side current, maintaining a desirable AC voltage. Because of the balanced power between AC and DC side, DC voltage can be regulated when AC voltage and current are stable.

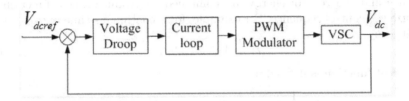

Fig. 2. Structure of VSC voltage controller

Using Kirchhoff's voltage laws (KVL) and commonly used abc/dq transformation, state equations on the AC side can be deduced as follows,

$$\begin{cases} L\frac{di_d}{dt} + Ri_d = V_{Nd} - V_{Cd} + \omega Li_q \\ L\frac{di_q}{dt} + Ri_q = V_{Nq} - V_{Cq} - \omega Li_d \end{cases} \tag{1}$$

where i_d, i_q are the dq components of the VSC AC side current, V_{Cd}, V_{Cq} and V_{Nd}, V_{Nq} are the dq components of the VSC output voltage, and AC network voltage respectively. L is the smoothing reactor including the inductance in the transformer between VSC and AC network. R is the reactor's line resistance. The dq transformation is aligned with AC side voltage, that is $V_{Nq} = 0$.

Current components i_d and i_q need to follow varying references coming from the voltage controller. It can be implemented using a PI controller to eliminate voltage drops on the inductor. In practical applications, due to the existence of current loop and PWM modulation, the output voltage can not update immediately at current sampling point, so there will be an output delay for voltage controller. By properly choosing current controller parameters, its output can normally follow the reference in one switching cycle. Then the current loop and PWM modulator can be modelled as an unit delay $e^{-\tau s}$ at one switching period $\frac{1}{f_s}$. Ignoring the disturbance in Fig. 2, the system model can be written as

$$G_p(s) = \frac{1}{Ts}e^{-\tau s} \tag{2}$$

where T is the time constant, and in a VSC it is equal to the capacitor value. To maintain the DC voltage, current reference can be set using a droop (P) controller.

$$I_{rref} = K_p(V_{ref} - V_{dc}) \tag{3}$$

where I_{rref} represents current reference for the inner current loop. $K_p > 0$ is the droop parameter. To simplify the analysis, the first-order Pade approximation is accepted in the following form, which gives an initial ranges of the controller parameter to make the system stable.

$$e^{-\tau s} \approx \frac{-\tau s + 2}{\tau s + 2} \tag{4}$$

In some modern control technologies, such as LQR, both the tracking performance and the control energy are combined to produce a cost function to optimize the control sequence. For example, for a continuous-time linear system described by

$$\dot{x}(t) = Ax(t) + Bu(t) \tag{5}$$

The cost function is defined as

$$J = \frac{1}{2} \int_0^\infty [x^T(t)Qx(t) + u^T(t)Ru(t)]dt \tag{6}$$

where $x(t)$ is the state, $u(t)$ is the control, while Q and R are the weighting factors, which are defined by human (engineers).

While the optimal control has provided a design framework to adjust the system performance as well as limit the control signals, in this paper the following system performance, such as stability, transient response specified in terms of peak overshoot, settling time, the tracking error and control energy are all considered.

3   Droop(P) Controller Synthesis

3.1   Range of K_p

In this part, we analyse the range of K_p from two aspects, system stability and damping ratio.

The stability is the basic requirement in controller design. The control parameters should first ensure the stability of the system for a given control structure. It is easy to obtain the closed-loop transfer function when the controller is a P controller. And it could be normalized into the following standard form:

$$\Phi(s) = \frac{\omega_n^2(-\frac{\tau}{2}s + 1)}{s^2 + 2\xi\omega_n s + \omega_n^2} \tag{7}$$

where ω_n and ξ are system natural frequency and damping ratio respectively.

$$\omega_n^2 = \frac{2K_p}{T\tau} \qquad 2\xi\omega_n = \frac{2T - K_p\tau}{T\tau} \tag{8}$$

To make the system stable, K_p should satisfy the following inequality relationship according to Routh criteria:

$$0 < K_p < \frac{2T}{\tau} \tag{9}$$

Considering the underdamped ratio case, that is $0 < \xi < 1$, we could get another range for K_p.

$$\frac{(6 - 4\sqrt{2})T}{\tau} < K_p < \frac{(6 + 4\sqrt{2})T}{\tau} \tag{10}$$

Then the final range of K_p is the intersection of (9) and (10).

$$\frac{(6 - 4\sqrt{2})T}{\tau} < K_p < \frac{2T}{\tau} \tag{11}$$

Note: this range of K_p is obtained through 1st order Pade approximation. In the simulation, the range will be narrowed down.

3.2 Dynamic Performance Analysis

From (7), the unit step response of the system is

$$y(t) = 1 - \frac{e^{-\xi\omega_n t}}{\sqrt{1 - \xi^2}} \Psi sin(\omega_d t + \theta - \psi) \tag{12}$$

where

$$\omega_d = \sqrt{1 - \xi^2}\omega_n \qquad \theta = arctan\frac{\sqrt{1 - \xi^2}}{\xi}$$

$$\Psi = \sqrt{2} \qquad \psi = arctan\frac{\frac{\tau}{2}\omega_d}{1 + \frac{\tau}{2}\xi\omega_n} \tag{13}$$

For the characteristic of the capacitance charging-discharging, there exists a integral link in the controlled object. P controller is enough to avoid the steady state error. Therefore, we only need to analyse the system dynamic performance indices, such as tracking performance, energy consumption and the sum of them.

Tracking error is the deviation of actual measured output from the reference set point which is a unit step in this paper.

$$e(t) = \frac{e^{-\xi\omega_n t}}{\sqrt{1 - \xi^2}} \Psi sin(\omega_d t + \theta - \psi) \tag{14}$$

The tracking performance is defined as the 2-norm of tracking error, and the calculated value is given by

$$E_{yc} = \| e(t) \|_2 = \sqrt{e_{yc}} \tag{15}$$

where

$$e_{yc} = \int_0^\infty e^2(t)dt = \frac{1 + \xi sin(3\theta - 2\psi - \frac{\pi}{2})}{2\xi\omega_n(1 - \xi^2)} \tag{16}$$

Controller output $u(t) = K_p e(t)$, then the control energy (that is, the 2-norm of $u(t)$) is given as

$$E_{uc} = \parallel K_p e(t) \parallel_2 = K_p E_{yc} \tag{17}$$

Now we are going to analyse the relationship between dynamic performance and proportional parameter K_p. Firstly, we give a proposition as follow which is very easy to be proved.

Proposition: Let $g = f(x)$, $h = f^2(x)$, if g is always bigger than zero within the range of x, then the monotonicity of g is the same as h. Therefore, e_{yc} and e_{uc} will be discussed in the following part instead of E_{yc} and E_{uc}.

Tracking Performance Analysis

For simplicity, we do not want to have the trigonometric function presenting in the performance indices. Then the upper boundary envelope line e_{ym}, that is the maximum value of e_{yc}, will be used to analyse the effect of K_p on the system performance.

$$e_{ym} < \frac{1}{(1 - \xi^2)\xi\omega_n} = \frac{16(T\tau)^2 K_p}{den} \tag{18}$$

where

$$den = \tau^3 K_p^3 - 14T\tau^2 K_p^2 + 28T^2\tau K_p - 8T^3 \tag{19}$$

Derivative of e_{ym} with respect to K_p is

$$de_{ym}/dK_p = num_y/den^2 \tag{20}$$

where

$$num_y = -32(T\tau)^2(\tau^3 K_p^3 - 7T\tau^2 K_p^2 + 4T^3) \tag{21}$$

Derivative of num_y is

$$dnum_y/dK_p = -32T^2\tau^4 K_p(3\tau K_p - 14T) \tag{22}$$

Obviously $num_y(K_p)$ is continuous. From (11), for the lower boundary value of K_p, $num_y(K_{p(min)}) < 0$ and for the upper boundary value of K_p, $num_y(K_{p(max)}) > 0$, there should be at least one K_p which makes $num_y = 0$.

For $dnum_y/dK_p > 0$ within the range of K_p, num_y grows with the increase of K_p. That is, there exists only one K_p which makes $de_{ym}/dK_p = 0$. Therefore, e_{ym} decreases first and then increases as K_p increases.

Energy Consumption Analysis

Similarly

$$e_{um} = \frac{16(T\tau)^2 K_p^3}{den} \tag{23}$$

Derivative of e_{um} with respect to K_p is

$$de_{um}/dK_p = num_u/den^2 \tag{24}$$

where

$$num_u = -32T^3\tau^2 K_p^2 (7\tau^2 K_p^2 - 28T\tau K_p + 12T^2) \tag{25}$$

Derivative of num_u is

$$dnum_u/dK_p = -128T^3\tau^2 K_p(7\tau^2 K_p^2 - 21T\tau K_p + 6T^2) \tag{26}$$

The roots of the above second order equation in the brackets are

$$r_{1,2} = \frac{(21 \pm \sqrt{13 * 21})T}{14\tau} \quad (r_1 < r_2) \tag{27}$$

By calculation, we know that the roots are out of the range of K_p. That is, the value of the derivative of num_u is always positive. It is the same for the analysis of e_{ym}, $num_u(K_{p(min)}) < 0$ and $num_u(K_{p(max)}) > 0$, so the change of e_{um} should be the same as the change of e_{ym}.

4   Simulation Results

In the simulation, the reference voltage is $400kV$ with power ratings of $500MW$. Controller parameters are chosen to be $T = 0.02$ and $\tau = 7.4 \times 10^{-4}$, which correspond to a capacitor value of $C = 62.4\mu F$ and PWM switching frequency $f_s = 1350Hz$ respectively.

According to (11), the range of K_p is from 9.3 to 54. In this simulation, it is narrowed down to 11.2 to 43.2 which are the 120 percent and 80 percent of the lower and upper bounds respectively. And it is increased with a step value of 0.1. The dynamic performance with different controller parameters are shown in Fig. 3, in which E_{ym} and E_{um} are the square roots of e_{ym} and e_{um} respectively, E_{ys} and E_{us} are the integral values of tracking error and control energy respectively, which are obtained by simulation within $20ms$. Input is DC voltage in per unit.

Fig. 3 gives the following observations:

Observation 1: All of the tracking performance curves decrease first and then increases with the growing of K_p. However, the energy consumption curves of E_{us} and E_{uc} keep growing, which are different from the theoretical analysis result E_{um} around the lower boundary.

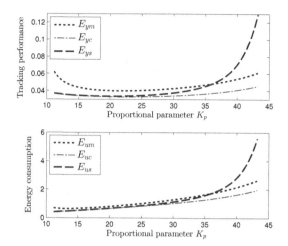

Fig. 3. Dynamic performance with different proportional parameters

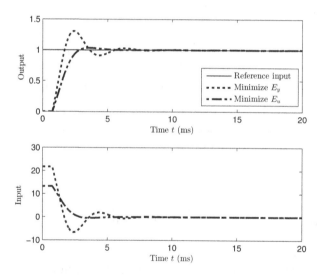

Fig. 4. Dynamic Response with different proportional parameters

Observation 2: Near the lower boundary, the calculated values E_{yc} and E_{uc} are close to the simulation results E_{ys} and E_{us}. However, the simulation values go up quickly with the growing of K_p in the last third part of its range, which means the upper boundary of K_p obtained according to the Pade approximation can not guarantee the stability of system.

Observation 3: The minimum E_{ym}, E_{yc} and E_{yr} are almost at the same K_p. And the lower boundary of K_p may not guarantee the damping ration of system, so we won't use the smallest K_p value either. Therefore, the original K_p values could be set to be the roots of (21) and (25).

The dynamic response results with different control parameters obtained from (21) and (25) which make system performance indices minimum separately are given in Fig. 4. The detailed numerical results are also summarized in Table 1, where $\sigma\%$ is the overshoot and t_s is obtained by simulation measurement. $E_{\_max}$, $E_{\_cal}$ and $E_{\_sim}$ means the K_p values are obtained by minimizing the maximum approximation values, calculated values and simulation values respectively.

From Fig. 4 and Table 1, it shows that we could give an initial proportional parameter according to (21) and (25), and then find an optimal value around them with the consideration of balancing the tracking error and control energy consumption.

Table 1. System performance for different K_p

Type	K_p	E_y	E_u	$\sigma\%$	$t_s(ms)$
$E_{\_max}$	21.7	0.0399	0.8657	30.49	6.76
	13.2	0.0486	0.6410	3.30	4.42
$E_{\_cal}$	22.4	0.0328	0.7355	33.02	6.80
	11.2	0.0369	0.4130	0.23	3.73
$E_{\_sim}$	20.0	0.0336	0.6720	24.43	5.45
	11.2	0.0369	0.4135	0.23	3.73

5   Conclusions and Future Works

5.1   Conclusions

This paper has investigated the P controller parameter optimization in the vector control of voltage source converters. Firstly, the proportional parameter is restricted in a range from two aspects, system stability and damping ratio. Then, the effect of K_p on tracking performance and energy consumption are investigated in detail. The energy consumption grows with the increase of K_p, but the tracking error decreases first and then increases. Finally, we have derived an equation to obtain the K_p to make the tracking error minimum. And the optimum value of K_p could be found around it by balancing the tracking performance with the control energy consumption.

5.2   Future Works

Based on the results obtained in this paper, future work will include the research on the parameter optimization of the PI controller based on a popular industrial model. The close loop transfer function will become much more complicated with the increase in the complexity of both the controller and the object model. The approaches will also be applied in some real-world systems, such as power electronics control in power transmission and distribution.

Acknowledgments. The author Y Wu and X Zhao would like to acknowledge the sponsorship from Chinese Scholarship Council (CSC) for their PhD study at Queen's University Belfast. This work was partially supported by UK Engineering and Physical Sciences Research Council (EPSRC) under grants EP/G042594/1 and EP/L001063/1, and the National Natural Science Foundation of China under Grant 61273040.

References

1. Camacho, E.F., Alba, C.B.: Model predictive control. Springer (2013)
2. Joe Qin, S., Badgwell, T.A.: A survey of industrial model predictive control technology. Control Engineering Practice 11(7), 733–764 (2003)
3. Teleke, S., Baran, M.E., Bhattacharya, S., Huang, A.Q.: Optimal control of battery energy storage for wind farm dispatching. IEEE Transactions on Energy Conversion 25(3), 787–794 (2010)
4. Astrom, K.J.: Control system design. Lecture notes (2002)
5. Li, C., Teo, K.L., Li, B., Ma, G.: A constrained optimal pid-like controller design for spacecraft attitude stabilization. Acta Astronautica 74, 131–140 (2012)
6. Armaroli, N., Balzani, V.: The future of energy supply: challenges and opportunities. Angewandte Chemie International Edition 46(1-2), 52–66 (2007)
7. Davey, E.: Green growth group ministers' statement on climate and energy framework for 2030 (2014),
   https://www.gov.uk/government/news/green-growth-group-ministers-statement-on-climate-and-energy-framework-for-2030
8. Cortes, P., Kazmierkowski, M.P., Kennel, R.M., Quevedo, D.E., Rodríguez, J.: Predictive control in power electronics and drives. IEEE Transactions on Industrial Electronics 55(12), 4312–4324 (2008)
9. Kaźmierkowski, M.P., Krishnan, R.: Control in power electronics: selected problems. Academic Press (2002)
10. Holtz, J., Qi, X.: Optimal control of medium-voltage drives an overview. IEEE Transactions on Industrial Electronics 60(12), 5472–5481 (2013)
11. Wiechmann, E.P., Aqueveque, P., Burgos, R., Rodriguez, J.: On the efficiency of voltage source and current source inverters for high-power drives. IEEE Transactions on Industrial Electronics 55(4), 1771–1782 (2008)
12. Li, K., Wu, Y., Li, S., Xi, Y.: Energy saving and system performance-an art of trade-off for controller design. In: 2013 IEEE International Conference on Systems, Man, and Cybernetics (SMC), pp. 4737–4742. IEEE (2013)

Research on Multivariable MPC Controller Design and Simulation of Interconnected Power Systems LFC

Hong Qian[1,2], Wei-xiao Jin[2], Jian-bo Luo[2], Min-rui Fei[1]

[1] Shanghai Key Laboratory of Power Station Automation Technology,
School of Mechatronic Engineering and Automation,
Shanghai University, Shanghai, China, 200072
qianhong.sh@163.com, mrfei@staff.shu.edu.cn
[2] School of Automation Engineering, Shanghai University of Electric Power, Shanghai 200090

Abstract. In order to make the control of interconnected grid frequency have better load adaptability and maintain the economic stability and reliable operation of the grid, a design method of multivariable predictive controller and its algorithm are proposed. They are based on the state-space model of the interconnected grid while considering the regional capacity constraint. To solve the problem with the control relying on K (the frequency error coefficient) in most such questions, the quadratic programming (QP) solving is employed. And so the optimal control is ensured even when there is interconnected grid support. According to the modeling and simulating of the two-area grid frequency control system, the proposed predictive algorithm is compared with the conventional algorithm via PI regulator in single area and dual zone load step disturbance, random white noise disturbance and model mismatch, etc. The results indicate that the proposed predictive control algorithm can significantly improve the recovery performance of the grid system frequency.

Keywords: frequency deviation factor, state-space model, predictive controller, QP problem solving.

1   Introduction

The advent of modern power grid characterized with grid interconnection makes it possible that regional grid can support each other. To restrain the influence of regional disturbance on the grid frequency, currently the traditional automatic generation control (AGC) uses the load frequency control (LFC). It is based on the calculation of the area control error (ACE) with traditional PI algorithm to obtain power adjustment commands. So the adjustment quality depends on K (the frequency error coefficient) directly. Because of the randomness and time variation of the feature and size of the grid load, the frequency error coefficient K is difficult to be measured and set [1]. The traditional control of the grid frequency is time-consuming and overshooting. The frequency error coefficient K is more difficult to be determined, when the region

K. Li et al. (Eds.): LSMS/ICSEE 2014, Part III, CCIS 463, pp. 97–108, 2014.

regulation capacity is insufficient. That will influence the recovery rate of the power quality and economic performance indicators of the regions [2-3].

In order to improve the control performance of the traditional AGC, many advanced control strategy had been used on the research of the LFC system controller, such as the sliding mode control [4], the neural network control [5] and the fuzzy control [6], etc. And the control effect is good. Model predictive control (MPC) is an advanced control technology based on model. It is ease-modeling and rolling optimization, etc. It can be used to solve the control defect of the traditional AGC. Using the MPC [10] and fuzzy control strategy, Documents [7-9] designed the AGC controller and achieved some consequence via simulation. But the research on this predictive control theory did not consider the limits of the regions active power output. It still relies on the frequency coefficient K. That cannot suit the regions load change and the support functions in the actual grid.

In this paper, the multivariable constrained predictive control based on dual-zone interconnected power grid is considered. It is a nonlinear constrained control problem based on state-space model. To solve the problem with the control relying on K, the quadratic programming (QP) solving is employed. Also the optimal control is ensured when there is interconnected grid support. The effectiveness of the multivariable constrained predictive control as the LFC controller with two regions is exemplified by simulation.

2   State-Space Model of Interconnected Grid LFC Controlled Object

The two regions interconnected grid has been used as controlled object of the interconnected grid LFC. Fig. 1 is the dynamic model of the controlled object of the two area interconnected grid LFC. It is composed by the transfer functions of each link, including generator objects and grid objects. $G_{a1}(s)$, $G_{a2}(s)$, $G_{b1}(s)$ and $G_{b2}(s)$ are generator objects of region-one and region-two; $G_{a3}(s)$ and $G_{b3}(s)$ are grid objects of region-one and region-two; Δf_1 is the variation of the frequency; P_{ci} is the dispatching instruction(the generator control variable) ; ΔP_{ti} is the output power increment of each region; ΔP_{Li} is the load variation of each region; ΔP_{12} is the variation of the tie-exchange power; T_{gi} is the governor time constant; T_{ti} is the generator time constant; R_i is the unit difference adjustment coefficient; ΔX_{gi} is the governor position increment; K_{li*} is the frequency adjustment effect coefficient of the load; T_{pi} is the frequency adjustment inertia time constant of the load; T_{12} is the tie-exchange power synchronous coefficient of region-one and region-two.

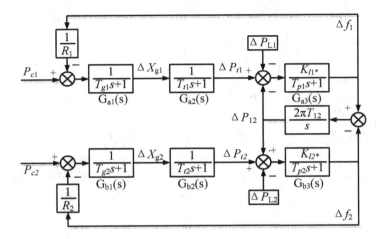

Fig. 1. Transfer function dynamic model of dual-zone interconnected grid LFC controlled objects

The discrete state space model of the controlled objects can be written in form of

$$\begin{cases} \dot{\mathbf{X}}(k+1)=\mathbf{A}\mathbf{X}(k)+\mathbf{B}\mathbf{U}(k)+\mathbf{R}\mathbf{W}(k)\,, \\ \mathbf{Y}(k)=\mathbf{C}\mathbf{X}(k)\,. \end{cases} \qquad (1)$$

Where $\mathbf{X}$, $\mathbf{U}$ $\mathbf{W}$ and $\mathbf{Y}$ are state variables, input variables, disturbance variables and output variables; $\mathbf{A}$, $\mathbf{B}$ $\mathbf{R}$ and $\mathbf{C}$ are the state matrix, input matrix, disturbance matrix and output matrix of the discrete system.
Where

$$\mathbf{U}=[P_{c1}\,P_{c2}]\,;\mathbf{W}=[\Delta P_{l1}\,\Delta P_{l2}]\,;\mathbf{Y}=[\Delta f_1\,\Delta f_2\,\Delta P_{12}]^T\,;$$

$$\mathbf{X}=[\Delta f_1\,\Delta P_{t1}\,\Delta X_{g1}\,\Delta P_{12}\,\Delta f_2\,\Delta P_{t2}\,\Delta X_{g2}]\,.$$

The sampling time constant is T_s, then

$$A = \begin{pmatrix}
1-\dfrac{T_s}{T_{p1}} & \dfrac{T_s K_{l1}*}{T_{p1}} & 0 & -\dfrac{T_s K_{l1}*}{T_{p1}} & 0 & 0 & 0 \\
0 & 1-\dfrac{T_s}{T_{t1}} & \dfrac{T_s}{T_{t1}} & 0 & 0 & 0 & 0 \\
-\dfrac{T_s}{T_{g1}R_1} & 0 & 1-\dfrac{T_s}{T_{g1}} & 0 & 0 & 0 & 0 \\
2pT_{12}T_s & 0 & 0 & 1 & -2pT_{12}T_s & 0 & 0 \\
0 & 0 & 0 & \dfrac{T_s K_{l2}*}{T_{p2}} & 1-\dfrac{T_s}{T_{p2}} & \dfrac{T_s K_{l2}*}{T_{p2}} & 0 \\
0 & 0 & 0 & 0 & 0 & 1-\dfrac{T_s}{T_{t2}} & \dfrac{T_s}{T_{t2}} \\
0 & 0 & 0 & 0 & -\dfrac{T_s}{T_{g2}R_2} & 0 & 1-\dfrac{T_s}{T_{g2}}
\end{pmatrix} ,$$

$$B = \begin{pmatrix}
0 & 0 & \dfrac{T_s}{T_{g1}} & 0 & 0 & 0 & 0 \\
0 & 0 & 0 & 0 & 0 & 0 & \dfrac{T_s}{T_{g2}}
\end{pmatrix}^T , \quad R = \begin{pmatrix}
-\dfrac{T_s K_{l1}*}{T_{p1}} & 0 & 0 & 0 & 0 & 0 & 0 \\
0 & 0 & 0 & 0 & 0 & 0 & -\dfrac{T_s K_{l2}*}{T_{p2}}
\end{pmatrix}^T ,$$

$$C = \begin{pmatrix}
1 & 0 & 0 & 0 & 0 & 0 & 0 \\
0 & 0 & 0 & 0 & 1 & 0 & 0 \\
0 & 0 & 0 & 1 & 0 & 0 & 0 \\
0 & 1 & 0 & 0 & 0 & 0 & 0 \\
0 & 0 & 0 & 0 & 0 & 1 & 0
\end{pmatrix} .$$

3 Design of the LFC Predictive Controller

Fig. 2 is the design structure of the LFC predictive controller, including estimation of the state by the observer and the optimization controller.

Because of the existence of the unknown disturbance of the grid load, the dynamic system is complicated. The observer can be used to estimate these state $\hat{x}(k)$. The design equations of the controller can be written in the form of

$$y(k) = \Psi\hat{x}(k) + \Upsilon u(k-1) + \Theta\Delta u(k) . \tag{2}$$

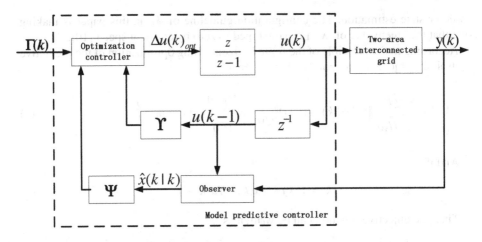

Fig. 2. LFC predictive controller structure

Where $\mathbf{y}(k) = \begin{bmatrix} \hat{y}(k+H_w|k) \\ \vdots \\ \hat{y}(k+H_p|k) \end{bmatrix}$; $\Delta\mathbf{u}(k) = \begin{bmatrix} \Delta\hat{u}(k|k) \\ \vdots \\ \Delta\hat{u}(k+H_u-1|k) \end{bmatrix}$; $\hat{\mathbf{x}}(k) = \begin{bmatrix} \hat{x}(k+1|k) \\ \vdots \\ \hat{x}(k+H_p|k) \end{bmatrix}$;

$\mathbf{\Psi} = \mathbf{H}\mathbf{\Psi}'$; $\Upsilon = \mathbf{H}\Upsilon'$; $\Theta = \mathbf{H}\Theta'$; $\mathbf{H} = \begin{bmatrix} \mathbf{C} & 0 & \cdots & 0 \\ 0 & \mathbf{C} & \cdots & 0 \\ \cdots & \cdots & \cdots & \cdots \\ 0 & 0 & \cdots & \mathbf{C} \end{bmatrix}$; $\Upsilon' = \begin{bmatrix} \mathbf{B} \\ \vdots \\ \sum_{i=0}^{H_u-1} \mathbf{A}^i\mathbf{B} \\ \sum_{i=0}^{H_u} \mathbf{A}^i\mathbf{B} \\ \vdots \\ \sum_{i=0}^{H_p-1} \mathbf{A}^i\mathbf{B} \end{bmatrix}$;

$\mathbf{\Psi}' = \begin{bmatrix} \mathbf{A} \\ \vdots \\ \mathbf{A}^{H_u} \\ \mathbf{A}^{H_u+1} \\ \mathbf{A}^{H_p} \end{bmatrix}$; $\Theta' = \begin{bmatrix} \mathbf{B} & \cdots & 0 \\ \mathbf{AB}+\mathbf{B} & \cdots & 0 \\ \cdots & \cdots & \cdots \\ \sum_{i=0}^{H_u-1} \mathbf{A}^i\mathbf{B} & \cdots & \mathbf{B} \\ \sum_{i=0}^{H_u} \mathbf{A}^i\mathbf{B} & \cdots & \mathbf{AB}+\mathbf{B} \\ \cdots & \cdots & \cdots \\ \sum_{i=0}^{H_p-1} \mathbf{A}^i\mathbf{B} & \cdots & \sum_{i=0}^{H_p-H_u} \mathbf{A}^i\mathbf{B} \end{bmatrix}$.

In this paper, the designed state observer model is the same as the model of the interconnected grid controlled objects. And the output deviation of the actual power and the state observer multiplied by an observer gain matrix $\mathbf{L}$. It is used as the observer state estimation. The configuration guideline of $\mathbf{L}$ in this paper is making sure that the eigenvalue of $\mathbf{A} - \mathbf{LC}$ converged to zero by constant speed [10].

According to the principle of the finite horizon rolling optimization of the predictive control, the objective function can be written as

$$V(k) = \sum_{i=H_W}^{H_P} \left\| \hat{y}(k+1|k) - r(k+i) \right\|^2_{Q(i)} + \sum_{i=0}^{H_u-1} \left\| \Delta\hat{u}(k+i|k) \right\|^2_{R(i)} . \tag{3}$$

And if

$$\varepsilon(k) = \Gamma(k) - \Psi x(k) - \Upsilon u(k-1) . \tag{4}$$

Then the objective function can be written as

$$V(k) = \left\| y(k) - \Gamma(k) \right\|^2_{Q(i)} + \left\| \Delta u(k) \right\|^2_{R(i)} . \tag{5}$$

Where $\mathbf{Q}$ and $\mathbf{R}$ are weighting matrix.

Then

$$\begin{aligned}
V(k) &= \left\| \Theta\Delta u(k) - \varepsilon(k) \right\|^2_Q + \left\| \Delta u(k) \right\|^2_R \\
&= \left[\Delta u(k)^T \Theta^T - \varepsilon(k)^T \right] Q[\Theta\Delta u(k) - \varepsilon(k)] + \Delta u(k)^T R\Delta u(k) \\
&= \varepsilon(k)^T \Theta\varepsilon(k) - 2\Delta u(k)^T \Theta^T Q\varepsilon(k) + \Delta u(k)^T \left[\Theta^T Q\Theta + R \right] \Delta u(k)
\end{aligned} \tag{6}$$

It can be written as

$$V(k) = const - \Delta u(k)^T q + \Delta u(k)^T r\Delta u(k) . \tag{7}$$

Where

$$q = 2\Theta^T Q\varepsilon(k) .$$

The solution of the optimal controller output $\Delta u(k)$ can be equivalent to the solution of

$$\min(\Delta u(k)^T r\Delta u(k) - q^T \Delta u(k) . \tag{8}$$

Then it is turned to be a typical quadratic programming (QP) problem. In this paper the optimal control law $\Delta u(k)_{opt}$ is resolved by using the internal point solution method. The constraint conditions can be written as

$$\begin{cases} u(1)=P_{c1} \leq u_{\max 1}, \\ \Delta u(1)=\Delta P_{c1} \leq \Delta u_{\max 1}, \\ u(2)=P_{c2} \leq u_{\max 2}, \\ \Delta u(2)=\Delta P_{c2} \leq \Delta u_{\max 2}. \end{cases} \quad (9)$$

Where

$u_{\max 1}$ and $\Delta u_{\max 1}$ are the maximum adjustment capacity value and limits of adjustment capacity rate of region-one. $u_{\max 2}$ and $\Delta u_{\max 2}$ are the maximum adjustment capacity value and limits of adjustment capacity rate of region-two.

4   Simulation

Using the Matlab/simulink, the two-area interconnected grid AGC system simulation model is built with simulink tool (PI control) and MPC toolbox (MPC control), respectively. The control objectives of the two-area interconnected grid AGC system simulation model are the tie-line exchange power and the frequency deviation. The MPC controller is designed by the algorithm aforementioned. Let the sampling period $T_s = 0.1s$, prediction horizon $P=15$, control horizon $M=10$, error weighting matrix $Q=10I$, control weighting matrix $R=0.01I$. The simulation of system dynamic performance (the self-capacity of each area is adequate and inadequate), the performance of resisting the random disturbance and the robustness are researched respectively.

4.1   Simulation Research on System Dynamic Performance (Self-capacity of Each Area Is Adequate)

When two areas have sufficient capacity (without capacity limit value), the AGC simulation model parameter of the two-area interconnected power system is shown in Table 1. The parameters are configured according to the 1:6 relationship of two regions load. At the time t=1s, step disturbances of the load of two areas are added respectively ($\Delta P_{l1} = 0.01 p.u.$, $\Delta P_{l2} = 0.03 p.u.$). Comparatively analyzing the system response curves of the MPC controlled system and the PI controlled system, Fig. 3- Fig. 5 are the AGC response curves of each area.

Table 1. Controlled objects parameter of two areas (model 1)

parameter	$T_{gi}(s)$	$T_{ti}(s)$	K_{li*}(Hz/MW)	$T_{pi}(s)$	R_i(Hz/ MW)	$T_{12}(s)$
Region-one	0.08	0.3	1.5	20	0.05	0.545
Region-two	0.08	0.3	0.25	20	0.008	

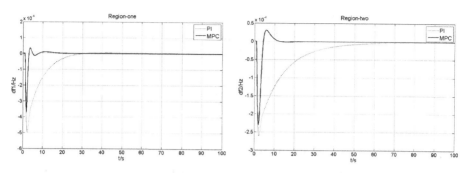

Fig. 3. Frequency deviation response curve of region-one with different controller action

Fig. 4. Frequency deviation response curve of region-two with different controller action

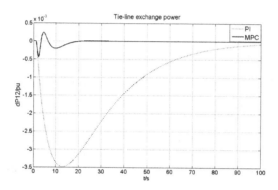

Fig. 5. Tie-line exchange power response curve of different controller action

By comparing the simulation output curves, the results show that the tie-line exchange power response overshoot of the MPC controller controlled system is much smaller than the PI controller (controlled system, and the regulation time is much smaller too. The frequency deviation response overshoot of these two regions with the MPC controller is smaller. And the regulation time is also much smaller.

4.2   Simulation Research on System Dynamic Performance (Self-capacity of Each Area Is Inadequate)

The method of research on system dynamic performance when the self-capacity of each area is inadequate is added by capacity limits on the simulation model. The AGC simulation model parameter of two-area interconnected power system is shown in Table 1, and 0.008p.u. capacity limit is added to the unit output of each region, which make $P_{ti} \leq 0.008p.u.$. At the time t=1s, step disturbances of the load of region-one are added ($\Delta P_{l1} = 0.01p.u.$). Comparatively analyzing the system response curves of the MPC controlled system and the PI controlled system, Fig. 6- Fig. 8 are the AGC response curves of each area.

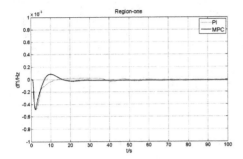
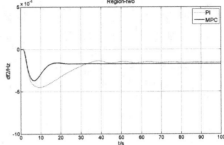

Fig. 6. Frequency deviation response curve of region-one with different controller action (with limits)

Fig. 7. Frequency deviation response curve of region-two with different controller action (with limits)

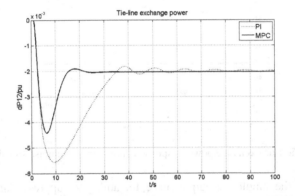

Fig. 8. Tie-line exchange power response curve of different controller action (with limits)

By comparing the simulation output curves, it can be concluded that regarding the tie-line exchange power and the frequency deviation response overshoot, the system with MPC controller is better than the system with PI controller. MPC controller still has the superiority on the regulation time and the response overshoot.

4.3   Simulation Research on the Performance of Anti-random Noise

The parameters of the AGC simulation model of the two-area interconnected power system are still the parameter of table.1. The white noise signal of the load is added to each region at the time t=1s (ΔP_{l1} and ΔP_{l2} are white noise signal with the amplitude 0.01p.u. and 0.03p.u.). The ability to suppress white noise of the grid load of the MPC controller is compared with the PI controller, Fig. 9- Fig. 11 are the AGC response curves of each area.

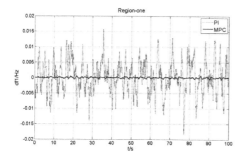

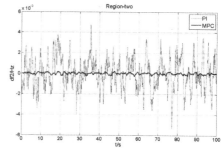

Fig. 9. Frequency deviation response curve of region-one with white noise signal as load

Fig. 10. Frequency deviation response curve of region-two with white noise signal as load

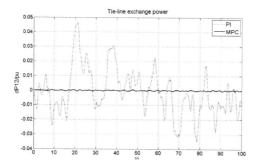

Fig. 11. Tie-line exchange power response curve with white noise signal as load

According to the simulation output curve, the ability to suppress white noise of the grid load of the MPC controller is much better than the PI controller.

4.4   Simulation Research on Model Mismatch

In order to compare the robustness between the MPC controlled AGC system and the PI controlled AGC system, the dynamic characteristics of the controlled objects are changed. The changed parameters are shown in Table 2. The system model of Table 1 is called model-one, and the system model of Table 2 is called model-two. The step disturbance of the load is added to model-one and model-two at the time t=1s ($\Delta P_{l1}=0.01\,p.u.$ and $\Delta P_{l2}=0.03\,p.u.$). The output curves of model-one and model-two with the MPC and PI control mode are shown in Fig. 12- Fig. 14.

Table 2. Controlled objects parameter of two areas (model-two)

parameter	$T_{gi}(s)$	$T_{ti}(s)$	K_{li*}(Hz/MW)	$T_{pi}(s)$	R_i(Hz/MW)	$T_{12}(s)$
Region-one	0.1	0.3	2	30	0.025	0.545
Region-two	0.09	0.375	1.25	20	0.01	

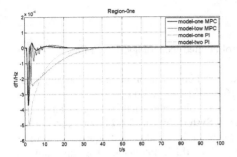 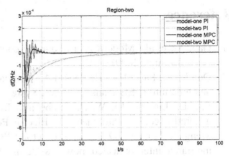

Fig. 12. Frequency deviation response curve of region-one with different model and different controller action

Fig. 13. Frequency deviation response curve of region-two with different model and different controller action

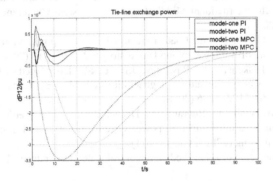

Fig. 14. Tie-line exchange power response curve with different model and different controller action

According to the output curve, the robustness of the system with both control modes is good. Compared to the PI controller, the MPC controller can get closer simulation output curve when the system model is changed. That shows the system with MPC controller has better robustness than the system with PI controller.

5   Conclusions

Constrained multivariable predictive control of the load frequency of the interconnected grid aforementioned overcomes the defect that the frequency control algorithm generally has. The method has practical engineering value with the constraints of the regional adjustment capacity limits. According to the simulation of the load step disturbance, the random white noise disturbance and the robustness, the control effect of the predictive control algorithm proposed in this paper has obvious advantages over the traditional methods for the control of the interconnected grid.

Acknowledgments. This work is supported by Shanghai Key Laboratory Power Station Automation Technology Laboratory under Grant 04DZ05901.

References

1. Gao, X., et al.: Modern grid frequency control applied technology. China Electric Power Press, Beijing (2009) (in Chinese)
2. Donde, V., Pai, M.A., Hiskens, I.A.: Simulation and optimization in an AGC system after deregulation. IEEE Trans. on Power System 16(3), 481–489 (2001)
3. Liu, L., Liu, R., Li, W.-D.: Probe into frequency bias coefficient setting in automatic generation control. Automation of Electric Power System. 30(6), 42–47 (2006) (in Chinese)
4. Meng, X.P., Gong, Q.X., Feng, L., et al.: PI fuzzy sliding mode load frequency control of multiarea interconnected power systems. In: 2003 IEEE International Symposium on Intelligent Control, pp. 1023–1027 (2003)
5. Sabahi, N.M.A.: Load Frequency control in inter-connected power system using modified dynamic neural networks. In: Proceedings of Mediterranean Conference on Control and Automation, Athens, Greece, pp. 1–5 (2007)
6. El-Metwally, K.A.: An adaptive fuzzy logic controller for a two area load frequency control problem. In: 12th International Middle-East Power System Conference, MEPCON 2008, pp. 300–306 (2008)
7. Yang, G., Liu, G.-M., Qu, Z.: Load frequency control of power systems based on MPC algorithm. Journal of Beijing Jiaotong University 36(2), 105–110 (2012) (in Chinese)
8. Venkat, A.N., Hiskens, I.A., Rawlings, J.B., et al.: Distributed MPC strategies with application to power system automatic generation control. IEEE Trans. on Control Systems Technology 16(6), 1192–1206 (2008)
9. Shafiee, Q., Morattab, A., Bevrani, H.: Decentralized model predictive load-frequency control for multi-area interconnected power systems. In: 2011 19th Iranian Conference on Electrical Engineering (ICEE), vol. 1 (2011)
10. Quan, J.X., Zhao, J., Xu, Z.H.: Predictive control. Chemical Industry Press, Beijing (2007) (in Chinese)

The Method to Establish the Simulation Model
of Internal Feedback Motor
Based on Software of Matlab/Simulink

Xiaodong Zhang and Hui Shi

JIBEI Electric Power Company Limited Skills Training Center, Baoding 071003, China
luckdd@163.com

Abstract. This paper presents the mothed to establish the simulation model of the internal feedback motor based on the dynamic model of the internal feedback winding motor, which fills the gaps in the simulik module library. The method makes use of the S Function in the tool-list of matlab/simulink to complete the simulation model. Simulation cases and field tests are used to testify the model, and it's found that the simulation results correspond to the data of field tests. Therefore the model which is established by the way, can be used correctly.

Keywords: internal feedback motor, cascade speed control system, simulation.

1 Introduction

The internal feedback winding motor, which is produced with the development of the cascade control technology in recent years, is a new type of winding motor. It adds a set of windings to the stator core of the asynchronous motor, which is named regulating winding, in order to receive the energy reacted from the rotors. And now the original stator winding is called the main winding[1-2]. The cascade control speeding system of the motor is showed in fig.1. The main winding of the stator connects with the grid, and the rotor winding connects with the cascade speeding control system which consists of rectifier and inverter, and the slip power is feedbacked to the regulating winding of the stator[3].

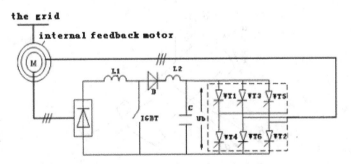

Fig. 1. The topology circuit of chopping cascade control system

K. Li et al. (Eds.): LSMS/ICSEE 2014, Part III, CCIS 463, pp. 109–116, 2014.

With the purpose of studying the principle of the internal feedback cascade control technology, the simulation method is used to do the analysis. Nowadays matlab/simulink is a kind of famous simulation software. However, there is not the available internal feedback winding motor model, which is a difficult thing for the project. With the purpose, this paper tells the principle of the simulation model, which is theoretically based on the program of matlab/simulink and the mathematic model of internal feedback motor, and established according to the S Function. Simulation cases and field tests are used to testify the model, and it's found that the simulation results correspond to the data of field tests. Therefore the model, which is established by the way, can be used practically and feasibly[4-5].

2   Thematic Model of the Internal Feedback Motor

According to the theory of the AC motor, after change the system of axes, the three-phase asynchronous motor can be considered as the equivalence of the two-phase motor model on the d–q axis in the synchronous revolution axes system. Suppose the axes are put on the rotor, with d axis as the straight one, q axis is advanced 90', and the included angel between the reference axis and the stator winding one (showed as figure 2), then we get a set of matrix by changing the abc coordinates to the dq0:

$$C_{sk} = \frac{2}{3} \left\{ \begin{array}{ccc} \cos(\theta-\theta_k) & \cos(\theta-\dfrac{2\pi}{3}-\theta_k) & \cos(\theta-\dfrac{4\pi}{3}-\theta_k) \\ \sin(\theta-\theta_k) & \sin(\theta-\dfrac{2\pi}{3}-\theta_k) & \sin(\theta-\dfrac{4\pi}{3}-\theta_k) \\ \dfrac{1}{\sqrt{2}} & \dfrac{1}{\sqrt{2}} & \dfrac{1}{\sqrt{2}} \end{array} \right\} \tag{1}$$

$$C_r = \frac{2}{3} \left\{ \begin{array}{ccc} \cos(\theta-\theta_r) & \cos(\theta-\dfrac{2\pi}{3}-\theta_r) & \cos(\theta-\dfrac{4\pi}{3}-\theta_r) \\ \sin(\theta-\theta_r) & \sin(\theta-\dfrac{2\pi}{3}-\theta_r) & \sin(\theta-\dfrac{4\pi}{3}-\theta_r) \\ \dfrac{1}{\sqrt{2}} & \dfrac{1}{\sqrt{2}} & \dfrac{1}{\sqrt{2}} \end{array} \right\} \tag{2}$$

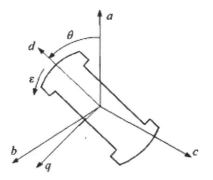

Fig. 2. Relation between abc and dq0 coordinates

Because the internal feedback motor adds a set of windings to the stator core of the asynchronous machine, then its structure and principle is very different from the normal asynchronous machine. The mathematic model in the static axes of system can be described in the following equations[6]:

Electric tension equations:

$$\left. \begin{array}{l} u_{d1} = r_{s1}i_{sd1} + \dfrac{d\psi_{d1}}{dt} \\[2mm] u_{q1} = r_{s1}i_{sq1} + \dfrac{d\psi_{q1}}{dt} \\[2mm] u_{d2} = r_{s2}i_{sd2} + \dfrac{d\psi_{d2}}{dt} \\[2mm] u_{q2} = r_{s2}i_{sq2} + \dfrac{d\psi_{q2}}{dt} \\[2mm] u_{rd} = r_{r}i_{rd} + \dfrac{d\psi_{rd}}{dt} + \omega_{r}\psi_{rq} \\[2mm] u_{rq} = r_{r}i_{rq} + \dfrac{d\psi_{rq}}{dt} - \omega_{r}\psi_{rd} \end{array} \right\} \tag{3}$$

Magnetic linkage equations:

$$\left. \begin{array}{l} \psi_{d1} = L_{11}i_{sd1} + L_{sm}(i_{sd1} + i_{sd2}) + L_{m}(i_{sd1} + i_{sd2} + i_{rd}) \\[2mm] \psi_{q1} = L_{11}i_{sq1} + L_{sm}(i_{sq1} + i_{sq2}) + L_{m}(i_{sq1} + i_{sq2} + i_{rq}) \\[2mm] \psi_{d2} = L_{22}i_{sd2} + L_{sm}(i_{sd1} + i_{sd2}) + L_{m}(i_{sd1} + i_{sd2} + i_{rd}) \\[2mm] \psi_{q1} = L_{22}i_{sq2} + L_{sm}(i_{sq1} + i_{sq2}) + L_{m}(i_{sq1} + i_{sq2} + i_{rq}) \\[2mm] \psi_{dr} = L_{1r}i_{rd} + L_{m}(i_{sd1} + i_{sd2} + i_{rd}) \\[2mm] \psi_{qr} = L_{1r}i_{rq} + L_{m}(i_{sq1} + i_{sq2} + i_{rq}) \end{array} \right\} \tag{4}$$

In the equations: r_{sk} ---- the resistance of regulating winding and the main winding, (k=1、2) ; r_{r} —the resistance of rotor winding; L_{11}, L_{22}, L_{1r} —the self-inductance of the main winding, regulating winding and rotor winding; L_{sm} —the mutual inductance of the stator two-phase winding; the resistance of the main winding and regulating winding; L_{m} — the self-mutual inductance of the stator and rotor; u — voltage; ψ —magnetic linkage; i —electric current.

Torque equations:

$$T_{e} = \frac{2}{3}P_{n}L_{m}(i_{rd}i_{sq1} + i_{rd}i_{sq2} - i_{rq}i_{sd1} - i_{rq}i_{sd2}) \tag{5}$$

Motion equations:

$$T_e - T_L = \frac{J_m}{P_n} \frac{d\omega_r}{dt} \tag{6}$$

In the equations: T_e −Torque of electromagnetism; T_L −loading torque; P_n −pole log; J_m −rotary inertia; ω_r −angular velocity of rotor. Then put the magnetic linkage equation (4) into the electric tension equation and erase the magnetic linkage item, then, we have the matrix equation of internal feed electromotor:

$$\dot{I} = A^{-1}BI + A^{-1}U \tag{7}$$

3   The Composition and Model Establishment of the S Function in M File

According to the equation (7) and the defined grammar of S function in matlab/simulink, the following simulation model of internal feedback motor is composed by the use of S Function form board:

```
Function [sys,x0,str,ts] =motor (t, x, u, lag,R1,R2,Rr, L11,L22, L1r,Lsm,Lm,Pn,Jm)
%motor's parameters
Switch flag
case 0    %initialization
sys=[6,   %number of continuous states
     0,   %number of discrete states
     7,   %number of outputs
     7,   %number of inputs
     1,   %direct feed through flag
     1];  %sample time
case 1    %Derivatives
A11=L11+Lsm+Lm; A13=Lsm+Lm; A15=Lm; A33=L22+Lsm+Lm; A55=L1r+Lm;
A=[A11 0 A13 0 A15 0;0 A11 0 A13  0  A15; A13 0 A33 0 A15 0;A13 0 A33
0 A15; A15  0  A15  0  A55  0;A15  0  A15  0  A55];
B52 = ωrLm =u(7)*Lm;

B56 = ωr(Lm+Lr) =u(7)*(Lm+L1r);

B=[-R1 0 0 0 0 0;
   0  -R1 0 0 0 0;
   0   0 -R2 0 0 0;
   0   0  0 -R2 0 0;
   0  -B52 0 -B52 -Rr -B56;
   B52 0 B52 0 B56 -Rr];
G=inv(A);
Sys=G*B*x+G* [u(1);u(2);u(3);u(4);u(5);u(6)];
case 2    %Discrete state update
sys=[];   %do nothing
```

```
case 3       %outputs
Te=1.5*Pn*Lm*(x(5)*x(2)+x(5)*x(4)-x(6)*x(1)-x(6)*x(3))
sys=[x(1);x (2);x(3);x(4);x(5);x(6);Te]
case 9       %Terminate
sys=[];      % do nithing
otherwise
sys=[];
end
```

The simulation mode of internal feedback motor is built according to the s function, shows as fig.3. And the cascade speeding control system of internal feedback motor is established by software of simulink.

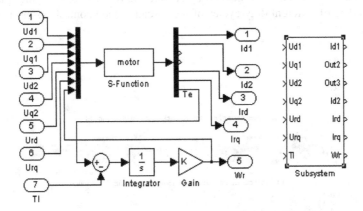

Fig. 3. Simulation mode of internal feedback motor based on s function in matlab/simulik

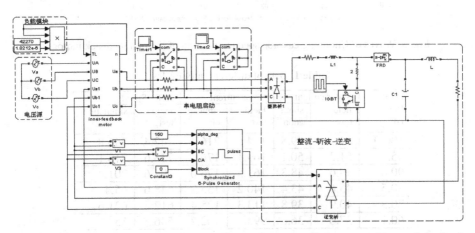

Fig. 4. Simulation model of chopping cascade speeding controlling system of internal feedback motor

4   Simulation Trial

In order to test the simulation model of the internal feedback motor's feasibility, the simulation model is applied in the internal feedback cascade control system.

The parameters of the internal feedback motor: P_e =570kw,   stator: U_e =6kv,

I_e =65A; rotor: U_2 =978v, I_2 =355A; feedback winding U_3=626v, I_3=125A,

η=0.9421, $\cos \varphi$=0.804, J_m =270 $kg \cdot m^2$, $\triangle$/Y Rejoining method.

Fig.5(a) is the speed wave of the internal feedback motor when it is running, and the fig.5(b) is the simulation wave of stator current, and the fig.5(c) is the simulation wave of rotor current,

The speed of motor can be regulated by change the IGBT chopping duty ratio. The simulation data of different duty-cycles of the cascade speed control system is showed as table.1.

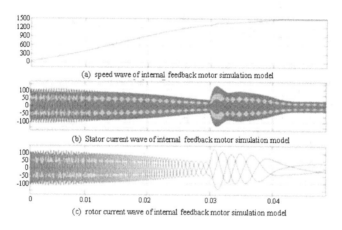

(a) speed wave of internal feedback motor simulation model

(b) Stator current wave of internal feedback motor simulation model

(c) rotor current wave of internal feedback motor simulation model

Fig. 5. Waves of Chopping cascade speed regulating of internal feedback motor

Table 1. Data of different duty-cycles

Duty Ratio (%)	Stator current		Rotor current		Rotor Speed	
	field test	simul ation	Field test	simulati on	Field test	simulation
100.0	51.7	49.8	194.6	216.4	742.4	746.8
95.3	51.2	50.1	169.1	176.4	713.6	718.1
90.0	49.6	48.3	161.5	176.0	694.6	701.8
80.5	45.6	43.4	140.2	155.2	648.6	65.4
70.3	40.9	39.8	113.3	125.4	612.4	621.6
60.1	37.6	36.6	99.6	102.7	546.3	554.6

Figure 6,7 are the simulation waves and experimental waves. Fig.5 is the rectifier voltage waves of internal feedback motor, fig5(a) is the experimental wave and fig.5(b) is the simulation wave.

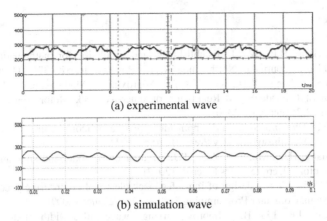

(a) experimental wave

(b) simulation wave

Fig. 6. Rectifier voltage wave of inner-feedback motor

The capacitance voltage wave of invertor DC side is showed as fig.6, fig.6(a) is the experimental wave and fig.6(b) is the simulation wave.

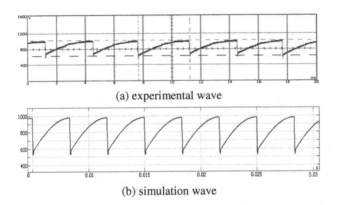

(a) experimental wave

(b) simulation wave

Fig. 7. Capacitance voltage wave of invertor dc side

5   Conclusion

Through the comparison of simulation data and field test data, it's found that the electric current of the model and its simulation curve nearly correspond to the field test data. And with the increasing duty ratio of chopped wave, the rotation speed of the electromotor increases gradually. Therefore, it proves the model is corrected and the way of establishing the internal feed electromotor introduced in this paper is feasible and practical.

References

1. Liu, J.Q., Gong, B.F., Wa, J.: Establishment and simulation for mathematical model of inner-feed motor. Journal of North China Electric Power Unicersity 32(1) (January 2005)
2. Wang, B.S., Zhang, X.D., Guo, X.Y.: Approximate Mathematical Modeling and Simulation of Cascade Speed Regulation of Inner-feedback Motor. Journal of System Simulation 22(2) (February 2010)
3. Wei, Z.G.: The principle and application of the silicon control cascade, vol. 8, pp. 138–181. Metallurgy Industry Press, Beijing (1985)
4. X, M., Huang, H.T., Mo, Y.P. Zhao, P.J.: Mathematical model and simulation of inner-feedback motor. Electrotechnics Electric, (2012)
5. Song, Y.J., Qi, X.Y.: A design of inner feedback electrical engineering speed regulating control system. Computer Programming Skill & Maintenance (2009)
6. Ma, H., Liu, J.F., Liu, B.S.: Improved triangle-wave pulse width modulation current control method for active power filter. Electric Power Automation Equipment 21(3), 30–33 (2004)

Agent-Based Simulation and Data Mining Analysis for Effect of Purchase Price in Households' Solar Energy Adoption Process

Yuanyuan Guo[1,2], Hong Zhang[3], Jiangshan Dong[1], Di Shen[1], and Jingyuan Yin[1]

[1] School of Computer Engineering and Science,
Shanghai University, Shanghai 200444, China
jyyin@staff.shu.edu.cn
[2] School of Communication & Information Engineering,
Shanghai University, Shanghai 200444, China
snowwhite@shu.edu.cn
[3] College of Envrionmental & Resource Sciences,
Shanxi University, Taiyuan 030006, China
zhanghong@sxu.edu.cn

Abstract. For promotion of solar energy popularization, study on effect of purchase price can facilitate proper product decisions making. In this paper, an agent based model is built to simulate the households' dynamic adoption process of solar energy. With varying purchase price, scenario analysis is taken to investigate the relevant market share changes. Random forest is used to measure the effect of purchase price by data collected from the model changes. The results show that impacts of purchase price differ with different types of energy using by households. By energy subsidy, product with higher purchase price can still attract market share of solar energy effectively. Thus, promotion strategies should be variable according to local using conditions of solar energy in residential consumer market.

Keywords: solar energy diffusion, agent-based simulation, households' dynamic adoption process, random forest.

1   Introduction

Using of conventional energies is facing with problems as energy depletion and carbon dioxide (CO_2) emission. Since these phenomena happen, the adoption of renewable energy sources are becoming vital for management of energy saving and pollution reduction.

In residential consumer market, electricity and nature gas are mostly used energy resources. Launching product of solar energy into the market can benefit the utilization of solar energy. As conventional energies dominate the market, promotion of solar energy introduce a considerable amount of risk for manufactures and the government. In order to reduce the risk, it is significant to analyze how product decisions give impact on the market shares of solar energy.

K. Li et al. (Eds.): LSMS/ICSEE 2014, Part III, CCIS 463, pp. 117–124, 2014.
© Springer-Verlag Berlin Heidelberg 2014

For manufactures of solar energy product, purchase price is the important factor in their product decisions making. Most manufactures always believe that only low purchase price can brings high market shares. But during the past years, competition in low purchase price didn't take obvious effects. Hence, it is essential to evaluate the effect of purchase price in households' solar energy adoption process.

Statistic analysis with surveys or historical data is the commonly used method in previous studies for promotion of renewable energy [1-3]. All these researches are designed to select the impact factors for the promotion. They can't reveal the dynamic adoption process of solar energy. In residential consumer market, the adoption process comes from micro (households) to macro (consumer market) level. Purchase price which impact on the households' energy adoption behavior will finally make the market share changes of solar energy.

It is difficult to conduct controlled experiments on the dynamic adoption process due to the lack of experimental control on many critical variables [4]. Fortunately, agent based model provide a tool to systematically conduct experiments automatically and decide how the impact factors affect the whole dynamic adoption process.

Several studies can be found on agent based modeling (ABM) for diffusion of technology. Shinde et al. [5] created an ABM framework to simulate and analyze the effect of multiple business scenarios on the adoption behavior of a group of technology products. T. Zhang et al. [6] studied the diffusion of alternative fuel vehicles. M. Günther et al. [7] built an agent based simulation for the new product diffusion of a novel biomass fule. G. Sorda et al. [8] used the agent based simulation to evaluate the promotion of electricity from agricultural biogas plants in Germany. D. J. Veit et al. [9] developed an agent-based model to study the dynamics in two-settlement electricity markets. T. Zhang et al. [10] presented an agent based model to evaluate the effects of different government policies on promoting new electricity technologies in complex systems.

In this paper, an agent-based model is developed to simulate the dynamic solar energy adoption process of households. Purchase price of solar energy product is selected from the impact factors to study its effect on the process. Scenario analysis is taken to see the changes of residential consumer market share due to purchase price changes. By the changing results, Random forest [11] is used as the data mining method to investigate the effect of purchase price. This paper is structured as follows. Section 2 introduces the methods used in this study. Section 3 presents scenario analysis and discusses the effect of purchase price. Conclusions are given in Section 4.

2   Method

2.1   Agent-Based Model for Households' Solar Energy Adoption Process

Considering the energy consuming reality, the households' adoption process of solar energy is a competitive process among conventional energies. There are three single kinds of energies for households to choose, they are electricity (A), natural gas (B) and solar energy (C). Solar water heater is used as the solar energy product for households' daily life use. In this model, agents represent households and differ with each other by kinds of energies they adopted. Types of agents are divided into 4 categories: 1. Only electricity consuming agents (A), 2. Electricity and natural gas consuming agents (AB), 3. Electricity and solar energy consuming agents (AC), 4. Electricity, natural gas and solar energy consuming agents (ABC). The scale-free network is used as the social network [12]. The whole scale-free network represents the residential consumer market.

Architecture of Agents. As agents represent households, they are smart agents with human intelligent behavior in decision making on adoption energies. In the real consumer market, the sale information and policy about energies starts from energy suppliers and government via mass media, and travels through the households' social network via word of mouth. In this model, agents have the ability to get all these kinds of information. They can also interact with their households and communicate with each other to understand their neighborhoods' energy choice. At the time of choice making, agents update their satisfaction level of energy adoption. They will compare the current single kind of energy in feature to feature way (such as price, average using flow and so on). Value proposition is used to make comparisons for all the single kind of energies. Calculation of value proposition includes all the impact factors that will influence the households' energy choice making. After comparison with value proposition, agents consider their economic condition and then make decisions about which kinds of energies to choose.

Impact factors which influence the energy adoption behavior are concluded as: i) nature features of energy; ii) sociological factors: social influence on households (the social network, word of mouth mechanism and the neighborhood effect); iii) environmental factors: regional factor (climate condition in the area), mass media factor (spread information of energy and the government encouraging policies), government policies; v) psychological factors: features of households.

Evaluations for Nature Features of Energy in the Model. Each kind of energy has certain nature features that distinguish from others. Agents compare them in feature to feature way. For electricity and nature gas, nature features are shown in Table1. For solar energy, the features of solar water heater are shown in Table 2.

Table 1. Nature features for electricity and nature gas

Name	Value	Unit
Price	[0, 3.0]	Yuan
Saner (Saving energy per month)	[0, 40]	Dimensionless quantity
Flow (Average use flow per day)	[0,180] (electricity) [0,50] (nature gas)	KWH (Electricity) Cubicmeter/hour (Natural gas)
Con (Using Convenience of each kind of energy)	[1, 6]	Dimensionless quantity
Purchase price	[0, 6000]	Yuan
Repair Cost (per month)	[0,500]	Yuan

Table 2. Features for solar water heaters

Name	Value	Unit
Price	[0, 3.0]	Yuan
Saner	[0, 40]	Dimensionless quantity
Capacity	[130,160]	L
Con	[1, 6]	Dimensionless quantity
Purchase price	[0, 6000]	Yuan
Repair Cost (per month)	[0,500]	

Calculation for Value Proposition of Energy. Agents update their satisfaction level of energy adoption at each time increment. The satisfaction level is also the threshold for comparison of energy value proposition. They compare the value proposition of their current adopted energies with that of other available energies. After comprehensive consideration including their economic conditions, agents will decide whether to keep or to change their current energy structure with certain probability.

For value proposition of energies features, utility theory [5,13] is used to describe it. Eq.1 is used when the agents target a maximum value of features (to calculate desirability function of saner and con). Eq. 2 is used when the target is the minimum value (to calculate desirability function of price and purchase price and repair cost). In the desirability function of price, y is defined as: y=price * flow or y=price * capacity.

$$D_{in1} = (\frac{y_{n1} - \min_{n1}}{\max_{n1} - \min_{n1}})^{R_{n1}} \tag{1}$$

$$D_{in2} = (\frac{\max_{n2} - y_{n2}}{\max_{n2} - \min_{n2}})^{R_{n2}} \tag{2}$$

Where, n_1 is energy features: saner, con; n_2 is energy features: price and flow/capacity, purchase price, repair cost; R_{n1}, R_{n2} are the risk variable for the corresponding features; i is the type of energy: electricity (A), nature gas (B), solar energy (C)

For solar water heater, the price of solar energy ($yuan / L$) has to be calculated. Eq.3 is used to calculate the heating cost of solar water heater per year.

$$C_y = (K * L * \Delta T / M / \mu) * T_c * Y_e + Y_w * (L / 1000) * T_y \tag{3}$$

C_y : heating cost of solar water heater per year

K : $4200J / kg \bullet °C$ for per kg water, the consuming electricity for rising the temperature per centigrade

L : the capacity of the solar water heater

ΔT : the increasing temperature for water in the solar water heater turns into hot

M : 36000 J / kwh

μ : the heating efficiency for solar water heater

T_c : the raining and cloudy days per year

Y_e : the unit price for electricity : $yuan / kwh$

Y_w : the unit price for water : $yuan /$ t

T_y : 365 the days per year: $days / year$

Eq. 4 is used to calculate the price of solar energy ($yuan / L$) per month

$$\text{Price}_{solar} = C_y / 12L \tag{4}$$

For each household, the function of value proposition for energy features is computed as a weighted average of the desirability of the features as shown in Eq.5.

$$U_{in} = \frac{\sum_n W_n D_{in}}{\sum_n W_n} \tag{5}$$

In addition to having different desirability functions, households can assign different importance (weight, W_n) to different features that make up the Eq.5. Where, n is energy features from Table 1.

Eq.6 is used for the value proposition of agents for social and environmental factors.

$$D\alpha i = \alpha * S\alpha \tag{6}$$

$D\alpha i$ is the value proposition of each household for sociological and environmental factors of energy i . In addition to having different value proposition functions, households can assign different sensitivities (sensitivity, $S\alpha$) to different factors that make up the Eq.6. Where, α is the social or environmental factor, $S\alpha$ is the corresponding sensitivity.

Eq.7 is used for the calculation of value proposition of each agent for all kinds of energy

$$VP_i = U_{in} + \sum D\alpha i = \frac{\sum_n W_n D_{in}}{\sum_n W_n} + \sum \alpha * S\alpha \tag{7}$$

2.2   Data Mining Method

In the scenario analysis section, effect of purchase price is identified. Data of model changes due to varying purchase price are collected. Random Forests (RF) is used summarize and quantify the impact of purchase price. Partial dependence plots [14] are used to interpret the results from the Random Forest model.

3   Scenario Analysis for Effect of Purchase Price

At the beginning of the simulation, Household A and AB dominate the whole residential market share, a small percentage of early household adopt solar energy (household AC or ABC).

Scenario analysis is taken in which solar water heaters, having a variety value of feature levels, is introduced into the residential consumer market. In order to study the effect of purchase price, only simulation results for variety value of purchase price and repair cost is selected. The features of energy A and B are setting as constant. Under these conditions, solar energy (C) is introduced into the residential consumer market. Values of purchase price and repair cost are varied between {1280, 6280} and {0, 200}. The value combination is varied between runs to simulate the introduction of solar energy. A total of 15 replicates are run at each combination of the features. After each run, the maximum share of households AC and ABC in the residential consumer market are collected. By the simulation results, random forest is used to analyse the effect of purchase price.

Figure 1 shows the partial dependency plots after using the Random Forest model. From Figure 1, it can be seen that purchase price is effective for the residential consumer market share of energy C. As the value of purchase price varies, the corresponding residential consumer market share of AC and ABC varies from 0.5 to 0.25. Compared with Figure 1. (a) and 1. (b), there exists an interesting phenomenon. For residential consumer market share of ABC predicted, lower purchase price brings higher market share. While for residential consumer market share of AC predicted, higher purchase price brings higher market share instead. As Figure 1. (a) shows, the residential market share of AC is less than 0.17 when the purchase price is lower than 4000. But when the purchase price is higher than 4000, the market tendency of AC begins to rise. Figure 1. (b) shows that the residential market share of ABC is always beyond 0.1 when the purchase price is lower than 4000. When the purchase price is higher than 4000, the market tendency of ABC begins to fail. In order to have better understand of this phenomenon, original plots corresponding to Figure 1 are presented in Figure 2. Figure 2. (a) and 2. (b) are original plots of the agent based model corresponding to Figure 1 when the purchase price is 5280 and 1280 respectively. From Figure 2. (a), it can be seen that there exists a long time interval before curves AC, ABC begin to rise. The reason can be explained as a certain amount of money is adding to the household's account at each time. Some households who can't use energy C because of inadequate economic condition can get enough money during the long time interval. Figure 2. (b) shows one original plots corresponding to Figure 1when the purchase price is 1280. It can be seen that there exists no time interval before curves AC, ABC begins to rise. The consumer market share of ABC is higher than AC. The result is on the contrary as compared with Figure 2. (a). In order to prove the explanation for long time interval in Figure 2. (a), simulations are taken by adding energy subsidy into households' account. Figure 2. (c) shows the result. From Figure 2. (c), it can be seen that the time interval is shorten with energy subsidy.

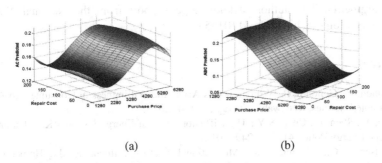

(a)   (b)

Fig. 1. Dependency plot showing the relative purchase price and repair cost effects on the residential consumer market of households AC (*a*), ABC (*b*)

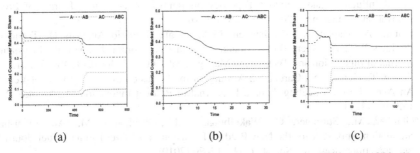

(a)   (b)   (c)

Fig. 2. Original plots of agent based model when purchase price is 5280 (*a*), 1280 (*b*) and purchase price is 5280 with energy subsidy is 50 per month (*c*)

4   Conclusion

In this paper, an agent based model is developed for simulation of households' solar energy adoption process. Random forest is used to identify the effect of purchase price for residential market share changes of solar energy. Original plots of the model are given to explain the analysis results from Random forest. It is shown that purchase price is very effective for promotion of solar energy. But the effect of purchase price differs between households AC and ABC. For households ABC, lower purchase price does brings higher residential consumer market share. But for households AC, higher purchase price brings higher residential consumer market share. Further study shows that this higher residential consumer market share is achieved after a long time interval. With energy subsidy, the time interval is shortening effectively. This indicates that different purchase price of solar energy product needs their exclusive promotion strategies. For areas without enough sunlight, the suitable solar energy product is always with higher purchase price. Hence, the promotion strategies must pay more attention on government policy. Subsidy should be given as to shorten the time interval before the residential consumer market share of solar energy rising.

While, for areas with enough sunlight, the solar energy product manager should pay more attention on the purchase price making. Lower purchase price is more competitive to attract market share of solar energy.

Acknowledgments. The authors acknowledge support from the National Natural Science Foundation of China (No.41101558).

References

1. Sardianou, E., Genoudi, P.: Which Factors Affect the Willingness of Consumers to Adopt Renewable Energies. Renewable Energy 57, 1–4 (2013)
2. Zhang, X., Shen, L., Chan, S.Y.: The Diffusion of Solar Energy Use in HK: What are the Barriers. Energy Policy 41, 241–249 (2012)
3. Michelsen, C.C., Madlener, R.: Motivational Factors Influencing the Homeowners' Decisions between Residential Heating Systems: An Empirical Analysis for Germany. Energy Policy 57, 221–233 (2013)
4. Delre, S.A., Jager, W., Bijmolt, T.H., Janssen, M.A.: Will it Spread or Not? The Effects of Social Influences and Network Topology on Innovation Diffusion. Journal of Product Innovation Management 27(2), 267–282 (2010)
5. Shinde, A., Haghnevis, M., Janssen, M.A., et al.: Scenario Analysis Of Technology Products With An Agent-Based Simulation And Data Mining Framework. International Journal of Innovation and Technology Management 10(05), 1340019–1340022 (2013)
6. Zhang, T., Gensler, S., Garcia, R.: A Study of the Diffusion of Alternative Fuel Vehicles: An Agent-Based Modeling Approach. Journal of Product Innovation Management 28(2), 152–168 (2011)
7. Günther, M., Stummer, C., Wakolbinger, L.M., Wildpaner, M.: An Agent-based Simulation Approach for the New Product Diffusion of a Novel Biomass Fuel. Journal of the Operational Research Society 62(1), 12–20 (2010)
8. Sorda, G., Sunak, Y., Madlener, R.: An Agent-based Spatial Simulation to Evaluate the Promotion of Electricity from Agricultural Biogas Plants in Germany. Ecological Economics 89, 43–60 (2013)
9. Veit, D.J., Weidlich, A., Yao, J., Oren, S.S.: Simulating the Dynamics in Two-settlement Electricity Markets via an Agent-based Approach. International Journal of Management Science and Engineering Management 1(2), 83–97 (2006)
10. Zhang, T., Nuttall, W.J.: Evaluating Government's Policies on Promoting Smart Metering Diffusion in Retail Electricity Markets via Agent-Based Simulation. Journal of Product Innovation Management 28(2), 169–186 (2011)
11. Svetnik, V., Liaw, A., Tong, C., Culberson, J.C., Sheridan, R.P., Feuston, B.P.: Random forest: a classification and regression tool for compound classification and QSAR modeling. Journal of Chemical Information and Computer Sciences 43(6), 1947–1958 (2003)
12. Inform, D., Hein, O., Ing, D.W., Schwind, M., König, W.: Scale-Free Network., From Math World–A Wolfram Web Resource, created by Weisstein, E.W.,
   http://mathworld.wolfram.com/Scale-FreeNetwork.html
13. Fishburn, P.C.: Utility theory for decision making. Research Analysis Corp. Mclean Va, New York (1970)
14. Friedman, J.H.: Greedy function approximation: a gradient boosting machine. Annals of Statistics, 1189–1232 (2001)

Thermal System Identification
Based on Double Quantum Particle Swarm Optimization

Pu Han, Shitong Yuan, and Dongfeng Wang

Hebei Engineering Research Center of Simulation & Optimzed Control for Power
Generation (North China Electric Power University), Baoding 071003, Hebei Province, China
hanpu102@263.net, yst.19@163.com, wangdongfeng@ncepubd.edu.cn

Abstract. In order to improve the convergence speed and precision of particle swarm optimization (PSO) and quantum PSO (QPSO), inspired by the idea of quantum physics, a new improved QPSO algorithm named double QPSO (DQPSO) is presented. The particle's encoding mechanism and the evolutionary search strategy are quantized in DQPSO algorithm, in which the evolution equation of the velocity vector is abandoned, thus the evolution equation is easier, and less parameter are used that makes the algorithm easier to control. Several benchmark multi-modal functions are used to test the proposed DQPSO algorithm, which verified that the new algorithm is superior to standard PSO and QPSO in search capabilities. Then, DQPSO is successfully used to the identification of a thermal system with pure time-delay and non-minimum phase. Finally, the algorithm is applied to the transfer function identification of thermal system based on field operation data.

Keywords: particle swarm optimization, QPSO, DQPSO, system identification, thermal system.

1   Introduction

Control system modeling is one of the common means of analyzing the characteristics of the system and the controller design and performance optimization. In recent decades, the intelligent optimization algorithms in the field of system identification and modeling has been widely used. Maria [1] proposed an improved genetic algorithm, and a stepwise methodology was applied to parameter identification of fed-batch cultivation of saccharomyces cerevisiae; Hu [2] used substructure ant colony optimization to the structure damage identification; ChiaNan Ko [3] presented an annealing dynamical learning algorithm (ADLA) to train wavelet neural networks (WNNs) for identifying nonlinear systems with outliers; An array synthesis based on quantum particle swarm optimization (QPSO) is proposed for the orthogonal MIMO radar [4]; Guemo [5] used the Gauss-Newton (GN) method, the Levenberg-Marquardt (LM) method and the Genetic Algorithms (GA) method to solve this optimization problem in order to identify 10 parameters of a lumped parameter thermal model for a permanent magnet synchronous machine (PMSM). In these algorithms, particle swarm optimization (PSO) algorithm for its simple, high efficiency and easy to

K. Li et al. (Eds.): LSMS/ICSEE 2014, Part III, CCIS 463, pp. 125–137, 2014.
© Springer-Verlag Berlin Heidelberg 2014

implement has been favored by many scholars. Alec Banks discussed the location of PSO within the broader domain of natural computing, considers the development of the algorithm, and refinements introduced to prevent swarm stagnation and tackle dynamic environments [6]; In another article, Alec Banks considered current research in hybridisation, combinatorial problems, multicriteria and constrained optimization, and a range of indicative application areas [7].

QPSO algorithm which evolved from PSO is a recently developed optimization method. In QPSO algorithm, the positions of particles are encoded by the probability amplitudes of quantum bits, the movements of particles are performed by quantum rotation gates, which achieve particles searching. As each quantum bit contains two probability amplitudes, each particle occupies two positions in space. Hence, the performance of the algorithm has been improved (e.g. [8], [9]). Same as basic PSO algorithm, the optimizing   performance of QPSO is largely dependent on the choice of initial parameters, and fall into local optimal value easily. In order to solve this problem, the mutations of particles are performed by quantum non-gate, which increase the diversity of particles. The performance of the algorithm has some improvements, but it does not fundamentally overcome premature convergence defects.

To improve search ability and optimization efficiency and to avoid premature convergence for QPSO, the double quantum particle swarm optimization (DQPSO) algorithm is proposed. In DQPSO, the evolutionary search strategy was improved. The particle is evolved by the median particle optimal position, local optimum position and global optimum position. The velocity vector is removed, the form of the evolution equation is simpler, smaller and easier to control parameters .

Through the simulation experiments, compare the three algorithms optimize results, we can prove that convergence speed and accuracy of the DQPSO algorithms have   achieved   satisfactory   results   in   most   function   optimization   problems. Meanwhile, we use this algorithm in a thermal system identification experiment, the results prove the validity of the identification algorithm .

2   PSO and QPSO Framework

2.1   Standard PSO Algorithm

Let there are N particles in D-dimensional search space. X and V denote the particle's position and its velocity in the search space. The term velocity represents the change in the position of each particle. Thus, the position of the i-th particle in D-dimensional space is represented as $x_i = (x_{i1}, x_{i2}, \cdots x_{id}, \cdots x_{iD})$. The velocity of the i-th particle in D-dimensional space is represented as $v_i = (v_{i1}, v_{i2}, \cdots v_{id}, \cdots v_{iD})$ .The best previous position explored by the ith particle is recorded and denoted as p_{id}. Another value that is tracked and stored by PSO is the best value obtained so far by any particle in the population. This best value is a global best and is denoted by p_{gd}. Each particle changes its position based on its current velocity and its distance. The modification can be represented by the concept of velocity and can be calculated as shown in the following formulas [10]:

$$v_{id}^{k+1} = wv_{id}^{k} + c_1 r_1 (p_{id} - x_{id}^{k}) + c_2 r_2 (p_{gd} - x_{id}^{k}) \tag{1}$$

$$\begin{cases} v_{id} = v_{max,d} & if \quad v_{id} > v_{max,d} \\ v_{id} = -v_{max,d} & if \quad v_{id} < -v_{max,d} \end{cases} \tag{2}$$

$$x_{id}^{k+1} = x_{id}^{k} + v_{id}^{k+1} \tag{3}$$

where k denotes iterations number of particles. The constants c_1, c_2 are acceleration factors; r_1, r_2 is random number from 0 to 1.

2.2   QPSO Algorithm

Most of the optimization problem can be viewed as m-dimensional space of the point or vector in the D-dimensional space optimization problem . And the optimization problem can be described as $\max f(x_1, \cdots x_n)$, $x_i \in [a_i, b_i]$, $i = 1, 2, \cdots n$, n is the number of variables to be optimized; $[a_i, b_i]$ is the definition domain of the variable x_i, f is the objective function, making its value as the particle fitness [8].

2.2.1   Initialize the Particle
In QPSO algorithm, the particle positions are encoded by the probability amplitudes of the corresponding states.

$$P_i = \left[\begin{array}{c|c|c|c} \cos(\theta_{i1}) & \cos(\theta_{i2}) & \cdots & \cos(\theta_{in}) \\ \sin(\theta_{i1}) & \sin(\theta_{i2}) & \cdots & \sin(\theta_{in}) \end{array} \right] \tag{4}$$

where $\theta_{ij} = 2\pi \times r$, r is random number from 0 to 1; $i = 1, 2, \cdots m$, $j = 1, 2, \cdots n$, n is the number of variables to be optimized, m is the population size of the particles. The probability amplitude of quantum state $|0>$ and $|1>$ corresponds to each particle as follows:

$$P_{ic} = (\cos(\theta_{i1}), \cos(\theta_{i2}) \cdots, \cos(\theta_{in})) \tag{5}$$

$$P_{is} = (\sin(\theta_{i1}), \sin(\theta_{i2}) \cdots, \sin(\theta_{in})) \tag{6}$$

Ep.(5) is particle's sine position, Ep. (6) is particle's cosine position.

2.2.2   Solution Space Transforming
P_i is ranged in [-1,1], we should map unit space to the optimization problem solution space. The solution space transformation as follows:

$$x_{ic}^{j} = \frac{1}{2} \left[b_i (1 + \alpha_i^{j}) + a_i (1 + \alpha_i^{j}) \right] \tag{7}$$

$$x_{is}^{j} = \frac{1}{2} \left[b_i (1 + \beta_i^{j}) + a_i (1 + \beta_i^{j}) \right] \tag{8}$$

Each particle corresponds to   two solutions of the optimization problem, $|0>$ quantum state's probability amplitude α_i^j corresponds to x_{ic}^j, $|1>$ quantum state's probability amplitude β_i^j corresponds to x_{is}^j.

2.2.3   Particle Position Update

The quantum-behaved particle swarm optimization algorithm updates its positions with the following equation:

$$\Delta\theta_{id}(t+1) = w\Delta\theta_{id}(t) + c_1 r_1 (\Delta\theta_l) + c_2 r_2 (\Delta\theta_g) \tag{9}$$

where c_1, c_2, r_1, r_2 are same as the PSO algorithm, $\Delta\theta_l$ is the angle difference between current angle of the individual and the individual optimal value, $\Delta\theta_g$ is the angle difference between current angle of the individual and the global optimal value, calculated as follows:

$$\Delta\theta_l = \begin{cases} 2\pi + \theta_{ilj} - \theta_{ij} & \theta_{ilj} - \theta_{ij} < -\pi \\ \theta_{ilj} - \theta_{ij} & -\pi < \theta_{ilj} - \theta_{ij} < \pi \\ \theta_{ilj} - \theta_{ij} - 2\pi & \theta_{ilj} - \theta_{ij} > \pi \end{cases} \tag{10}$$

$$\Delta\theta_g = \begin{cases} 2\pi + \theta_{igj} - \theta_{ij} & \theta_{igj} - \theta_{ij} < -\pi \\ \theta_{igj} - \theta_{ij} & -\pi < \theta_{igj} - \theta_{ij} < \pi \\ \theta_{igj} - \theta_{ij} - 2\pi & \theta_{igj} - \theta_{ij} > \pi \end{cases} \tag{11}$$

In the iteration it works with following equation:

$$\begin{bmatrix} \cos(\theta_{ij}(t+1)) \\ \sin(\theta_{ij}(t+1)) \end{bmatrix} = \begin{bmatrix} \cos(\Delta\theta_{ij}(t+1)) & -\sin(\Delta\theta_{ij}(t+1)) \\ \sin(\Delta\theta_{ij}(t+1)) & \cos(\Delta\theta_{ij}(t+1)) \end{bmatrix} \begin{bmatrix} \cos(\theta_{ij}(t)) \\ \sin(\theta_{ij}(t)) \end{bmatrix}$$
$$= \begin{bmatrix} \cos(\theta_{ij}(t)) + \Delta\theta_{ij}(t+1) \\ \sin(\theta_{ij}(t)) + \Delta\theta_{ij}(t+1) \end{bmatrix} \tag{12}$$

where $i = 1, 2, \cdots m$, $j = 1, 2, \cdots n$.

2.2.4   Particle Mutation

Mutation factor in the evolutionary algorithm increase the diversity of population, and overcome premature or local-best solution. Mutation operation is implemented by quantum NOT gate. Assign a random value for each quantum, the particles are randomly selected in 2 qubits if the value is less than the given mutation probability. Their optimal position and rotation vector remains unchanged. Particles mutated as:

$$\begin{bmatrix} 0 & 1 \\ 1 & 0 \end{bmatrix} \begin{bmatrix} \cos(\theta_{ij}) \\ \sin(\theta_{ij}) \end{bmatrix} = \begin{bmatrix} \cos(\theta_{ij} + \frac{\pi}{2}) \\ \sin(\theta_{ij} + \frac{\pi}{2}) \end{bmatrix} \tag{13}$$

where $i = 1, 2, \cdots m$, $j = 1, 2, \cdots n$。

3   DQPSO Algorithm and Process Comparison

3.1   DQPSO Algorithm

In quantum time-space framework, the quantum state of a particle is depicted by wave function $\psi(\vec{\theta},t)$, According to superposition state characteristics and probabilities expression characteristics of quantum theory, in three-dimensional, the probability density function satisfying   (e.g. [12], [13], [14], [15], [16]):

$$\int_{-\infty}^{+\infty} |\psi|^2 \, d\theta dydz = \int_{-\infty}^{+\infty} Q d\theta dydz = 1 \tag{14}$$

The state function $\psi(\vec{\theta},t)$ develops in time according to Schrodinger equation:

$$i\hbar \frac{\partial}{\partial t}\psi(\vec{\theta},t) = \hat{H}\psi(\vec{\theta},t) \tag{15}$$

$$\hat{H} = -\frac{\hbar^2}{2m}\nabla^2 + V(\vec{\theta}) \tag{16}$$

where $\hat{H}$ is the Hamiltonian operator; $\hbar$ is Planck Constant.
The Schrodinger equation for the model is

$$\frac{d^2\psi}{dy^2} + \frac{2m}{\hbar^2}[E + \gamma\delta(y)]\psi = 0 \tag{17}$$

we can represent the normalized wave function as

$$\psi(y) = e^{-\beta|y|} \tag{18}$$

Hence, the probability density function Q is given by

$$Q(y) = \frac{1}{L}e^{-2|y|/L} \tag{19}$$

We obtain $L = 1/\beta = \dfrac{\hbar^2}{m\gamma}$ is the characteristic length of Delta potential well.
Monte Carlo Method can simulate the process of measurement. Let s be the random number uniformly distributed on (0 , 1/L), that is

$$s = \frac{1}{L}u = \frac{1}{L}e^{-2|y|/L} \tag{20}$$

where u is random number from 0 to 1. Substitutes for Q in (20), we obtain

$$u = e^{-2|y|/L} \tag{21}$$

$$y = \pm \frac{L}{2} \ln(1/u) \qquad (22)$$

Hence, position of the particle can be measured by

$$\theta(t) = \theta + y \qquad (23)$$

L is decided by $L(t+1) = 2\beta|\theta best - \theta(t)|$. Finally, the evolution equation of DQPSO is

$$\theta best = \sum_{i=1}^{M} \theta_i / M = \sum_{i=1}^{M} \theta_{i1} / M, \sum_{i=1}^{M} \theta_{i2} / M, \cdots \sum_{i=1}^{M} \theta_{id} / M \qquad (24)$$

$$\theta_j = (\phi_{1j}\theta_{ij} + \phi_{2j}\theta_{gj}) / (\phi_{1j} + \phi_{2j}) \qquad (25)$$

$$\theta(t+1) = \theta \pm \beta|\theta best - \theta(t)| \cdot \ln(1/u) \qquad (26)$$

where θ_j is the local attractor; θ_{best} is the mean best position, θ_{best} is defined as the center of pbest positions of the swarm; M is the population size; ϕ_1, ϕ_2 is a random number uniformly distributed in (0, 1); β is called the contraction-expansion (CE) coefficient, which can be tuned to control the convergence speed of the algorithms. The most commonly used control strategy of β is to initially setting it to 1.0 and reducing it linearly to 0.5.

3.2   Process Comparison of Three Algorithms

The above three algorithms calculation process shown in Table 1.

Table 1. The comparison of algorithm flow

	PSO	QPSO	DQPSO
Step1	Initializes the swarm: sets the swarm size, the number of variables, the maximum number of iterations, and so on.	same as left	same as left
Step2	Initialize positions of particles randomly.	Quantum coding positions of particles.	same as left
Step3	(blank)	Solution space transforming according to Eq (7),(8).	same as left
Step4	Calculate the fitness value for each particle and determined individual's best p_{id} and swarm's best p_{gd}	same as left	same as left
Step5	(blank)	(blank)	Calculate the mean best position θ_{best} according to Eq (24)

Table 1. (*continued*)

Step6	Update the position of particle as Eq (1),(2),(3).	Update the position of particle as Eq (9),(10),(11),(12).	Update the position of particle as Eq (25),(26).
Step7	(blank)	Particles mutated as Eq (13).	same as left
Step8	Repeat step3–8 until stop criteria are satisfied.	same as left	same as left

4 Functions Optimization Test

In this paper, four benchmark functions are used to test the efficiency of DQPSO. To compare the performance of DQPSO with some other methods, PSO and QPSO are also simulated in the paper. The benchmark functions of the test as shown in Table 2.

Table 2. Benchmark test functions

Function	Formulation	Initialization range
f1	$\min f_1(x, y) = (1+(x+y+1)^2(19-14x+3x^2-14y+6xy$ $+3y^2))*(30+(2x-3y)^2(18-32x+12x^2$ $+48y-36xy+27y^2))$	$x、y \in [-2,2]$
f2	$\min f_2(x, y) = \left\{\sum_{i=1}^{5} i\cos[(i+1)x+i]\right\}\left\{\sum_{i=1}^{5} i\cos[(i+1)y\right.$ $\left.+i]\right\}+0.5\left[(x+1.42)^2+(y+0.8)^2\right]$	$x, y \in [-10,10]$
f3	$\max f_3(x) = 0.5-\dfrac{\left[\sin\sqrt{x^2+y^2}\right]^2-0.5}{\left[1+0.001*(x^2+y^2)\right]^2}$	$x、y \in [-100,1\cdot$
f4	$\min f_4(x) = \dfrac{1}{4000}\sum_{i=1}^{n}(x_i-100)^2 - \prod_{i=1}^{n}\cos(\dfrac{x_i-100}{\sqrt{i}})+$	$x_i \in [100,600]$

f1 is Goldstein-Price function, there are four local minimum points: (1.2,0.8), (1.2,0.2), (−0.6,−0.4), (0,−1). The global minimum point is (0,−1). The global minimum value is 3. If the result is less than 3.005, it is judged to be qualified, and terminate the iterative calculation.

f2 is Shubert function, there are 760 local minimum points, and only one global minimum point (−1.42513,−0.80032), The global minimum value is −186.7309. This function is very easy to fall into local minima value −186.34. If the result is less than −186.34, it is judged to be qualified, and terminate the iterative calculation.

f3 is Shaffer's function, there is an infinite number of local maxima, and only one global maxima point (0,0). The global maxima value is 1. If the result is greater than 0.99, it is judged to be qualified, and terminate the iterative calculation.

f4 is Griewank function. It is multi-dimensional functions, The global minimum value is 0. If the result is less than 5, it is judged to be qualified in 10-dimensional function. 50-dimensional function's qualifying conditions is less than 150.

Set the algorithm parameters as: swarm size is 50; the maximum number of iteration is 500; In DQPSO algorithm, the contraction-expansion coefficient β is initially setting to 1.0 and reducing it linearly to 0; In QPSO algorithm, inertia weight $w=0.5$, acceleration factors $c_1=2.0$, $c_2=2.0$, mutation probability is 0.05; In QPSO algorithm, $w=0.5$, $c_1=2.0$, $c_2=2.0$. Repeat the test 100 times, Table 3 is the comparison of optimization results.

From Table 3, DQPSO algorithm presented in this paper has been improved on the number of qualified and optimization convergence precision. The optimization results are closer to the optimal value, especially with high accuracy in a multidimensional optimization problems.

Table 3. Comparison of optimization results

Functions	Algorithm	Best result	Worst result	Average value	Theory value	Qualified number
f1	DQPSO	3	3.0050	3.0028	3	49
	QPSO	3.0001	3.0048	3.0031	3	41
	PSO	3.0003	3.0581	3.0198	3	26
f2	DQPSO	-186.7230	-166.9959	-185.5856	-186.7309	44
	QPSO	-186.5971	-170.3618	-184.0189	-186.7309	27
	PSO	-186.7097	146.8709	-175.2368	-186.7309	22
f3	DQPSO	1	0.9903	0.9966	1	50
	QPSO	1	0.9856	0.9929	1	44
	PSO	0.9990	0.9847	0.9899	1	22
f4 10-dim	DQPSO	0.0015	6.3827	0.8782	0	45
	QPSO	1.8227	7.6705	4.6208	0	37
	PSO	1.1656	1.2333e+03	646.03	0	6
50-dim	DQPSO	24.2259	295.1235	128.2356	0	40
	QPSO	156.538	402.5672	256.78634	0	6
	PSO	219.3874	1.1359e+03	1001.5684	0	0

5   Thermal System Identification

5.1   Main Steam Pressure System Identification of CFB Unit

When combustion rate disturb in circulating fluidized bed (CFB) unit, main steam pressure dynamic characteristics model in certain conditions can be expressed as

$$G(s) = \frac{5(1-20s)}{(225s+1)^2} e^{-80s} \tag{27}$$

Fig.1 is the principle of system identification. Equation (28) is the initial model structure. The fitness function is mean square error, according to equation (29).

$$G^*(s) = \frac{K(1-as)}{(Ts+1)^n} e^{-Ls} \tag{28}$$

$$Err = \sqrt{\frac{1}{LP} \sum_{k=1}^{LP} \left[y(k) - y(\,k\,) \right]^2} \tag{29}$$

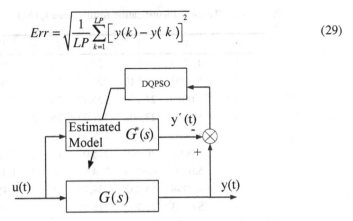

Fig. 1. Identification diagram for main steam pressure-combustion rate

Estimate K,a,T,n,L use the PSO、QPSO and DQPSO. Optimization range of parameters: $K \in [0,10]$, $a \in [0,50]$, $T \in [0,500]$, $n \in \{1,2,3\}$, $L \in [0,200]$. Other parameters settings are same with the classic function test experiment.

Table 4, Table 5 and Table 6 are identification results. The result is the average of 100 times test repeat, Their mean square errors are 4.10e-3, 1.20e-3, 2.09e-4. In PSO identification, the mean square error is largest, the results deviate significantly from the true value, the identification fall into the local optimum; In QPSO identification, the mean square error is large, the result is unsatisfactory; In DQPSO identification, the best result is the true value, it is better than the other two algorithms. Thus it can be seen that DQPSO algorithm is feasible and effective to identify the transfer function.

Table 4. Identification results using PSO

N	$K^{1)}$	$a^{1)}$	$T^{1)}$	$L^{1)}$	$n^{1)}$	Err
1	5.0014	36.669	226.67	61	2	5.53e-3
2	5.0029	47.282	228.37	48	2	1.08e-2
3	5.0019	40.824	227.27	56	2	7.43e-3
4	5.0045	55.821	230.13	37	2	1.61e-2
5	5.0022	45.106	227.77	51	2	9.51e-3
6	5.0069	75.103	234.06	12	2	3.02e-2
7	5.0026	45.692	228.08	50	2	9.93e-3
8	5.0019	41.645	227.39	55	2	7.83e-3
9	5.0029	47.282	228.37	48	2	1.08e-2
10	5.0013	35.828	226.55	62	2	5.17e-3
Average	5.0028	47.125	228.47	48	2	4.10e-3[2)]
Best	5.0013	35.828	226.55	62	2	5.17e-3[3)]

Note 1): K,a,T reserved five significant digits; Delay time L reserved integer;
Note 2): The mean square error of average result;
Note 3): The mean square error of best result.

Table 5. Identification results using QPSO

N	K	a	T	L	n	Err
1	4.9996	6.6955	224.49	94	2	1.92e-3
2	5.0003	22.910	225.23	77	2	9.64e-4
3	5.0010	21.296	225.49	78	2	1.08e-3
4	5.0002	23.602	225.27	76	2	9.24e-4
5	5.0005	24.495	225.38	75	2	1.19e-3
6	5.0001	20.900	225.07	79	2	2.22e-4
7	5.0000	6.6118	224.59	94	2	1.95e-3
8	4.9998	15.252	224.79	85	2	9.45e-4
9	5.0003	21.759	225.20	78	2	4.98e-4
10	5.0001	20.896	225.08	79	2	2.23e-4
Average	5.0002	18.442	225.06	81	2	1.20e-3
Best	5.0001	20.900	225.07	79	2	2.22e-4

Table 6. Identification results using DQPSO

N	K	a	T	L	n	Err
1	5.0001	20.909	225.07	79	2	2.21e-4
2	5.0000	20.000	225.00	80	2	<eps[4)]
3	5.0001	20.455	225.03	79	2	1.20e-4
4	5.0001	21.825	225.12	78	2	3.49e-4
5	5.0000	20.000	225.00	80	2	<eps
6	5.0000	20.000	225.00	80	2	<eps
7	4.9999	19.225	224.84	83	2	4.70e-4
8	4.9999	19.158	224.89	82	2	4.04e-4
9	5.0000	20.000	225.00	80	2	<eps
10	4.9999	19.158	224.89	82	2	4.04e-4
Average	5.0000	20.073	224.98	80	2	2.09e-4
Best	5.0000	20.000	225.00	80	2	<eps

Note 4): eps is the minimum value of simulation software used in this article can be represented.

5.2 History Data Identification of Thermal System

Keep the boiler combustion economy and optimality is an important task of the boiler combustion control. The oxygen content in flue gas is one of the boiler parameters to improve the combustion efficiency in coal power plants. If the oxygen content in flue gas is high, the combustion efficiency of the boiler will be reduced, and increase the heat loss of exhaust gas; If the oxygen content in flue gas is low, incomplete burning of pulverized coal will increased chemical and mechanical incomplete combustion heat loss, and the combustion efficiency of the boiler will be also reduced. Hence, the oxygen content in flue gas must be kept in a reasonable range. Best flue gas oxygen content equivalent to the best air fuel ratio [17], the transfer function identification of oxygen control system is important to controller design and optimization.

This experiment chose a group data of the boiler operation from a 1000MW ultra supercritical power plant. The data obtained from the DCS system in 900MW load conditions. The data obtained from the DCS system in 900MW load conditions. Sampling period is 6s, Sampling time is one hour. Fig.2 is the historical data curve of the oxygen content in flue gas, total air and total fuel. Fig.3 is the principle of oxygen system identification. In this model, total air and total fuel were considered as inputs and oxygen content was considered as outputs.

Using QPSO to identify the transfer function based on field operation data, the algorithm parameters setting as: swarm size is 50; the maximum number of iteration is 500; the contraction-expansion coefficient β is initially setting to 1.0 and reducing it linearly to 0; mutation probability is 0.05. Optimization range of parameters: $K_1 \in [0,1]$, $K_2 \in [-1,0]$, T_1、 $T_2 \in [0,1000]$, L_1、 $L_2 \in [0,1000]$, n_1、 $n_2 \in \{1,2,3,4,5\}$.

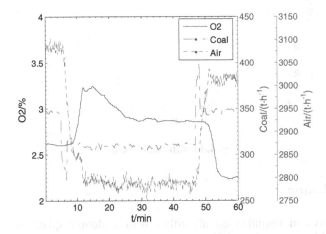

Fig. 2. The curves of original data

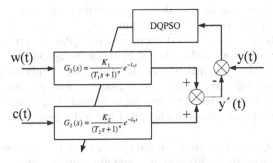

Fig. 3. Identification diagram

The original data deal with zero mean and five-spot triple smoothing. The optimized results of system transfer function are

$$\begin{cases} G_1(s) = \dfrac{0.0013}{(105.63s+1)^3} e^{-480s} \\ G_2(s) = \dfrac{-0.0166}{(70.72s+1)^3} e^{-216s} \end{cases} \tag{30}$$

Fig.4 is the comparison of identification model output and the actual output. The identification mean square error is 7.6608e-4. It can be seen from the figure that the two curves are very close and the identification error can be accepted.

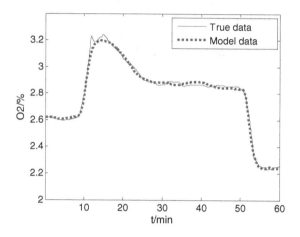

Fig. 4. Identification results

6   Conclusion

The particle swarm optimization algorithm with a double quantum behavior is proposed in this paper. The particle's encoding mechanism and the evolutionary search strategy are quantized in DQPSO algorithm, in which the evolution equation of the velocity vector is abandoned, thus the evolution equation is easier, and less parameter are used that makes the algorithm easier to control. Several benchmark multi-modal functions are used to test the proposed DQPSO algorithm, which verified that the DQPSO is superior to other algorithm in search capabilities.

DQPSO is applied in thermal process identification in this paper. The dynamic characteristic between the oxygen content in flue gas, total air and total fuel in ultra supercritical boiler is identified, and gets perfect effect. The identification algorithm put forward in this paper has generality, and it can be used in other identification systems.

Like PSO and QPSO algorithm, DQPSO is sensitive to parameters. In this paper, the contraction-expansion coefficient is initially setting to 1.0 and reducing it linearly to 0. The adaptive method of parameter control may lead to a more efficient algorithm and our next work is focusing on this problem.

Acknowledgments. The authors thank the editors and the anonymous referees for their valuable comments and suggestions, which improved the original manuscript. This work is supported by the Fundamental Research Funds for the Central Universities of China (13ZP13).

References

1. Angelova, M., Atanassov, K.F., Pencheva, T.: Purposeful model parameters genesis in simple genetic lgorithms. Computers & Mathematics With Applications 64(3), 221–228 (2012)
2. Hu, M., Zhang, L.: Application of optimization analysis on structure damage identification based on ACO algorithm. Advanced Design And Manufcture III 450(3), 506–509 (2011)
3. Ko, C.N.: Identification of nonlinear systems with outliers using wavelet neural networks based on annealing dynamical learning algorithm. Engineering Applications of Artificial Intelligence 25(3), 533–543 (2011)
4. Liu, H., Yang, G., Song, G.: MIMO radar array synthesis using QPSO with normal distributed contraction-expansion factor. Procedia Engineering 15, 2449–2453 (2011)
5. Guemo, G.G., Chantrenne, P., Jac, J.: Parameter identification of a lumped parameter thermal model for a permanent magnet synchronous machine. In: Electric Machines & Drives Conference 2013 IEEE International, pp. 1316–1320 (2013)
6. Banks, V.J., Anyakoha, C.: A review of particle swarm optimization Part I: background and development. Natural Computing 6(4), 467–484 (2007)
7. Banks, V.J., Anyakoha, C.: A review of particle swarm optimization Part II: hybridisation, combinatorial, multicriteria and constrained optimization, and indicative applications. Natural Computing 7(1), 109–124 (2008)
8. Li, S.Y., Li, P.C.: Quantum particle swarms algorithm for continuous space optimization. Chinese Journal of Quantum Electronics 24(5), 569–574 (2007)
9. Zhang, Z.: Quantum-behaved particle swarm optimization algorithm for economic load dispatch of power system. Expert Systems with Applications 37(2), 1800–1803 (2010)
10. Clerc, M., Kennedy, J.: The particle swarm: Explosion stability and convergence in a multi-dimensional complex space. IEEE Transactions on Evolutionary Computation, 58–73 (2002)
11. Sun, J., Feng, B., Xu, W.B.: Particle swarm optimization with particles having quantum behavior. In: Proceedings of 2004 Congress on Evolutionary Computation, pp. 325–331 (2004)
12. Sun, J., Xu, W.B.: A global search strategy of quantum-behaved particle swarm optimization. In: Proceedings of the IEEE Congress on Cybernetics and Intelligent System, pp. 111–116 (2004)
13. Fang, W., Sun, J., Ding, Y., Wu, W., Xu, W.: A Review of Quantum-behaved particle swarm optimization. IETE Technical Review 27(4), 336–348 (2010)
14. Luitel, B., Venayagamoorthy, G.K.: Particle swarm optimization with quantum infusion for system identification. Engineering Applications of Artificial Intelligence 23(5), 635–649 (2010)
15. Horng, M.H.: Vector quantization using the firefly algorithm for image compression. Expert Systems with Applications 39(1), 1078–1091 (2012)
16. Luitel, B., Venayagamoorthy, G.K.: Quantum inspired PSO for the optimization of simultaneous recurrent neural networks as MIMO learning systems. Neural Networks 23(5), 583–586 (2010)
17. Sun, L.F., Wang, Y.C.: Soft-sensing of Oxygen Content of Flue Gas Based on Mixed Model. Energy Procedia 17, 221–226 (2012)

Simulation of Energy Efficiency
in Wireless Meter Reading System by OPNET

Ping Huang, Shiwei Ma[*], Lin Lin, and Bilal Ahmad

School of Mechatronic Engineering & Automation,
Shanghai Key Laboratory of Power Station Automation Technology,
Shanghai University, No. 149, Yanchang Rd.
200072 Shanghai, China
masw@shu.edu.cn

Abstract. A method using OPNET for the energy efficiency analysis is proposed to save the energy cost of the wireless meter reading system in this paper. Firstly, a node energy module is generated to reveal the energy consumption accurately. Then, a channel noise is added to the network to generate the practical model. Finally, OPNET simulation is conducted to analyze the influence of a super frame structure on energy consumption by comparing the performance of the system. In comparison with other methods, the energy consumption of the system is minimized. Simulation results show that a longer lifetime of the node is acquired in the proposed method which can be well applied in the real meter reading system.

Keywords: energy efficiency, wireless meter reading system, OPNET.

1 Introduction

The IEEE 802.15.4 protocol, combined with the ZigBee Alliance closely, constitutes the so-called ZigBee Stack network protocol. ZigBee technology is widely used in wireless meter reading system because of its protocol simplicity, low power consumption, high reliability, low cost, large capacity of network, data security and free use of frequency [1]. Two kinds of nodes are used: full function devices (FFD), and reduced devices (RFD). The former can be configured as a coordinator and the latter can be collectors.

The Zigbee protocol is fully qualified for residential areas because the terminals are usually distributed closely and the data transmission rate is low [2]. For each meter reading terminals and concentrators, the demand of ultra-low-power, low-cost (equipment and operating costs are included) and the power supply by the batteries is desired. However two AA batteries must be ensured for a 6 to 24 months of use in the system because of the high cost of the battery replacement. So, saving power consumption is an important topic in the study of wireless meter reading system. OPNET, an excellent commercial software which has a strong model library and device model library, contains full edge of both the NS and OMNET simulation

---

[*] Corresponding author.

K. Li et al. (Eds.): LSMS/ICSEE 2014, Part III, CCIS 463, pp. 138–146, 2014.
© Springer-Verlag Berlin Heidelberg 2014

software. It is more conducive to a true simulation of the actual system. In this paper, the wireless meter reading system scenarios using the IEEE 802.15.4 protocol are built by OPNET. The beacon-enabled mode is introduced to extend the battery life or save power in data transmission process. So the performance of IEEE802.15.4 protocol is analyzed and improved mainly by limiting the turn-on time of the device or the transceiver of the coordinator, or by setting them into a passive state in the absence of data transmission. It depends on the superframe structure which can be chosen by the Beacon order and the Superframe order.

The paper is organized as follows. In Section 2, the IEEE 802.15.4 standard MAC layer is described and the beacon enable mode is proposed. In section 3, an energy module and an interference channel are introduced to the system. In Section 4, the simulation system is built and results are evaluated. The conclusions are presented in section 5.

2   IEEE 802.15.4 Standard MAC Layer

It is highly advantageous in the realization of ubiquitously networked societies, and demand has been also increased nowadays in view of significance in energy management usage. MAC layer may loss unnecessary energy when there is data transmission in Wireless Sensor Network (WSN). They are mainly shown in the following aspects [3]: idle listening, failure of transmitting, the control overhead and the crosstalk.

All nodes of the IEEE 802.15.4 / ZigBee network are working on the same channel. So it may cause conflicts if the neighbor nodes send data at the same time. Therefore the technology of CSMA/CA is adopted for MAC layer. Simply, the node needs to listen to the channel before it sends the data. If the channel is idle, it can send data, otherwise, a random time delay has to be carried out, followed by another monitoring. The retreat time has an exponential increase with a maximum value. If the node detects busy channel again after the last backoff, the backoff time will be doubled. So the nodes should wait for a longer time to avoid the busy monitoring. The waste of energy exists in the process of competition for acquiring the channel.

There are two kinds of channel access mode in MAC layer [4].

One is non-beacon enable mode; it allows the terminal node (ZE) only for periodic sleep, the coordinator and all the routers must be in a working condition for a long time. The ZigBee standard adopts the mechanism, which realizes the periodic dormancy of ZE. The parent node caches data for the child node and the ZE node extracts data from its parental node.

The other is beacon enable mode, in which a format of super-frame is implemented. The beacon frame, which contains a number of timing and network information, is transmitted at the beginning of super frame, followed by the competitive access period. During the time, each node accesses channel is in a competitive mode. Then in the inactive period, the node goes into sleep mode and waits for the next super frame cycle to send the beacon frame. If the neighboring ZEs send a beacon frame together, conflicts must cause the child devices failing to receive data from the parent node. Furthermore, the synchronization will be affected.

According to superframe structure which is defined by the coordinator, we can see it is comprised of an active part and an optional inactive part from Fig. 1. The length of the superframe (a.k.a. beacon interval, BI) and the length of its active part (a.k.a. superframe duration, SD) are defined as follows:

$$BI = aBaseSuperframeDuration * 2^{BO}$$
$$SD = aBaseSuperframeDuration * 2^{SO}$$

Where,

$$BO = Beacon\ Order$$
$$SO = Superframe\ Order$$

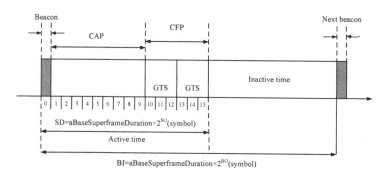

Fig. 1. An Example of the Superframe Structure

In the MAC layer, aBaseSuperframeDuration is a constant, which means the numbers of symbols occupied in a superframe when SO is zero. It is always defined about 960 symbols. The number of CAPs contained in a super-frame period are $2^{(BO-SO)}$ (as BI/CAP), another ($2^{(BO-SO)}-1$) CAPs can be contained in the inactive period. The period time may be set more reasonably by sending a stagger beacon frame so as to avoid conflict. In Section 4, the Beacon Order and Superframe Order will be discussed by simulating the system.

3   Design of Energy Module and Channel Interference

Among the network simulation tools, OPNET is widely used in the simulation of wireless network because of its rich protocol model libraries and the capability of wireless transmission characteristics modeling [5]. However, a lack of energy model reduces the perfection of OPNET tools to a certain extent. The energy model is not provided by the existing network simulation tool or too simple (using a linear model of the NS2 for example). The initial energy value is set in each node of the system. A specified value of energy is subtracted when each node sends or receives a data packet. The existing energy module is failing to be calculated under simulating the periodic dormancy.

A WSN network energy modeling method is presented in this paper based on OPNET. At first, a battery model with the non-linear characteristics is introduced, which named Rakhmatov Battery Model (RM) and can be more accurately simulate the energy consumption. Then an energy module is added to the IEEE 802.15.4 model by using the real-time interaction mechanism to get the model parameters. The battery module will compute the consumed and remaining energy levels. Firstly, the process is initialized by reading the attributes and set the global variables. Secondly, the ID of the node is got. The module gets the value to check if this node is coordinator or not. At last, the maintained and the consumed energy is updated by the formula follows as the product of current, time and voltage. This node module is adopted by both the concentrator and the collector, with a few different parameter settings in a simple proposed star network [6]. The survial time of wireless meter reading network can be calculated.

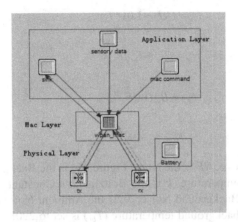

Fig. 2. Node Model

In Figure 2, the sensor data or the mac command in IEEE802.15.4 model produces a specified length of data packet and sends the packet to the MAC layer at fixed time intervals. The sink module is responsible for the generation of super frame and data statistics such as total packet from the network layer. The backoff algorithm is used to determine sequences when all nodes try to communicate by the arrival of beacon frame. The MICAZ sensor nodes, which support IEEE 802.15.4 MAC layer, wireless network providing services or similar system layers, are used in the RM cell model by default. The initial power is 200J and the supply voltage is 3V. The detailed parameter settings are shown in Table 1.

Table 1. Current draw of battery

State	Current
Receive Mode	19.7mA
Transmission Mode	17.4mA
Idle Mode	20μA
Sleep Mode	1μA

In the actual simulation, parameters of energy module are modified through the chip date sheet, and this can greatly improve the efficiency of network simulation. On the other hand, since the wireless meter reading network is in the open wireless space, the signal propagation is uncertain. The system works on free ISM frequency band, the signal propagation will be inevitably disturbed by other signals on the link, including interference from the equipments in the communication system and the external equipment of communication system [7]. Internal equipment interference is mainly studied in this paper. Interference within the device is determined primarily by the performance of the radio transceiver, and the interference within the model mainly concentrated in the background noise, of which is made up by the total thermal noise and environmental noise. Environmental noise power is the product of bandwidth and power spectral density. The formulas of the background noise power are listed as follows:

$$T_{rx} = (NF - 1.0) * T_{bk}$$ (1)

$$N_b = (T_{rx} + T_{bk})B_{rx}k$$ (2)

$$N_a = B_{rx}(1.0E^{-26})$$ (3)

$$N = N_b + N_a$$ (4)

In the equations above, NF is the Noise Figure, T_{bk} is the Background Temperature, T_{rx} is the Receiver Temperature, k is the Boltzmann's Constant, B_{rx} is the Receiver Bandwidth, N_b is the Background Noise, N_a is the Ambient Noise and N is the Noise. The default effective background temperature (T_{bk}) is set to 290K, which is a constant and is not consistent with the real environment. Since the module is either connected to the meter or placed in the meter, and the temperature varies at the range of 0°C to 40°C, the attribute is modified to change randomly between 275 and 315 so that the randomness of interference at each time can be simulated.

4   Simulation and Results

The entire wireless meter reading system is simplified as the network simulation for a single floor. The scene of 100,000 square meters (100×100) is shown in Figure 3. The node turns into a concentrator or a collector depending on the parameters of the MAC layer. The battery process is added to the system to complete the statistics of the energy consumption of the whole network or a single node.

The performance of a wireless network can be viewed through an average traffic received, average network delay, packet drop network or average energy consumption of network nodes. In a network with different values of BO or SO, the node is dynamic throughout the duration of the beacon. This means that the node will be active so hopeful synchronization between nodes for all the simulation time. The simulation for the BO and SO is changed between 1 and 14 on $BO{\geq}SO$. We obtained the figure of the average traffic received and energy consumed as shown in Figure 4. In this figure, the traffic received on $BO{=}8$ and $SO{=}2$ is almost better than the others. And its consumed energy is lower. Then, a reasonable value of BO and SO is chosen by comparing the performance of one node.

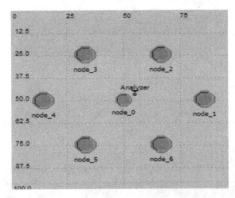

Fig. 3. Simple Model of Meter Reading System

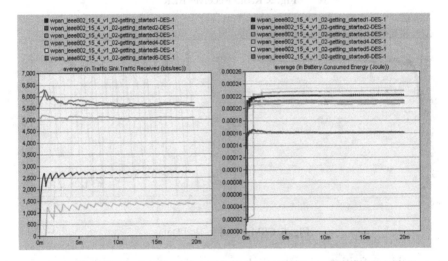

Fig. 4. Traffic Receive and Global Consumed Energy with different *BO* or *SO*

Furthermore, error code (signal) in transmission is generated due to the destruction of the signal caused by the change of the signal voltage. Noise, impulse caused by AC power or lightning, transmission equipment failures and other factors can lead to errors. Various specifications of equipment have strict definition of the bit error rate. In Figure 5, due to various reasons, the error is generated inevitably in the transmission of digital signal. An error is generated when the interference strengthens to a certain extent. The scenario with random noise interferences has a higher bit error rate which can well simulate the real wireless meter reading system. And it is stable at a reasonable value about 0.0009 which conforms to the requirements of the system.

CSMA-CA channel access mechanism with slots is used in the case of the beacon-enable network [8]. The backoff time slot time is exactly aligned with the beacon transmission in such network. The CSMA end-to-end delay (Figure 6) also changes cyclically with the average delay of about 1.3 seconds. An error is generated when signal distortion or interference strengthens to a certain extent. It can be seen that beacon mode is not suitable for the network of demanding high synchronicity [9].

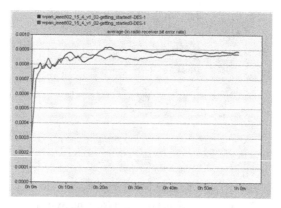

Fig. 5. Radio Receiver BER

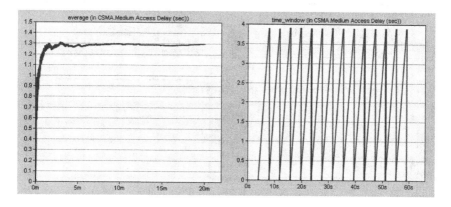

Fig. 6. CSMA End-to-End Delay

The inside definition of protocol shows that the energy consumption is mainly concentrated in four modes: Receive Mode, Transmission Mode, Idle Mode, and Sleep Mode. The sleep period is 3.87072s (241920 symbols), which is calculated according to the definition of the super frame : *Sleep period = BI-SD = aBaseSuperframeDuration (2 ^ BO-2 ^ SO)*. The transmitting time (tx_time) is the attribute passed from the parent function after calling the function. The transmitting process takes 0.004096s. Ideally, the energy consumption is 2.138112×10-4 J in each session of work and is about 1.116316×10^{-6} J in the sleeping time.

The physical layer, which is ideal and has no energy loss, is a free space mode for the transmission of data. The local battery statistics of concentrator and collector is shown as Figure 7, 8. Energy consumption of the collector is significantly higher than that of the concentrator. So the central issue of the research is the energy consumption of the collector. The initial energy of each collector is 200 joules. The nodes are waked up 4s periodically and sending a data packet. An hour later, energy consumption of the collector is 1.8J. So 200 joules of energy can probably be used for five days. Two AA batteries (1.5V, 1600mAh, about 34,560 joules) will ensure the working time of about 2.5 years for the system according to the last energy consumption.

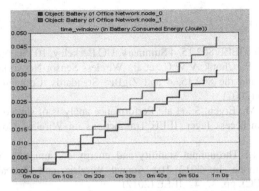

Fig. 7. Collector of Local Energy Consumption Statistics (Unit: Joule)

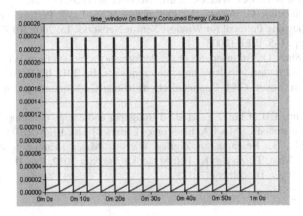

Fig. 8. Network of Global Energy Consumption Statistics (Unit: Joule)

5   Conclusions

An energy module is introduced to the MAC layer of the IEEE 802.15.4 mechanism by using OPNET. Analytical model is being used to predict energy consumption of the star networks under the beacon-enable environment. The channel interference has been exploited in this model to establish a real environment. Then, a relatively appropriate value of Beacon Order and Superframe Order is proposed. Moreover, the energy consumption of a single node is lower within the stable beacon-enabled model from the experiment. However, low system energy consumption does not guarantee good delay performance. In general, the beacon-enable model is suitable for real meter reading system and can save the energy consumption well.

Acknowledgments. This work was financially supported by the Shanghai Science and Technology Foundation (13510500400).

References

1. Ergen, S.C.: ZigBee/IEEE 802.15. 4 Summary, UC Berkeley, 10 (September 2004)
2. Baronti, P., Pillai, P., Chook, V.W.C., et al.: Wireless Sensor Networks: A Survey on the State of the Art and the 802.15. 4 and ZigBee Standards. Computer Communications 30(7), 1655–1695 (2007)
3. Pollin, S., Ergen, M., Ergen, S., et al.: Performance analysis of slotted carrier sense IEEE 802.15. 4 medium access layer. IEEE Transactions on Wireless Communications 7(9), 3359–3371 (2008)
4. Yin, D., Lee, T.T.: Throughput stability and energy consumption of IEEE 802.15. 4 beacon enabled mode. In: 2012 21st Annual Wireless and Optical Communications Conference (WOCC), pp. 37–42. IEEE (2012)
5. Zhou, H.Y., Luo, D.Y., Gao, Y., et al.: Modeling of Node Energy Consumption for Wireless Sensor Networks. Wireless Sensor Network 3(1) (2011)
6. Otal, B., Verikoukis, C., Alonso, L.: Optimizing MAC Layer Performance Based on a Distributed Queuing Protocol for Wireless Sensor Networks. In: 2008 IEEE GLOBECOM Workshops, pp. 1–5. IEEE (2008)
7. Bhatti, S.A., Shan, Q., Atkinson, R., et al.: Performance simulations of WLAN and Zigbee in electricity substation impulsive noise environments. In: 2012 IEEE Third International Conference on Smart Grid Communications (SmartGridComm), pp. 675–679. IEEE (2012)
8. Park, T.R., Kim, T.H., Choi, J.Y., et al.: Throughput and energy consumption analysis of IEEE 802.15. 4 slotted CSMA/CA. Electronics Letters 41(18), 1017–1019 (2005)
9. Wang, Y., Ma, S.: Research on ZigBee Wireless Meter Reading System in Opnet Simulator. In: Xiao, T., Zhang, L., Fei, M. (eds.) AsiaSim 2012, Part II. CCIS, vol. 324, pp. 45–53. Springer, Heidelberg (2012)

Predictive Maintenance for Improved Sustainability — An Ion Beam Etch Endpoint Detection System Use Case

Jian Wan[1], Seán McLoone[1,2], Patrick English[3], Paul O'Hara[3], and Adrian Johnston[3]

[1] Department of Electronic Engineering, National University of Ireland, Maynooth, Ireland
[2] School of Electronics, Electrical Engineering and Computer Science,
Queen's University Belfast, Belfast, UK
s.mcloone@ieee.org
[3] Seagate Technology®, Derry, UK

Abstract. In modern semiconductor manufacturing facilities maintenance strategies are increasingly shifting from traditional preventive maintenance (PM) based approaches to more efficient and sustainable predictive maintenance (PdM) approaches. This paper describes the development of such an online PdM module for the endpoint detection system of an ion beam etch tool in semiconductor manufacturing. The developed system uses optical emission spectroscopy (OES) data from the endpoint detection system to estimate the RUL of lenses, a key detector component that degrades over time. Simulation studies for historical data for the use case demonstrate the effectiveness of the proposed PdM solution and the potential for improved sustainability that it affords.

Keywords: PM, PdM, OES, RUL, Ion Beam Etch.

1   Introduction

Sustainability has emerged as a result of significant concerns about the unintended social, environmental, and economic consequences of rapid population growth, economic growth and excessive consumption of natural resources. The consideration of sustainability has become an integral part of many industrial activities [1]. The benefits of sustainable energy and environmental management include better accountability, better control and allocation of cost, improved performance and reduction in waste. For the semiconductor manufacturing industry reliable and efficient maintenance schemes play an important role in improving sustainability as they increase the plant yield and reduce downtime and waste of energy and materials significantly [3]. Currently, time-based preventive maintenance (PM) strategies are widely used in the semiconductor manufacturing industry where maintenance is carried out periodically according to prior or historical knowledge of the process or equipment. However, PM is quite a conservative and yet insecure strategy as maintenances are usually performed well before the relevant failure while the true failure development is not monitored. Frequent PM activities also increase the cost as more energy, materials and uptime are wasted by the maintenance activities, which impacts negatively on the environment.

Considering the disadvantages of PM, the concept of predictive maintenance (PdM) was proposed where maintenance actions are taken only when necessary and maintenance tasks can be optimally scheduled so as to improve efficiency and reduce waste

K. Li et al. (Eds.): LSMS/ICSEE 2014, Part III, CCIS 463, pp. 147–156, 2014.

[4]. PdM utilizes all available data sources from the process to develop predictive models for remaining useful life (RUL) estimation of key equipment components which need to be maintained. These data sources for PdM can be from existing sensors, test sensors and test signals [5]. A broad range of data mining and machine learning methods can be used for data pre-processing, feature extraction/selection and health model development for PdM [2,3]. For example, regularization methods are used to identify health predictive models for ion-implantation in [6] and Bayesian networks, random forest and linear regression modelling methods are compared in [2] for PdM on an implanter system. Generally speaking, there exists an ongoing shift from traditional PM approaches to PdM schemes in the semiconductor manufacturing industry [7].

Echoing the advances in PdM technologies and the need to improve sustainability by reducing the waste of energy and materials, this paper studies the development of an online PdM module to replace the existing PM scheme for the lens used in the endpoint detection system of an ion beam etch tool used in semiconductor manufacturing. The developed PdM module uses the existing optical emission spectroscopy (OES) data from the endpoint detection system to estimate the RUL of lenses, a key component that degrades over time. The rest of the paper is organized as follows: Section 2 briefly introduces the ion beam etch process and the corresponding maintenance task; Section 3 describes the proposed online PdM module for the lens used in the endpoint detection system; Section 4 details the experiments and simulations used to evaluate the proposed PdM lens RUL estimation module; Finally, some conclusions are drawn in Section 5.

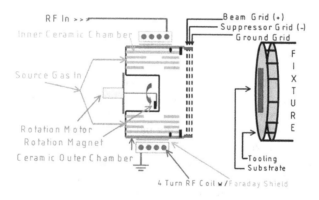

Fig. 1. Overview of an ion bean etch tool

2   Ion Beam Etch and Maintenance Task

Ion beam etch is a versatile etch process for pattern delineation and material modification in which the substrate to be etched is placed in a vacuum chamber in front of a broad-beam ion source. The diagram of the considered ion beam etch tool is shown in Fig. 1. The magnetic field for the ion source is created by a cylindrical solenoid RF coil. High ionization efficiency is achieved as the electrons generated by the source follow a circular path that has been designed so that the electrons have a high probability of collision with the process gas molecules that fill the plasma chamber. Three grid plates which are separated with ceramic insulators are used to extract ions from the source and accelerate them towards the wafer as beams. The wafer is held in place by clamp claws and the fixture can rotate or tilt to change the mill angle in order to optimize the smoothness of the etch.

An endpoint detection system is often fitted as an integral part of ion beam etch tools. This system uses an optical sensor to capture light emission from the chamber, performs OES to obtain the spectral decomposition of the light, and then analyzes the resulting spectrum to determine the endpoint of each etch run. Fig. 2 shows the main components of the interface between the endpoint detection system and the chamber. Here one key component is the lens which acts as the light pathway.

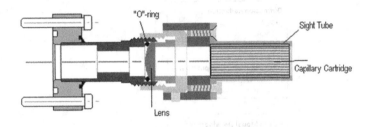

Fig. 2. Components used in the endpoint detection system

The quality of the collected OES data influences the accuracy of endpoint detection and thus it is vital to guarantee the reliability of the measured OES data for the ion beam etch process. However, the lens as well as the capillary used in the endpoint detection system becomes dirty/degrade over time. In particular, as dirt builds up on a lens its opacity increases and this reduces the amount of light reaching the OES sensor and hence the intensity of the recorded OES data. Thus, it follows that if we can track these changes over time we can generate a health index for lenses that can potentially be used to predict their RUL. This is enabled by collecting and analyzing OES data from monitor wafters, which are blank aluminium wafers processed periodically in the chamber as a pre-conditioning etch step before processing of production wafers. These monitor wafers undergo a fixed processing recipe and hence identical input conditions. Therefore, changes observed in monitor wafer OES signals over time are largely driven by changes in tool health.

Dirty lenses need to be replaced before they degrade to a point where they impact on tool performance. Currently this is done as part of a PM scheme, with the frequency of replacements determined by process engineers based on past experience of lens-related process failures. As such, many lenses are replaced prematurely due to the use of a conservative PM strategy. In the following an online PdM module is proposed that enables much better utilisation of lenses.

3 The Online PdM Module

Making use of the OES data from monitor wafers and the computing capabilities from the endpoint detection system, the proposed online PdM module is shown in Fig. 3. At each time instant, the OES data consists of the chamber light emission intensity recorded at 1201 distinct wavelengths. Thus the complete OES data for each monitor wafer etch run is an $m \times 1201$ matrix, where m is the number of sample points.

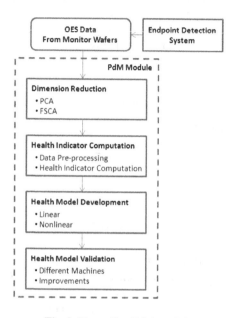

Fig. 3. The online PdM module

Using dimensionality reduction techniques the raw high-dimensional OES data can be reduced to one dimension. Here, principal component analysis (PCA) and forward selection component analysis (FSCA) are investigated for dimension reduction [8,9]. The reduced data is further pre-processed and a single health indicator for the lens is computed for subsequent health model development. Based on the trend of the computed health indicator over multiple etch runs, linear and nonlinear models can be identified for real-time lens RUL estimation. The simplicity of the resulting lens health model allows it to be easily applied to different machines. In addition, the online PdM module can continuously update the prediction model for better RUL estimation using additional information from the expanding production history and maintenance records.

4 Experiments and Simulations

The experiments were conducted using OES data collected for 1746 monitor wafers processed on a Seagate® ion beam etch process from February to June 2013. The OES data for a typical monitor wafer (each one a 170×1201 matrix) is plotted in Fig. 4. Each peak corresponds to a chemical species present in the plasma. As can be seen, only a limited number of channels have peak values that are significantly greater than zero, which is a reflection of the relatively simple chemistry of the ion beam etch process during monitor runs.

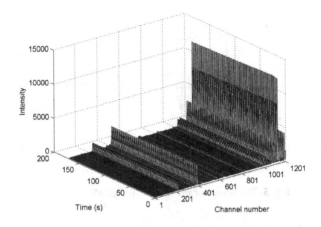

Fig. 4. The channel intensity over time for a typical run

4.1 Dimension Reduction and Health Indicator Computation

As seen in Fig. 4, the collected high-dimensional OES data for the typical monitor wafer is highly redundant enabling dimensionality reduction to be applied with minimum information loss. There are two main approaches for reducing dimensionality [8]: feature extraction methods such PCA which find a new set of k dimensions that are combinations of the original d dimensions and variable selection methods such as FSCA which select k of the d dimensions that best represent the information in the full d dimensional dataset. Here PCA is performed on the dataset to demonstrate the suitability of using OES data for estimating lens RUL and FSCA is performed on the dataset to identify the key channels for health indicator computation in practice.

Combining the dataset together temporally and performing PCA on the resulting 2968210×1201 dimension matrix yields the scores for the first principal component (PC) as plotted in Fig. 5. Here, the color of the scores changes from black to red to reflect the evolution of time (used later). The first PC accounts for 98.49% of the data variability confirming that the original OES data are highly redundant and that PCA can successfully reduce the dimension with little loss of information. Cross-checking the patterns in this plot against maintenance logs revealed that the two biggest jumps in score corresponded to the maintenance events where a capillary change along with lens change occurred, while all the other jumps highlighted by blue lines corresponded to lens changes. Thus two patterns are evident. The first is a long term trend linked to the aging of the capillary. Superimposed on this is a short term trend linked to lens deterioration. Thus, it can be concluded that the PCA score plot effectively captures the evolution of the OES data over time and the score contains a clear lens and capillary health signature.

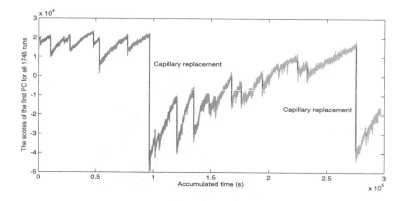

Fig. 5. The scores of the first PC for all 1746 runs

In order to facilitate the health indicator computation and link the health indicator with physical signals rather than the scores from PCA, FSCA is performed on the dataset to identify the key channels that account for most data variability. FSCA is an extension of forward selection regression for selecting a subset of variables that best represent the original full set of variables [9]. This is equivalent to selecting successive components whose combination with the previously selected components explain the most variance across all the data. The components selected by FSCA are the most important features. Performing FSCA on the 2968210×1201 temporally combined dataset, channel 995 is selected as the most representative channel accounting for 98.48% of the data variability. The intensity evolution for the selected top channel is plotted in Fig. 6 for all 1746 runs. Similarly to Fig. 5, maintenance activities as well as lens deterioration are clearly reflected in the intensity changes of the channel. Therefore, only the OES data from this channel is needed for health indicator computation.

The OES data of channel 995 for each monitor wafer is a 170 sample time-series signal from which a single health indicator value needs to be computed. Here, this is achieved by defining the mean intensity of the time-series as the health indicator. Fig. 7 shows a plot of this health indicator for all 1746 processing runs. As can be seen it retains the lens and capillary health signature observed in the raw data.

4.2   Estimating RUL of Lenses

Based on the jumps in the computed health indicator values in Fig. 7, 20 lens changes can be identified and the identified locations are consistent with the maintenance logs. The evolution of the health indicator tends to be linear over the life of each lens. In order to confirm this characteristic, the 20 lenses are plotted in parallel together with their linear approximations in Fig. 8. The time sequence of the lenses is conveyed through

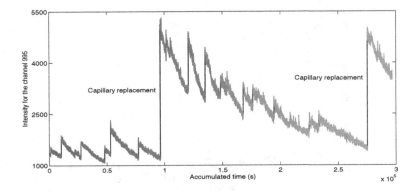

Fig. 6. The intensity evolution of the most representative channel as selected by FSCA

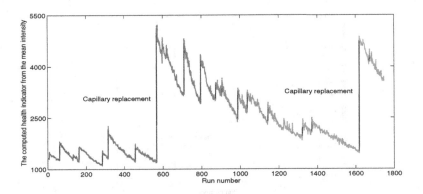

Fig. 7. The health indicator values for all 1746 runs

the color changing from black to red over time. It can be seen from Fig. 8 that the starting point and slope of every lens health indicator evolution are different and thus a single generic linear model will not work for all lenses. Instead a new model needs to be identified online for each lens to fit its specific linear trend. The model for each lens can be identified and updated at each run using the health indicator values from previous runs and the RUL of the lens can then be predicted, where the RUL is defined as the number of runs required for the health indicator to drop below a specified threshold.

Taking the 3rd lens in Fig. 8 as an example, the threshold is set to be the health indicator value at run 100, then the prediction using a linear model at run 50 and 75 are shown in Fig. 9, respectively. The predicted run number at 50 is 98 with an error of -2% while the predicted run number at 75 is 91 with an error of -9%. The RUL of this lens at other run numbers can be computed in a similar way and the computed profile starting from run 20 is shown in Fig.10 with the comparison to the actual RUL. Similarly, the RUL of all other lenses can be predicted at any run number. The prediction errors at run 50 and 75 in terms of run number and health indicator value for the 6 lenses with a life greater than 100 runs are listed in Table 1, where the threshold is also set to be the

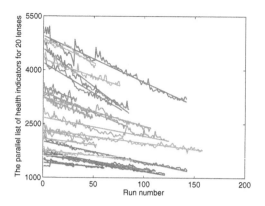

Fig. 8. The health indicator signal evolution for 20 lenses

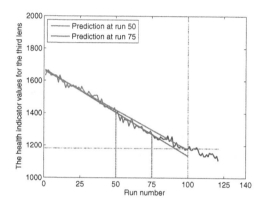

Fig. 9. The linear prediction for the 3rd lens

health indicator value at run 100. Specifying the lowest health indicator value of these 6 lenses as the common threshold to reach, the life of the 6 lenses can be extended by an average of 137% if a PdM rather than a PM strategy is employed (i.e. average lens utilisation is only 42% with the adopted PM strategy).

It can be seen from Fig. 10 that the estimated RUL of lenses tends to be conservative at later run numbers as the trajectory for the health indicator strays away from its linear approximation, as shown in Fig. 9. Nonlinear models such as a polynomial model can be used instead of the linear model to improve the accuracy of the estimation. For example, using a third-order polynomial model to estimate the RUL of this lens at run 75 gives an error of 3% compared to the -9% obtained with the linear model. In practice, multiple models can be applied for online estimation and the most suitable model selected at each run number based on the former prediction errors and prior knowledge on the evolution of health indicator values.

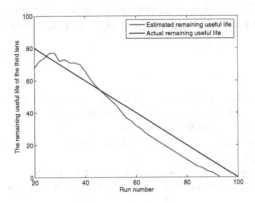

Fig. 10. Remaining useful life of the 3rd lens

Table 1. The prediction errors in terms of run number and health indicator value at run 50 and 75

Lens ID	At Run 50		At Run 75	
	Number_100	Indicator_100	Number_100	Indicator_100
3	98 (-2%)	1174 (-0.6%)	91 (-9%)	1135 (-3.9%)
5	83 (-17%)	1271 (-9.4%)	91 (-9%)	1344 (-4.2%)
6	79 (-21%)	1145 (-8.2%)	95 (-5%)	1127 (-1.6%)
7	90 (-10%)	3491 (-3.7%)	112 (12%)	3767 (3.9%)
13	124 (24%)	2543 (6.6%)	122 (22%)	2533 (6.2%)
17	83 (-17%)	1861 (-4.4%)	88 (-12%)	1894 (-2.7%)

5   Conclusions

This paper has demonstrated that an online PdM module can be developed for the ion beam etch endpoint detection system with limited extra effort by making full use of available data sources and existing computing capabilities. Data mining techniques such as PCA have been used to develop the health model for the PdM module. The developed PdM module can issue maintenance alerts based on real-time predictions of RUL of lenses derived from analysis of OES data collected during the processing of monitor wafers. This enables maintenance activities to be optimally scheduled to reduce the waste of energy and materials, thereby improving sustainability. As such lens utilisation approaching 100% can be achieved with PdM (compared to a utilisation level of 42% with PM).

In general, the development and integration of similar online PdM modules using existing data sources, such as those provided by supervisory control and data acquisition (SCADA) and condition monitoring systems (CMS), can be an economic and sustainable practice in the semiconductor manufacturing industry and in many other manufacturing sectors as well.

Acknowledgments. This work was supported by The Irish Center for Manufacturing Research (ICMR) and Enterprise Ireland (Grants CC/2010/1001 and CC/2011/2001).

References

1. Besnard, F., Bertling, L.: An Approach for Condition-Based Maintenance Optimization Applied to Wind Turbine Blades. IEEE Trans. Sustainable Energy 1(2), 77–83 (2010)
2. Mattes, A., Schopka, U., Schellenberger, M., Scheibelhofer, P., Leditzky, G.: Virtual Equipment for Benchmarking Predictive Maintenance Algorithms. In: Proc. 2012 Winter Simulation Conference, (WSC), pp. 1–12, 9–12 (2012)
3. Lee, J., Siegela, D., Lapiraa, E.R.: Development of a Predictive and Preventive Maintenance Demonstration System for a Semiconductor Etching Tool. ECS Transactions 52(1), 913–927 (2013)
4. Susto, G.A., Beghi, A., De Luca, C.: A Predictive Maintenance System for Epitaxy Processes Based on Filtering and Prediction Techniques. IEEE Trans. Semiconductor Manufacturing 25(4), 638–649 (2012)
5. Hashemian, H.M., Wendell, C.B.: State-of-the-Art Predictive Maintenance Techniques. IEEE Trans. on Instrumentation and Measurement 60(10), 3480–3492 (2011)
6. Susto, G.A., Schirru, A., Pampuri, S., Beghi, A.: A Predictive Maintenance System based on Regularization Methods for Ion-implantation. In: 23rd Annual SEMI Advanced Semiconductor Manufacturing Conference (ASMC), pp. 175–180 (2012)
7. International Technology Roadmap for Semiconductors (2014), http://public.itrs.net/
8. Alpaydin, E.: Introduction to Machine Learning, 2nd edn. The MIT Press (2010)
9. Prakash, P., Johnston, A., Honari, B., McLoone, S.F.: Optimal Wafer Site Selection using Forward Selection Component Analysis. In: 23rd Annual SEMI Advanced Semiconductor Manufacturing Conference (ASMC), pp. 91–96 (2012)

Temperature Synchronization in the Multi-cooling System

Beibei Wang, Liang Chen, Dehong Lian, and Zhengyun Ren

College of Information Science & Technology, Donghua University,
201620 Shanghai, China
{chenliang,renzhengyun}@dhu.edu.cn

Abstract. The multi-cooling system has a decentralized control structure. Practice and simulation show that the temperature controllers located in the cold rooms tend to synchronize, which decreases the storage quality and reduces lifetime of the compressors. The paper takes the supermarket refrigeration system, a typical multi-cooling system, as example, and investigates the temperature synchronous phenomenon. The thermal load of the surrounding air in the display cases are proposed as an important factor, which dramatically influences behaviors of the system. By adopting different thermal loads through adjusting air speeds of the air curtains, the system operates in the de-synchronization situation and achieves better performances. All simulations are based on a simulation model of the supermarket refrigeration system in Matlab/Simulink. Ideas and results in the paper are not limited into the supermarket refrigeration system and can be extended to other multi-cooling systems.

Keywords: Multi-cooling, supermarket refrigeration system, temperature, synchronization.

1 Introduction

The refrigeration system widely exists in the supermarket, buildings, ships, etc, where require the air condition or goods refrigeration. The refrigeration system is a closed system consisting of the compressor, condenser and cold storage rooms. By utilizing a refrigerant in a refrigeration cycle, heat is transported from the cold rooms to the outdoor surroundings. The multi-cooling system is the refrigeration system with many independent cold rooms. Each cold room is typically equipped with a temperature controller which adjusts the refrigerant flow such that the desired temperature is reached. All the cold rooms share a compressor rack and a condenser unit. The supermarket refrigeration system is a typical multi-cooling system, where foods are classified and stored in different display cases.

The multi-cooling system has a decentralized control structure. The temperature controllers located in the cold rooms perform their jobs without exchanging information with each other. The concept is flexible and simple; however, it neglects cross-coupling effects between the subsystems. It has been observed in the supermarket refrigeration

K. Li et al. (Eds.): LSMS/ICSEE 2014, Part III, CCIS 463, pp. 157–164, 2014.

system that the temperature in one display case influences the temperature in the neighboring ones. These interactions from time to time lead to a synchronous operation in the display cases, that causes inferior control performance and reduces the lifetime of the compressors, the heart of the multi-cooling system. The similar synchronous phenomenon is also found in a building with thermally-coupled rooms [1].

The synchronization problem is significantly complicated by the fact that the system features switched dynamics turning the supermarket refrigeration system into a hybrid one. Recently, a kind of hybrid model predictive control approaches has been developed [2, 3], that need deal with the nonlinear programming problem and may lead to highly complex solutions. In our previous work, the Poincaré map theory has been used to analyze the synchronization dynamics [4]. The Poincaré map has been used before, e.g. in [5, 6], for studying stability of the switched system. Ref. [4] further developed the method and the synchronization is interpreted as a stable two-periodic orbit. The numerical solution was given to the stability of the general decentralized hysteresis control system. The control structure of the decentralized hysteresis controllers is typically applied in today's supermarket refrigeration system and many other multi-cooling systems. In the paper, the supermarket refrigeration system is taken as an example, but the method can be applied in other multi-cooling systems. The thermal load of the surrounding air of the display case is proposed to be the key parameter to dramatically influence behaviors of the system. By changing its value actively, the synchronous phenomenon can be avoided effectively. The numerical simulations are performed in Matlab/Simulink and the control performance is compared with the traditional method used in today's supermarket system.

2   Synchronization Problem Description

Let's consider the mathematical model of the supermarket refrigeration system developed in [7] because the model focuses on the dynamics of the temperature control in the display cases and the suction pressure control in the compressors, which are greatly relevant to the synchronization phenomenon. The model is summarized as follows:

$$\frac{dT_{air,i}}{dt} = \frac{\dot{Q}_{goods\text{-}air,i} + \dot{Q}_{airload,i} - \delta_i \dot{Q}_{e,max,i}}{\left(1 + \dfrac{UA_{goods\text{-}air,i}}{UA_{air\text{-}wall,i}}\right) M_{wall,i} C_{pwall,i}} \tag{1}$$

$$\frac{dP_{suc}}{dt} = \frac{\sum\limits_{i=1}^{N} \delta_i \dot{m}_{0,i} + \dot{m}_{r,const} - \dot{V}_{comp}\left(\alpha_\rho P_{suc} + b_\rho\right)}{V_{suc} \cdot \nabla \rho_{suc0}} \tag{2}$$

where $T_{air,i}$ is the air temperature of the ith display case, P_{suc} is the suction pressure, δ_i indicates whether the valve of the ith display case is open or closed. It is controlled by a hysteresis controller in the following way:

$$\delta_i(k+1) = \begin{cases} 1, & \text{if} \quad T_{air,i} \geq T_{air,i,up} \\ 0, & \text{if} \quad T_{air,i} \leq T_{air,i,low} \\ \delta_i(k), & \text{if} \quad T_{air,i,low} < T_{air,i} < T_{air,i,up} \end{cases} \tag{3}$$

where k denotes the time index, $T_{air,i,up}$ is the upper bound of the air temperature and $T_{air,i,low}$ is the lower bound.

In the supermarket refrigeration system, most display cases are open type. Numerous experiments show that in the open display cases, 50% of heat come from the surrounding air through the air curtain; in the vertical open display cases, the percent is more than 70%. The heat is closely related to the temperature change in the display cases. Therefore, the thermal load of the surrounding air $\dot{Q}_{airload}$ is an important parameter that has dramatic effect on behaviors of the system. In addition, the dynamics of the system can be studied through the bifurcation and chaos theory. The aim of the theory is to investigate dramatic changes in the qualitative or topological structure of a system by changing smoothly a system parameter [8]. It can be dated back to 1975 when the first mathematical definition of 'chaos' was given [9]. Fig. 1 gives a bifurcation diagram to show how the temperature in the first display case $T_{air,1}$ behaves with the smooth change of the parameter, the thermal load of the second display case $\dot{Q}_{airload,2}$. Here, the supermarket refrigeration system model with two display cases and two compressors (2d-2c) are used and the thermal load of the first display case is kept to 3000w. It can be seen that the system has very complicated behaviors with any bifurcations occurring when small smooth changes are made to the values of $\dot{Q}_{airload}$.

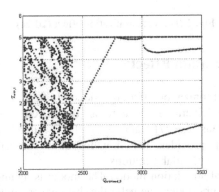

Fig. 1. The bifurcation diagram

Fig. 2 shows that when the thermal loads of two display cases are the same, or almost same, the temperatures in the display cases tends to agree (see Fig. 2a), that leads to their temperature controllers act synchronously, thus periodic high and low amount of vaporized refrigerant flow into the suction manifold. Hence, large fluctuation in the suction pressure P_{suc} is a consequence (see Fig. 2b) which then induces frequently switching on and off compressors in the compressor rack (see Fig. 2c). This subsequently leads to excessive wear on the compressors, and thus reducing lifetime of the compressors, the heart of the refrigeration system.

From above analysis, it can be seen that the thermal load $\dot{Q}_{airload}$ is an important parameter that affects behaviors of the system and when the values of two display cases are the same, or almost same, the synchronous phenomenon tends to happen. Therefore, if one of the thermal loads of the display cases is actively changed, the synchronization can be avoided. Because $\dot{Q}_{airload}$ is the result that heat passes into the display case through the air curtain, if one want to change its value, he needs to change the temperature difference between the inside and outside of the display case, or change the heat transfer coefficient according to the Newton's law of cooling. It can be easily accomplished by adjusting the air speed of the air curtain. When the speed decreases, the heat transfer coefficient increases.

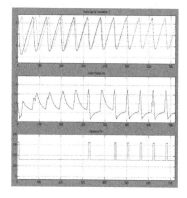

Fig. 2. The synchronization effect (2d-2c)

3   De-synchronization Effect

To evaluate the de-synchronization effect mentioned in Section 2, a simulation model of the supermarket refrigeration system is developed mainly based on the mathematical expressions in [7] and uses the Matlab/Simulink environment exclusively. Ref. [10] provides the downloadable simulation model in two environments: Windali and Matlab/Simulink.

Fig. 3 shows the total simulation model. The model is attained by connecting in- and outputs from four sub-models, namely, the display case module, the compressor control module, the suction manifold module and the performance evaluation module. The display case module can embed any number of the display cases, which is useful for studying the large-scale multi-cooling system. In the compressor control module, a suction pressure controller with a dead band is equipped in the compressor rack consisting of many compressors. The controller can turn one or more compressors on/off according to the refrigeration demand. The number of compressors also can be chosen for application in the large-scale multi-cooling system. The suction manifold module mainly describes dynamics of the suction manifold injected into the compressor rack. The performance evaluation module gives three indices to examine the control performance of the de-synchronization scheme. The indices are: the constraint satisfaction of the temperature controller γ_{con} , which measures the storage quality; the number of switches in the compressor rack γ_{sw}, which measures lifetime of the compressors; γ_{pow} and the power consumption of the compressors . The three indices are smaller, the control performance is better. In addition, all simulations in the paper use the parameter settings shown in the graphic user interface as Fig. 4.

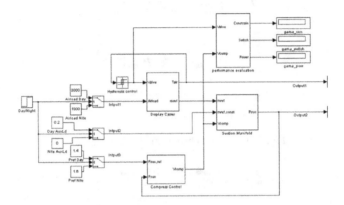

Fig. 3. The total simulation model of the supermarket refrigeration model

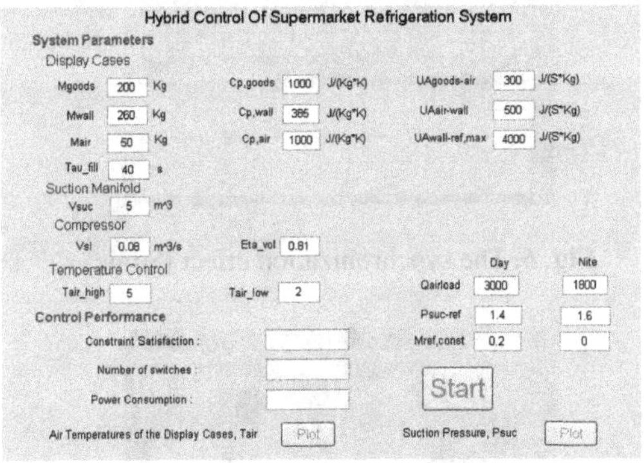

Fig. 4. Parameter settings of the simulation model

Fig. 5 shows the de-synchronization effect of the supermarket refrigeration system with two display cases and two compressors (2d-2c). Comparing with Fig. 2, it can be seen from Fig. 5 that the temperatures in the two display cases don't synchronize (see Fig. 5a), the suction pressure oscillates in a relative small range (see Fig. 5b) and the switching frequency of the compressors dramatically decreases (see Fig. 5c).

Furthermore, simulations are performed in the large-scale supermarket refrigeration system with eight display cases and four compressors (8d-4c). Fig. 6

shows the almost synchronous operation and Fig. 7 shows the de-synchronization effect by using the method proposed in the paper, where the values of $\dot{Q}_{airload,i}$ are 2700, 2800, ..., 3400, respectively. In Fig. 7, the temperatures' curves are much denser than those in Fig. 6, which implies the de-synchronous phenomenon.

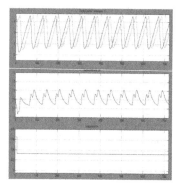

Fig. 5. The de-synchronization effect (2d-2c)

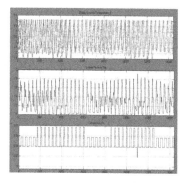

Fig. 6. The synchronization effect (8d-4c)

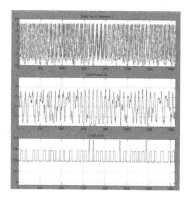

Fig. 7. The de-synchronization effect (8d-4c)

Table 1 summarizes the three performance indices by comparing the de-synchronous operation with the synchronous one. From Table 1, it can be concluded that for the performances of the storage quality γ_{con} and lifetime of the compressors γ_{sw}, the method proposed in the paper has obvious improvement over the tradition control used in today's supermarket refrigeration system. For example, for the system with two display cases and two compressors (2d-2c), the number of the temperature in the display case going out of the hysteresis control bounds reduces 65.77% by comparing the method in the paper with the tradition one; and the number of compressor switches decrease 89.09%. The performance improvement proves in some extent that the thermal load of the surrounding air $\dot{Q}_{airload}$ is more important as a key parameter to affect behaviors and performances of the system. In addition, for the index of the energy consumption, the values remain almost unchanged. The main reason may be because the major reduction in the energy consumption comes from increasing the reference value of the suction pressure in the suction pressure controller [11]. It can be easily achieved at the same time when using the above de-synchronization methods.

Table 1. Performance Indices

	γ_{con}	γ_{sw}	γ_{pow}
2d-2c	-65.77%	-89.19%	-0.97%
8d-4c	-52.67%	-42.47%	-0.6%

4   Conclusions

The paper proposes that the thermal load of the surrounding air is an important factor which can influence behaviors of the system. By adopting different thermal loads in the display cases through adjusting air speeds of the air curtains, the system operates in the de-synchronization situation and achieves better storage quality and longer lifetime of the compressors. All simulations are based on the simulation model of the supermarket refrigeration system in Matlab/Simulink. Ideas and results in the paper can be extended to other multi-cooling systems.

Acknowledgments. The work is supported by National Natural Science Foundation of China under Grant 61104113 and the Fundamental Research Funds for the Central Universities.

References

1. O'Brien, J., Sen, M.: Temperature synchronization, phase dynamics and oscillation death in a ring of thermally-coupled rooms. In: Proceedings of the ASME 2011 International Mechanical Engineering Congress & Exposition, pp. 1–10 (2011)
2. Sarabia, D., Capraro, F., Larsen, L.F.S., de Prada, C.: Hybrid NMPC of supermarket display cases. Control Engineering Practice 17(4), 428–441 (2009)

3. Ricker, N.L.: Predictive hybrid control of the supermarket refrigeration benchmark process. Control Engineering Practice 18, 608–617 (2010)
4. Chen, L., Lian, D.H., Wang, T.: Synchronization stability of a distributed hysteresis controlled system. In: 7th IEEE Conference on Industrial Electronics and Applications, Singapore, pp. 1613–1617 (2012)
5. Kousaka, T., Ueta, T., Kawakami, H.: Bifurcation of switched nonlinear dynamical systems. IEEE Transactions on Circuits and Systems II: Analog and Digital Signal Processing 46, 878–885 (1999)
6. Hiskens, I., Reddy, P.: Switching-induced stable limit cycles. Nonlinear Dynamics 50, 575–585 (2007)
7. Larsen, L.F.S.: Model Based Control of Refrigeration Systems, Ph.D. thesis, Department of Control Engineering - Aalborg University, Aalborg (2006)
8. Devaney, R.L.: An Introduction to Chaotic Dynamical Systems. Westview Press (2003)
9. Li, T.Y., Yorke, J.A.: Period three implies chaos. Amer. Math. Monthly 82, 481–485 (1975)
10. Larsen, L.F.S., Izadi-Zamanabadi, R., Wisniewski, R., Sonntag, C.: Supermarket refrigeration systems – A benchmark for the optimal control of hybrid systems, http://astwww.bci.unidortmund.de/hycon4b/wprelated/sr.pdf
11. Larsen, L.F.S., Thybo, C., Stoustrup, J., Rasmussen, H.: A method for online steady state energy minimization, with application to refrigeration systems. In: 43rd IEEE Conference on Decision and Control, pp. 4708–4713 (2004)

A Variant Gaussian Process for Short-Term Wind Power Forecasting Based on TLBO

Juan Yan[1], Zhile Yang[1], Kang Li[1], and Yusheng Xue[2]

[1] School of Electronics, Electrical Engineering and Computer Science
Queen's University Belfast, Belfast, BT9 5AH, UK
[2] State Grid Power Engineering Research Institute, Nanjing, China
jyan03@qub.ac.uk

Abstract. In recent years, renewable energy resources have drawn a lot of attention worldwide in developing a more sustainable society. Among various forms of renewable energies, wind power has been recognized as one of the most promising ones in many countries and regions including Northern Ireland and Ireland according to the National Renewable Energy Action Plans (NREAPs). However, due to the variability nature of wind power, the wind generation forecasting hours even days ahead proves to be imperative to enhance the flexibility of the operation and control of real-time power systems. In this paper, a variant Gaussian Process employing only nearby measured wind power data is proposed to make short term prediction of the overall wind power production for the whole island of Ireland. Multi Gaussian Process submodels are developed, and the model capability in reflecting the variability and uncertainty in the wind generation system is enhanced. In such method, local data could be utilized more efficiently and computation complexity is reduced at the same time. The forecasting results have been verified in comparison with standard Gaussian Process and persistence model, and improvements can be observed in terms of the model complexity and prediction accuracy. Moreover, a recently proposed teaching-learning based optimization algorithm (TLBO) is applied to build the Gaussian model, and simulations show its faster convergence speed and better global searching capability.

Keywords: Gaussian Process, TLBO, Wind Power Forecasting.

1 Introduction

In the era when the importance of renewable energy supply can not be underestimated any more, the European Union (E.U.) has set ambitious goals for its development and commits to devote a lot more to achieve the targets. Considering the special meteorological and topographic characteristics in the island of Ireland, the power system operation company Eirgrid has put a lot of effort to develop wind power to contribute to the target. However, the variability nature of wind makes the wind power forecasting imperative before it is integrated into grid.

Wind power forecasting methods could be grouped into several types with reference to different forecasting horizons. Numerical weather prediction (NWP) is usually employed for long term prediction from 6 hours to several days ahead. Future local weather information (wind speed, wind direction, temperature, air density etc.) is obtained first

K. Li et al. (Eds.): LSMS/ICSEE 2014, Part III, CCIS 463, pp. 165–174, 2014.
© Springer-Verlag Berlin Heidelberg 2014

with NWP method and then converted into wind power generation with statistical methods [1]. For short term and very short term wind power prediction, from few minutes to 6 hours ahead, time series model could be developed directly using solely historic measurement data [2]. The time-series model recursively predicts wind power by employing the output data in a fixed window. Many statistical methods have been applied in this area, such as ARMA (autoregression moving average) [3], neural network based models [4], and Gaussian Process [5] etc.

As a non-parametric method, Gaussian process shows great global property [6] and has been applied in system identification [7]. It supposes that the system outputs follow multi-variable joint Gaussian distribution with fixed mean function and covariance function. Such assumption could only be satisfied when the system is stable enough during the relevant horizon. However, due to the uncertainty of wind power, it becomes inappropriate to assume all the available data follow one joint Gaussian distribution when the data covers a large horizon up to several days. In such a situation, a novel multi local Gaussian Process is developed assuming that each prediction point follows a joint Gaussian distribution with only its former L outputs which loosens the assumption of the standard one. In this way N (referring to the number of available data) sub-models are built at each available point and the hyperparameters related to the covariance function and mean functions are tuned by minimizing the estimation error of training data set. This model is supposed to fit better to the actual situation.

For the optimization of the proposed model, non-linear complex optimization problems impose difficulties on conventional optimization techniques such as linear programming and quadratic programming, while the dynamic programming often has to endure its dimension curse. Meta-heuristic algorithms open a new window to such predicament. They are immune from the formulation behaviour of the objective function as well as the constraints and almost omnipotent to tackle all kinds of problems though not being able to guarantee to find the global optimum. The teaching-learning based optimization is a new member of this family [8, 9] and has be adopted to solve real-world problems [10–14] due to the fast convergence speed and good exploitation ability. In this paper, TLBO is used to optimize the proposed non-linear Gaussian Process variant model.

This paper is organized as follows. First, the reference model Gaussian Process for time series forecasting is illustrated in Section II. Then Section III proposes the prediction error adjusted variant Gaussian process. In Section IV, TLBO optimization method is introduced and analysed. Section V designs the experiment and presents the results. Finally, section VI concludes the paper.

2   Gaussian Process for Wind Power Forecasting

As shown in (1), a multi-input-single-output (MISO) nonlinear system output could be expressed as the sum of a fixed function of the input and the noise where $x(k), y(k)$ denotes available input-output measurements for the kth sample and $v(k)$ is an i.i.d random sequence of zero mean and finite variance σ_v^2.

$$y(t) = f(x(k)) + v(k) \tag{1}$$

Gaussian Process is a collection of random variables, any finite number of which have (consistent) joint Gaussian distributions [6]. During the implementation of Gaussian Process modelling, the available data $Y=(y(1), y(2), \ldots, y(N))^\top$ and the new output y_0 are assumed to follow one joint Gaussian distribution. The mean function could be set to zero and the covariance matrix could be defined by a covariance function, which is expressed as

$$\begin{bmatrix} y_0 \\ Y \end{bmatrix} \sim N(0, \begin{bmatrix} A_p & B_p \\ B_p^\top & C_p \end{bmatrix}) \tag{2}$$

where

$$A_p = cov_p(y_0, y_0)$$
$$B_p = (cov_p(y_0, y(1)), \ldots, cov_p(y_0, y(N))) \tag{3}$$
$$C_p(i, j) = cov_p(y(i), y(j))$$

$cov(.)$ represents the covariance between two variables, and p represents the hyper-parameters used to define the covariance. There are many different forms of covariance function, of which automatic relevance determination (ARD) shown in (4) is a very popular one due to its infinite differentiability. In (4), D refers to the dimension of the model input and δ_{ij} refers to the Kronecker delta representing the observation noise for each case. $p = [v_1, v_2, w_1, \ldots, w_D]$, and all the elements are non-negative.

$$cov_p(y(i), y(j)) = \Phi(x(i), x(j)) = v_1 \exp[-\frac{1}{2}\sum_{d=1}^{D} w_d(x_d(i) - x_d(j))^2] + v_2\delta_{ij} \tag{4}$$

From (2), it can be verified by the Bayesian Theorem that the conditional density of y_0 on Y also follows the Gaussian distribution with the mean value of $B_p C_p^{-1} Y$ and variance of $A_p - B_p C_p^{-1} B_p^\top$ shown in (5)(6). The hyper-parameters in cov could be tuned by maximizing the marginal density as shown in (7).

$$y_0|Y \sim N(B_p C_p^{-1}Y, A_p - B_p C_p^{-1} B_p^\top) \tag{5}$$

$$\hat{y}_0 = B_p C_p^{-1}Y \tag{6}$$

$$p* = arg \max_p \frac{1}{(2\pi)^{\frac{N}{2}}|C_p|^{\frac{1}{2}}} \exp(-\frac{1}{2}Y'C_p^{-1}Y) \tag{7}$$

No matter whether a system has exogenous input, Gaussian Process could be utilized. For a time-series systems, let L represent the time lag, then

$$y(t) = f(y(t-1), y(t-2), \ldots, y(t-L)) + v(t) \tag{8}$$

Denote $x(t) = (y(t-1), y(t-2), \ldots, y(t-L))$, representing the state space vector which decides the system state at time t, then (8) has the same form as (1). Given a set of data as training data, for any time t, the output could be predicted as follows, similar to (6).

$$\hat{y}(t) = B(t)C^{-1}Y$$
$$B(t) = (\Phi(x(t), x(1)), \Phi(x(t), x(2)), \ldots, \Phi(x(t), x(N))) \tag{9}$$
$$C(i, j) = \Phi(x(i), x(j)) \quad i, j \in [1, N]$$

Here Y represents the training data and N represents the number of training data. It is obvious that when N is large enough, the computation complexity of C^{-1} would be huge which is a burden for the application of Gaussian Process.

3 Proposed Modelling Method

Standard Gaussian Process assumes that all the available outputs follow one joint Gaussian distribution with the new prediction, however, this might be too strict for the wind power system which shows strong uncertainty and variability. In this paper, a variant of Gaussian Process is proposed, assuming that each sample follows one Gaussian distribution with its previous L samples, and prediction can be made based on this kind of smaller sized Gaussian Process submodels. Instead of maximising the marginal likelihood method, the sum of squared prediction errora is minimized to identity the hyper-parameters involved in covariance function. So this model could be regarded as a mixture of multi local Gaussian Processes. In this way, the influence of faraway data point could be reduced and the computation complexity for matrix inversion could be reduced.

As illustrated above, when making a new prediction $y(t)$, only the L outputs before t, denoted as $Y(t) = (y(t-1, y(t-2), \ldots, y(t-L))^\top$, follow one joint Gaussian distribution with $y(t)$, and only those L data are used to estimate it. In such circumstance, the computation demand is greatly decreased because the size of $Y(t)$ is much smaller than that of Y in (9). In spite of the reduction in used dataset dimension, the accuracy of this method could be guaranteed due to the efficient use of highly correlated data. The proposed method could be expressed in (10).

$$\hat{y}(t) = B_p(t)C_p^{-1}(t)Y(t)$$

$$= B_p(t)C_p^{-1}(t) \begin{pmatrix} y(t-1) \\ y(t-2) \\ \vdots \\ y(t-L) \end{pmatrix} \tag{10}$$

$B_p(t)$ and $C_p(t)$ are shown in (11) where $\boldsymbol{x}(t-i)$ is the corresponding input to $y(t-i)$ $(i = 1, \ldots, L)$.

$$B_p(t) = (\Psi(\boldsymbol{x}(t), \boldsymbol{x}(t-1)), \ldots, \Psi(\boldsymbol{x}(t), \boldsymbol{x}(t-L))$$

$$C_p(t) = \begin{pmatrix} \Psi(\boldsymbol{x}(t-1), \boldsymbol{x}(t-1)) & \ldots & \Psi(\boldsymbol{x}(t-1), \boldsymbol{x}(t-L)) \\ \vdots & \ddots & \vdots \\ \Psi(\boldsymbol{x}(t-L), \boldsymbol{x}(t-1)) & \ldots & \Psi(\boldsymbol{x}(t-L), \boldsymbol{x}(t-L)) \end{pmatrix} \tag{11}$$

For a time sequence of data, the proposed method works like a moving window. Each point is estimated with its former L data, and the window move forward step by step to make further prediction. The hyperparameters could be tuned by minimizing the sum of square error in (12), and all the submodels employs the same covariance function. The validation procedure is similar.

$$p* = arg \min_{p} \sum_{t=1}^{N} (\hat{y}_t - y_t)^2$$
$$= arg \min_{p} \sum_{t=1}^{N} (B_p(t)C_p(t)^{-1}Y(t) - y_t)^2 \tag{12}$$

For multi-step ahead prediction, iterative prediction is employed [15], so that the forecasting horizon could reach multi sampling intervals and the model does not have to be trained separately for different prediction steps [16].

4   Teaching-Learning-Based Optimization

Teaching-learning based optimization is a novel population based meta-heuristic algorithm proposed by Rao et.al [8, 9] in 2011. The general process is partly inspired by its ancestors particle swarm optimization (PSO) and differential evolution (DE) but combines both of their strengths. A teaching phase is designed first to select a teacher (e.g the best student) and share knowledge for the other students, trying to lead the class towards the better direction. Such scheme is similar with the social learning of PSO that solutions learn from the global best solution. The teaching phase is associated by a second stage named as learning phase, where students share knowledge by randomly selecting a classmate and learning from him/her. The idea is basically taken from DE in which particles learn from the interaction. One of the most salient features of TLBO lies on the non-tuning-parameter process. The social learning parameter in PSO and mutation and crossover parameters in DE are all removed, and particles mutation all depend on the statistics information of the whole population and solutions interaction. The procedure of TLBO is illustrated as follows.

4.1   Teaching Phase

As aforementioned, a teacher will be first selected by sorting all the fitness function values of the whole population. Moreover, the mean values of each dimension in the population is calculated and treated as the mathematical statistics for providing valuable information to the teacher. By utilizing these data, the teacher will try to share knowledge to the class. The knowledge is quantitatively measured by the difference between the teacher and the mean values. A fixed teaching factor is utilized to decide the difference value which denoted by the following equation,

$$DM_i = rand_1 \times (T_i - T_F Mean_i) \tag{13}$$

where DM_i is the value difference of i-th iteration while $Mean_i$ denotes the column-wise mean value in all dimensions in the solutions of i-th iteration. T_i denotes the selected teacher, and T_F is the teaching factor. According to the original paper of TLBO, the T_F can be either 1 or 2 denoted as:

$$T_F = round(1 + rand_2(0,1)) \tag{14}$$

where the length of the learning step (value difference) is various. Students will subsequently gain knowledge from the difference as:

$$p_{ij}^{new} = p_{ij}^{old} + DM_i \tag{15}$$

where p_{ij}^{new} and p_{ij}^{old} denote the old and fresh learners before and after gaining knowledges respectively in the j-th solution of i-th iteration . The fresh learners will compete with his/her predecessor and replace him/her if a better fitness value is achieved.

4.2 Learning Phase

Besides learning from the teacher, each student has a chance to learn from a classmate in a second stage named learning phase. In this phase, each solution would randomly select another solution to compare the fitness, and update the knowledge storage according to the interaction. The phase is denoted as follow,

$$p_{ij}^{new} = \begin{cases} p_{ij}^{old} + rand_3(p_{ik} - p_{ij}) & if \ f(p_{ik}) < f(p_{ij}) \\ p_{ij}^{old} + rand_3(p_{ij} - p_{ik}) & if \ f(p_{ij}) < f(p_{ik}) \end{cases} \tag{16}$$

where the j-th learner p_{ij} and k-th learner p_{ik} are randomly selected from the population in the i-th iteration. Through a competition, the initial learner p_{ij} will refresh his/her knowledge based on the deviation of the two learners. Similarly with teaching phase, the refresh student will have to compete with its predecessor and the better one will be selected.

It could be observed that none of any pre-set tune-able parameters has been introduced into the optimization process besides the common initialization parameters: the size of particles and the number of iterations. Solutions update their values entirely based on the interactions between the teacher and other students. Experiments on well known benchmarks have been conducted and the prominent performance of TLBO has been proved [8, 9].

5   Experimental Results and Discussions

In this section, the overall wind power generation in the island of Ireland was predicted with the introduced time series models which could benefit the economic dispatch among different forms of energy and different generating countries. The power generation of 15th and 16th of April 2013 were acquired every 15 minutes, and that of 17th and 18th were forecast, therefore there were 384 data points in total, half of which were used for training and the other half for validation. These data were normalized first and then the automatic relevance determination (ARD) covariance function shown in (4) was employed. According to some trial-and-error experiment results, the length

Table 1. The Normalized RMSE of three different models and the accuracy improvement for 8 steps prediction (2 hours)

RMSE(m=10)	1	2	3	4	5	6	7	8
Persistence model	0.0752	0.1071	0.1383	0.1727	0.2069	0.2385	0.2744	0.3063
Standard Gaussian Process (SGP)	0.0678	0.1051	0.1342	0.1689	0.2012	0.2335	0.2673	0.2980
Proposed variant of Gaussian Process	0.0676	0.0997	0.1295	0.1622	0.1942	0.2239	0.2542	0.2836
Improvement over Persistence	10%	6.9%	6.3%	6.0%	6.1%	6.1%	7.3%	7.4%
Improvement over SGP	0.3%	5%	3.5%	4.0%	3.4%	4.3%	5.0%	4.8%

of the state vector was set as L=10, generating 12 parameters in all to be tuned. TLBO was employed to identify those hyper-parameters and the global optima could be approximately approached with multi simulation and proper setting of initial points. The validation result can be seen from Fig.1 indicating the effectiveness of the model with the maximum error of 10%.

The effectiveness of the proposed variant of Gaussian Process for multi-step prediction could be seen from Fig.2. The proposed method beats the persistence model in the whole forecasting horizon and the maximum accuracy improvement can reach as much as 10% according to Table I.

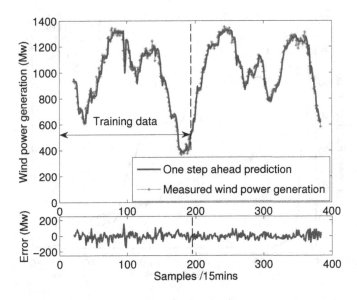

Fig. 1. One step ahead prediction result of the proposed method

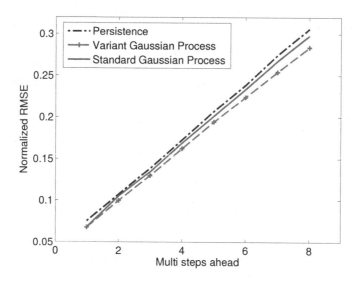

Fig. 2. The normalized RMSE of different models

Further, from Fig.3, it can be seen that TLBO shows much faster convergence rate and better capability to find the global optima when the optimisation starts with initial particles distributing all over the working space. Although it is generally considered

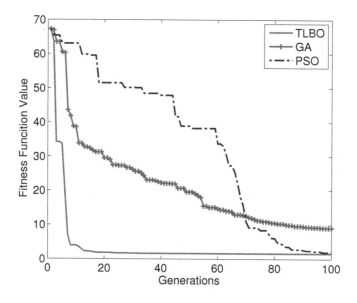

Fig. 3. The optimization procedure of TLBO compared with PSO and GA

that in each generation, the fitness function gets evaluated twice that number for PSO and GA, the convergence rate of TLBO is still much faster with respect to the number of function evaluations according to Fig.3.

6 Conclusion

This paper proposes a variant of standard Gaussian Process for short term wind power forecasting with time series models. It weakens the assumption that all the available data have to follow one joint Gaussian distribution, and therefore reduces the influence of less correlated samples which is faraway. Local information are employed more efficiently, and more accuracy results can be achieved even for multi step prediction up to 2 hours.

The hyper-parameters in the covariance function are optimized by minimizing the sum of square error of the estimated value for the training data. TLBO is employed and proven to be an efficient tool to optimize non-linear problems with faster convergence rate and better exploitation capability. The method could help in economic load dispatch with large penetration of wind power and other renewable resources.

Acknowledgment. This work was funded by Engineering and Physical Sciences Research Council (EPSRC) under grant number EP/L001063/1 and EP/G042594/1. Many thanks to the review's great suggestions and comments.

References

1. Kariniotakis, G., Pinson, P., Siebert, N., Giebel, G., Barthelmie, R., et al.: The state of the art in short-term prediction of wind power-from an offshore perspective. In: Proceedings, Ocean Energy Conference ADEME-IFREMER (2004)
2. Giebel, G., Landberg, L., Kariniotakis, G., Brownsword, R., et al.: State-of-the-art methods and software tools for short-term prediction of wind energy production. In: Proceedings 2003 EWEC (European Wind Energy Conference and Exhibition) (2003)
3. Milligan, M., Schwartz, M., Wan, Y.: Statistical wind power forecasting models: results for us wind farms. National Renewable Energy Laboratory, Golden, CO (2003)
4. Sideratos, G., Hatziargyriou, N.D.: An advanced statistical method for wind power forecasting. IEEE Transactions on Power Systems 22(1), 258–265 (2007)
5. Mori, H., Kurata, E.: Application of gaussian process to wind speed forecasting for wind power generation. In: IEEE International Conference on Sustainable Energy Technologies, ICSET 2008, pp. 956–959. IEEE (2008)
6. Rasmussen, C.E.: Gaussian processes in machine learning. In: Bousquet, O., von Luxburg, U., Rätsch, G. (eds.) Machine Learning 2003. LNCS (LNAI), vol. 3176, pp. 63–71. Springer, Heidelberg (2004)
7. Gregorčič, G., Lightbody, G.: Nonlinear system identification: From multiple-model networks to gaussian processes. Engineering Applications of Artificial Intelligence 21(7), 1035–1055 (2008)
8. Rao, R., Savsani, V., Vakharia, D.: Teaching–learning-based optimization: A novel method for constrained mechanical design optimization problems. Computer-Aided Design 43(3), 303–315 (2011)

9. Rao, R.V., Patel, V.: An improved teaching-learning-based optimization algorithm for solving unconstrained optimization problems. Scientia Iranica (2012)
10. Crepinsek, M., Liu, S.-H., Mernik, L.: A note on teaching–learning-based optimization algorithm. Information Sciences 212, 79–93 (2012)
11. Waghmare, G.: Comments on a note on teaching–learning-based optimization algorithm. Information Sciences 229, 159–169 (2013)
12. Niknam, T., Kavousi Fard, A., Baziar, A.: Multi-objective stochastic distribution feeder reconfiguration problem considering hydrogen and thermal energy production by fuel cell power plants. Energy 42(1), 563–573 (2012)
13. Niknam, T., Azizipanah-Abarghooee, R., Aghaei, J.: A new modified teaching-learning algorithm for reserve constrained dynamic economic dispatch. IEEE Transactions on Power Systems 28(2), 749–763 (2013)
14. Yang, Z., Li, K., Foley, A., Zhang, C.: A new self-learning tlbo algorithm for rbf neural modelling of batteries in electric vehicles. In: 2014 IEEE World Congress on Computational Intelligence. IEEE (2014)
15. Girard, A., Rasmussen, C.E., Quinonero-Candela, J., Murray-Smith, R.: Gaussian process priors with uncertain inputs? application to multiple-step ahead time series forecasting (2003)
16. Yan, J., Li, K., Bai, E.-W.: Prediction error adjusted gaussian process for short-term wind power forecasting. In: 2013 IEEE International Workshop on Intelligent Energy Systems (IWIES), pp. 173–178 (November 2013)

Modeling Electric Vehicles in Equilibrium Analysis of Electricity Markets*

Ming Xie, Xian Wang, and Shaohua Zhang

Key Laboratory of Power Station Automation Technology,
Department of Automation, Shanghai University, Shanghai, 200072, China
xieming@shu.edu.cn

Abstract. High penetration of electric vehicles (EVs) in power system will significantly bring a large number of load and storage capacities. Different competition modes of EVs will have different effects on electricity markets. In this paper, Cournot competition equilibrium models of electricity market with different competition modes of EVs are developed and the theoretical analysis of market prices become leveled during off-peak and peak period is made. Numerical examples are presented to verify the validity of the theoretical analysis and the effectiveness of the proposed models. It is shown that EVs fleet can smooth market prices curve with arbitrage charging and discharging, especially when EVs fleet acts as a price-taker, it can reach the effect of "peak load shifting". In addition, the battery depreciation costs will largely affect the price-smoothing effect.

Keywords: Electric Vehicles, Electricity Market, Cournot Competition, Equilibrium Analysis.

1 Introduction

With the development of power industry reform in the 1980s, the competitive electricity markets are restructured in many countries [1]-[3]. The fossil fuels scarcity, higher environmental awareness as well as technological improvements in battery manufacturing have brought extensive use of EVs in the electricity markets [4]-[5]. Using vehicle-to-grid (V2G) technology [6], a fleet of EVs will bring a large number of load and storage capacity to power systems, which will affects the strategic behaviors of each player. Equilibrium analysis of oligopoly electricity market is often used to research the strategic behaviors of generators and assessment of market power [7]. In this context, researching in the equilibrium models of electricity markets with EVs are identified as an important issue.

Until now, a considerable amount of research about EVs is mainly concentrated on the impacts of electric network security and dispatching. As reported in [8]-[10], a large number of EVs' uncontrolled charging will increase the power

---

* This project is supported by National Natural Science Foundation of China (No. 70871074).

K. Li et al. (Eds.): LSMS/ICSEE 2014, Part III, CCIS 463, pp. 175–185, 2014.
© Springer-Verlag Berlin Heidelberg 2014

distribution network loss and voltage deviation. Instead, the controlled charging can reduce the peak load demand and improve the above negative effects effectively. The different optimal dispatching strategy of EVs are presented and analyzed in [11]-[14]. The research shows that using the storage characteristics of EVs battery in optimal operation will contribute to improve the security and economy of power systems. It can be noted that to date, the researches concerning the impacts of EVs' charging and discharging on each participant's behaviors are still less.

According to the different competition modes of EVs, Cournot-Nash equilibrium models of electricity markets with EVs are presented in this paper [7],[15]. The influence of market prices curve is theoretically examined in detail, as EVs are price-taker. Numerical example is presented to verify the validity of the theoretical analysis and examine the impacts of EVs' charging and discharging on generators' strategic behaviors and market equilibrium.

2   Theoretical Model

2.1   Assumptions

Assuming that in a power market, there are n traditional strategic generators and a certain number of EVs. Then a given day is divided into 24 hours. The market demand in period t is expressed by the following linear demand function:

$$D_t = a_t - b_t p_t .$$ (1)

Where, p_t is the hourly market prices; D_t is the hourly power demand; a_t and b_t $(a_t, b_t \geq 0)$ are constant coefficients with given values greater than 0.

The conventional generators have the following quadratic cost functions in period t:

$$C_{i,t}(x_{i,t}) = \alpha_i x_{i,t} + 0.5\beta_i x_{i,t}^2, i = 1, 2, ..., n .$$ (2)

Where, $x_{i,t}$ is the generation of generator i in period t; α_i and β_i are cost parameters with given values greater than 0.

Assuming that a fleet of EVs connecting to power systems by V2G in period of low prices and discharges in period of high prices. The output of EVs fleet in period t is expressed by the following function:

$$x_{v,t} = -L_t + SO_t - SL_t .$$ (3)

Where L_t is the EVs' charging for driving in period t; SO_t and SL_t represent EVs' charging and discharging for arbitrage in period t respectively.

Since the market demand D_t in period t satisfies:

$$D_t = \sum_{i=1}^{n} x_{i,t} + x_{v,t} .$$ (4)

The demand function (1) can be reformulated as follows:

$$\sum_{i=1}^{n} x_{i,t} - L_t + SO_t - SL_t = a_t - b_t p_t .$$

(5)

During any given hour of the day, the amount of each hourly charging and discharging cannot exceed the battery available capacity. There is a certain energy loss in EVs fleet storage for arbitrage. Symbol η presents charging efficiencies of EVs, and $SL_t \eta$ is used to represent the actual energy for arbitrage. Further assume that utilizing EVs battery for arbitrage may lead to battery depreciation costs $vc \cdot SO_t$, where symbol vc is the variable costs of storage operation.

According to the different competition modes of EVs involved in electricity markets, three cases are used to discuss in this paper: Case 1, EVs fleet is a price-taker to participate in market competition; Case 2, EVs fleet is owned by a monopolist player without any generation capacity; Case 3, EVs fleet is owned by a monopolist traditional generator.

2.2   The Equilibrium Model

The following program formulates the optimization problem of each player:

$$\max_{\substack{x_{i,t} \\ L_t \\ SL_t \\ SO_t}} \sum_{t \in T} [p_t(x_{i,t} - L_t + SO_t - SL_t) - (\alpha_i x_{i,t} + 0.5\beta_i x_{i,t}^2) - vc \cdot SO_t] .$$

(6a)

$$s.t. \qquad\qquad 0 \le x_{i,t} \le \bar{x} \qquad\qquad (\lambda_{i,t}, \forall i, t) . \qquad (6b)$$

$$x_{i,t} - x_{i,t-1} - \xi_i^{up} \bar{x}_i \le 0 \qquad\qquad (\lambda_{i,t}^{rup}, \forall i, t) . \qquad (6c)$$

$$x_{i,t-1} - x_{i,t} - \xi_i^{down} \bar{x}_i \le 0 \qquad\qquad (\lambda_{i,t}^{rdo}, \forall i, t) . \qquad (6d)$$

$$\sum_{t \in d} L_t - Ld_d = 0 \qquad\qquad (\lambda_d^{Ld}, \forall d) . \qquad (6e)$$

$$L_t + L_t - \overline{st}^{in} \le 0 \qquad\qquad (\lambda_t^{si}, \forall t) . \qquad (6f)$$

$$SO_t - \overline{st}^{out} \le 0 \qquad\qquad (\lambda_t^{so}, \forall t) . \qquad (6g)$$

$$\sum_{\tau=1}^{t} SO_\tau - \sum_{\tau=1}^{t-1} SL_\tau \eta \le 0 \qquad\qquad (\lambda_t^{stlo}, \forall t) . \qquad (6h)$$

$$\sum_{\tau=t-1}^{t} SL_\tau \eta \le \overline{st}^{cap} - [\sum_{\tau=1}^{t-1} SL_\tau \eta - \sum_{\tau=1}^{t-1} SO_\tau](\lambda_t^{scap}, \forall t) \quad (\lambda_t^{scap}, \forall t) . \qquad (6i)$$

$$L_t, SL_t, SO_t \ge 0 \qquad\qquad \forall t . \qquad (6j)$$

Eq.(6a) represents the profit function faced by each player in the model. It includes revenues from selling electricity, which are either generated by firms $(p_t x_{i,t})$ or previously stored $(p_t SO_t)$. It also includes quadratic cost for generator, variable costs of storage operation (vc), costs of storage loading $(p_t SL_t)$, and costs of storage recharging $(p_t L_t)$.

Condition (6b) demands that player's electricity generation never exceeds its available generation capacity $\bar{x}_i$. Eq.(6c) is a 'ramping up'restriction: between two subsequent time periods, electricity generation can only be increased to a certain degree, depending on the total available capacity and a parameter ξ^{up}, which takes on values between 0 and 1. Likewise, Eq.(6d) represents a 'ramping down'restriction. Condition (6e) specifies the daily vehicle fleet requires a certain amount of energy for driving purpose: It can take place during any given hour of the day, or it could be split up over all 24 hours. Eq.(6f) ensures that the EVs battery loading rate never exceeds the available storage charging capacity $\overline{st}^{in}$ (in MW). Likewise, (6g) ensures that selling electricity back to the market from storage never exceeds the available storage discharging capacity $\overline{st}^{out}$. Eq.(6h) ensures that discharging never exceeds the net of previous storage charging and discharging. Eq. (6i) ensures that the amount of electricity used to be stored in period t cannot exceed the total available capacity of battery $\overline{st}^{cap}$ (in MWh), minus the net amount of previous storage. Eq.(6h) and (6i) include efficiency losses: only a share η of previously stored electricity can be sold back to the market. Finally, Eq.(6j) ensures non-negativity of the decision variables. $\lambda_{i,t}$, $\lambda_{i,t}^{rup}$, $\lambda_{i,t}^{rdo}$, λ_d^{Ld}, λ_t^{si}, λ_t^{so}, λ_t^{stlo}, λ_t^{scap} are dual variables.

The equilibrium model of electricity markets with EVs is composed of the optimization problem of each player (6a)-(6j) and the condition of market clearing (5).

2.3   Model Solution

From condition (5), p_t is solved and inserted into (6a). The resulting equation does no longer include the market prices, but only the production decisions of each player. Then each player's first-order optimality (Karush-Kuhn-Tucker, KKT) conditions are developed. The KKT conditions for each player's optimization problem can be expressed as follows:

$$
0 \leq -p_t + \lambda_{i,t} - \lambda_{i,t}^{rup} - \lambda_{i,t+1}^{rup} - \lambda_{i,t}^{rdo} + \lambda_{i,t+1}^{rdo} + \alpha_i + (\beta_i + \frac{1}{b_t})x_{i.t}+ \tag{7a}
$$
$$
\mu_i \frac{1}{b_t}[-L_t + SO_t - SL_t] \perp x_{i,t} \geq 0 \qquad \forall i,t .
$$

$$
0 \leq p_t + \lambda_d^{Ld} + \lambda_t^{si} - \psi\frac{1}{b_t}[-L_t + SO_t - SL_t] - \mu_i\frac{1}{b_t}x_{i,t} \perp L_t \geq 0 \quad \forall i,t . \tag{7b}
$$

$$
0 \leq vc + \lambda_t^{so} + \sum_{\tau=t}^{T}\lambda_\tau^{stlo} - \sum_{\tau=t}^{T-1}\lambda_{\tau+1}^{scap} - p_t + \psi\frac{1}{b_t}[-L_t+SO_t - SL_t]+ \tag{7c}
$$
$$
\mu_i\frac{1}{b_t}x_{i,t} \perp SO_t \geq 0 \qquad \forall i,t .
$$

$$
0 \leq p_t + \lambda_t^{si} - \sum_{\tau=t}^{T-1}\lambda_{\tau+1}^{stlo}\eta + \sum_{\tau=t}^{T}\lambda_\tau^{scap}\eta - \psi\frac{1}{b_t}[-L_t + SO_t - SL_t]- \tag{7d}
$$
$$
\mu_i\frac{1}{b_t}x_{i,t} \perp SI_t \geq 0 \qquad \forall i,t .
$$

$$0 \leq -x_{i.t} + \bar{x}_i \perp \lambda_{i,t} \geq 0 \qquad\qquad \forall i, t \,. \qquad (7\text{e})$$

$$0 \leq -x_{i.t} + x_{i.t-1} + \xi_i^{up}\bar{x}_i \perp \lambda_{i,t}^{rup} \geq 0 \qquad\qquad \forall i, t \,. \qquad (7\text{f})$$

$$0 \leq -x_{i.t-1} + x_{i.t} + \xi_i^{down}\bar{x}_i \perp \lambda_{i,t}^{rdo} \geq 0 \qquad\qquad \forall i, t \,. \qquad (7\text{g})$$

$$\sum_{t \in d} L_t - Ld_d = 0 \qquad\qquad \lambda_d^{Ld} free \,. \qquad (7\text{h})$$

$$0 \leq -SL_t - L_t + \overline{st}^{in} \perp \lambda_t^{si} \geq 0 \qquad\qquad \forall t \,. \qquad (7\text{i})$$

$$0 \leq -SO_t + \overline{st}^{out} \perp \lambda_t^{so} \geq 0 \qquad\qquad \forall t \,. \qquad (7\text{j})$$

$$0 \leq -\sum_{\tau=1}^{t} SO_\tau + \sum_{\tau=1}^{t-1} SL_\tau\eta \perp \lambda_t^{stlo} \geq 0 \qquad\qquad \forall t \,. \qquad (7\text{k})$$

$$0 \leq -\sum_{\tau=1}^{t} SL_\tau\eta + \sum_{\tau=1}^{t-1} SO_\tau + \overline{st}^{cap} \perp \lambda_t^{scap} \geq 0 \qquad\qquad \forall t \,. \qquad (7\text{l})$$

$$a_t - b_t p_t = \sum_i x_{i,t} - L_t + SO_t - SL_t \qquad\qquad p_t \ free \,. \qquad (7\text{m})$$

Using a special nonlinear complementarity function, these KKT conditions are reformulated as a set of nonlinear algebraic equations. Then the market equilibrium can be obtained by solving these equations.

These conditions include market parameters ψ, and μ_i. $\psi = 0$ indicates that the EVs operator is a price-taker, otherwise $\psi = 1$. $\mu_i = 0$ indicates that generator i dose not operate EVs fleet, otherwise $\mu_i = 1$.

Through theoretical analysis in the appendix, it can find that market prices curve will reach the effect of "peak load shifting", when EVs fleet is a price-taker for arbitrage.

3   Numerical Example

3.1   Date

A day is divided into 24 hours in this example. The coefficients of demand function are given to simulate the demand curve in electricity markets: $b_t = b = 0.4(\text{MWh}/\$)$, a_t is shown in Table 1 and Fig. 1.

Table 1. Parameter a_t

Time	1	2	3	4	5	6	7	8	9	10	11	12
a_t(MW)	40	35	32	30	35	40	60	75	90	100	110	115
Time	13	14	15	16	17	18	19	20	21	22	23	24
a_t(MW)	123	127	125	123	120	118	120	125	120	100	72	60

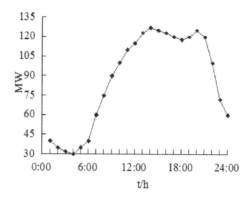

Fig. 1. Parameter a_t

Table 2. The Parameters of Generator

Generator	$\bar{x}_i$(MW)	ξ^{up}	ξ^{down}	α_i($/MWh)	β_i($/(MW)^{2}h)
1	150	0.22	0.18	10	1.5
2	170	0.22	0.18	15	1.0

The electricity market includes three players, among them two are oligopolistic generators. The remaining is an EVs operator without any generation capacity. Table 2 lists the parameters of oligopolistic generator.

The amount of average daily recharging requirement (Ld_d) is 20MWh. The available storage charging capacity is around 20MW ($\overline{st}^{in}$), as well as discharging capacity ($\overline{st}^{out}$). The total available capacity of battery is around 40MW ($\overline{st}^{cap}$). Furthermore assume a round-trip efficiency of $\eta = 0.8$ for EVs battery.

In Case 1, EVs fleet is a price-taker; In Case 2, EVs fleet is owned by EVs operator; In Case 3, EVs fleet is owned by generator 1.

The Baseline which does not include any EVs is defined in order to serve as a point of reference.

3.2   The Effect of EVs on Market Equilibrium

Fig.2 and Fig.3-Fig.5 indicate the impacts of EVs on market prices and the strategic behaviors of each player in each case respectively. EVs fleet charges as consumer during off-peak period and discharges as suppliers during peak period.

It can be seen from Fig.2 that compared to the Baseline, the smoothness of market prices is significantly improved after introduction of EVs. Besides, the degree of improvement is related to the competition modes of EVs fleet.

Combined with Fig.3-Fig.5, it can find that the remaining battery capacity is fully used when EVs operator is a price-taker (Case 1). In this case, the off-peak prices are the highest because of the highest additional electricity demand.

Conversely, the peak prices are lowest owing to the highest electricity supply. Besides, the market prices curve achieves the effect of "peak load shifting".

In Case 2, the EVs operator withholds some battery capacity during off-peak period. As a result, the off-peak prices are lower, and the peak prices are higher than Case1.

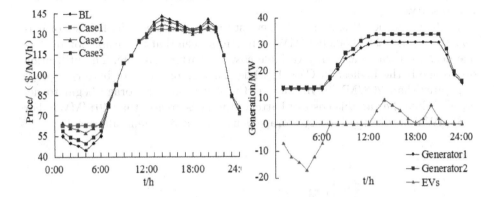

Fig. 2. Market prices in equilibrium under different case

Fig. 3. Players strategy in equilibrium under case 1

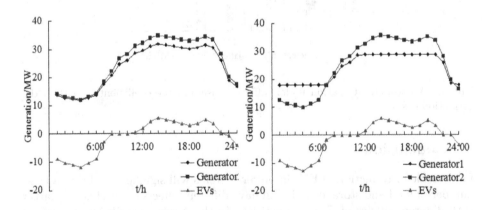

Fig. 4. Players strategy in equilibrium under case 2

Fig. 5. Players strategy in equilibrium under case 3

In Case 3, the generator 1 substantially increases its generation in order to strategically decrease market prices during the period of EVs charging. Furthermore generator 1 decreases its generation for obtaining more arbitrage profits during the period of EVs discharging. For this reason, the smoothness of its market prices curve is the worst.

3.3 Sensitivity Analysis of Battery Degradation

All of the previous analyses is to assume zero variable costs of battery storage ($vc = 0$). In actuality, it cannot be ignored. However, it is difficult to find a fixed value to describe it, because of battery depreciation costs depend on battery technology, the depth of discharging, and the kind of charging and discharging cycles heavily[16],[17]. Therefore, the sensitivity analysis of battery degradation is necessary.

Fig.6 shows the utilization of battery storage while the variable costs of battery storage is 0,20,40,60,80($/MWh). It can be seen that the utilization of storage decreases with increasing variable costs of battery storage. The utilization of storage is the highest in Case 1, while the variable costs of battery storage is greater than 20$/MWh. In Case 1, the utilization of storage begin to decrease, while the variable costs of battery storage is greater than 40$/MWh. For 80$/MWh, hardly any battery storage is used for arbitrage in all cases.

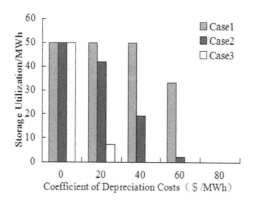

Fig. 6. EVs storage utilization under different case and the coefficient of battery depreciation costs

4 Conclusion

Large-scale introduction of EVs in power systems will significantly bring a large number of load and storage. This will have different effects on electricity market with different modes of EVs competition. In this paper, equilibrium models of electricity market with EVs are established and theoretical analysis is made to prove that the electricity prices become leveled during off-peak and peak periods while EVs fleet is a price-taker. Numerical examples are presented to verify the validity of the theoretical analysis and the effectiveness of the proposed models. The main work and key contributions of this paper are as follows:

1. Introduction of EVs fleet with controlled charging and discharging on electricity market can smooth the price curve. In addition, its smooth effect depends on the competition mode of EV.

2. If EVs fleet is a price-taker, the smoothness of market prices curve is best by making full use of storage, and achieves the effect of "peak load shifting". If EVs fleet is an oligopoly without any generation capacity, it may withhold some storage capacity in order to obtain more profits. If EVs fleet is owned by an oligopolistic generator, it not only withholds some storage capacity but also increases generation during off-peak period and reduces generation during peak period, which makes the worst price-smoothing effect.

3. The utilization of battery decreases with increasing degradation costs. High battery degradation costs can hardly bring the price-smoothing effect.

References

1. Oksanen, M., Karjalainen, R., Viljainen, S., Kuleshov, D.: Electricity markets in Russia, the US, and Europe. In: 6th International Conference on the European Energy Market, pp. 1–7. IEEE Press, Leuven (2009)
2. Kirschen, D.S.: Demand-side view of electricity markets. IEEE Trans on Power Systems 18(2), 520–527 (2003)
3. Yu, E.K., Xie, K., Han, F., et al.: A Preliminary Introduction to Electric Power Market. Power System Technology 19(3), 58–62 (1995)
4. Lopes, J.A.P., Soares, F.J., Almeida, P.M.R.: Integration of electric vehicles in the electric power system. Proceedings of the IEEE 99(1), 168–183 (2011)
5. Pantos, M.: Exploitation of Electric-Drive Vehicles in Electricity Markets. IEEE Trans. on Power Systems 27(2), 682–694 (2012)
6. Liu, X.F., Zhang, Q.F., Cui, S.M.: Review of Electric Vehicle V2G Technology. Transactions of China Electro Technical Society 27(2), 121–127 (2012)
7. Zhang, S.H., Wang, X., Kang, X.N., Li, Y.Z.: Integration of Gencos Risk Preference in Cournot Equilibrium Analysis of Electricity Markets. Power System Technology 35(9), 176–180 (2010)
8. Clement-Nyns, K., Haesen, E., Driesen, J.: The Impact of Charging Plug-In Hybrid Electric Vehicles on a Residential Distribution Grid. IEEE Trans. on Power Systems 25(1), 371–380 (2010)
9. Schneider, K., Gerkensmeyer, C., Kintner-Meyer, M., Fletcher, R.: Impact assessment of plug-in hybrid vehicles on pacific northwest distribution systems. In: Power and Energy Society General Meeting-Conversion and Delivery of Electrical Energy in the 21st Century, pp. 1–6. IEEE Press, Pittsburgh (2008)
10. Gong, X., Lin, T., Su, B.H.: Impact of Plug-in Hybrid Electric Vehicle Charging on Power Distribution Network. Power System Technology 36(11), 30–35 (2012)
11. Wang, X.F., Shao, C.C., Wang, X.L., Du, C.: Survey of Electric Vehicle Charging Load and Dispatch Control Strategies. Proceedings of the CSEE 33(1), 1–10 (2013)
12. Yao, W.F., Zhao, J.H., Wen, F.S., Xue, Y.S., et al.: A Charging and Discharging Dispatching Strategy for Electric Vehicles Based on Bi-level Optimization. Automation of Electric Power Systems 36(11), 30–37 (2012)
13. Luo, Z.W., Hu, Z.C., Song, Y.H., Xu, Z.W., et al.: Coordinated Charging and Discharging of Large-scale Plug-in Electric Vehicles with Cost and Capacity Benefit Analysis. Automation of Electric Power Systems 36(10), 19–26 (2012)
14. Yang, J.Q.: The Prediction of Electric Vehicles Charge and Discharge Capacity and The Study on the Control Strategy of V2G. Beijing Jiao Tong University, Beijing (2012)

15. Yang, J.L.: The Study for the Equilibrium Models of the Electricity Market and its Related Algorithms. Shanghai Jiao Tong University, Shanghai (2010)
16. Sovacool, B.K., Hirsh, R.F.: Beyond batteries: An examination of the benefits and barriers to plug-in hybrid electric vehicles (PHEVs) and a vehicle-to-grid (V2G) transition. Energy Policy 37(3), 1095–1103 (2009)
17. Peterson, S.B., Whitacre, J.F., Apt, J.: The economics of using plug-in hybrid electric vehicle battery packs for grid storage. Journal of Power Sources 195(8), 2377–2384 (2010)

Appendix

EVs fleet is a price-taker. Suppose that T is the set of 24 hours. $T1$ is the set of periods of charging. $T2$ is the set of period of discharging. $T3$ is the set of period of no charging and discharging. So $T1 + T2 + T3 = T$.

a).Period of charging

The expression of market prices can be obtained by Eq.(7b) and Eq.(7d):

$$p_t = \lambda_d^{Ld} + \lambda_t^{si} \tag{8}$$

$$p_t = -\lambda_t^{si} + \sum_{\tau=t}^{T-1} \lambda_{\tau+1}^{stlo}\eta - \sum_{\tau=t}^{T} \lambda_{\tau}^{scap}\eta \tag{9}$$

From the Eq.(7i), it can be found that if $\overline{st}^{in} > SL_t + L_t$, λ_t^{si} equals to 0. This means that battery hourly charging capacity is so large that cannot be reached, when $t \in T$.

From the Eq.(7k), it can find that when

$$-\sum_{\tau=1}^{t} SO_\tau + \sum_{\tau=1}^{t-1} SL_\tau\eta > 0 \tag{10}$$

λ_t^{stlo} equals to 0.

The Eq.(10) indicates that discharging never exceeds the net amount of previous storage, which only occurs to the last discharging. That is to say when t is before the period of last discharging, $\lambda_t^{stlo} = 0$, otherwise $\lambda_t^{stlo} > 0$. Owing to η is a constant, $\sum_{\tau=t}^{T-1} \lambda_{\tau+1}^{stlo}\eta$ equaling to $\sum_{\tau=Tl}^{T-1} \lambda_{\tau+1}^{stlo}\eta$ (Tl is the last period before discharge) is a constant, when $t \in T1$.

From the Eq.(7l), it can find that when

$$-\sum_{\tau=1}^{t} SL_\tau\eta + \sum_{\tau=1}^{t-1} SO_\tau + \overline{st}^{cap} > 0 \tag{11}$$

λ_t^{scap} equals to 0.

The Eq.(11) indicates that the amount of electricity used to be stored in period t cannot exceed the total available capacity of battery, minus the net amount of previous storage. That is to say when $t \in T3$, $\lambda_t^{scap} > 0$, otherwise $\lambda_t^{scap} = 0$. So $\sum\limits_{\tau=t}^{T} \lambda_\tau^{scap} \eta$ is a constant when $t \in T1$.

As λ_d^{Ld} is a constant and market prices is a continuous function, the market prices is a constant in period of EVs' charging, when $\overline{st}^{in}$ is so large that quantity of discharging cannot reach it. The market prices will completely be filled during this period.

b).Period of discharging

The expression of market prices can be obtained by Eq.(7c):

$$p_t = vc + \lambda_t^{so} + \sum_{\tau=t}^{T} \lambda_\tau^{stlo} - \sum_{\tau=t}^{T-1} \lambda_{\tau+1}^{scap} \tag{12}$$

From the Eq.(7j), it can find that if $\overline{st}^{out} > SO_t$, λ_t^{so} equals to 0. This situation will happen when battery hourly discharging capacity is so large that cannot be reached, when $t \in T$.

According to the above, it can be found that when t is in period of discharging, $\sum\limits_{\tau=t}^{T} \lambda_\tau^{stlo}$ equals to $\sum\limits_{\tau=Td}^{T} \lambda_\tau^{stlo}$ (Td is the last period of discharging), which is a constant, $\sum\limits_{\tau=t}^{T-1} \lambda_{\tau+1}^{scap}$ and equals to 0. Besides, vc is a constant, the market prices is a constant in period of EVs' discharging,when $\overline{st}^{out}$ is so large that quantity of discharging cannot reach it. The peak of market prices will be clipped.

To sum up, the "peak load shifting" will happen when EVs fleet is a price-taker and $\overline{st}^{in}$ and $\overline{st}^{out}$ are large enough in this model.

A Novel Quantum Particle Swarm Optimization for Power Grid with Plug-In Electric Vehicles in Shanghai

Jinwei Gu[1], Manzhan Gu[2], and Quansheng Shi[1]

[1] School of Economics and Management, Shanghai University of Electric Power,
Yangpu District, Shanghai, 20090, China
[2] Department of Applied Mathematics, Shanghai University of Finance
and Economics, Yangpu District, Shanghai, 200433, China

Abstract. This paper studies the plug-in electric vehicles charging/discharging mode under the intelligent power grid in Shanghai with the objective of minimizing the total mean square of load curve of charging and discharging electricity. Considering constrains on battery capacity, electricity power and available time, an electric vehicles charging/discharging optimization model is build for power grid in Shanghai. Based on the parasitic and anti-parasitic behaviors in the nature, we propose a Novel Quantum Particle Swarm Optimization (NQPSO) to solve the problem. Two populations - host group and parasitic group are generated to dynamically changing the population size within the parasitic mechanism so as to improve the population genetic. A quantum particle encoding method is designed according the characteristics of the problem. Finally we apply NQPSO upon instances to explore the performance of our algorithm, and the results have showed the computational evidence for its effectiveness.

Keywords: smart power grids, electric vehicle, charge and discharge mode, quantum particle swarm optimization.

1   Introduction

Recently, the countries all over the the world are facing the common problems of energy shortage and environmental deterioration, and the energy saving and emission reduction are in sight and vocal. Under this background, the plug-in electric vehicles (PEV) come on stage with the properties of saving energy and protecting environment. For this reason, the industry has kept developing and growing in recent years, and the number of PEV increases year by year.

The PEV are connected to the power grid by the distributed energy storage element - automobile battery. The electricity needed by the vehicles is the power load in the grid, and it will bring additional enormous pressure to the power grid and transformers. On the other hand, PEV also act as power storage "batteries", which will improve the capacity and reliability of power grid. Therefore, it is of realistic

K. Li et al. (Eds.): LSMS/ICSEE 2014, Part III, CCIS 463, pp. 186–197, 2014.

significance and research value to realize the intelligent management of charging/ discharging for PEV, i.e. charging the vehicles in the "Valley" and discharging in the "Peak".

Some intelligent algorithms have been proposed for the PEV charging/ discharging problems. In 2011, based on real-time price Zou et al. [2] defined a PEV centralized charging mechanism. Activated by the idea of dynamic estimation interpolation, they proposed an algorithm to fill valley on electricity-supply side and balance the cost paid by users. In addition, Zhao et al. [3] researched the electric vehicle scheduling problem and designed an algorithm for the problem. They also assessed the economic benefits of electric vehicles joining in the power grid, and gave their advices. From the perspective of economy, the paper [4] studies the setting where we could cut peak and fill valley for the power grid with PEV. Our paper focuses on the power grid with PEV in Shanghai, and research how to use the pricing of electricity to manage the charging/discharging of PEV, and finally optimize the PEV charging/discharging mode in power grid.

We are interested in applying Particle Swarm Optimization (PSO) to solve the problem in this paper. PSO is an effective tool to solve many optimization problems. Because of its properties: simplicity and effectiveness, the algorithm attracts the attention of many researchers, and is applied upon problems in many fields. In literature [5], [6], the authors proposed modified PSO to forecast the electricity load of PEV. Based on numerical experiments, they proved the effectiveness of the algorithm. Hutson et al. [7] researched the plug-in hybrid electric vehicles problem with the objective to maximize the profit of vehicle owners. They designed a binary-based PSO and found that the algorithm accurately finds near optimal solutions and significantly increases the potential profits for the vehicle owners.

However, the experimental results show that traditional PSO is unsatisfactory in some situations. One of the reasons is that too much attention is focused on the strategies of single population improvement, while the mutual influence within populations and the influence between populations and external environment are ignored. Built on the natural parasitic behavior mechanism where multiple populations influence each other, we propose a Novel Quantum Particle Swarm Optimization (NQPSO). We define two populations named as host group and parasitic group, and define a population evolution model for the problem. In this model the sizes of the two populations are updated dynamically by simulating parasitic and anti-parasitic relationships in the nature. This idea could help to improve the population diversity, speed up the convergence and overcome the premature phenomenon. In the end of the paper, we verify the feasibility and effectiveness of NQPSO by numerical experiments.

2   Problem and Model

2.1   Objective Function

Define each time period consists of 24 time pieces, and for each vehicle the power of charging/discharging is a variable. By considering PEV in power grid, the peak of the load curve could be cut and the valley could be filled, which helps to smooth the fluctuation of load curve. This paper considers the optimization problem with PEV in power grid with the objective to minimize the total mean square of load curve, i.e.

$$\min \ S = \sum_{j=1}^{24} \left(P_j + \sum_{i=1}^{n} P_{ij} - \sum_{j=1}^{24} P_j \Big/ 24 \right)^2 \tag{1}$$

where S denotes the total cost, n is the number of vehicles. P_j is the power in j-th time piece, i.e. the original value of power before vehicles join in. $\sum_{j=1}^{24} P_j \Big/ 24$ denotes the average value of power of the 24 time pieces, and P_{ij} is the power capacity of the j-th time for the i-th vehicle.

2.2   Constraints

(1) *Power Load*
For the j-th time piece, we define the maximum current for charging is $I_i/3$, and $2I_i$ for discharging. Circuit power is no more than 15KW. So we have

$$\min\left(15, V_{ij} \times 2I_i\right) \le P_{ij} \le \max\left(-15, -V_{ij} \times I_i/3\right) \tag{2}$$

where V_{ij} is the real-time voltage of battery, I_i is the rated current, depending on the type of vehicle i.

(2) *Quantity of electricity*
Denote by ΔQ_{ij} the modified quantity of the electricity in the battery of vehicle i in the j-th time piece. We define

$$\left(S_{socij} - S_{\min}\right) Q_{i\max} \le \Delta Q_{ij} \le \left(S_{socij} - S_{\max}\right) Q_{i\max} \tag{3}$$

where $Q_{i\max}$ is the maximum capacity of the battery of vehicle i. S_{socij} denotes the state of charge (SOC) of vehicle i in the j-th time piece, where SOC is the ratio of remaining amount of electricity over maximum capacity. $S_{\min}$ and $S_{\max}$ respectively denote the minimum and maximum values of SOC, usually taking values 0.2 and 1.

(3) *Fluctuation of the Power*

$$Y = \sum_{i=1}^{n} \sum_{j=1}^{24} \left[\tilde{I}_{ij} P_{ij} p_j - \overline{I}_{ij} P_{ij} \left(p_j + \gamma - \eta \right) \right] \le H \tag{4}$$

Where Y denotes the total cost. We define two indicators $\tilde{I}_{ij}, \overline{I}_{ij}$. $\tilde{I}_{ij} = 1$ if the i-th vehicle is charging in the j-th time piece, otherwise $\tilde{I}_{ij} = 0$. $\overline{I}_{ij} = 1$ if the i-th vehicle is discharging in the j-th time piece, otherwise $\overline{I}_{ij} = 0$. p_j represents the price of electricity in j-th time piece, γ denotes the government subsidy for each unit of electricity discharged by vehicles, and η denotes the loss ratio of batteries. H is an upper bound on the cost of power, which is a given constant.

The price of the electricity for household use in Shanghai is shown in Table 1. According to statistics, a vehicle in Shanghai travels averagely 39 km each day. Assume each vehicle consumes electricity 0.215KW·h per mile, then 251.55 KW·h is needed for a vehicle each month. The amount of electricity used for other purposes is assumed to be 110-150 KW·h for each family. Then the total consumption of electricity must be out of the first power range [0, 260). Based on the principle of saving energy, the total electricity consumption each month lies in interval [261, 400). Therefore, we have $p_j = 0.677$ for peak period and $p_j = 0.647$ for valley period.

Table 1. Fee standards for Shanghai household electricity

Quantity of electricity (Monthly)	Peak Period (6:00 am-22:00 pm)	Valley Period (22:00 pm-6:00 am)
0-260	0.617 RMB/ KW·h	0.307 RMB/ KW·h
261-400	0.677 RMB/ KW·h	0.647 RMB/ KW·h
≥ 400	0.977 RMB/ KW·h	0.797 RMB/ KW·h

For the fee involving charging and discharging, let $\gamma - \eta = 0.6$, i.e. by discharging 1 KW·h electricity, vehicle owners could get 0.6 RMB from the government subsidy after deducting battery cost. The data 0.6 RMB is almost like the price of electricity in the valley period of the interval (261,400).

(4) *Available Time*

For PEV owners, it is common that charging/discharging operation is carried out after work hours. We assume each family charges/discharges one time each day. Since government gives supply subsidy during peak period, we define PEV owners choose to discharge electricity during the peak period, i.e. from 18 pm to 22 pm, and charge electricity in the valley period from 22 pm to 6 am. We give the following inequalities:

$$18 \leq T_{s1} \leq T_{f1} \leq 22 \tag{5}$$

$$22 \leq T_{s2} \leq T_{f2} \leq 30 \tag{6}$$

where T_{s1} and T_{f1} respectively denotes the start and completion times of discharging operation, and T_{s2} and T_{f2} are start and completion times of charging operation.

3   The Parasitic Mechanism

3.1   Growth Model for Parasite Population

During the long history of nature evolutionary, the relationship between different populations is complex. The phenomenon that two populations live together is generally referred to as *symbiosis*. This paper studies a type of symbiosis named *parasitic*, which refers to two populations of birds living in the same environment. One population obtains benefit from the other one, and the other (called *host*) provides nutrition and residences for the former (called *parasite*).

Generally, the parasite try to put their eggs into the host's nest, and the young will also damage eggs of host in the nest, reducing the survival rate of host, or even completely killing chicks of host. On the other hand, the host will try to improve their immunity to eliminate eggs of parasite. The relationship between host and parasite is a dynamic development process, similar to a long "arms race". After a long period of evolution, some of the features are to be preserved within the interaction between parasite and the host, and then reflected on genes of populations.

In the field of mathematical biology, the parasitic relationship between host and parasite has always been a hot topic, but few papers consider applying parasitic mechanism into optimization algorithms. This paper will try to simulate the parasitic relationship to improve the population genetic of the optimization algorithm. The parasitic behavior of two populations can be defined by the following differential equations model, denoted by the following model (7).

$$
\begin{cases}
\dfrac{dx_1}{dt} = r_1(t)x_2(t) - \dfrac{r_1(t)x_1(t)^2}{K_1} - k(t)x_1(t)y_2(t)x_1(t)^2 \\[3mm]
\dfrac{dx_2}{dt} = k(t)x_1(t)y_2(t) - \dfrac{r_2(t)x_1(t)^2}{K_2} \\[3mm]
\dfrac{dy_1}{dt} = r_3(t)y_2(t) - \dfrac{r_3(t)y_1(t)^2}{K_3} - c(t)y_1(t) - p(t)x_1(t)y_1(t) \\[3mm]
\dfrac{dy_2}{dt} = c(t)y_1(t) - \dfrac{r_4(t)y_2(t)^2}{K_4}
\end{cases}
\tag{7}
$$

Here, x_1, x_2 respectively denote the numbers of young individuals of parasite (YIP) and adult individuals of parasite (AIP). y_1, y_2 are the numbers of young individuals of host (YIH) and adult individuals of host (AIH) respectively. K_1, K_2, K_3, K_4 are the largest possible numbers of the corresponding individuals, and r_1, r_2, r_3, r_4 denote the growing rates. k denotes the proportion of the number of parasite individuals fed by host, and c is the proportion of number of host individuals growing into adult. p is the proportion of the number of YIH killed by parasite.

3.2   The Nest Parasitism Schema

In this section, we will introduce the mechanisms how the number of host individuals and parasite individuals are updated. Firstly, we define the fitness of an individual (chromosome) is the reciprocal of the objective function value. For a fight between two individuals of host and parasite, we define the one with lower fitness is defeated. Next we introduce the four mechanisms, and describe how the numbers of individuals are updated.

(1) *Parasitic mechanism*

Assume parasite is in the dominant position. If some AIHs are defeated, then we will put YIPs host's nest, and the YIPs will absorb host's nutrition and create damage to the host.

If $(x_2 \le y_2)$

For each AIP, assign $N_s = \lceil x_2 / y_2 \rceil$ AIHs to fight against it. Denote by n_i the number of AIHs who are defeated. Obviously, the x_2-th AIP may have less than N_s AIHs opponents. Define the defeated rate of host is $C_1 = \sum_{i=1}^{x_2} n_i / y_2$, and the successful rate of parasite is $C_2 = \sum_{i=1}^{x_2} (n_i / N_s) / x_2$. In this case, we choose $\lceil C_2 x_1 \rceil$ YIPs to migrate into the population of YIHs.

Else

Noting the number of AIHs is limited, we let some AIPs left with no opponent while each of the remaining AIPs has one AIH to fight. Let $C_1 = \sum_{i=1}^{x_2} n_i / y_2$, $C_2 = \sum_{i=1}^{x_2} (n_i / N_s) / x_2$, where denotes the number of AIPs, each of which has an opponent. We still choose $\lceil C_2 x_1 \rceil$ YIPs to migrate into the population of YIHs.

End

(2) *Killing young host mechanism*

After YIPs migrate into the population of YIHs, YIPs may create damage to YIHs, which is called the killing young host mechanism. We define a half of YIHs with lower fitness are killed by YIPs, and the same amount of new YIHs are initialized in order to keep the same amount of YIHs in the current generation.

(3) *Counter parasitic mechanism*

We define each AIH has the ability or to clean up YIPs or to reject the migrating of YIPs, in order to enhance its immunity. We allow removing of a YIP under the situation where the fitness of an AIH is greater than that of the YIP, then the AIH could remove the YIP with the probability C_3, where $C_3 = \sum_{i=1}^{x_2} (N_s - n_i) / y_2$. Thus we can remove about $\lceil C_3 x_1 \rceil$ YIPs.

(4) *Growing mechanism*

The growing mechanism exists both in parasite populations and the host population. In the parasite population, the youth individuals are all grow up to be adult individuals.

In the host population, we regard the growing mechanism as a migration process. Traditionally, the ratio of the YIHs growing up to be AIHs is fixed. But this approach is not flexible and cannot follow the dynamic migration situation. In order to improve the efficiency of our algorithm, we propose a new migration strategy. Let fit_1, fit_2 respectively be the minimal fitness values of AIHs and YIHs, and $\overline{f_1}, \overline{f_2}$ be the average fitness values of AIHs and YIHs. In the evolution of every generation, the important work is to determine the growing up ratio λ, which can be calculated as follows:

$$\lambda = \begin{cases} \dfrac{\min\left\{\left|fit_1 - fit_2\right|, \left|\overline{f_1} - \overline{f_2}\right|\right\}}{\max\left\{\left|fit_1 - fit_2\right|, \left|\overline{f_1} - \overline{f_2}\right|\right\}}, & fit_1 > fit_2 \ \& \ \overline{f_1} > \overline{f_2} \\[3mm] \dfrac{\left|fit_1 - fit_2\right|}{\max\left\{fit_1, fit_2\right\}} & , fit_1 > fit_2 \ \& \ \overline{f_1} \le \overline{f_2} \ or \ fit_1 \le fit_2 \ \& \ \overline{f_1} > \overline{f_2} \\[3mm] 0 & , else \end{cases} \qquad (8)$$

This mechanism has the property of spreading good genes to AIHs, which increases the host's ability to fight against parasitic individuals.

4   The Novel Quantum Particle Swarm Optimization

4.1   The Quantum Particle Coding

In order to express more states of the particle, we learn the characteristics of superposition and probability expression of quantum theory, and introduce the qubit representation (see Fig. 1). Each dimension for quantum particles is $24n$, where n represents the number of electric vehicles. $q_{ij} = \left[\alpha_{ij}, \beta_{ij}\right]^T (i = 1, 2, ..., n, j = 1, 2..., 24)$ is a qubit. We only consider non-working hours, i.e. from 18 pm to 6 am in the next day. Therefore, the quantum coding dimension of the particle can be shorten to 12 n. Map the quantum particles to space domain of the problem, then mapping relation can be expressed as $q_{ij} = q_{ij}^2 \cdot (ub - lb) + lb$.

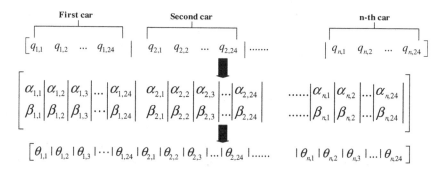

Fig. 1. Coding method of quantum particle

4.2   Updating Operation of Quantum Particle

Base on the speed updating method in the basic PSO, we could calculate the amount of *updating phase shift* of quantum particles by the following formula:

$$\Delta\theta_{ij}^{t+1} = \omega \times \Delta\theta_{ij}^t + C_1 \cdot r_1^t \cdot (\theta_{Gbest} - \theta_{ij}) + C_2 \cdot r_2^t \cdot (\theta_{Pbest} - \theta_{ij}) \qquad (9)$$

where $\Delta\theta_{ij}^t$ represent the amount of *updating phase shift* of the j-th dimension in i-th particle quantum at the t-th iteration. θ_{ij} is the current phase. θ_{Gbest} and θ_{Pbest} respectively denote the best phases in the current generation and in history. ω is an inertia weight coefficient and C_1 and C_2 are acceleration coefficients. Then update the quantum particles according to the formula below:

$$\begin{bmatrix} \alpha_{ij}^{t+1} \\ \beta_{ij}^{t+1} \end{bmatrix} = \begin{bmatrix} \cos\Delta\theta_{ij}^{t+1} & -\sin\Delta\theta_{ij}^{t+1} \\ \sin\Delta\theta_{ij}^{t+1} & \cos\Delta\theta_{ij}^{t+1} \end{bmatrix} \begin{bmatrix} \alpha_{ij}^{t} \\ \beta_{ij}^{t} \end{bmatrix}, \tag{10}$$

where $\left(\alpha_{ij}^{t}, \beta_{ij}^{t}\right)$ represents the j-th probability amplitude of the t-th iteration.

4.3   The Main Procedure of NQPSO

The process of novel quantum particle swarm optimization (NQPSO) is shown in Fig. 2.

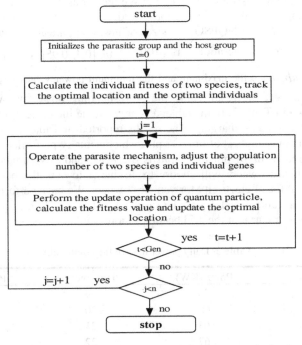

Fig. 2. The flow chart of novel quantum particle swarm optimization (NQPSO)

5   Computational Results

Assume the number of PEV is limited, the load of electricity which could be adjusted is small, and it has little influence on the whole Shanghai area, but still can improve the load curve of a community area. In contrast, for the problem with many vehicles, if each vehicle has its own charging/discharging power capacity, and the available time varies, then it is a complicated, difficult and high-demission problem. Therefore, we assume the whole Shanghai area can be divided into several research centers, and the management of charging/discharging could be carried out in each center, which could finally realize the optimization of PEV charging/discharging in power grid in Shanghai area. Hence this experiment is divided into two parts: a community, an area consisting of several communities.

5.1   Charging/Discharging PEV in a Community

For the small PEV, let $I_i = 80A$, $V_{ij} = V_{i\min}$ and $Q_{i\max} = V_{i\min} \cdot I_i$. Considering 5 vehicles, define the initial capacity of battery $S_{SOCi} = 0.3(i = 1, 2, ...)$. The daily power load of the community is shown in Table 2. We only consider the time period from 18 pm to 6 am, and the time period is divided into 12 time pieces (No. 1,2,…,6,19,20,…,24). In **NQPSO**, the sizes of host and parasite populations are 100, and the number of iterations is 1000. $\omega \in [0.4, 0.95]$, $C_1 = C_2 = 2$ and $r_1^t, r_2^t \in [0,1]$. ub, lb are the maximum and the minimum power values respectively. Based on NQPSO, we get the results, and the corresponding power per vehicle in each period is shown in Table 3. In the table, a positive value indicates discharging while a negative value indicates charging. The results reported in Table 3 shows that the all vehicles discharging highest at time 22 pm, which actually performs "fill the valley". The time period from 18 pm to 22 pm is the peak, and government give subsidy. Therefore, vehicles owners will choose to discharge to "cut the peak", which also reduces the total fee of electricity for vehicles owners. We have performed an instance with the data in Table 2 and 3, and the results are shown in Fig. 3. In the figure, we find that the power curve of the grid becomes more stable than before.

Table 2. Daily power load of the community

Time piece	Power(KW)	Time piece	Power(KW)
1	66	99	106
2	66	20	98
3	63	21	104
4	67	22	93
5	69	23	70
6	68	24	66

Table 3. The charging/discharging power load of the PEV

No. of Vehicle	Power (W) of time pieces											
	1	2	3	4	5	6	19	20	21	22	23	24
1	-3	-5	-3.9	-5	-4	-1	5	2	3	0	-4.1	-2.9
2	-6	-6	-9	-2.7	-3	-3	8.3	3.3	1.3	0.3	0	-3
3	-7	-4	-1	-3	-2	-1	2.3	2	6	0	-5.5	-5
4	-2.7	-5	-6	-3	-3	-1	3	3	0	0	0	-4
5	0	0	0	0	0	0	0	2	6.1	2.7	-1	-7.5

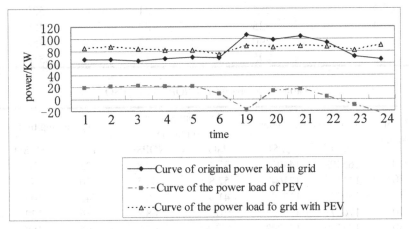

Fig. 3. Curves of power load in a community

5.2   A Charging/Discharging PEV in an Area Consisting of Communities

In this section, we consider the management of charging/discharging PEV in an area consists of several communities, and the values of power in time pieces are given in Table 4. Obviously, compared to the situation in a community, the power load in this section is larger. In this way, more vehicles are needed, and the improvement of power curve is bettered. But the cost of charging/discharging will also be increased. We perform NQPSO and particle swarm optimization (PSO) upon instances with different number of vehicles (10vehicles, 20 vehicles, 30 vehicles, 40 vehicles and 50 vehicles). The results are reported in Table 5.

For each instance, the two algorithms are ran 10 times, and we obtain the average objective value and average convergence generation. The results in Table 5 imply that compared to PSO, NQPSO performs better on the perspectives of average objective function value and average convergence generation. The reason is when dealing with instance with high dimension, PSO is easy to fall into a local optimum and get prematurity. But the algorithm NQPSO always has smaller fitness value and a faster convergence speed. By parasitic mechanism, NQPSO carries out the explicit key selection in the neighborhood of each solution, which improves the speed and performance of NQPSO. In addition, we use quantum and particle updating operation to search solutions, so the whole population will give sub-populations a momentum when the whole population is trapped in a local optimum, and help to escape from the local optimum. In conclusion, our algorithm NQPSO is more competitive than the basic PSO.

Table 4. Daily power load of an area consisting of communities in Shanghai

Time piece	Power(KW)	Time piece	Power(KW)
1	330	19	530
2	330	20	540
3	285	21	520
4	325	22	450
5	290	23	350
6	295	24	330

Table 5. Comparison of NQPSO with PSO

No. of vehicles	Avg Objective value		Avg convergence generation		Avg running time	
	PSO	NQPSO	PSO	NQPSO	PSO	NQPSO
10	165308	144910	35.4	29.8	23	51
20	141170	121647	52.8	32.5	36	78
30	150271	117703	41.4	40.7	42	103
40	176271	120323	167.2	68.9	56	129
50	183260	135208	123.4	62.8	96	188

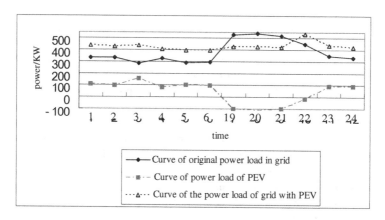

Fig. 4. Curves of power load of an area consisting of communities

6   Conclusion

This paper studies the power grid with PEV charging/discharging problem in power grid in Shanghai with the objective to minimize the total mean squares of load curve of charging and discharging. We consider constrains on the battery capacity, electric power and available time, and establish the mathematical model regarding the price of electricity in Shanghai. To overcome the defect that PSO is likely to trap in a local optimum when dealing with problems with high dimension, we propose NQPSO. Through implementation of our algorithm, we show that NQPSO dominates PSO including average objective value and average convergence generation. Of course, this paper studies the problem only with household PEV. Other vehicles such as bus, tax and so on could also be considered. Therefore, the charging/discharging problem with other types of vehicles could be one of the future topics.

Acknowledgments. The authors are grateful to the anonymous referees for their valuable comments and suggestions. This research is supported by National Natural Science Foundation of China (11201282 and 61304209), the Ministry of Education of Humanities and Social Science Fund Project (10YJCZH032), Innovation Program of Shanghai Municipal Education Commission (14YZ127).

References

1. Yang, J., Wang, W., Zhang, Y., Wang, D., Ye, F.: Applying Power Battery of Electric Vehicles for Regulating Peak in Grid. East China Electric Power 38(11), 1685–1687 (2010)
2. Zou, W., Wu, F., Liu, Z.: Centralized charging strategies of plug-in hybrid electric vehicles under electricity markets based on spot pricing. Automation of Electric Power Systems 35(14), 62–67 (2010)
3. Zhao, J., Wen, F., Yang, A., Xin, J.: Impacts of Electric Vehicles on Power Systems as Well as the Associated Dispatching and Control Problem. Automation of Electric Power Systems 35(14), 2–10 (2011)
4. Peterson, S.B., Whitacre, J.F., Apt, J.: The economics of using plug-in hybrid electric vehicle battery packs for grid storage. J. of Power Sources 195(8), 2377–2384 (2010)
5. Wu, C., Wang, F., Dong, Z., Suo, R.: Electric power load prediction model Using Particle swarm algorithm. Electric System Technology 33(2), 27–30 (2009)
6. Saber, A.Y., Venayagamoorthy, G.K.: Unit commitment with vehicle-to-grid using particle swarm optimization. In: Proceedings of IEEE Bucharest Power Technology Conference, Bucharest, Romania, pp. 1–8 (2009)
7. Hutson, C., Venayagamoorthy, G.K., Corzine, K.A.: Intelligent scheduling of hybrid and electric vehicle storage capacity in a parking lot for profit maximization in grid power transaction. In: IEEE Energy 2030 Conference, Atlanta, GA, US, pp. 1–8 (2008)
8. Sun, B., Shi, Q., Xie, P.: Charge/Discharge mode of electric vehicle home users in Shanghai. East China Electric Power 40(4), 576–579 (2012)
9. He, J.: Co-evolution of nest parasitism. Bulletin of Biology 43(10), 7–9 (2008)
10. Feng, J., Zhong, W., Qiang, F.: An Scale Chaos Quantum Particle Swarm Algorithm. J. of East China University of Science and Technology (Natural Science Edition) 34(5), 714–718 (2008)

Strategy and Loading-Test of Servo Electro-Hydraulic Fatigue Testing Machine Based on XPC System

Zhonghua Miao[1], Xiaodong Hu[2], Chuang Li[2], Mingchao Lu[2], and Keli Han[3]

[1,2] Department of Automation, College of Mechatronics Engineering and Automation, Shanghai University; Shanghai Key Laboratory of Power Station Automation Technology, 200072 Shanghai, China
[3] Chinese Academy of Agricultural Mechanization Sciences, 100083 Beijing, China
zhhmiao@shu.edu.cn,
{hxd0805010424,lichuang_boy,lmc635572715,hanky2008}@163.com

Abstract. Fatigue testing plays a more and more important role in modern industry. In order to obtain the general features and the optimal algorithms of closed-loop control systems of the electro-hydraulic fatigue testing machines, a mathematical model of servo electro-hydraulic fatigue testing machine based on xPC real-time control system is developed and simulated in Simulink. In this model, the compound control system with feed-forward (speed feed-forward and acceleration feed-forward) is introduced. And then the static and dynamic loading testing for both position and force control are done to get the control laws and strategies. The research and results not only reveal the control laws and suitable control strategies of the fatigue-testing machine but also prove that the system has high control accuracy, fast signal processing and response ability. It can also realize the position and force control of the fatigue testing machines perfectly.

Keywords: servo electro-hydraulic fatigue testing machines, xPC real-time control system, compound control system, feed-forward control.

1 Introduction

Fatigue testing of specimens is playing a more and more important role in modern industry through which people can obtain the fatigue lives of the specimens under different forces. This will contribute greatly to the structure design of the instruments [1]. Nowadays, fatigue testing is conducted mainly by electromagnetic resonant fatigue testing machines and servo electro-hydraulic fatigue testing machines [2]. And there are also thousands of control algorithms for fatigue testing which are with low real-time response characteristics. As a result, designers should do much more study and research on choosing the optimal algorithm to make sure that the testing machines are in their best conditions and so they can get their best performances. This paper mainly explores the control characteristics and strategies of closed-loop control systems of the electro-hydraulic servo fatigue testing machine based on the xPC real-time control platform. The feed-forward control algorithm with speed-forward and

K. Li et al. (Eds.): LSMS/ICSEE 2014, Part III, CCIS 463, pp. 198–207, 2014.

acceleration-forward control is introduced to compare with the traditional closed-loop control algorithm without feed-forward control [3], [4]. The testing results also showed that the algorithm this paper introduced have great advantage over the existing PID control algorithms on servo electro-hydraulic fatigue testing machines.

2   Principle of Feed-Forward Control

2.1   Introduction of Feed-Forward Control

Feed-forward control takes advantage of the information that the controller gives out which is mainly aimed at setting the target's motion curve [5]. Generally speaking, the standard of high-quality controller is that the controller set the target's motion curve and the actuator moves along with the curve precisely. In traditional closed-loop PID control system, the target generator will calculate the position, speed and acceleration of the target to be pre-generated [6], [7] after the PID controller updates the controlling information. Therefore, feed-forward control doesn't need to wait to update after the PID controller has made a comparison of the target position and the real position. It can directly control and set the speed and acceleration of the target through feed-forward which makes the control action more prompt and less-influenced by the system delay. The output gain of traditional PID controller is the product of PID parameters and the feed-back error. While the output of the feed-forward PID control system is the sum of two products [8]. One is the product of the parameters and speed. The other is the product of the parameters and acceleration as is shown in equation (1). The structure diagram of compound PID control with feed-forward control is presented in Fig.1.

$$Feedforward output = K_v \times T \arg et speed + K_a \times T \arg et acceleration \tag{1}$$

In equation (1), K_v refers to the speed feed-forward gain; and K_a is the acceleration feed-forward gain.

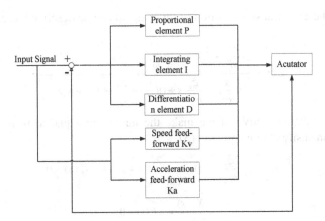

Fig. 1. Structure diagram of compound PID control

The advantage of feed-forward control is its quick response feature while the weak point is the low accuracy compared with normal PID closed-loop control. There is a contradiction between improving the accuracy and the response speed. The compound control system keeps the whole system in an open-loop state while the system is in transient state. And it turns to the closed-loop state when the system goes to steady-state. In conclusion, the compound control system can handle the contradiction between the control accuracy and the system stability perfectly.

2.2   Establishment of Feed-Forward Control Model

In this paper, we make great efforts to study the servo electro-hydraulic system and Fig.2 shows the block diagram of the compound control system with differential feed-forward control. This block diagram introduces the speed feed-forward and acceleration feed-forward control on the basis of traditional PID control [9], [10], [11].

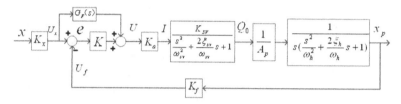

Fig. 2. Block diagram of compound control with forward feedback

Set the main controller as proportional control whose gain is K_1. We can obtain the transfer function shown as equation (2).

$$\frac{X_p}{X_s} = \frac{(K_1 + G_F(s))K_v}{s(\frac{s^2}{w_{sv}^2} + \frac{2\xi_{sv}}{w_{sv}}s + 1)(\frac{s^2}{w_h^2} + \frac{2\xi_h}{w_h}s + 1) + K_1 K_v} \tag{2}$$

If $K_f = 1$, the error transfer function of input signal can be described as equation (3).

$$E(s) = \frac{s(\frac{s^2}{\omega_{sv}^2} + \frac{2\xi_{sv}}{\omega_{sv}}s + 1)(\frac{s^2}{\omega_h^2} + \frac{2\xi_h}{\omega_h}s + 1) - K_V G_F(s)}{s(\frac{s^2}{\omega_{sv}^2} + \frac{2\xi_{sv}}{\omega_{sv}}s + 1)(\frac{s^2}{\omega_h^2} + \frac{2\xi_h}{\omega_h}s + 1) + K_1 K_V} \tag{3}$$

From the equation above, if we make the numerator equal to 0 as equation (4) shows we can easily obtain equation (5).

$$s(\frac{s^2}{\omega_{sv}^2} + \frac{2\xi_{sv}}{\omega_{sv}}s + 1)(\frac{s^2}{\omega_h^2} + \frac{2\xi_h}{\omega_h}s + 1) - K_V G_F(s) = 0 \tag{4}$$

$$\frac{X_p(s)}{X_s} = 1, E(s) = 0 \tag{5}$$

The result means the output can completely "track" the input signal $\theta_i(t)$, which also can be understood that the system has infinite response bandwidth and zero track error. Considering the fact that the parameters of servo hydraulic systems are time-variant and uncertain, we know that it is nearly impossible to achieve full-compensation. So in engineering practice, people use approximate compensation method to imitate full-compensation. So we use approximate compensation to get equation (6).

$$G_F(s) = \lambda_1 s + \lambda_2 s^2 \tag{6}$$

λ_1, λ_2 are two non-dimensional coefficients. That makes the compound system to be a type "3" system. The steady-state error of the system for the ramp function input and parabolic function input are nearly zero. From equation (3) and equation (6) we can see if λ_1 and λ_2 are set properly, we can get equation (7). a, b and c are constants.

$$E(s) = \frac{as^5 + bs^4 + cs^3}{s(\frac{s^2}{w_{sv}^2} + \frac{2\xi_{sv}}{w_{sv}}s + 1)(\frac{s^2}{w_h^2} + \frac{2\xi_h}{w_h}s + 1) + K_1 K_v} \tag{7}$$

In conclusion, the compound control system with speed and acceleration feed-forward can greatly improve the magnitude frequency characteristics of the whole system comparing with the traditional closed-loop control system.

3   Simulation of Compound Control System in Matlab

Based on the transfer function and the block diagram of the compound control system, this paper establishes the simulation model in Matlab as is shown in Fig.3 and it completes the time domain simulation analysis [12], [13], [14].

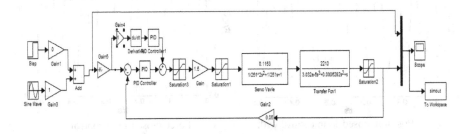

Fig. 3. Simulink simulation model of position control system

Here are four figures that show the simulation results of the system with and without compound control. Fig.4.a and Fig.4.b present the traditional PID closed-loop control system. Fig.4.c and Fig.4.d are the compound control of position and force with feed-forward control.

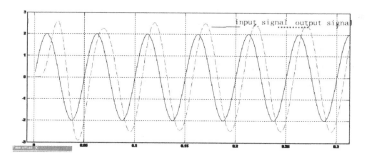

(a) Traditional position closed-loop control

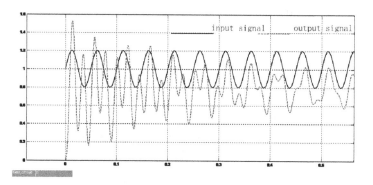

(b) Traditional force closed-loop control

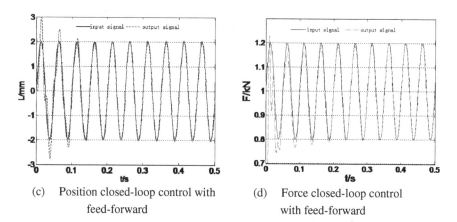

(c)   Position closed-loop control with   (d)   Force closed-loop control
            feed-forward   with feed-forward

Fig. 4. Sine response curves of traditional PID control and compound control with forward feedback (20Hz)

From the comparison of the simulation results we can see that traditional PID closed-loop control of position and force has its own limitations. There are certain phase-lag and track error in traditional PID closed-loop control system. There is also an oscillation in the early stage which definitely is not required in a control system. While from Fig.4.c and Fig.4.d we can see the output signal tracks the reference input in a very accurate way and there are slight errors between the input and output signals. So the system with compound PID control can make the output signal "track" the input signal.

4   Loading Testing on Fatigue-Testing Machine

Test experiments of the compound system with feed-forward control on the fatigue-testing machine are divided into two parts which are position-loading test and force-loading test. Detailed analysis is done to reveal the characteristics and control methods of the servo electro-hydraulic fatigue testing machine.

4.1   Loading Test of Position Closed-Loop Control

Position closed-loop control is the basis of force closed-loop control which makes it more important for us to put much more efforts on position closed-loop control. Loading-test of position control includes static and dynamic loading testing. Apparently, we will get different outputs when we adjust the parameters of the compound PID control system, such as P, I or D. What this paper wants to find out is how the compound PID control system improves performance.

Static Loading Test of Position Control
The static loading test of position is divided into two parts. One is only with P control and the other is with P, I and D control. The loading-test results are shown in Fig.5.

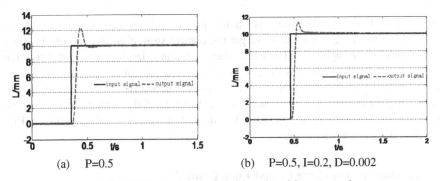

(a)   P=0.5   (b)   P=0.5, I=0.2, D=0.002

Fig. 5. The static test of position closed-loop control

From Fig.5.a we can see that the rise time is 0.056s, the maximum overshoot is 23.5%, and the steady-state error is 0.06. While from Fig.5.b we can get that the rise time is 0.040s, the maximum overshoot is 13.9% and the steady-state error is 0.03.

Because the closed-loop control system of fatigue testing machine is type "1" system, the theoretical steady-state error is zero. We only need P and D control in position closed-loop control to meet the system demand of dynamic and static characteristics.

Dynamic Loading Test of Position Control
The amplitude of the input sinusoidal signal in Fig.6 is 2mm and the frequency is 10Hz. From Fig.6 we can get that the control system with traditional PID control has a lag-behind which means that traditional PID control is not able to meet the command in real-time signal tracking. While just as Fig.6.b reveals the performance of system has been greatly improved when the feed-forward control is introduced into the closed-loop control system. As the frequency of the input signal increases, the skewing aggravates and the median line of the response curve deviates downward.

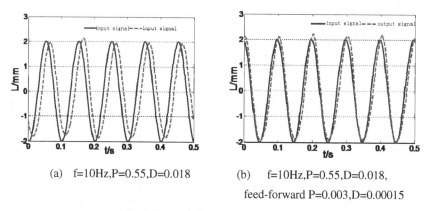

(a)   f=10Hz,P=0.55,D=0.018   (b)   f=10Hz,P=0.55,D=0.018,
                                            feed-forward P=0.003,D=0.00015

Fig. 6. Dynamic Loading-test of Force

In this part we did some analysis on the features of the response curve of the fatigue testing machine from which we get the control strategies. In conclusion, when the frequency of the input signal is 10Hz, there is a phase-lag in the output. From the figures we can see the system with speed feed-forward control and acceleration feed-forward control can greatly improve the dynamic characteristics. When the system goes in a higher frequency, the dynamic characteristics of the system become worse which mainly presents in the amplitude attenuation and the adjustment in PID parameters does not work anymore.

4.2   Loading Test of Force Closed-Loop Control

The loading test of force closed-loop control is the main function that the fatigue testing machine is to realize. Force closed-loop control includes static loading test and dynamic loading test. The static loading test is to give the fatigue testing machine a constant force input and then get the related data to check the static characteristics and dynamic characteristics of the system. While the dynamic loading-test is to load sinusoidal signals on the fatigue testing machine and adjust the parameters of PID controller to check the output signals of the whole system.

Static Loading Test of Force Control

This paper divides static loading-test into two groups. One is only with P control and the other is with normal PID control namely P, I and D control. The testing results are shown in Fig.7.

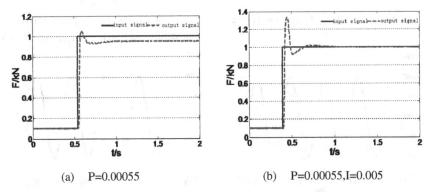

(a)   P=0.00055   (b)   P=0.00055,I=0.005

Fig. 7. Static test of force closed-loop control

From Fig.7.a we can get that the rise time is 0.026s, the maximum overshoot is 5%, and the steady-state error is 50N. While we can see the rise time of the response curve in Fig.7.b is 0.017s, the maximum overshoot is 33.0% and the steady-state error is 3N which is greatly reduced compared to Fig.7.a. Due to the fact that the system of fatigue testing machine is a type "0" system, the theoretical steady-state error is zero when the input signal is a step signal. So we add the integral (I) control in the control system in the force loading test.

Dynamic Loading Test of Force Control

We load sinusoidal signals on the fatigue-testing machine to obtain the dynamic characteristics of the force closed-loop control. The amplitude of the input sinusoidal signals is 200N and the same offset which is 1000N. Fig.8 shows the test curves of the force loading-test with simple PID control and compound PID control when the input signals are with frequencies of 10Hz and 15Hz. With the frequency of the input signal increases, systems with simple PID control perform not well. Just as Fig.8.a and Fig.8.b imply, the phase-lag increases and the amplitude attenuates but "flattop" is gone gradually. With the increase of the input frequency, the phase-lag of the dynamic response is larger and the median line of the amplitude starts to move downwards. From Fig.8.c and Fig.8.d we can see the improvement of the performance when the speed feed-forward control and the acceleration feed-forward control are imported into the control strategy.

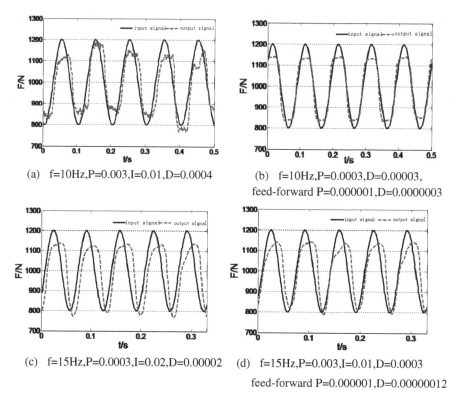

(a) f=10Hz,P=0.003,I=0.01,D=0.0004

(b) f=10Hz,P=0.0003,D=0.00003,
feed-forward P=0.000001,D=0.0000003

(c) f=15Hz,P=0.0003,I=0.02,D=0.00002

(d) f=15Hz,P=0.003,I=0.01,D=0.0003
feed-forward P=0.000001,D=0.00000012

Fig. 8. Dynamic Loading-test of Force

In this part we did some analysis on the features of the response curves of the fatigue testing machine and from which we discuss the control strategies. When the system is with simple PID control algorithms, the system performs badly and the response curve is not smooth. The dynamic performance improves as the feed-forward control is imported and the "flattop" is gone. When the system is in middle-frequency band, the "flattop" is fading away gradually but the median line of the response curve is shifting downward. When the system goes in a higher frequency band, the dynamic characteristics of the system become even worse and the phase-lag increases.

5   Conclusion

Through the study on the control system of servo electro-hydraulic fatigue testing machine, this paper figures out the transfer functions of traditional PID control system and compound PID control system. This paper establishes the Simulink model of two systems. Simulation results have shown that compound PID control with feed-forward can decrease the error between the input and output of the system and increase the dynamic characteristics of the system. And then it implements the static and dynamic loading test of position and force on the fatigue testing machine control system.

This paper also does analysis on the loading test results and obtains the features and control strategies of the whole system. It points out that the compound PID control system with speed feed-forward and acceleration feed-forward can greatly improve the dynamic characteristics and the stability of the servo electro-hydraulic fatigue testing machine control system. But compared with speed feed-forward control the acceleration feed-forward control needs much more calculation and it needs to consider more comprehensive factors. It recommends that we use adjustable modern controller to implement the auto calculation and design of the acceleration feed-forward element which is just what we need to study and research on in the future.

Acknowledgments. The manuscript is supported by the National Natural Science Foundation of China under Grant 51375293, and in part supported by Shanghai Education Commission Innovation Project (No. 12JC1404100), and Key industrialization projects (No. 12YZ010).

References

1. Luo, H.X.: Servo electro-hydraulic fatigue testing control system. Machinery Industry Press, Beijing (1991)
2. Zhang, N.: Research on high frequency dynamic fatigue testing machine. Zhejiang University of Technology, Hangzhou (2009)
3. Smuts, J.: Process control for practitioners, OptiControls Inc. (2011)
4. Shoukun, W.: A new kind of electro-hydraulic proportion loading method for actuator testing based on integral sliding mode. International Journal of Control, Automation and Systems (2013)
5. Nachtwey, P.: Feed forwards make closed-loop tuning easier. Practical Design for Fluid Power Motion Control, Washington (2006)
6. Shoukun, W.: Open-closed-loop iterative learning control for hydraulically driven fatigue test machine of insulators. Journal of Vibration and Control (2013)
7. Tsao, T.C.: Optimal feed-forward digital tracking controller design. ASME J. of Dynamic System, Measurement and Control 116, 583–591 (1994)
8. Hao, L., Lihua, D., Zhong, S.: Adaptive dynamic surface based nonsingular fast terminal sliding mode control for semi-strict feedback system. Journal of Dynamic System, Measurement and Control (2012)
9. Modeling in Simulink,
   http://www.mathworks.cn/products/simulink/description2.html
10. Shuai, Z., Jinbing, C., Tao, W., Wei, F.: Nonlinear modeling and simulation of pneumatic servo postion system of rodless cylinder. Applied Mechanics and Materials (2011)
11. Zhian, S.: Analysis and design of servo-electrical control system based on MATLAB. National Defence Industry Press, Beijing (2007)
12. Qibai, L.: Servo-electrical proportional and mathematical control system. China Machine Press, Beijing (2001)
13. Desmarais, B.: Simulation in simulink. Odum Institute for Research in Social Science (2009)
14. Simulation in simulink,
   http://ece.wpi.edu/courses/es3011/sim/simulation.html

Iterative Learning Control Design with High-Order Internal Model for Permanent Magnet Linear Motor

Wei Zhou[1, 2], Miao Yu[1,*], and Donglian Qi[1]

[1] College of Electrical Engineering, Zhejiang University, 310027 Hangzhou, China
[2] College of Engineering, Jiangsu Institute of Commerce, 211100 Nanjing, China
pink_20020351@163.com, zjuyumiao@gmail.com, qidl@zju.edu.cn

Abstract. In this paper, an iterative learning control algorithm was proposed for improving the permanent magnet linear motor (PMLM) velocity tracking performance under iteration-varying desired trajectories. A high-order internal model (HOIM) was utilized to describe the variation of desired trajectories in the iteration domain. By incorporating the HOIM into P-type ILC, the convergence of tracking error can be guaranteed. The rigorous proof was presented to show that the system error converge well. The simulation results indicate that the proposed high-order internal models based approach yields a good performance and achieves perfect tracking.

Keywords: iterative learning algorithm, high-order internal models, discrete-time plant, permanent magnet linear motors.

1   Introduction

Along with the rapid development of manufacturing industry, high speed and high precision motion control systems are playing an increasingly important role. Permanent magnet linear motors (PMLMs) are the basis of the high speed and positioning precision machines in motion control system due to its mechanical simplicity. The main features of a PMLM include high force density achievable, low thermal losses, high dynamic performance and high positioning precision, etc [1].

PMLM operates repeatedly in a finite interval. This makes iterative learning control (ILC) particularly useful in such a circumstance. ILC is a well-known control method in applications with repetitive tasks [2]. The basic objective of ILC is to overcome the imperfect knowledge of the plant using previous tracking information and to achieve output tracking through repetition [3-5].

However, in the conventional ILC, the desired trajectories are assumed to be iteration-invariant. And it is difficult to ensure in the control practice. Sometimes, tracking tasks vary in different iterations. The approaches to the problems of nonrepetitiveness always are hot topics in the field of ILC. All of the iteration-varying problems, such as variant disturbances, uncertainties and desired trajectories have

---

*Corresponding author.

K. Li et al. (Eds.): LSMS/ICSEE 2014, Part III, CCIS 463, pp. 208–217, 2014.
© Springer-Verlag Berlin Heidelberg 2014

been discussed in [6], and two methods were designed to make use of the nonrepetitiveness in ILC. Recently, a discrete-time adaptive ILC approach is presented in [7] for the system with random initial condition and iteration-varying reference trajectory. The scheme can achieve the point-wise convergence over a finite time interval along the iterative learning axis. In [8], an adaptive ILC method was proposed for a class of nonlinear strict-feedback discrete-time systems with random initial conditions and iteration-varying desired trajectories. And by using high-order internal model (HOIM), an ILC approach has been studied for iteratively varying reference trajectories for continuous-time linear time-varying systems, and the initial resetting condition, extension to nonlinear cases are also explored [9].

ILC algorithm has been developed widely from continuous-time system to discrete-time system for the last two decades associated with the real implementation [10]. [11] deals with the uncertainties in system parameters and disturbance of discrete-time ILC algorithm. A unified learning scheme was designed for a class of discrete-time nonlinear systems with the initial shift problem [12]. A 2-D system theory based ILC has been studied in [13] for discrete linear time-invariant systems with variable initial conditions.

In this paper, we deal with the ILC problem for PMLM system tracking iteration-varying desired trajectories. Furthermore, a sufficient condition is derived under which the convergence of the learning process is guaranteed. Simulation results are documented to illustrate the effectiveness of the proposed robust ILC algorithm for PMLM velocity tracking control.

2   Problem Formulation

Consider a discrete linear time-invariant system with the relative degree of one as follows:

$$\begin{cases} x_k(t+1) = Ax_k(t) + Bu_k(t) \\ y_k(t) = Cx_k(t) \end{cases}, \tag{1}$$

where k indicates the number of operation cycle and t is the discrete-time index; $x_k(t) \in \mathbf{R}^n$, $u_k(t) \in \mathbf{R}^m$ and $y_k(t) \in \mathbf{R}^m$ denote the state, input and output vector, respectively; $A \in \mathbf{R}^{n \times n}$, $B \in \mathbf{R}^{n \times m}$ and $C \in \mathbf{R}^{m \times n}$ are real matrices; and without loss of generality, it is assumed that matrices B and C are of full rank.

Let $e_{k+1}(t) = y_{k+1}^r(t) - y_{k+1}(t)$ be the tracking error at the time interval $t \in [0, T]$ of the $(k+1)$-th iterations. The objective of ILC is to find an appropriate control input sequence $u_{k+1}(t)$ such that the system output $y_k(t)$ follows the reference trajectory $y_k^r(t)$. Consider the iteration-varying reference trajectories $y_k^r(t)$ which are related to the reference trajectories of past iterations. The variations of desired trajectories along the iteration axis can be expressed by a high-order internal model (HOIM) as follows:

$$y_{k+1}^r(t) = h_1 y_k^r(t) + h_2 y_{k-1}^r(t) + h_3 y_{k-2}^r(t) + \cdots + h_m y_{k-(m-1)}^r(t), \tag{2}$$

where h_i ($i = 1, 2, ..., m$) are the coefficients of a stable polynomial

$$H(z) = z^m - h_1 z^{m-1} - h_2 z^{m-2} - \cdots - h_m. \tag{3}$$

From (2) and (3) we can see that, the trajectory $y_{k+1}^r(t)$ generated by m-order internal model is related to the trajectories of past m iterations. Then introduce a shift operator ω^{-1} [14] with the property that $\omega^{-1} y_{k+1}^r(t) = y_k^r(t)$, $\forall t \in [0, T]$. We can rewrite (2) as

$$y_{k+1}^r(t) = H(\omega^{-1}) y_k^r(t), \tag{4}$$

where polynomial $H(\omega^{-1}) = h_1 + ... + h_m \omega^{-m+1}$ is used to describe m-order internal model.

Definition: The λ-norm [15] is defined for function $e_k(t)$ as $\left\| e_k(t) \right\|_\lambda = \sup_{t \in [0,T]} e^{-\lambda t} \left\| e_k(t) \right\|$. And the λ-norm for function $H(\omega^{-1}) e_k(t)$ is defined as

$$\left\| H(\omega^{-1}) e_k(t) \right\|_\lambda = \left\| H(\omega^{-1}) \right\| \left\| e_k \right\|_\lambda = |h_1| \left\| e_k(t) \right\|_\lambda + ... + |h_m| \left\| e_{k-m+1}(t) \right\|_\lambda. \tag{5}$$

With respect to the system dynamics (1) and desired trajectories (4), we have the following assumptions:

Assumption 1: Initialization is satisfied throughout repeated trainings, $e_k(0) = 0$, $k = 1, 2, ...$.

Assumption 2: The matrix Q is defined with bound $b_Q = \sup_{t \in [0,T]} Q$ ($Q \in \{A, B, C\}$).

Assumption 3: $H(z)$ is stable or critical stable which means that all roots of $H(z) = 0$ are within the unit circle or at least one root is lying on the unit circle.

According to the internal model principle [16], the generator of nonrepetitiveness must be involved in the ILC controller to improve tracking performance. For the known law of iteration-varying desired trajectories (4), the HOIM must be incorporated into the design of learning control update law. Use HOIM of (2) and we can design an iterative learning control update law, after k-th operation cycle,

$$\begin{aligned}
u_{k+1}(t) &= h_1 u_k(t) + h_2 u_{k-1}(t) + ... + h_m u_{k-m+1}(t) + \gamma_1 e_k(t+1) + ... + \gamma_m e_{k-m+1}(t+1) \\
&= H(\omega^{-1}) u_k(t) + \Gamma(\omega^{-1}) e_k(t+1),
\end{aligned} \tag{6}$$

where learning gain polynomial $\Gamma(\omega^{-1}) = \gamma_1 + \gamma_2 \omega^{-1} + ... + \gamma_m \omega^{-m+1}$.

3   Learning Convergence Analysis

In this section, we discuss the ILC convergence.

Theorem 1: For the discrete linear time-invariant system (1), given the HOIM-based desired trajectories (4), assume that the Assumptions 1-3 are satisfied. If the learning gains γ_k are chosen such that the asymptotic stability of the following polynomial is guaranteed

$$P(z) = z^m - \zeta_k z^{m-1} - \ldots - \zeta_{k-m+1},\qquad(7)$$

where $\zeta_j = \left\| h_{k+1-j} - CB\gamma_{k+1-j} \right\|$, $j \in [k,\ldots,k-m+1]$, then the output error $e_k(t)$ converges to zero in $[0,T]$ as $k \to \infty$ under the HOIM-based ILC law (6), that is, $\lim_{k\to\infty} e_k(t) = 0$.

Proof: Substituting desired trajectories (4) to the $(k+1)$-th tracking error $e_{k+1}(t+1)$, we obtain

$$e_{k+1}(t+1) = H\left(\omega^{-1}\right) y_k^r(t+1) - y_{k+1}(t+1).\qquad(8)$$

Then substituting system dynamics (1) to (8) yields

$$\begin{aligned}
e_{k+1}(t+1) &= H\left(\omega^{-1}\right) e_k(t+1) - Cx_{k+1}(t+1) + H\left(\omega^{-1}\right)\left[Cx_k(t+1)\right] \\
&= H\left(\omega^{-1}\right) e_k(t+1) - C\left[Ax_{k+1}(t) + Bu_{k+1}(t)\right] \\
&\quad + H\left(\omega^{-1}\right) C\left[Ax_k(t) + Bu_k(t)\right].
\end{aligned}\qquad(9)$$

From the HOIM-based ILC law (6), we get

$$e_{k+1}(t+1) = \left[\dot{H}\left(\omega^{-1}\right) - CB\Gamma\left(\omega^{-1}\right)\right] e_k(t+1) - CA\left[x_{k+1}(t) - H\left(\omega^{-1}\right) x_k(t)\right].\qquad(10)$$

Taking the norms of (10) and considering Assumption 2, it can be derived that

$$\left\| e_{k+1}(t+1) \right\| \leq \left\| H\left(\omega^{-1}\right) - CB\Gamma\left(\omega^{-1}\right) \right\| \left\| e_k(t+1) \right\| + b_c b_A \left\| x_{k+1}(t) - H\left(\omega^{-1}\right) x_k(t) \right\|.\qquad(11)$$

From plant (1), we have

$$x_{k+1}(t+1) - H\left(\omega^{-1}\right) x_k(t+1) = Ax_{k+1}(t) + Bu_{k+1}(t) - H\left(\omega^{-1}\right)\left[Ax_k(t) + Bu_k(t)\right].\qquad(12)$$

Taking the norms of (12) yields

$$\begin{aligned}
\left\| x_{k+1}(t+1) - H\left(\omega^{-1}\right) x_k(t+1) \right\| &\leq \left\| A\left[x_{k+1}(t) - H\left(\omega^{-1}\right) x_k(t)\right] \right\| \\
&\quad + \left\| B\left[u_{k+1}(t) - H\left(\omega^{-1}\right) u_k(t)\right] \right\|.
\end{aligned}\qquad(13)$$

Substituting (6) to (13) yields

$$\begin{aligned}
\left\| x_{k+1}(t+1) - H\left(\omega^{-1}\right) x_k(t+1) \right\| &\leq \left\| A\left[x_{k+1}(t) - H\left(\omega^{-1}\right) x_k(t)\right] \right\| \\
&\quad + \left\| B\Gamma\left(\omega^{-1}\right) e_k(t+1) \right\|.
\end{aligned}\qquad(14)$$

Applying Assumption 2 and we get

$$\begin{aligned}
\left\| x_{k+1}(t+1) - H\left(\omega^{-1}\right) x_k(t+1) \right\| &\leq b_A \left\| x_{k+1}(t) - H\left(\omega^{-1}\right) x_k(t) \right\| \\
&\quad + b_B \left\| \Gamma\left(\omega^{-1}\right) e_k(t+1) \right\|.
\end{aligned}\qquad(15)$$

Now we can write the above inequalities from $t=0$ to $t=T$. When $t=0$, we have

$$\left\|x_{k+1}(1)-H\left(\omega^{-1}\right)x_k(1)\right\|\le b_A\left\|x_{k+1}(0)-H\left(\omega^{-1}\right)x_k(0)\right\|+b_B\left\|\Gamma\left(\omega^{-1}\right)e_k(1)\right\|. \quad (16)$$

From Assumption 1, we get

$$
\begin{aligned}
y_k^r(0) &= C(0)x_k^r(0) \\
&= y_k(0) \\
&= C(0)x_k(0).
\end{aligned} \quad (17)
$$

Thus we have $x_k^r(0)=x_k(0)$. And noting that

$$
\begin{aligned}
y_{k+1}^r(0) &= H\left(\omega^{-1}\right)y_k^r(0) \\
&= H\left(\omega^{-1}\right)C(0)x_k(0) \\
&= y_{k+1}(0) \\
&= C(0)x_{k+1}(0),
\end{aligned} \quad (18)
$$

we get $x_{k+1}(0)=H\left(\omega^{-1}\right)x_k(0)$. Substituting it into (16), we can obtain that

$$\left\|x_{k+1}(1)-H\left(\omega^{-1}\right)x_k(1)\right\|\le b_B\left\|\Gamma\left(\omega^{-1}\right)e_k(1)\right\|. \quad (19)$$

Similarly, when $t=1$ we get

$$\left\|x_{k+1}(2)-H\left(\omega^{-1}\right)x_k(2)\right\|\le b_A\left\|x_{k+1}(1)-H\left(\omega^{-1}\right)x_k(1)\right\|+b_B\left\|\Gamma\left(\omega^{-1}\right)e_k(2)\right\|. \quad (20)$$

Following the same procedure, we can now conclude that for $t\in[0,T]$,

$$\left\|x_{k+1}(t)-H\left(\omega^{-1}\right)x_k(t)\right\|\le\sum_{j=0}^{t-1}b_A^{t-1-j}b_B\left\|\Gamma\left(\omega^{-1}\right)e_k(j+1)\right\|. \quad (21)$$

Substituting (21) into (11), we have

$$\left\|e_{k+1}(t+1)\right\|\le\left\|H\left(\omega^{-1}\right)-CB\Gamma\left(\omega^{-1}\right)\right\|\left\|e_k(t+1)\right\|+b_C\sum_{j=0}^{t-1}b_A^{t-j}b_B\left\|\Gamma\left(\omega^{-1}\right)e_k(j+1)\right\|. \quad (22)$$

Multiplying both sides of (22) by $e^{-\lambda(t+1)}$, we can derive from the definition of λ-norm that

$$\sup_{t\in[0,T]}e^{-\lambda(t+1)}\left\|e_{k+1}(t+1)\right\|\le\left\|H\left(\omega^{-1}\right)-CB\Gamma\left(\omega^{-1}\right)\right\|\left\|e_k\right\|_\lambda+a^2\delta\left\|\Gamma\left(\omega^{-1}\right)\right\|\left\|e_k\right\|_\lambda, \quad (23)$$

where $\delta=\dfrac{1-a^{(1-\lambda)T}}{a^{(\lambda-1)}-1}$ and $a=\max\{e,b_A,b_B,b_C\}$. Choose λ large enough so that δ is arbitrarily small. By expressing the HOIM in the above inequality, we can derive

$$\left\|H\left(\omega^{-1}\right)-CB\Gamma\left(\omega^{-1}\right)\right\|\left\|e_k\right\|_\lambda=\left\|h_1-CB\gamma_1\right\|\left\|e_k\right\|_\lambda+...+\left\|h_m-CB\gamma_m\right\|\left\|e_{k-m+1}\right\|_\lambda, \quad (24)$$

and

$$a^2\delta\left\|\Gamma\left(\omega^{-1}\right)\right\|\left\|e_k\right\|_\lambda=a^2\delta|\gamma_1|\left\|e_k(t)\right\|_\lambda+...+a^2\delta|\gamma_m|\left\|e_{k-m+1}(t)\right\|_\lambda. \quad (25)$$

Substituting (24) and (25) into (23), we get

$$\|e_{k+1}\|_\lambda \leq \|h_1 - CB\gamma_1\|\|e_k\|_\lambda + ... + \|h_m - CB\gamma_m\|\|e_{k-m+1}\|_\lambda + a^2\delta|\gamma_1|\|e_k\|_\lambda + ...$$
$$+ a^2\delta|\gamma_m|\|e_{k-m+1}\|_\lambda .$$

(26)

Now rewrite (26) from $t = 0$ to $t = T$. We can deduce that

$$\begin{bmatrix} \|e_{k+1}(1)\|_\lambda \\ \vdots \\ \|e_{k+1}(T+1)\|_\lambda \end{bmatrix} \leq \mathbf{F}_k \begin{bmatrix} \|e_k(1)\|_\lambda \\ \vdots \\ \|e_k(T+1)\|_\lambda \end{bmatrix} + \cdots + \mathbf{F}_{k-m} \begin{bmatrix} \|e_{k-m}(1)\|_\lambda \\ \vdots \\ \|e_{k-m}(T+1)\|_\lambda \end{bmatrix},$$

(27)

where $\quad \mathbf{F}_j = \begin{bmatrix} \rho_j & 0 & \cdots & 0 \\ 0 & \rho_j & \cdots & 0 \\ \vdots & \vdots & & \vdots \\ 0 & 0 & \cdots & \rho_j \end{bmatrix}$, $\quad \rho_j = \|h_{k+1-j} - CB\gamma_{k+1-j}\| + a^2\delta|\gamma_{k+1-j}|$,

$j \in [k, k-1,..., k-m+1]$. Finally, from (27) we can see that the convergence of $\|e_{k+1}\|_\lambda$ is determined by diagonal matrix $\mathbf{F}_j$. Considering definition of δ, we can furthermore deduce that the convergence of $\|e_{k+1}\|_\lambda$ is determined by polynomial $P(z) = z^m - \zeta_k z^{m-1} - ... - \zeta_{k-m+1}$, where $\zeta_j = \|h_{k+1-j} - CB\gamma_{k+1-j}\|$, $j \in [k,..., k-m+1]$. Clearly, we have $\lim_{k\to\infty} e_k(t) \to 0$ as result.

4   Simulation Example

The simulation results are given to verify the effectiveness of the presented ILC design with HOIM.

Consider the PMLM system[17, 18] as follows:

$$\begin{cases} x_1(t+1) = x_2(t)\Delta + x_1(t) \\ x_2(t+1) = -\Delta\dfrac{k_1 k_2 \psi_f^2}{Rm} x_2(t) + x_2(t) + \Delta\dfrac{k_2 \psi_f}{Rm} u(t), \\ y(t) = x_2(t) \end{cases}$$

(28)

where $x_1(t)$ and $x_2(t)$ are the motor position and rotor velocity, respectively; $u(t)$ is the voltage of stator; $k_1 = \pi/\tau$, $k_2 = 1.5\pi/\tau$, τ is pole pitch; ψ_f, R and m are the flux linkage, resistance of stator and rotor mass, respectively; the discrete time interval $\Delta = 10$ ms, the operation cycle is $N = \{0,1,...100\}$ which means that plant (28) will operate repeatedly in time interval $[0,1s]$. The parameters of PMLM are described in Table 1.

The iteration-varying desired trajectory can be described by a second-order internal model:

$$H(\omega^{-1}) = 2\cos(10\Delta) - \omega^{-1}.$$

(29)

And the desired trajectory of first iteration and second iteration are

$$y_1^r(t) = -0.3\Delta^2 \left(60\Delta t^3 - 30\Delta^2 t^4 - 30t^2\right),$$
$$y_2^r(t) = -0.2\Delta^2 \left(50\Delta t^3 - 31\Delta^2 t^4 - 27t^2\right).$$

(30)

Table 1. Parameters of PMLM

Parameters	τ / m	R / Ω	m / kg	ψ_f / Wb
	0.03	8.6	1.635	0.35

Fig. 1 illustrates the reference trajectories in time and iteration domains. We can see the variation of the desired trajectories. Fig. 2 gives iteration-varying tracking trajectories in $t = 0.2s$ and $t = 0.6s$ to show the variation in iteration domain.

From (28) we can see that the product of the output/input coupling matrix $CB = 0.0378$ is full column rank. The ILC update law with HOIM is designed as

$$u_{k+1}(t) = 1.99u_k(t) - u_{k-1}(t) + 34.8e_k(t+1) - 27.5e_{k-1}(t+1).$$

(31)

We can calculate the coefficients of polynomial $P(z)$ that $\zeta_1 = \|1.99 - CB\gamma_1\| = 0.67$, and $\zeta_2 = \|-1 - CB\gamma_2\| = 0.04$, respectively.

The maximum absolute tracking results of HOIM-based ILC are illustrated in Fig. 3. It can be found that the convergence is achieved as the iteration number increases. To show the time-domain behavior, Fig. 4 gives the tracking profiles at the 4th and 15th iterations. And we can clearly see that the convergence of 15th iteration can give better performance than that of 4th.

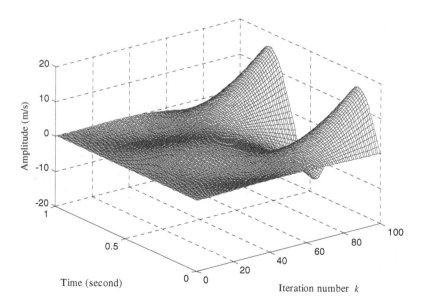

Fig. 1. Iteration-varying tracking trajectories.

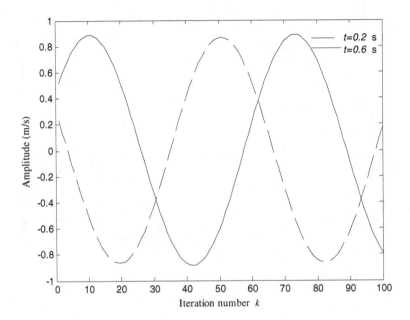

Fig. 2. Iteration-varying tracking trajectories in $t = 0.2$s and $t = 0.6$s

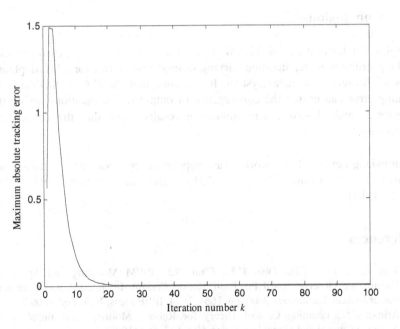

Fig. 3. The maximum absolute tracking error along the iteration axis

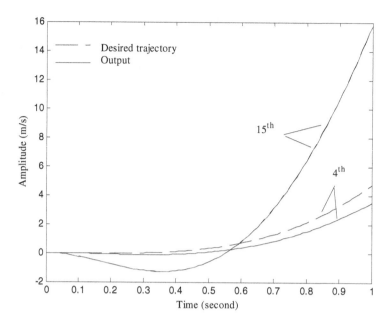

Fig. 4. Tracking profiles of the HOIM-based ILC for the 4th and 15th iterations

5   Conclusions

A high-order internal model (HOIM) based ILC scheme was proposed to tackle the tracking problem under iteration-varying desired trajectories for PMLM plant for a class of discrete-time linear system. It is shown that the ILC method uses P-type tracking error can ensure the convergence of output to the iteration-varying desired trajectories with $k \to \infty$. The simulation results show that the convergence is satisfactory.

Acknowledgments. This work was supported by National Natural Science Foundation of China (No. 61171034) and the National 973 Program (2012CB316400).

References

1. Tan, K.K., Lee, T.H., Dou, H.F., Chin, S.J.: PWM Modeling and Application to Disturbance Observer-based Precision Motion Control. In: International Conference on Power System Technology 2000, pp. 1669–1674. IEEE Press, New York (2000)
2. Arimoto, S.: Learning Control Theory for Robotic Motion. International Journal of Adaptive Control and Signal Processing 4(6), 543–564 (1990)

3. Chen, Y.Q., Gong, Z.M., Wen, C.Y.: Analysis of a High-Order Iterative Learning Control Algorithm for Uncertain Nonlinear Systems with State Delays. Automatica 34(3), 345–353 (1998)
4. Yu, M., Zhang, J., Qi, D.L.: Discrete-Time Adaptive Iterative Learning Control with Unknown Control Directions. International Journal of Control, Automation and Systems 10(6), 1111–1118 (2012)
5. Huang, D.Q., Xu, J.X., Li, X.F., Xu, C., Yu, M.: D-type Anticipatory Iterative Learning Control for a Class of Inhomogeneous Heat Equations. Automatica 49(8), 2397–2408 (2013)
6. Chen, Y.Q., Moore, K.L.: Harnessing the Nonrepetitiveness in Iterative Learning Control. In: 41st IEEE Conference on Decision and Control, pp. 3350–3355. IEEE Press, New York (2002)
7. Chi, R.H., Hou, Z.S., Xu, J.X.: Adaptive ILC for a Class of Discrete-Time Systems with Iteration-Varying Trajectory and Random Initial Condition. Automatica 44(8), 2207–2213 (2008)
8. Yu, M., Wang, J.S., Qi, D.L.: Discrete-Time Adaptive Iterative Learning Control for High-Order Nonlinear Systems with Unknown Control Directions. International Journal of Control 86(2), 299–308 (2013)
9. Liu, C.P., Xu, J.X., Wu, J.: On Iterative Learning Control with High-Order Internal Models. International Journal of Adaptive Control and Signal Processing 24(9), 731–742 (2010)
10. Jang, T.-J., Ahn, H.-S., Choi, C.-H.: Iterative Learning Control for Discrete-Time Nonlinear Systems. International Journal of Systems Science 25(7), 1179–1189 (1994)
11. Wang, D.W.: Convergence and Robustness of Discrete Time Nonlinear Systems with Iterative Learning Control. Automatica 34(11), 1445–1448 (1998)
12. Sun, M.X., Wang, D.W.: Initial Shift Issues on Discrete-Time Iterative Learning Control with System Relative Degree. IEEE Transactions On Automatic Control 48(1), 144–148 (2003)
13. Fang, Y., Chow, T.W.S.: 2-D Analysis for Iterative Learning Controller for Discrete-Time Systems with Variable Initial Conditions. IEEE Transactions On Circuits And Systems—I: Fundamental Theory And Applications 50(5), 722–727 (2003)
14. Moore, K.L.: A Matrix Fraction Approach to Higher-Order Iterative Learning Control: 2-D Dynamics Through Repetition-Domain Filtering. In: 2nd International Workshop on Multidimensional (nD) Systems, pp. 99–104. IEEE Press, New York (2000)
15. Saab, S.S.: A Discrete-time Learning Control Algorithm for a Class of Linear Time-Invariant Systems. IEEE Transactions on Automatic Control 40(6), 138–1142 (1995)
16. Tayebi, A., Zaremba, M.B.: Internal Model-Based Robust Iterative Learning Control for Uncertain LTI Systems. In: 39th IEEE Conference on Decision and Control, pp. 3439–3444. IEEE Press, New York (2000)
17. Jin, S.T., Hou, Z.S., Chi, R.H., Li, Y.Q.: Discrete-Time Adaptive Iterative Learning Control for Permanent Magnet Linear Motor. In: 5th International Conference on Cybernetics and Intelligent Systems, pp. 69–74. IEEE Press, New York (2011)
18. Tan, K.K., Huang, S.N., Lee, T.H., Lim, S.Y.: A Discrete-Time Iterative Learning Algorithm for Linear Time-Varying Systems. Engineering Applications of Artificial Intelligence 16(3), 185–190 (2003)

Electron Beam Welding of Dissimilar Materials and Image Acquisition

Shun Guo, Qi Zhou[*], Yong Peng, and Meiling Shi

Nanjing University of Science & Technology, Nanjing,
Jiangsu Province, China 210094

Abstract. Use of electron beam welding method for welding dissimilar materials of nitrogen steel and armor steel is proposed in this article. Used high energy electron beam as heat source, by means of welding experiment, the welding performance, joint microstructure, hardness which all had been tested, experimental results show that main organization of the weld zone is dendrite, main components between dendrite is low melting phase and impurities, hardness of the weld zone is higher than two kinds of steels, heat affected zone of armor steel is highest. Experiments had also established a visual collection system, image of electron beam welding of two dissimilar steel had been collected and processed, which can provide a reference for future research in this area.

Keywords: nitrogen steel, armor steel, electron beam welding, image acquisition.

1   Introduction

In general, ferritic steel with nitrogen content more than 0.08% (mass fraction) and austenitic steel with nitrogen content more than 0.4% (mass fraction), are called high nitrogen steel [1]. The use of nitrogen has many advantages: promote refinement of grain, reduce ability of formation of the ferrite and deformation martensite, and greatly improve material resistance to pitting and crevice corrosion.

Armor steel is one of the main materials for armor protection, which has higher requirements for crack resistance. Armor steel commonly used is 616 armor steel and manual arc welding is used to weld it. Due to low efficiency of manual arc welding, developing advanced welding methods is necessary [2-4]. At the same time, due to the high strength of armor steel, welding which is a rapid heating and cooling process will cause joints crack. The main types of cracks are cold crack and hot crack [5]. So it has a very important significance to use an efficient and reasonable welding method to weld armor steel.

In some cases, any kind of metallic material cannot fully meet the requirements of practical application. Therefore, combination of dissimilar materials will be a good attempt. High nitrogen steel and armor steel both of them have outstanding

---

[*] Corresponding author.

K. Li et al. (Eds.): LSMS/ICSEE 2014, Part III, CCIS 463, pp. 218–226, 2014.
© Springer-Verlag Berlin Heidelberg 2014

performance, so achieving docking of these two kinds of steels will have a good meaning in fully reflecting its good superiority.

Electron beam welding is easy to achieve deep penetration, small heat affected zone and is under condition of vacuum which is cut off the adverse effects of air to welded joints. Therefore, the use of electron beam welding to weld these two dissimilar materials is a good method.

Weld appearance is a major factor which affects weld quality. By visual inspection to discover geometry of the weld pool, it can play an important role on controlling of weld quality. However, vision sensing of the electron beam welding is not mature, carrying out research on visual sensing of electron beam welding will have an important reference on its future development.

2   Experimental

616 armor steel and 1Cr22Mn16N high nitrogen steel are selected as experimental materials. Size of two plates is determined to be 130mm×50mm×9mm, docking without groove is adopted as the joint assembly method. Chemical components and mechanical properties of 616 armor steel and 1Cr22Mn16N high nitrogen steel are as Tab.1a-b:

Table 1a Chemical components （Wt.%）

Element	C	Si	Mn	S	P	Ti	N	P
616	0.19 - 0.25	0.7- 1.0	1.5- 1.8	0.025	0.025	0.02- 0.05	/	/
1Cr22Mn 16N	0.14 8	0.49	16.00	0.002	0.029	/	0.56	0.02 9

Table 1b Mechanical properties

Standard range	σ_b/MPa	σ_s/MPa
616	1580-1590	1470-1540
1Cr22Mn16N	1100-1200	800-900

Table 2. Main equipment parameters

Model	Accelerating voltage U_a/kV	Filament current I_f/mA	Focusing current I_l/mA	Electron beam I_b/mA	Vacuum chamber space /m^3
ZD60- 6A 500L	60	0~850	0~1000	0~100	0.5

ZD60-6A CV500L which is selected as experimental welding machine is manufactured by AVIC Beijing Aeronautical Manufacturing Technology Research Institute. The equipment mainly consists of vacuum chamber, a vacuum pump, the electron gun, a monitor, cooling water system, power control component. The main technical parameters are as Tab.2.

Before experiment is carried out, focus of electron beam is needed to be measured. In the absence of constraints of external electric or magnetic field, electron beam would be dispersed due to the action of the electric field of its space charge, In order to maintain a certain cross section of a long transmission distance in the axial, electron beam must be focused by effect of electric or magnetic. The focused electron beam is generally first converged and then diverged. In this process, a minimum cross-section of the beam trajectory is called electron beam focus [6-7].

In this experiment, a test of parameters to weld butt steel well has been done on Q235 steel, parameters of preliminary current and voltage are obtained. When penetration is almost equal to the thickness of 10mm Q235 steel, focus current is 710mA; current of electron beam is 54mA. According to the current, focus position used a similar AB method is measured. Test results is shown as Tab.3, Referring to the above data, range of I_f is 700-720mA, I_b is 60-90mA.

Reasonable tooling of welding steel is needed. a fixture for the docking is designed as Fig.1a. The fixture can ensure a good butt, improve weld quality and reduce distortion. In addition, a magnetic fixture which makes device of image acquisition a good placement inside the vacuum chamber is designed as Fig.1b.

Table 3. relationship between focus current and focus position (distance of working)

Number	Focus current I_f/mA	Electron beam I_b/mA	Welding speed V/mm/s	Distance of arc H/mm	Distance of closing arcH/mm	Distance of working H/mm
1	700	20.9	6	202	322	Unobvious change
2	705	20.9	6	202	322	246
3	710	20.9	6	208	323	244
4	715	20.9	6	202	322	239
5	716	20.9	6	203	320	238

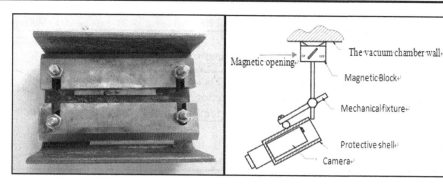

Fig. 1. a fixture for the docking   **b** magnetic fixture for camera

Fig. 2. a Weld formation of surface   b Section condition

Electron beam welding and image acquisition are carried out on the butt of high nitrogen steel and armor steel. When the penetration is close to thickness of steel as Fig.2b, distance of working is 233mm, I_f is 718mA, $I_{b:}$ is 90mA, welding speed is 3mm/s. At the same time, weld formation is relatively good as Fig.2a, but there are some splashes. The main reason is oversaturation of nitrogen which forms N_2 escaped from the weld pool.

3   Analysis of Micro Organizational Characteristics of Electron Beam Joint

Metallurgical analysis is an important means of metallic materials. It can help us understand internal organization of material and determine material properties. Metallographic sample is cut from the sample of Fig.2. It's taken some pretreatment, including polish and corrosion, and Axiover 40MAT metallurgical optical microscope is selected to observe its organization. Pictures are shown as Fig.3a-f.

From Fig 3a,results of metallographic analysis show microstructure of armored steel is low carbon lath martensite and bainite.

From Fig 3b, heat affected zone from inside to outside can be divided into four parts,the fusion line, coarse grain zone, fine crystal zone and transition zone. In process of welding, temperature of fusion area is very high which reaches a semi-molten state,and the temperature of coarse crystal nearby is also a little high. overheating makes austenite grains grow quickly and convert to coarse martensite after rapid cooling. A little far from the fusion area, grains are relatively small fine grain, in this fine grain zone, austenite transformation temperature is moderate, therefore, austenite is transformed to fine martensite.

From Fig 3c, the organisation of HAZ (heat affected zone) is lots of martensite and ferrite structure ,near the weld pool. the grain size of martensite and ferrite is smaller ,central fusion zone is composed of thick lath martensite, martensite of interface near HAZ becomes small, the growth direction of columnar crystals and the interface of HAZ are perpendicular, this region has experienced process of rapid melting and solidification, after crystallization, coarse austenite is transformed into coarse martensite. Due to the large temperature gradient, fast thermal conduction and better dendrite growth conditions which is perpendicular to the interface of fusion line, in this direction,columnar crystals preferentially which is extended to the fusion center is formed and then a transitional organization has been resulted.

From Fig3.d, results of metallographic analysis show that the main organization of weld area is dendrite. between the dendrite, there are some precipitate coagulated finallys, defect, or air mass which formed by some alloy elements gathered. There are

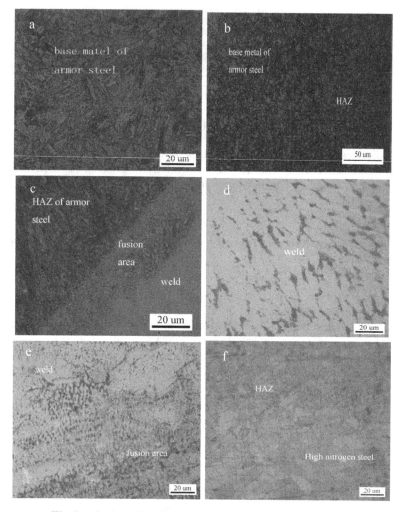

Fig. 3. a-f: phase diagram of microstructure of welding joint

enrichment of low melting phase and inclusions between the dendritic. Material contains a large amount of alloy elements of N, Cr. In weld zone, Nitride are formed by the interaction between these elements,which is deposited finlly before solidification.

From Fig.3e, the main organization of high nitrogen steel near the fusion line is austenite and ferrite, due to excessive welding heat input, size of austenite grain close to weld pool is quite thick, because of lower input, the size of austenite grain close to heat affected zone is relatively small.

From Fig3.f, organizations of heat affected zone are austenitic and small amounts of ferrite which is distributed on the boundary of austenitic grain. Ferrite of coarse grained region grows along the grain boundaries. The main organizations of high nitrogen steel components are austenite and a small amount of ferrite.

4   Analysis of Hardness of Joints

Mechanical properties of metals are an important reference of material application, thus, it's strongly necessary to test the mechanical properties. In the experiment, tests of hardness are carried out.

Select appropriate sample from the joint area and take some pretreatment of the sample, including polish and corrosion. HVS-1000Z digital micro hardness tester is selected to test the hardness. Preload and dwell time are designed as 0.5kg and 10s. Sampling point of specimen is shown as Fig.4a-b schematically.

When dot Pitch is 0.3mm.Accessing points along dashed line of Fig.4b, hardness distribution chart is gotten as Fig.5. Weld centerline is shown as vertical line of Fig 4a. The right side of the weld centerline is high nitrogen steel, coordinate value is negative. The left side of the weld centerline is armor steel, coordinate value is positive.

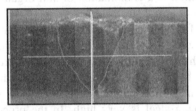 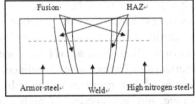

Fig. 4. a Sampling point of specimen   b Schematic diagram of sampling point of specimen

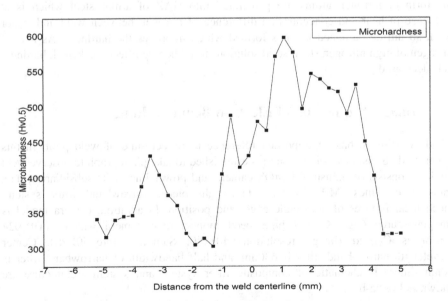

Fig. 5. hardness distribution chart

From Fig.5, hardness of HAZ of high nitrogen steels is higher than base metal. In the process of electron beam welding, due to the large heat input and quick welding speed, HAZ of high nitrogen steels are heat-treated, which is like to quench. Fusion zone near the high nitrogen steel, the hardness decrease, and slightly lower than base metal. By metallographic observation, the organizations of this location are coarse ferrite and austenite, hardness of both are very low, In addition, due to the loss of nitrogen, effect of solid solution strengthening weakened, which reduce the hardness.

In weld zone, hardness increases with the location of sampling points closer to the armor steel. The reason may be that distribution of solute is not uniform, organization near high nitrogen steel is mainly formed by prior austenite of high nitrogen steel, and organization near armor steel is mainly formed by prior lath martensite of armor steel, thus, hardness is higher.

Hardness of weld is higher than both base metals. Two dissimilar materials have been achieved metallurgical bonding in weld zone. On figure of hardness, weld zone is like to a transition layer, which organization is a combination of soft organization of high nitrogen steel and hard organization of armor steel.

The highest hardness is the junction of fusion zone and HAZ of armor steel. The lowest hardness is the junction of fusion zone and HAZ of high nitrogen steel. In process of welding, these areas have gone through rapid heating and cooling, previous lath martensite of armor steel is refined, which is similar to quench. Thus, hardness gets increased, but, previous austenite becomes thick, hardness gets decreased. At the same time, nitrogen content of high nitrogen steel is very high. In the process of welding, nitrogen will be reallocated in part of joint. Part of nitrogen in form of nitrogen gas overflow from weld pool or form nitrogen hole. Also some nitrogen in the form of nitrogen atoms are penetrated into HAZ of armor steel which is at influence of heat and concentration difference of nitrogen between weld and armor steel, so nitride hardened layer is formed which improves the hardness. As loss of nitrogen of high nitrogen steel, solid solution strengthening effect weakened, hardness gets decreased

5   Image Acquisition of Electron Beam Welding

Weld pool image has an important reference to reflect state of weld pool. In this paper, CMOS vision sensing system is established to take photograph to observe weld pool. It consists of industrial CMOS camera and power line of its subsidiaries, data transmission lines, MP3514 lens, less light piece, narrowband filter system, mechanical fixtures of adjustable angle and position. In addition, camera model is ImagingSource GigE CMOS high-speed monochrome camera which MT9V024 sensor is Carried (image resolution≥640*480, Sampling rate≥50FPS). Center wavelength narrowband filter is 630nm, and half bandwidth of narrowband filter is 10nm. Internal schematic of composite filter system and system architecture are shown as Fig.6a-b.

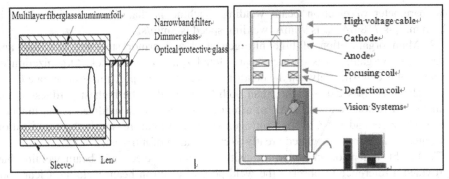

Fig. 6. a Internal schematic of composite   b line connection
filter system

Exposure and gain of camera is auto, acquisition frame rate and storage format are set as 115FPS and AVI. One of collected image is shown in Fig 7a. Image processing software has been prepared to deal with the boundary of weld pool. Fig 7b-d are shown which are treated by the software with different threshold of 210,220,230 gray.

a Original image b Thresholding210gray   c Thresholding220gray d Thresholding230gray

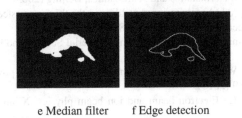

e Median filter   f Edge detection

Fig. 7. extraction of boundary of weld pool

Observed by different thresholds, a reasonable threshold is set as 230 gray. The image is processed by median filter which image is gotten as Fig.7e, then Fig.7e is extracted the boundary which result is as Fig.7f. Boundary of keyhole and weld pool is gotten.

6   Conclusion

1. Electron beam welding of dissimilar materials of high nitrogen steel and armor steel is carried out. When penetration is close to the thickness of 9mm steel,

the parameter of electron beam welding machine is that distance of working is 233mm, I_f is 718mA, I_b is 90mA, welding speed is 3mm/s.

2. Main organization of high nitrogen steel and armor steel are austenite and martensite. Organizations of each heat affcted zone become thick. Organization of weld pool is dendrite, between the dendrite, there are some precipitate coagulated.

3. Hardness of weld is higher than both base metals. The highest hardness is the junction of fusion zone and HAZ of armor steel. The lowest hardness is the junction of fusion zone and HAZ of high nitrogen steel. The main reason is the organization change and also may be related greatly to reallocation of nitrogen.

4. Establishment of an image acquisition system of electronic beam welding has provided a reference to observe the weld pool of electron beam welding. Clear pool images have been collected. By this system and software, boundary of keyhole and weld pool are shown in figure.

Acknowledgement. Thanks for the National Natural Science Foundation of China (No.51375243), High-end CNC machine tools and Basic manufacturing equipment of National Science and Technology Major Project (2010ZX04007-041) and the nature science foundation of Jiangsu province (No.BK20140784) supporting.

References

1. Speidel, M.O.: Properties and applications of high nitrogen steels. High Nitrogen Steels–HNS 88, 92–96 (1988)
2. Tan, W.: Study of amphibious armored vehicle body cracks and corrosion comprehensive. Beijing University of Aeronautics and Astronautics, Beijing (2003)
3. Yang, B.: Tests on 707 (A) ferrite cored welding in the thin armor steel. Ordnance Material Science and Engineering 21(2), 27–32 (1998)
4. Guo, B.H.: Armor material GMAW. Ordnance Material Science and Engineering J., 27–32 (1981)
5. Hu, K., Wang, P.: Welding of parts copper and copper alloy of power plant equipment. Welder 40(8), 86–89 (2010)
6. Tang, T.T., Liu, C.L.: Electron beam and ion beam physics. Xi'an Jiao tong University Press, Xi'an (2001)
7. Lawson, J.D.: Charged particle beam physics. Atomic Energy Press, Beijing (1988)

A Hybrid Algorithm for Reversible Toffoli Circuits Synthesis

Xiaoxiao Wang[1,2], Licheng Jiao[1], and Yangyang Li[1]

[1] Key Laboratory of Intelligent Perception and Image Understanding of Ministry
of Education of China,
International Research Center for Intelligent Perception and Computation,
Xidian University, Xi'an, Shaanxi Province 710071, China
[2] School of Computer Science, Xi'an Shiyou University,
Xi'an Shaanxi Province, 710065, China
xxwang@xsyu.edu.cn, {lchjiao,yyli}@mail.xidian.edu.cn

Abstract. In this paper, we propose a hybrid algorithm aimed at optimally synthesizing reversible Toffoli circuits in terms of the quantum cost for 4-bit and 5-bit reversible benchmarks. The hybrid algorithm alternates a variable-length evolutionary process with a heuristic factor subtraction algorithm based on Positive Polarity Reed Muller (PPRM) expansion. Further more, the variable length evolutionary algorithm employs a new constraint solving method, which introduces a trade-off factor to control a pair of contradictions: the decreasing of constraint violation and the increasing of quantum cost. The experimental results show that the hybrid algorithm outperforms existing combinations of a definite synthesis approach and a post-optimization method on some commonly used 4-bit and 5-bit benchmarks in point of quantum cost, and obtain some better results than the best known ones.

Keywords: reversible logic circuit, hybrid algorithm, variable-length evolutionary algorithm, constraint solving.

1 Introduction

Reversible logic gained extensive attention in the area of low-power design, optical computing and quantum computing. Hence, the synthesis of reversible logic has become a flourishing research area in the recent years.

Mathematically, the synthesis of reversible logic circuits can be formulated as decomposing the expected reversible matrix to some smaller reversible matrices which corresponding to the general reversible gates. In contrast to synthesis with traditional irreversible gates there are two main restrictions for reversible gates: fan-out and feed-back are not allowed. Consequently a circuit or network for a reversible logic consists of a cascade of reversible gates. Apart from satisfying the reversible specification, the obtained circuit should be minimized in the gate count (GC) or quantum cost (QC).

K. Li et al. (Eds.): LSMS/ICSEE 2014, Part III, CCIS 463, pp. 227–239, 2014.
© Springer-Verlag Berlin Heidelberg 2014

Reversible synthesis algorithms can be classified into three categories: definite algorithm, heuristic algorithm and hyper heuristic algorithm.

Some definite algorithms employ different representations of a reversible logic and can always find solutions in less time even for a large scale problem, such as transformation method [1], cycle based algorithm [2][3], Binary Decision Diagram (BDD) based synthesis [4][5]. However, the appropriate optimization techniques are required to reduce the cost of obtained circuits, such as rewriting rules based approach [6], templates based techniques [7] or detection and elimination of non-trivial reversible identities [8]. Other definite algorithms use exhaustive search method which can find the minimized result on GC for 3-bit [9] and 4-bit circuit [10][11][12] or on QC for 3-bit circuit [13]. However due to the super-exponential increase on the memory requirement, the exhaustive method can not apply to the reversible functions with more than 4 bit so far.

Reed-Muller Reversible Logic Synthesizer (RMRLS) [14] is an innovative heuristic algorithm. It proceeds to subtract factors from the Positive Polarity Reed-Muller (PPRM) expansion of a reversible function, then to construct a priority-based search tree and employ heuristics to rapidly prune the search space. Many other algorithms employ the idea to improve the performance of their algorithms, such as [3] and [7]. While the greedy nature of the heuristics method may sometimes lead the algorithm to trap in local optimum.

Some evolutionary algorithms are already used in reversible circuit synthesis [15][16]due to its global search ability. However, owing to some innate aporia, such as the unknown length of the optimum, the difficulty of construction of feasible solutions and the absence of infeasible reparation mechanism, only some small scale problems with low complexity were tested.

Considering the above discussion, we introduce a new hybrid algorithm aimed at synthesizing reversible circuits with generalized Toffoli gates and reducing the quantum cost meanwhile. We formulate the synthesis of reversible logic circuit (RLC) as a minimization problem with equality constraint. The hybrid algorithm alternately conducts a variable-length evolutionary process and the factor subtraction operation inspired by Gupta's RMRLS algorithm [14]. The two process are complementary each other. On the one hand, the factors subtracted from the PPRM of a selected individual can provide information about potential search space for the subsequent evolutionary process. On the other hand, the evolutionary process can conquer the myopia of greedy heuristic and avoid mistake pruning according to the priority of factors. Further more, a new constraint solving method is proposed and employed in the variable-length evolutionary process. It introduces a trade-off factor who can balance the decreasing of constraint violation (CV) and the increasing of objective value, or quantum cost, and at the same time avoids the chromosome bloat, the common problem in variable-length evolution.

The tests are mainly focused on 4-bit and 5-bit functions which exceed the processing capability of current computer system using exhaustive algorithm. The experimental results show that we can obtain some superior results than the definite algorithm combined with post-optimization method in terms of QC.

Section II gives the background knowledge about reversible logic, variable-length evolutionary algorithm and equality constraint solving. Section III describes the details of the hybrid algorithm. Section IV shows the experimental result.

2   Preliminaries

Before detailing the hybrid algorithm, we introduce some basics about reversible logic, variable-length evolution algorithm and equality constraint solving. Hence, the idea behind the algorithm can be understood better.

2.1   Reversible Logic

Definition 1. A *gate is reversible* if the function it computes is bijective. If it has n input and out wires, it is called $n \times n$ gate, or a gate on n wires.

Definition 2. Let $X:=\{x_1, x_2, ..., x_n\}$ be the set of variables. A *generalized Toffoli gate* has the form $TOF(C, t)$, where $C=\{x_{i1}, x_{i2}, ..., x_{ik}\} \subset X$ is the set of control lines and $t=\{x_j\}$ with $C \cap t = \varnothing$ is the target line. It maps the Boolean pattern $\{x_1^0, x_2^0, ..., x_n^0\}$ to $\{x_1^0, x_2^0, ..., x_{j-1}^0, x_j^0 \oplus x_{i1}^0 x_{i2}^0 ... x_{ik}^0, x_{j+1}^0, ..., x_n^0\}$. For $|C|=0$, $|C|=1$ and $|C|=2$, the gates are NOT, CNOT and Toffoli respectively. See Fig. 1.

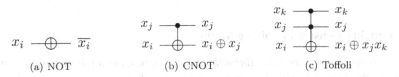

(a) NOT   (b) CNOT   (c) Toffoli

Fig. 1. NOT, CNOT and Toffoli gates

Reversible circuit is to find a series of reversible gates from a universal reversible gate set, which can realize the specification of a reversible function. For example, the specification of 3-bit reversible function 3_17 is [7, 1, 4, 3, 0, 2, 6, 5]. One of reversible circuits of function 3_17 is shown in Fig. 2.

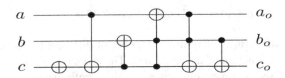

Fig. 2. One of reversible circuits for 3_17 on GTs

Definition 3. The *quantum cost of a reversible circuit rc* is the sum of the quantum cost of each reversible gate g_i consisting of the circuit, and can be represented as

$$gqc\big(T(C,t)\big) = \begin{cases} 1 & |C| = 0 \\ 2^{|C|+1} - 3 & |C| \geq 1 \end{cases} \qquad (1)$$

Where $gqc(g_i)$ represent the quantum cost of a general Toffoli gate g_i and $len(rc)$ is the number of gates consisting rc.

The quantum cost (QC) of a general Toffoli gate is calculated according to the mechanics reported at http://webhome.cs.uvic.ca/~dmaslov/definitions.html. If the number of control bit of a general Toffoli gate is less than or equal to 1, its quantum cost is 1, else is $2^{|c|+1}$-3.

Definition 4. *Positive Polarity Reed-Muller Expansion of Reversible Function is the form of representation using an EXOR sum-of-products (ESOP) expansion in which only uncomplemented variables are allowed. The PPRM expansion of the function 3_17 is as follow:*

$$
\begin{aligned}
a_o &= 1 \oplus b \oplus ab \oplus c \oplus bc \\
b_o &= 1 \oplus a \oplus b \oplus c \\
c_o &= 1 \oplus a \oplus c \oplus ac \oplus bc
\end{aligned}
\tag{2}
$$

References [12] gave efficient techniques for transformation from the truth table to the PPRM of a reversible function.

2.2   Equality Constraint Solving Method

Existing evolutionary algorithms formulate RLC synthesis as a minimization problem and define the fitness function as a weighted sum of the circuit cost and the error (reciprocal relation to correctness) [15], [16]. It is difficult to adjust the weighting coefficient and the correctness of the circuit is not guaranteed. In order to find correct circuit, we formulate the problem as a minimization problem with equality constraint.

The constraint violation (CV) is often inversely proportional to the correctness. Here we propose a new way to calculate the CV drawn inspiration from RMRLS [14], which is defined as (3).

$$
cv(rc) = \text{diffTerm}(reduced\ PPRM)
\tag{3}
$$

The function diffTerm (p) returns the number of different terms between p and the PPRM of identical function of the same size. The *reduced PPRM* is obtained through substituting the series reversible gates of the synthesized reversible circuit rc successively into the *PPRM* of the reversible function.

The handling of equality constraints has long been a difficult issue for evolutionary optimization methods, on account of feasible space being very small compared to the entire search space. Recently appeared algorithms for equality constraint handling are specialized for continuous functions.

For RLC synthesis, its feasible solutions are difficult to build and there are not feasible reparation methods. Consequently, the evolving population is full of infeasible ones and the ranking of infeasible solutions needs to be paid attention to particularly. This paper employs the separation of objectives and constraints mechanism to solve equality constraints and emphasis the comparison of infeasible ones. The new constraint solving method is proposed based on stochastic ranking (SR) [21]. SR ranks the individuals in the population according to the objective value or the CV stochastically through a probability p_f. It may sometimes ranks the individual with a small cost and overlarge CV or the individual with a small CV and overlarge cost front and hence result in evolution degradation or chromosome bloat. The new constraint solving method introduces a trade-off factor which is used when ranking two infeasible ones. It means that one unit of the decreasing of CV worth at most ρ unit of the increasing of objective value or quantum cost.

2.3   Variable-Length Evolutionary Algorithm

Variable-length representation falls into three categories according to the size of solutions: fixed size, bounded size and unbounded size. Unbounded size is referred to the situation where the length of optimum solution is unknown, which has open-ended complexity. RLC synthesis belongs to the third category.

Evolutionary algorithms solving the unbounded size problem usually initialize the population with chromosomes of small sizes and naturally evolve populations whose mean chromosome length grows short over time by the effect of combination operator and selection bias [17]. What is the matter is how to avoid bloat - chromosome redundancy and undue growth.

Methods for controlling bloat are numerous in the GP literature. The review and comparison of different GP model to avoid bloat are given in [18] [19]. We use three mechanisms to control bloat, including size limit, nondestructive crossover-SVLC [20] and the new constraint solving method proposed in this paper. We set the maximum length of chromosomes according to the complexity of RCL problem, see Sec. 3. If recombination produces an individual larger than the maximum length, truncation is conducted. Nondestructive crossover-SVLC ensures that genome size within a population be increased gradually over time. The small variation in size over time affords the GA the opportunity to search for solutions in more correlated landscape.

3   Hybrid Algorithm for RLC Synthesis

The framework of the hybrid algorithm for RLC synthesis is seen in Fig. 6. There are two main differences compared with common GA framework. Firstly, if the best solution does not been improved for more than t generations, we conduct the operation of factor subtraction to the tournament-selected individuals, and update them by randomly appending selected factors to the corresponding individuals. Otherwise, the variable-length evolutionary algorithm grows the length of individuals gradually under the effect of SVLC crossover. Secondly, the revised stochastic ranking (RSR) method is used to ranking the individuals in the population through a bubble-like process and the subsequent tournament selection is conducted based on the ranking results. The RSR is not only a constraint-solving method but also an implicit bloat-avoiding method.

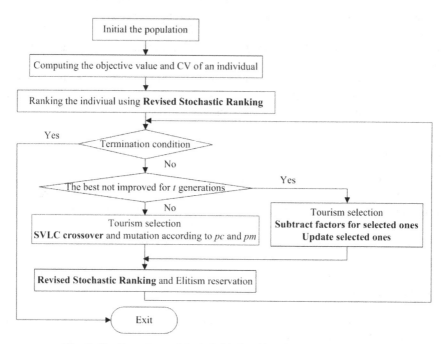

Fig. 3. The flow chart of the hybrid algorithm for RLC synthesis

3.1   Chromosome Coding

A chromosome is encoded by a series of structures, each of which represent a GT gate and consists of two parts: control bits and a target bit. For a circuit of size n, the control bits are represented by a 0-1 bit string of length n. The target bit is represented by an integer which denotes the position of the target bit. For example, if a 3-bit circuit includes three gates: CNOT(a, b), NOT(0), Toffoli(a, b, c), the encoding is as follow:

Control bit: (0 0 1) (0 0 0) (0 1 1)
Target bit: 1   0   2

3.2   Factor Subtraction and Individual Update

Eligible factors can be subtracted from the PPRM of a reversible function. For each variable v_i, we search for eligible factors in the PPRM expansion of $v_{o,i}$ that do not contain v_i.

During the initialization phase, the eligible factors generated from the original PPRM of the reversible function can be used to construct an individual. For example, we can extract 9 eligible factors from (2), among them there are 4 factors for variable a, including $a=a\oplus1$, $a=a\oplus b$, $a=a\oplus c$ and $a=a\oplus bc$, corresponding NOT(a), CNOT(b, a), CNOT(c, a) and T(b, c, a) respectively.

When the evolution falls into stagnation, for each tournament-selected individual, we can subtract factors from the reduced PPRM which is obtained through substituting each gate of the individual into the original PPRM of the reversible function. Then a fixed number of factors randomly selected were appended to the individual. The heuristic operation can detect the reduced search space and pull out the evolution from stagnation.

```
Program Update(Ij)
    Output: an extended individual I_j' based on I_j.
    const Unit=10;
    var MaxLen,RealLen,DiffLen,AddLen
    begin
      get the maximum length of an individual MaxLen;
      get the real length of an individual RealLen;
      DiffLen:= MaxLen-RealLen;
      if DiffLen > Unit then
          AddLen:= Unit;
      else
          AddLen:= DiffLen;
      if  rand()%5==1 then
        for i:=RealLen to RealLen+AddLen do
              append a gate from subtracted factors to
   I_j';
        else
            for i:=RealLen to RealLen+AddLen do
                append a gate random generated to I_j';
        compact and evaluate updated I_j';
    end.
```

3.3   Revised Stochastic Ranking

```
Program RSR
Const
var P_f,  , population(I_1,...,I_n)
begin
    I_j is the jth chromosome in the population
    for i:=1 to n do
        for j:=1 to n do
            sample u• U(0,1)
            if  cv(I_j)> =cv(I_{j+1})and f(I_j)>=f(I_{j+1})then
                swap(I_j,I_j+1);
            else if  cv(I_j)>cv(I_{j+1})&& f(I_j)<f(I_{j+1}) then
                if((f(I_{j+1})-f(I_j))/(cv(I_j)-cv(I_{j+1}))<   then
                    swap(I_j,I_{j+1})
                else if u<P_f  then
```

```
                    swap(I_j,I_{j+1})
        else if cv(I_j)<cv(I_{j+1})&&f(I_j)>f(I_{j+1})   then
            if((f(I_{j+1})-f(I_j))/(cv(I_j)-cv(I_{j+1}))>   then
            if u>P_f  then
                    swap(I_j,I_{j+1})
    end for
    if no swap done break
  end for
end.
```

The pseudo code of RSR is listed above. cv() is used to compute the CV of an individual according to (3) and f() to compute the quantum cost of an individual in accordance with (1). If two adjacent individuals I_j and I_{j+1} have the dominate relationship in light of their objective value and CV. The dominating one will rank first. If they are non-dominated each other, we compute the ratio of the discrepancy of CVs to the discrepancy of costs. If the result is small than the trade-off factor ρ, the individual with lower CV should rank first, otherwise, the individual with lower CV and overlarge cost will rank first with small probability.

4   Experimental Studies

We conduct four experiments to dissect the performance of our hybrid algorithm. First, we compare our algorithm with a definite algorithm with a post-optimization on a group of 4-bit benchmark functions [8]. Then, we test another group of 4-bit problems from the exhaustive method for GC minimization [10], due to the absence of QC minimization algorithm. In the next experiment, we aimed at explain the function of heuristic process, that is, the factor subtraction and individual update. The comparison are made between the results obtained from the hybrid algorithm and those from pure evolutionary process on some commonly used 4-bit and 5-bit problems coming from http://webhome.cs.uvic.ca/~dmaslov/. The last experiment tests some randomly generated 5-bit functions.

Table 1 shows the value of experimental parameters. Apart from the parameters in evolutionary process, such as the size of population p_s, the maximum generation g, the ratio of crossover p_c and the ratio of permutation p_m, there are another three parameters: the trade-off factor ρ and the probability p_f used in RSR and t which decides the opportunity when individual update is conducted. The value of ρ is 9 for 4-bit funciotn and 21 for 5-bit function respectively.

Table 1. Parameter values for the hybrid algorithm

p_s	g	p_c	p_m	ρ	p_f	t
300	3000, 9000	0.8	0.2	9, 21	0.2	100

4.1   Comparison the Hybrid Algorithm with the Post-Optimization Algorithm

The results coming from [8], our hybrid algorithm and the pure evolutionary process without the update of individuals over 30 runs are listed in Table 2. We can see that the hybrid algorithm can acquire the superior QC than 0 on 12 problems, and superior GC on 9 problems. If the operation of individual update is eliminated, we will obtain the lower feasible ratio Fr on 5 problems and the larger average convergence generation Gm on 13 problems.

Table 2. Comparison between the hybrid algorithm and [8]

Func.	[7]	Hybrid algorithm			Hybrid-individual update		
	(g, c)	(g, c)	Fr	Gm	(g, c)	Fr	Gm
App2.1	(10,30)	(11,31)	1	248	(11,31)	1	330
App2.2	(18,102)	(12,40)	1	523	(13,41)	0.9333	828
App2.3	(13,43)	(12,32)	1	427	(12,32)	1	550
App2.4	(9,36)	(10,34)	1	378	(10,34)	1	448
App2.5	(10,50)	(9,29)	1	510	(15,35)	0.9333	535
App2.6	(6,14)	(4,12)	1	14	(4,12)	1	23
App2.7	(15,59)	(9,29)	1	360	(9,29)	1	410
App2.8	(15,53)	(11,43)	1	500	(12,44)	0.9000	558
App2.9	(11,47)	(11,33)	1	360	(10,34)	0.9667	470
App2.10	(13,57)	(10,38)	1	525	(13,41)	1	630
App2.11	(12,80)	(11,31)	1	320	(11,35)	1	650
App2.12	(17,35)	(8,32)	1	550	(10,34)	1	720
App2.13	(12,52)	(13,45)	1	920	(13, 45)	0.9333	1200

4.2   Comparison the Hybrid Algorithm with the Exhaustive Search Method

Table 3 lists the results coming from the exhaustive search method for GC minimization on 4-bit problems and those from our hybrid algorithm for QC minimization. We have obtain dominate solution on oc8 in terms of GC and QC, un-dominated solutions on oc5, oc6 and oc7 and the same solution on 4_49, 4bit-7-8 and decode42. This experiment can reflect the search capability of our hybrid algorithm. It can find the circuits with smaller QC.

4.3   Synthesis of Random Generated 5-Bit Reversible Functions

In this test, we first choose 10 complex permutations of 5-bit (r5_1 to r5_10) from the permutations randomly generated. The maximum number of control bit of each function is 4 and the specifications are listed in the appendix. Then we choose two commonly used 5-bit benchmark functions [14]. The value p is 21 for r5 and 9 for 5one013 and 5one245. We synthesize their circuits using our hybrid algorithm over 30 runs. The maximum generation is 9000.

We can see in the Table 4, the minimum feasible ratio is 0.7667 and the average quantum cost is 195 for random permutations. We obtain the best QC on 5one013 and 5one245 by far.

Table 3. Comparison between the hybrid algorithm and [10]

Func.	[9] (g, c)	Our (g, c)	Our optimal circuit
4_49	(12,32)	(12,32)	C(3,0)C(1,2)N(0)T(3,2,1)C(0,2)T(1,0,2)C(2,0)T4(2,1,0,3)C(3,2) C(3,1)T(2,1,3)C(0,2)C(2,1)
4bit-7-8	(7,19)	(7,19)	C(3,1)C(3,0)C(2,3)T4(0,1,3,2)C(2,3)C(3,1)C(3,0)
decode42	(10,30)	(10,30)	C(2,1)C(3,0)C(1,3)C(0,2)T4(3,2,0,1)T(3,1,0)C(1,3)T(3,2,1)C(0,2) C(0,1)
hwb4	(11,23)	(11,23)	C(2,0)C(3,1)T(3,1,2)C(1,2)C(0,1)T(2,1,3)T(3,0,2)C(1,0)C(3,1) C(0,3)C(2,0)
oc5	(11,39)	(13,37)	C(3,0)C(1,2)T(3,2,1)T4(2,1,0,3)N(0)T(1,0,2)C(3,0)C(3,1)C(0,2) T(2,1,3)C(0,1)C(2,0)C(0,2)
oc6	(12,44)	(13,41)	C(2,0)T4(2,1,0,3)C(3,2)T(2,1,3)C(2,1)T(3,0,2)C(1,3)C(2,0) T(2,1,0)T(3,0,2)N(0)C(0,3)C(1,0)
oc7	(13,41)	(13,41)	T(1,0,2)C(1,3)C(2,0)N(1)T(3,1,2)T(3,0,1)C(0,3)C(2,3)N(0)C(1,0) T4(2,1,0,3)T(3,2,1)C(1,2)
oc8	(12,48)	(12,40)	C(0,1)T(3,2,1)N(0)T4(3,1,0,2)C(1,3)T(3,0,1)C(1,0)C(0,2)N(1)T(3,2,1)T(2,1,3)C(0,2)

Table 4. Synthesis results on number of 5-bit permutations

Func.	Gm	Fr	Best (g, c)
R5_1	6532	0.8000	(35,195)
R5_2	5315	0.8333	(43,199)
R5_3	6822	0.8000	(40,188)
R5_4	5980	0.8667	(42,190)
R5_5	5202	0.8333	(45,185)
R5_6	5026	0.8000	(43,187)
R5_7	6231	0.7667	(39,203)
R5_8	5189	0.7667	(31,207)
R5_9	5369	0.8000	(44,196)
R5_10	5062	0.7667	(39,195)
5one013	2968	1	(19,59)
5one245	1643	1	(17,61)

5   Conclusion

This paper studied on a hybrid algorithm for reversible synthesis using Toffoli gates. The hybrid algorithm combines two complementary methods together. The heuristic factor subtraction operation can provide the information for the variable search space of the variable-length evolution and improve the convergence speed and the feasible ratio. The evolution process can conquer nearsightedness of the heuristic method [14] in some degree.

The experimental results show that it outperforms the definite algorithm with post-optimization [8] and can compete with the exhaustive search [10] on some 4-bit benchmarks. Moreover, the hybrid algorithm can synthesize and optimize 5-bit circuits perform well on 5one013 and 5one245. It may be a solution between the exhaustive method and the definite algorithm.

Acknowledgments. This effort was sponsored by the National Basic Research Program 973 of China, under grant number 2013CB329402, the National Natural Science Foundation of China, under Grants 61 273 317 and 61 271 301.

References

1. Miller, D.M., Maslov, D., Dueck, G.W.: A Transformation Based Algorithm for Reversible Logic Synthesis. In: Proceedings of ACM Design Automation Conference, pp. 318–323. IEEE Press, New York (2003)
2. Saeedi, M., Zamani, M.S., Sedighi, M., Sasanian, Z.: Reversible Circuit Synthesis Using a Cycle-based Approach. ACM Journal on Emerging Technologies in Computing Systems 6, 1–26 (2010)
3. Saeedi, M., Sedighi, M., Saheb, Z.M.: Library-based Synthesis Methodology for Reversible Logic. Microelectronics Journal 41, 185–194 (2010)
4. Wille, R., Drechsler, R.: BDD-based Synthesis of Reversible Logic for Large Function. In: Proceedings of ACM Design Automation Conference, pp. 26–31. IEEE Press (2009)
5. Wille, R., Drechsler, R.: Effect of BDD Optimization on Synthesis of Reversible and Quantum Logic. In: Proceedings of the Workshop on Reversible Computation, pp. 57–70 (2009)
6. Arabzadeh, M., Saeedi, M., Saheb, Z.M.: Rule-based Optimization of Reversible Circuits. In: Proceedings of the 15th Asia and South Pacific Design Automation Conference (ASPDAC), pp. 849–854 (2010)
7. Maslov, D., Dueck, G.W., Miller, D.M.: Techniques for the Synthesis of Reversible Toffoli Networks. ACM Trans. on Design Automation of Electronic Systems (TODAES) 12, 42 (2007)
8. Younes, A.: Detection and Delimitation of Non-trivial Reversible Identities. International Journal of Computer Science Engineering and Applications (IJCSEA) 2, 49–61 (2012)
9. Shenda, V.V., Aditya, K., Igor, L., Markov, J.P., Hayes, M.S.: Synthesis of Reversible Logic Circuits. IEEE TCAD 22, 710–722 (2003)
10. Golubiscky, O., Maslov, D.: A Study of Optimal 4-bit Reversible Toffoli Circuits and Their Synthesis. IEEE Transactions on Computers 61, 1341–1353 (2012)
11. Yang, G., Song, X., Hung, W.N., Perkowski, M.: Bi-Directional Synthesis of 4-Bit Reversible Circuits. Computer J. 51, 207–215 (2008)
12. Li, Z., Chen, H., Xu, B., Liu, W., Song, X., Hue, X.: Fast Algorithm for 4-Qubit Reversible Logic Circuits Synthesis. In: Proceedings of IEEE Congress on Evolutionary Computation, pp. 2202–2207. IEEE Press, New York (2008)
13. Maslov, D., Miller, D.M.: Comparision of the Cost Metrics for Reversible and Quantum Logic Synthesis. arXiv preprint quant-ph/0511008 (2005)

14. Gupta, P., Agrawal, A., Jha, N.K.: An Algorithm for Synthesis of Reversible Logic Circuits. IEEE Trans. Computer-Aided Design and Integration Circuits Syst. 25, 2317–2330 (2006)

15. Zhang, M., Zhao, S., Wang, X.: Automatic Synthesis of Reversible Logic Circuit Based on Genetic Algorithm. In: Proc. of ICISS, Shanghai, pp. 542–546 (2009)

16. Lukac, M., Pekowski, M., Goi, H., Pivtoraiko, M., Yu, C.H., Chung, K.: Evolutionary Approach to Quantum and Reversible Circuit Synthesis. Artificial Intelligence Review, 361–417 (2003)

17. Stringer, H., Wu, A.S.: Bloat is Unnatural: an Analysis of Changes in Variable Chromosome Length Absent Selection Pressure. University of Central Florida. Orlando, FL, http://neuro.bstu.by/ai/To-dom/My_research/Paper-0-again/For-courses/Messy-GA/cs-tr-04-01.pdf

18. Panait, L., Luke, S.: Alternative Bloat Control Methods. In: Deb, K., Tari, Z. (eds.) GECCO 2004. LNCS, vol. 3103, pp. 630–641. Springer, Heidelberg (2004)

19. Whigham, P.A., Dick, G.: Implicitly Controlling Bloat in Genetic Programming. IEEE Trans. Evolutionary Computing 14, 173–190 (2010)

20. Hutt, B., Warwick, K.: Synapsing Variable-length Crossover Meaningful for Variable Length Genomes. IEEE Trans. Evolutionary Computing 11, 118–131 (2007)

21. Runarsson, T.P., Yao, X.: Stochastic Ranking for Constrained Evolutionary Optimization. IEEE Trans. Evolutionary Computing 4, 284–294 (2000)

Appendix: Specifications of Random 5-Bit Permutations

R5_1
[21, 7, 5, 3, 18, 19, 12, 23, 31, 4, 27, 16, 22, 10, 26, 20, 2, 25, 30, 1, 11, 15, 14, 28, 24, 13, 0, 6, 29, 9, 17, 8]
R5_2
[23, 19, 26, 30, 9, 24, 12, 1, 28, 21, 11, 5, 4, 27, 16, 15, 18, 31, 20, 25, 13, 7, 6, 8, 29, 2, 3, 10, 22, 0, 14, 17]
R5_3
[31, 6, 9, 3, 10, 0, 8, 27, 2, 1, 4, 16, 26, 22, 14, 7, 20, 13, 11, 24, 12, 19, 30, 29, 23, 17, 25, 28, 5, 15, 21, 18]
R5_4
[0, 8, 5, 10, 13, 28, 1, 9, 22, 21, 16, 20, 19, 17, 30, 7, 2, 31, 23, 29, 14, 24, 6, 18, 4, 27, 25, 12, 11, 3, 15, 26]
R5_5
[2, 9, 6, 8, 11, 16, 3, 21, 19, 26, 18, 0, 12, 30, 24, 29, 27, 22, 25, 23, 5, 14, 10, 20, 4, 13, 17, 1, 31, 7, 28, 15]
R5_6
[29, 30, 27, 0, 1, 20, 31, 22, 15, 28, 9, 7, 3, 11, 10, 23, 21, 12, 2, 13, 17, 24, 26, 4, 8, 5, 16, 14, 18, 25, 6, 19]
R5_7
[28, 22, 3, 7, 18, 29, 30, 8, 27, 13, 16, 24, 17, 14, 10, 0, 20, 5, 31, 2, 11, 19, 21, 4, 1, 25, 26, 12, 23, 6, 9, 15]
R5_8

[31, 0, 8, 21, 23, 4, 30, 25, 15, 27, 18, 24, 2, 5, 22, 17, 11, 26, 28, 19, 3, 10, 13, 7, 29, 1, 14, 16, 6, 9, 12, 20]
R5_9
[24, 1, 20, 17, 5, 25, 14, 16, 9, 31, 8, 23, 21, 7, 26, 27, 30, 11, 2, 22, 19, 10, 3, 0, 29, 13, 12, 18, 28, 4, 6, 15]
R5_10
[21, 5, 12, 2, 16, 4, 23, 29, 19, 20, 26, 1, 22, 15, 17, 31, 7, 18, 11, 6, 28, 24, 14, 10, 30, 13, 9, 25, 8, 3, 27, 0]

Research of the Optimal Algorithm in the Intelligent Materials

Enyu Jiang [1], Xiaojin Zhu[2], Zhiyan Gao[2], and Weihua Deng[1,*]

[1] College of Electrical Engineering, Shanghai University of Electric Power
Shanghai, 200090, P. R. China
enyu_1981@163.com
[2] School of Mechatronics Engineering and Automation, Shanghai University
Shanghai, 200072, P. R. China

Abstract. Adopting piezoelectric material as sensors and actuators, active vibration control of a cantilever beam is studied in this paper based on Linear Quadratic Regulator (LQR). Firstly, the actuator equation, sensor equation, and the vibration equation is constructed, and then the vibration equation is converted to modal state equation using modal analysis method. Secondly, the optimal control law is given by LQR method, with the detailed control flow. Finally, the active vibration control simulation is done for the vibration suppression of a piezoelectric beam. The results show that the control performance for the step response of the first and second vibration modal is good, as well as the coupled modal of the first two modal. And the effectiveness of the proposed LQR method is verified.

Keywords: piezoelectric material, smart structure, LQR control, active vibration control.

1 Introduction

Piezoelectric materials have been widely used as sensors and actuators for the smart structures [1-2]. Meanwhile active vibration control for flexible structures with attached piezoelectric materials becomes a hot spot [3-5]. And many control strategies have been proposed, Glugla and co-workers[6] presents a modification of the LMS algorithm by adjusting the underlying gradient descent alogrithm in active vibration control; Jae Hanghan and Keun Horew[7] utilizing a linear quadratic Gaussian control algorithm using the laminated composite beam with piezoelectric sensors and actuators reduce the beam vibration; Yan Ruhu and Alfred Ng[8] developed active robust vibration control method for active vibration control of flexible structures; Gustavo et al[9] proposes an on-line self-organizing fuzzy logic controller design applied to the control of vibration in flexible structures containing distributed piezoelectric actuator patches.

---

[*] Corresponding author.

K. Li et al. (Eds.): LSMS/ICSEE 2014, Part III, CCIS 463, pp. 240–249, 2014.
© Springer-Verlag Berlin Heidelberg 2014

With optimal configuration of the piezo patches according to the same location principle, the controlled system is guaranteed to be minimum phase system, and the observation and control spillover can be prevented from modal truncation[10]. Using modal analysis method, the vibration equation of cantilever beam is converted to modal state space equation. With LQR based independent modal control method adopted, this paper takes a cantilever piezo beam as an example, and gives simulation results of the step response and control output voltage for the first two modal frequency. And the effectiveness of the method has been approved by the simulation results.

2   Kinetic Equation for Smart Piezoelectric Structures

2.1   Character of the Piezoelectric

The piezoelectric characteristics can be described as a constitutive relation which characterizes the coupling effect between mechanical and electrical properties as follows:

$$\varepsilon_p = S_{pq}^E \sigma_q + d_{ip} E_i , \quad p,q = 1, \cdots, 6 \tag{1}$$

$$D_i = d_{ip} \sigma_q + \varepsilon_{ik}^\sigma E_k , \quad i,k = 1,2,3 \tag{2}$$

Where σ_q and ε_p represent the stress and strain, respectively, E_k and D_i represent the electric field and electric displacement, respectively. Also S_{pq}^E, d_{ip} and ε_{ik}^σ represent the elastic compliance, piezoelectric strain/charge coefficient, and electric permitivity, respectively. In (1) and (2) a lamina can be either a piezoelectric material or a conventional composite lamina. Take the relations (1) and (2) transformed into the relations in the geometric axes, and, recalling that the stress component except σ_x are negligible, the induced strain of an unconstrained piezoelectric actuator can be written as

$$\varepsilon_x = d_{31} E_z = d_{31} u / t_{pe} \tag{3}$$

Where d_{31} is the transformed piezoelectric constant, u is the input voltage of the piezoelectric voltage, and t_{pe} is the thickness of piezoelectric actuator.

2.2   Piezoelectric Actuator Equation

The moment of actuator force for the cantilever beam when control voltage is input by the control circuit is got from integrating (3)

$$m(x,t) = K_a u [h(x - x_2) - h(x - x_1)] \tag{4}$$

$m(x,t)$ is the moment force, and $h(x)$ is the Heaviside step function. x_1 and x_2 is the distance between the edges of the piezoelectric patch and the cantilever fixed end. K_a is the proportional constant and it can be got by (5).

$$K_a = \frac{1}{2}bd_{31}E_{pe}\left(t+t_{pe}\right) \tag{5}$$

Where b is the width of the cantilever beam and the piezoelectric patch. E_{pe} is the elastic modulus of the piezoelectric actuator. And t is the thickness of the beam.

By taking the derivative of (4) with respect to x, the actuator force is

$$\frac{\partial m(x,t)}{\partial x} = k_a u \left[\varphi_i(x_2) - \varphi_i(x_1)\right] \tag{6}$$

Where $\varphi_i(x)$ is the mode shape of free vibration.

2.3   Piezoelectric Sensor Equation

While the beam is subjected to symmetric bending stress and small deformation and the applied electric field is zero, the output voltage of the i-th piezoelectric sensor amplified by the charge amplifier is shown in (7).

$$U_i = \frac{kd_{31}b_{pe}r_{pe}E_{pe}}{C_{pe}}\int_{x_1}^{x_2}\frac{\partial^2 w(x,t)}{\partial x^2}dx \tag{7}$$

Where $i = 1, \cdots, r$, and r is the number of the sensors, k is the amplification factor of the charge amplifier. $w(x,t)$ is deflection function. b_{pe} is the width of the sensors. r_{pe} is the distance between the sensor's neutral plane and the flexible beam's neutral plane in z direction. C_{pe} is the capacity of the piezoelectric sensors.

2.4   Vibration Equation of the Cantilever Beam

Assuming the principal inertia axis of all the cross section and the external load are in xoy plane, the transverse vibration of the beam is also in the same plane, then the major transformation of the beam is bend transformation. And if the shear deformation and the rotational inertia influence is ignored, the beam is called Bernoulli-Euler beam. And its vibration mode can be expressed by the deflection $w = w(x,t)$, a function of coordinate x and time t. $f(x,t)$ is the external force of unit length beam.

$m(x,t)$ is the applied moment, $\rho(x)$ is the linear density of the beam, A is the cross section area. E is the elastic modulus of the cantilever beam. $I(x)$ is the inertia moment of the cross section and the neutral surface.

$$EI(x)\frac{\partial^4 w(x,t)}{\partial x^4} + A\rho(x)\frac{\partial^2 w(x,t)}{\partial t^2} = f(x,t) - \frac{\partial m(x,t)}{\partial x} \tag{8}$$

Equation (8) is the transverse vibration differential equation of the Bemoulli-Eule beam. And the boundary conditions for fixed end beam is,

$$w(0,t)=0, \quad EI\frac{\partial w(0,t)}{\partial x}=0 \tag{9}$$

According to the related theory of the vibration mechanics and the given boundary conditions, to use the assumed modes method, the function $w(x,t)$ is expanded as an infinite series in the form:

$$w(x,t)=\sum_{i=1}^{\infty}\varphi_i(x)q_i(t) \tag{10}$$

In the above (10), $q_i(t)$ is the generalized displacement.

Substitute (10) into (8), the transverse vibration differential equation can be gained as follows:

$$EI(x)\sum_{i=1}^{n}\phi_i^{(4)}(x)q_i(t)+A\rho(x)\sum_{i=1}^{n}\phi_i(x)q_i^{(2)}(t)=f(x,t)-\frac{\partial m(x,t)}{\partial x} \tag{11}$$

Both side of the (11) multiply the mode function $\varphi_i(x)$, and calculate the integral along with the whole length of the beam, with the perpendicular of the mode function, the decoupled mode coordinate equation can be expressed as:

$$\ddot{q}_i(t)+\omega_i^2 q_i(t)=k_a u[\varphi_i(x_2)-\varphi_i(x_1)] \tag{12}$$

Assume: $B_i = k_a[\varphi_i(x_2)-\varphi_i(x_1)]$, (12) change into:

$$\ddot{q}_i(t)+\omega_i^2 q_i(t)=B_i u \tag{13}$$

Substitute the (10) into (7), the output voltage of the ith piezoelectric sensor can be gained as follows:

$$U_i = \frac{kd_{31}b_{pe}r_{pe}E_{pe}}{C_{pe}}\sum_{i=1}^{\infty}q_i(t)\int_{x_{1j}}^{x_{2j}}\frac{\partial^2\phi_i(x,t)}{\partial x^2}dx$$

$$= \frac{kd_{31}b_{pe}r_{pe}E_{pe}}{C_{pe}}\sum_{i=1}^{n}[\phi_i'(x_2)-\phi_i'(x_1)]q_i(t) \tag{14}$$

Assume: $C_i = \frac{kd_{31}b_{pe}r_{pe}E_{pe}}{C_{pe}}[\varphi_i'(x_2)-\varphi_i'(x_1)]$

So (14) can be written as:

$$U_i(t)=\sum_{i=1}^{n}C_i q_i(t) \tag{15}$$

3   State Space Equation for Cantilever Piezoelectric Beam

Where $x(t)=\{q_1(t),q_2(t),\cdots,q_n(t),\dot{q}_1(t),\dot{q}_2(t),\cdots,\dot{q}_n(t)\}^T$ is the state space vector, then, (12) and (14) can be converted to state-space form.

$$\begin{cases} \dot{x}(t)=Ax(t)+Bu(t) \\ \quad y(t)=Cx(t) \end{cases} \tag{16}$$

$y(t)$ is output of the sensor and $u(t)$ is the input of the actuator. A is state matrix, B is control matrix, D is output matrix.

$$A=\begin{bmatrix} 0_{n\times n} & I_{n\times n} \\ -\Omega^2 & 0 \end{bmatrix}, \quad B=\begin{bmatrix} 0_{n\times 1} \\ B_1 \\ B_2 \\ \vdots \\ B_n \end{bmatrix}, \quad C=\begin{bmatrix} C_1 & C_2 & \cdots & C_n & 0_{1\times n} \end{bmatrix} \tag{17}$$

where

$$\Omega=\begin{bmatrix} \omega_1 & & & \\ & \omega_2 & & \\ & & \ddots & \\ & & & \omega_n \end{bmatrix} \tag{18}$$

As the low frequency response is the main concern, only the first two modal is considered in this paper. The natural frequency of the flexible beam can be obtained by (19).

$$\omega_i=\lambda_i^2\sqrt{\frac{EI}{m}} \tag{19}$$

$I=\dfrac{bt^3}{12}$ is the cross-section inertia moment about y axis, m is the mass per unit length of the cantilever beam.

$$\varphi_i(x)=\sin(\lambda_i x)+D_i\cos(\lambda_i x)+E_i\sinh(\lambda_i x)+F_i\cosh(\lambda_i x) \tag{20}$$

$$\varphi_i'(x)=\lambda_i[\cos(\lambda_i x)-D_i\sin(\lambda_i x)+E_i\cosh(\lambda_i x)+F_i\sinh(\lambda_i x)] \tag{21}$$

For the first vibration modal, $\lambda_1=1.875/l$, $D_1=-1.3622$, $E_1=-1$, $F_1=1.3622$; For the second vibration modal, $\lambda_1=4.694/l$, $D_2=-0.9819$, $E_2=-1$, $F_2=0.9819$. l is the length of the cantilever beam.

4   Linear Quadratic Regulator

4.1   A Linear Quadratic Regulator Algorithm

While the control system is linear, the performance function of the state variables and control variables is quadratic function, and this optimal control problem is called linear quadratic optimal control problem. As the solution of linear quadratic problem is linear function of state variables, and closed loop optimal control can be achieved by state variable feedback, so the practical value of this method is rather high. The quadratic performance index of the state variables and control variables is defined as,

$$J_s = \frac{1}{2} \int_0^\infty [x^T Q x + u^T R u] dt \qquad (22)$$

Q is the weighting matrix of state variables and R is the weighting matrix of input variables. And Q is a positive definite matrix, R is a positive definite matrix.

As $u(t)$ is not restricted, the optimal control should satisfy

$$\frac{\partial H}{\partial u} = R^T u(t) + B^T \lambda \qquad (23)$$

As R is positive definite matrix and its inverse exists, thus

$$u^*(t) = -R^{-1} B^T \lambda \qquad (24)$$

As $\dfrac{\partial^2 H}{\partial u^2} = R > 0$, $u^*(t)$ makes the minimum value of H exist.

Substitute $u^*(t)$ into the Hamilton canonical equation

$$\dot{x}(t) = A x(t) - B R^{-1} B^T \lambda(t), x(t_0) = x_0 \qquad (25)$$

$$\dot{\lambda}(t) = -A^T \lambda(t) - Q x(t) \qquad (26)$$

It is the linear homogeneous equation of $x(t)$ and $\lambda(t)$, and the relation of $x(t)$ and $\lambda(t)$ is linear at any moment.

Take derivations of (26) about t, and put (25) together

$$\left(-Q - A^T P(t)\right) x(t) = \left(\dot{P}(t) + P(t) A - P(t) B R^{-1} B^T P(t)\right) x(t) \qquad (27)$$

As equation (27) works for any value of $x(t)$

$$\dot{P}(t) + P(t) A + A^T P(t) - P(t) B R^{-1} B^T P(t) + Q = 0 \qquad (28)$$

Equation (28) is the first order differential equation about $P(t)$, and it is called differential Riccati equations. And it can be approved that while the elements of A, B, Q, R is sectional-continuous function about t on $[t_0, \infty]$, equation (28) has the unique solution by discovering convergence condition of the boundary condition.

$$u^*(t) = -R^{-1} B^T P(t) x \qquad (29)$$

Let $K(t) = R^{-1} B^T P(t)$, thus

$$u^* = -K(t) x \qquad (30)$$

And closed linear feedback control can be realized by using state variable linear feedback.

Substituting (29) into (15), and the state equation of the closed loop system is

$$\begin{cases} \dot{x} = Ax - BR^{-1}B^T Px \\ \quad y = Cx \end{cases} \tag{31}$$

The closed loop control principle diagram of LQR control with state feedback is shown in Figure 1.

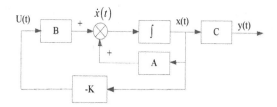

Fig. 1. LQR closed loop control principle diagram

As the low frequency vibration of the cantilever beam plays a major role, suppressing the first few order vibration energy is enough. Taking $x(t)$ as the modal control variable and using LQR method, state feedback control is designed to control $u(t)$ to drive actuators' control moment $m(x,t)$. During $[t_0, \infty]$, optimal control is used to find the optimal control force and the input voltage $u^* = -Kx$ to let the system performance function achieve the minimum value and transfer the system from initial state to zero state. And the output voltage of the piezo sensors reflects the vibration state of the piezo cantilever beam.

4.2   The Controllability of Control System

For linear time-invariant systems are infinite state regulator, request system can fully controllable. Because the control region in the infinite state regulator, when the time extend to infinity, so whatever control vector will be extend to infinity if the system can't be controlled. While for a limited time state regulator, Because the system performance in the upper limit of the integral term is finite, even if the system was not entirely controllable, but in the limited integration time, the integral value is limited, so for a limited time state regulators, can control the system from time to emphasize of the requirements

The size of the aluminium base plate and piezoelectric patches and the physical parameters is listed in Table 1. According to the closed loop state (33) and the derived system state matrix A and system input B the system matrix can be obtained,

$$A = \begin{bmatrix} 0 & 0 & 1 & 0 \\ 0 & 0 & 0 & 1 \\ 273.3 & 0 & 0 & 0 \\ 0 & 1712.8 & 0 & 0 \end{bmatrix}, B = \begin{bmatrix} 0 & 0 & 0.1565 & 0.1381 \end{bmatrix}^T$$

Thus $rank\left(\begin{bmatrix} B, A*B, A\wedge 2*B, A\wedge 3*B \end{bmatrix} \right) = 4$

Therefore, the system is controllable, as the rank of the system controllability is 4, and it is equal to the system dimension. And considering that the Q and R is symmetric positive definite matrix, optimal control exists and is unique.

Table 1. Parameters of the beam and piezo patch

	E (Gpa)	Density (kg/m³)	Length (mm)	Width (mm)	Thickness (mm)	Piezo Constant (m/v)	Capacity (μf)
Beam	59.7	2500	400	20	2.5		
Piezo Patch	18.95	1800	100	20	0.3	4.47*1010	0.06

5   Simulation Example

The smart cantilever beam illustrated in Fig. 2 is taken as a simulation example with the sensors and actuators in the same direction. Step signal is imposed on the system. While the system is undamped, the initial conditions of the modal variables are $q_1(0) = q_2(0)$, $\dot{q}_1(0) = \dot{q}_2(0)$, Linear quadratic regulator control is used to suppress vibrations of the first two modal frequencies, while $Q = \begin{bmatrix} 10e7 & 0 \\ 0 & 10e7 \end{bmatrix}$, $R = [1]$.

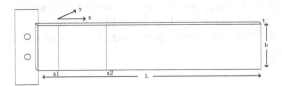

Fig. 2. Cantilever beam model with attached piezo patches

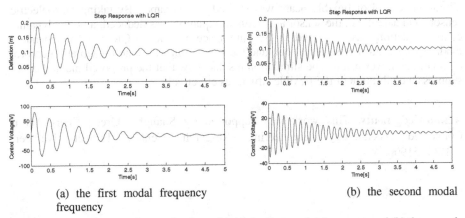

(a) the first modal frequency

(b) the second modal frequency

Fig. 3. Control response and control voltage for (a) the first modal frequency and (b) the second modal frequency

The vibration control response and the control voltage for the first modal frequency is illustrated in Fig. 3(a) while feedback control is on, and the results for the second modal frequency is illustrated in Fig. 3(b). For both of the first two modal frequencies, the mixed control response and control voltage is shown in Fig. 4(a), and the state response is shown in Fig. 4(b).

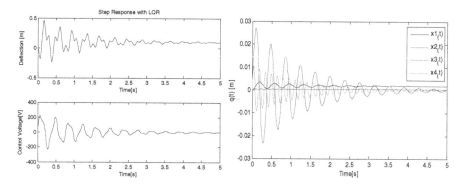

Fig. 4. Control response for the first two modal frequency

Fig. 5.   State response for the first two modal frequency

As shows in Fig. 3(a) and Fig. 3(b), the LQR control is effective and can suppress the vibration response significantly with less steady-state error. As shown in Fig. 3(a) and Fig. 3(b), the control voltage for the second modal frequency is lower than that is needed for the first modal frequency. Fig. 4 and Fig. 5 shows that the first two modal vibration is coupled, though the control performance is also good, the required control voltage is much bigger.

6   Conclusion

Vibration control of flexible beam was studied in this paper. By taking piezoelectric sensors and actuators , the sensor and actuator equation was deduced, and the dynamic control equation was converted to the state space equation. LQR based independent modal space control was carried out, and the simulation was done for vibration control of smart cantilever beam. The simulation results show that the proposed method could suppress the vibration of the flexible structures effectively.

Acknowledgments. This work was supported by Shanghai Green Energy Grid Connected Technology Engineering Research Center, Project Number: 13DZ2251900.

References

1. Carbonari, R.C., Paulino, G.H., Silva, E.C.N.: Intergral piezoatuator system with optimum placement of functionallu graded material-A topology optimization paradigm. Journal of Intelligent Material Systems and Structures 21, 1653–1668 (2010)
2. Vasques, C.M.A., Rodrigues, J.D.: Active vibration control of smart piezoelectric beams: comparison of classical and optimal feedback control strategies. Computers and Structures 84, 1402–141 (2006)
3. Banks, H.T., del Rosario, R.C.H., Tran, H.T.: Proper orthogonal decomposition-based control of transverse beam vibrations: experimental implementation. IEEE Transactions on Control Systems Technology 10, 717–726 (2002)
4. Moshrefi-Torbati, M., Keane, A.J., Elliott, S.J., et al.: Active vibration control (AVC) of a satellite boom structure using optimally positioned stacked piezoelectric actuators. Journal of Sound and Vibration 292, 203–220 (2006)
5. Ramesh, K., Narayanan, S.: Active vibration control of beams with optimal placement of piezoelectric sensor/actuator pairs. Smart Materials and Structures 17, 1–15 (2008)
6. MGlugla, M., Schulz, R.K.: Active vibration control using delay compensated LMS algorithm by modified gradients. Journal of Low Frequency Noise Vibration and Active Control 27, 65–74 (2008)
7. Hang, H.J., Ho, R.K., Lee: An experimental study of active vibration control of composite structures with a piezo-ceramic actuator and a piezo-film. Sensor, Smart Materials and Structures 6, 549–558 (1998)
8. Hu, Y.-R., Ng, A.: Active robust vibration control of flexible structures. Journal of Sound and Vibration 288, 43–56 (2005)
9. Wang, Y.F., Wang, D.H., Chai, T.Y.: Active control of frition-induced self-exited vibration using adaptive fuzzy systems 330, 4201–4210 (2011)
10. Zhang, J., He, L., Li, X., Gao, R.: The independent modal space control of piezoelectric intelligent structures based on LQG optimal control method. Chinese Journal of Computational Mechanics 27, 789–794 (2010)

Adaptive Window Algorithm for Acceleration Estimation Using Optical Encoders

Shuang Wang[1], Fei Hou[1], Surong Huang[*], and Pinghua Zhang[2]

[1] School of Mechatronic Engineering and Automation, Shanghai University, Shanghai, China
[2] Automotive Regional Center, Infineon Technologies Company, Shanghai, China
srhuang@shu.edu.cn

Abstract. Optical incremental encoders are extensively for position measurements used in servo control systems. The position measurements suffer from quantization errors. Acceleration is common obtained by numerical differentiation, which will largely amplify quantization errors. In order to improve the performance of acceleration estimations, an adaptive window algorithm is employed. It maximizes the accuracy of acceleration according to the window size. Besides, velocity is estimated by the mixed time and frequency method, which is utilized to estimate acceleration. Finally, Validity of the acceleration estimation is verified by the experiments.

Keywords: Acceleration, Adaptive Window, Estimation.

1 Introduction

Optical incremental encoders are extensively utilized for servo control systems. Because of the resolution limitation of optical incremental encoders, position measurements suffer from quantization errors which are largely amplified by velocity and acceleration signals which are acquired by the finite differentiation. The signals will affect the performance of the servo system. Therefore, direct differentiation mostly leads to signals useless. In order to improve the accuracy of the acceleration, high resolution encoders or acceleration sensors are employed, but costs are increased. Moreover, many researchers are dedicated to studying smart algorithms which are utilized to improve the accuracy of velocity and acceleration estimations.

Velocity measurement is realized generally by either a fixed-time (M method) or a fixed- position (T method) method [1]. In M method, velocity is acquired by counting the pulses between the fixed sampling time and then performing differentiation on the position signal obtained from the optical encoder. However, the quantization errors are not intolerable when the speed is low. On the contrast, velocity is obtained accurately by fixed-position method when the speed is low, but at high speed the quantization errors are increased extremely. Tsuji [2] [3] proposed a synchronous-measurement method which is called S method. In S method, velocity is calculated when the pulses are changed and the pulses are counted between the fixed sampling

---

[*] Corresponding author.

K. Li et al. (Eds.): LSMS/ICSEE 2014, Part III, CCIS 463, pp. 250–257, 2014.
© Springer-Verlag Berlin Heidelberg 2014

time. The S method may play a same role as an average filter. Compared with M method, it improves the accuracy of the velocity estimation, but it may cause great time delay. Hence, it is not feasible to estimate the acceleration. In order to estimate acceleration, Hiroyuki Nagatomi and Kouhei [4] proposed an acceleration estimation method based on the least squares algorithm, but the computation is complex.

R. Petrella and M. Tursini [5] proposed a novel mixed time and frequency measurement. The basic idea of the proposed method is to compensate for the quantization errors of pure fixed-time measurement due to the lack of synchronization between encoder signals and sampling time. Without compromising the time delay, accuracy of velocity acquired by mixed time and frequency are improved greatly. In order to estimate acceleration, an adaptive window algorithm is employed which maximizes the accuracy of acceleration according to the window width.

2   Velocity Measurement

The mixed time and frequency velocity measurement is aimed at compensating for the quantization errors caused by M method. This can be realized by measuring time intervals (T_1 and T_2 in Fig1) between the bonds of sampling window (T_s) and incoming encoder pulses. m_1 is the number of the pulses counting in every sampling time. The real time when the encoder turns m_1 pulses is expressed as T_d. In this way, the quantization errors (due to the asynchrony between encoder pules and sampling window) can be eliminated greatly. A high accuracy estimation value can be obtained within each sampling period.

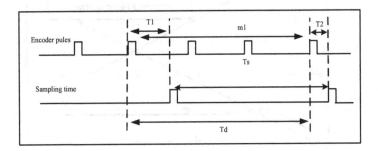

Fig. 1. The Schematic Diagram of Velocity Estimation

The basic calculating equation is,

$$n = \frac{60m_1}{Z(T_s + T_1 - T_2)} \qquad (1)$$

Where, Z represents the number of pules when the encoder rotates one cycle.

Velocity is obtained in each sampling time by formula (1). When the speed is low, there may be no encoder pulse within the sampling time. In this case, the speed information can be updated when a valid pulse is accepted. Though the velocity estimated by this method is accurate, there still exist errors. The measuring errors are in terms of synchronization between sampling window and the high frequency clock pulses. The error formula can be described as follows:

$$\sigma = \frac{2}{T_d * f_0} * 100\%$$
(2)

Here, f_0 is the clock frequency used to measure time intervals.

Steady-state error curves with different velocity estimation methods are presented in Fig.2. The analysis results are assumed that the system clock frequency is 30MHz, sampling time is 1ms and resolution of the incremental encoder is 2500 slits/r. In Fig.2, the advantages of using mixed time and frequency method are apparent. The velocity estimation obtained by this method is far more accurate than that of others. What is more important, this method can keep the accuracy at both low speed and high speed.

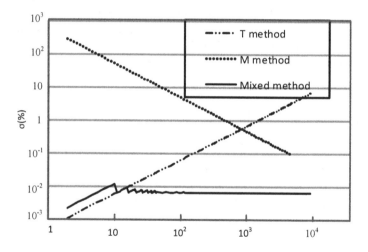

Fig. 2. Steady-state Errors of Velocity Estimation

3   Acceleration Estimation

Acceleration is conventionally calculated by numerical differentiation with low-pass filter. And the acceleration $a(i)$ can be obtained as,

$$a(i) = \frac{v(i) - v(i-1)}{T_s}$$
(3)

Here, T_s is the sampling time, $v(i)$ is the ith velocity estimation.

If $e(i)$ represents the quantization errors caused by the differential, the equation can also be described as,

$$a(i) = \frac{\tilde{v}(i) - \tilde{v}(i-1)}{T_s} + \frac{e(i) - e(i-1)}{T_s} \tag{4}$$

Where, $\tilde{v}(i)$ is the actual speed value.

From the above equation, it can be noted that quantization errors are amplified by the quotient, especially T_s is short. Therefore, low-pass filters are often utilized to get useful acceleration estimations, but it may cause great time delay.

In order to eliminate the quantization errors, the longer sampling time can be adopted. In other words, the sampling window should be increased as follows.

$$\hat{a}(i) = \frac{1}{n} \sum_{k=0}^{k=n-1} \hat{a}(i-k) = \frac{v(i) - v(i-n)}{nT_s}$$
$$= \frac{\tilde{v}(i) - \tilde{v}(i-n)}{nT_s} + \frac{e(i) - e(i-n)}{nT_s} \tag{5}$$

The window size is expressed as n. The more accurate acceleration can be obtained with larger window size. Nevertheless, the larger window size means the time delay and reduces the estimation reliability.

In order to trade precision against time delay, the window size n should be selected properly depending on the signal itself. The window size should be narrow when the acceleration is high, producing faster and reliable estimation. When the acceleration is low, the window size is large. And the more precise estimation is yielding.

Therefore, the allowable error σ is introduced to select the proper window size n. It is established to determine whether the slope of a straight line approximates the differentiation between two estimations $v(i)$ and $v(i-k)$. It can be employed to decide the longest window size which satisfied the accuracy requirement. The solution can be described as finding largest window size n where $\forall k \in \{1, 2, 3...\}$ satisfies,

$$\left| v(i-k) - \hat{v}(i-k) \right| \leq \sigma \tag{6}$$

Where, $\hat{v}(i-k) = a_n + b_n(i-k)T_s$ and b_n is equal to the acceleration estimation $\hat{a}(i)$, which can be calculated by

$$\hat{a}(i) = b_n = \frac{v(i) - v(i-n)}{nT_s} \tag{7}$$

Here, $n = \max\{1, 2, 3...\}$.

The slope of a straight line (b_n) is the acceleration estimation. σ is selected depending on the performance of velocity estimation. If the quantization errors of velocity estimation are large, a big value should be chosen. It can get a large window size. Thus, the quantization errors can be reduced effectively. If the quantization errors are small, the σ can be set as a small value. As a result, the accuracy of acceleration estimation is improved with short delay time.

The acceleration estimator can be described as follows:

1) Store the velocity estimations employed by mixed time and frequency method
2) Set $k=1$.
3) Set $v(i)$ is the last sample velocity estimation and $v(i-k)$ is the kth estimation before $v(i)$.
4) Calculate b_n, the slope of the straight line passing through $v(i-k)$ and $v(i)$.
5) Check whether the $v(i-j)$ satisfy the equation (6), where $j \in \{1,2,...,k\}$.
6) If so, set $k=k+1$ and GOTO 4). Else return the last acceleration estimation.

4   Experimental Results

A simple specific hardware system has been developed in order to emulate the performance of the proposed acceleration estimation. Quadrature signals are generated by a microcontroller and are fed into Infineon microcontroller XMC4500, which is employed to calculate the velocity and acceleration. The resolution of the incremental encoder is simulated as 2500 lines, which is common used in industrial applications. Fig.3 shows the experimental results where the speed is increasing from 0 to 500r/min and the sampling time is 1ms. It is noted that the velocity estimated by mixed time frequency method is much accurate. The quantization errors are within the 2r/min when it reaches the steady-state condition.

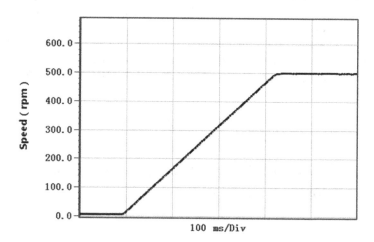

Fig. 3. Velocity Estimations Obtained by Mixed Time and Frequency Method

In order to compare the performance of acceleration estimations which are estimated by adaptive window method with that of numerical differentiation method, acceleration estimations are obtained from the same speed signal shown in Fig.3. Velocity is rising from 0 to 500r/min at a constant acceleration (1500r/(min·s). Acceleration estimations obtained by direct differentiation with a low-pass filter are presented in Fig.4. The quantization errors are about 15 r/(min·s) and the delay time (Δt) is 39ms. In Fig.5, the acceleration is obtained by the adaptive window method (σ=1). The accuracy of the acceleration estimations is almost the same with results shown in Fig. 4. However, the time delay is reduced to 24ms, which improved the rapidity. In order to improve the accuracy of the acceleration estimations, a larger value σ can be selected. Experimental results are shown in Fig. 6 where σ=4, but the time delay is increased to 30ms, where the delay time is also shorter than that of using filter method (in Fig.4). Therefore, compared to employing numerical differentiation with filter method, the adaptive window method has great advantages of obtaining better performance acceleration estimations. The acceleration estimations calculated by adaptive window method are more accurate with shorter time delay.

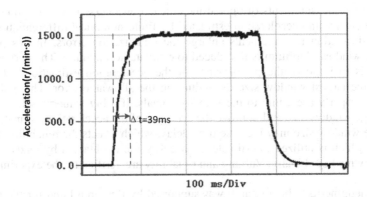

Fig. 4. Acceleration Estimations with Low-pass Filter

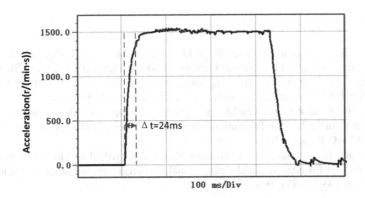

Fig. 5. Acceleration Estimation with Adaptive Window Method When σ=1

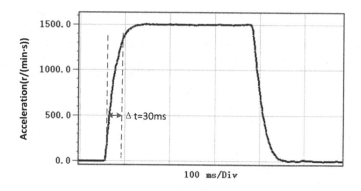

Fig. 6. Acceleration Estimations with Adaptive Window Method When σ=4

5   Conclusion

The position measurements of optical incremental encoders suffer from quantization errors. Velocity and acceleration estimated by finite numerical differentiation of the quantized position measurements enlarge the quantization errors. In this paper, an adaptive window algorithm is introduced to estimate acceleration. The purpose of the method is to improve the performance of the acceleration estimations. It aims at finding the largest window size depending on the allowable error. The σ should be selected properly according to the velocity signals. If a big valueσis selected, it can get a large window size and improve the accuracy of the acceleration estimations. But the large window size may increase time delay, which affects the reliability. Besides, velocity which is utilized to estimate the acceleration is obtained by mixed time and frequency method. Finally, Validity and reliability are verified by the experiments.

Acknowledgments. This research was supported by Research Fund for the Doctoral Program of Higher Education (20113108110008).

References

1. Merry, R.J.E., van de Molengraft, M.J.G., Steinbuch, M.: Velocity and Acceleration Estimation for Optical Incremental Encoders. J. Mechatronics 20(1), 20–26 (2010)
2. Tsuji, T., Hashimoto, T.: A Wide-Range Velocity Measurement Method for Motion Control. IEEE Transactions Industrial Electronics 56(2), 510–519 (2009)
3. Tsuji, T., Mizuochi, M., Ohnishi, K.: Technical Issues on Velocity Measurement for Motion Control. In: 12th International Power Electronics and Motion Control Conference, EPE-PEMC 2006, pp. 106–111 (2006)
4. Nagatomi, H., Ohnishi, K.: Acceleration Estimation Method of Motion Control System with Optical Encoder. In: IEEE International Conference on Industrial Technology, ICIT 2006, pp. 1480–1485 (2006)

5. Petrella, R., Tursini, M.: An Embedded System for Position and Speed Measurement Adopting Incremental Encoders. IEEE Transactions on Industry Applications 44(5), 1436–1444 (2008)
6. Ohmae, T., Matsuda, T., Kamiyama, K., Tachikawa, M.: A Microprocessor-controlled High-accuracy Wide-range Speed Regulator for Motor Drives. IEEE Transactions on Industrial Electronics 29(3), 207–221 (1982)
7. Petrella, R., Tursini, M., Peretti, L., Zigliotto, M.: Speed Measurement Algorithms for Low Resolution Incremental Encoder Equipped Drives: Comparative Analysis. In: Proc. of International Aegean Conference on Electrical Machines and Power Electronics, ACEMPELECTROMOTION Joint Conference, Bodrum, Turkey, pp. 780–787 (2007)

An Improved Algorithm of SOC Testing
Based on Open-Circuit Voltage-Ampere Hour Method

Ye Deng, Yueli Hu[*], and Yang Cao

Microelectronic R&D Center, Shanghai University, No. 149, Yanchang Rd. 200072
Shanghai, China
huyueli@shu.edu.cn

Abstract. An improved algorithm is proposed to estimate the SOC of the lithium-ion. Based on the combination of the open-circuit voltage and the Ampere-Hour integral measurements, aimed at the long standing time and error accumulation, improved the algorithm. Assessed manganese acid lithium battery, obtained the relationship between VOC and SOC and the optimal parameter, we can estimate the open-circuit voltage of the lithium battery in a short time by the model parameters, which greatly reduces the open circuit voltage method of incubation time. When charging or discharging the battery, compute the variation of the SOC through Ampere–Hour integral. Experiment confirmed the accuracy of the algorithm. This algorithm can accurately estimate the SOC thus has a certain reference value for the research of lithium battery management system.

Keywords: LiMn2O4 battery, SOC estimation, battery model, open-circuit voltage, ampere-hour integral method.

1   Introduction

As the deepening prominent of the energy crisis and environment pollution, Lithium-ion batteries gradually replace the traditional batteries, widely used in automotive, aerospace, Marine and other fields for its high energy density, high working voltage, high energy, low pollution, low self-discharge rate[1].Battery management system is required to predict the battery charged state (state of charge, SOC) accurately in real time to prevent the overcharge and overdischarge and prolong the lifetime of the battery, so as to improve the performance and security of the vehicle[2]. One of the important parameters that are required to ensure safe charging and discharging is SOC. It is an important parameter used to describe the capacity of the battery. But due to the complexity of its structure, the influence of the open-circuit voltage, charge and discharge current, temperature, self-discharge, cycle life and its nonlinear, it is difficult to estimate the SOC of the battery. Accurate estimation of SOC prevents battery damage or rapid aging by avoiding unsuitable overcharge and overdischarge. The importance of SOC determination in battery management is summarized as follows:

---

[*] Corresponding author.

K. Li et al. (Eds.): LSMS/ICSEE 2014, Part III, CCIS 463, pp. 258–267, 2014.

✧ For state of health (SOH) estimation applications, since there is no absolute definition of the SOH, SOC becomes an important indicator to estimate SOH.

✧ For cell protection, BMS determin whether the battery is over-charged or over-discharged based on the value of SOC, thereby to control the charge and discharge current.

✧ For cell equalization applications, it is only necessary to know the SOC of any cell relative to the other cells in the battery chain. BMS deals with unequal SOC by only charging the cell with low SOC.

At present, lots of research has been done at home and abroad in battery model and the algorithm of SOC, including discharge test method, ampere-hour integral method, open-circuit voltage method, neural network, Kalman Filtering and so on[3].To obtain fast and robust SOC estimation, in this paper, an improved Open-Circuit Voltage method is put forward, which is based on the equivalent circuit battery model of the lithium-ion battery. In addition, an On Open-Circuit Voltage-Ampere Hour Method is designed for the SOC estimation of the lithium-ion battery.

The paper is organized as follows. Section 2 introduced the traditional algorithm of SOC testing, the proposed algorithm is discussed in Section 3. Experimental results on the chosen algorithm is carried out in Section 4, followed by conclusions in Section 5.

2   Traditional Algorithm of SOC Testing

Many techniques have been developed for estimating the SOC of a battery. Some are specific to particular cell chemistries, and others depend on measuring the battery parameters that depend on the state of charge. In this section we give a brief overview of main SOC estimation methods, and discuss their pros and cons.

1)   Discharge Test Method
Discharge test is the most reliable method for determining the SOC of a battery. Battery is usually discharged in constant current and regards the product of the discharge current and the time as the residual capacity, the discharge test method is the most reliable method for estimating SOC, but due to the long time it needs, it is not suitable for a running vehicle since the charge/discharge current is varying during the vehicle running.

2)   Open-Circuit Voltage Method
The battery open circuit voltage(OCV) has a certain proportion relationship with SOC principle, we can estimate the SOC of the battery by measuring OCV continuously, it's easy and simple, but it needs a long time to get the over-circuit voltage, thus it's unable to accurately estimate the SOC online. Furthermore, OCV curve is sensitive to different discharge rates and temperature. Therefore this method is effective for measuring SOC only at the initial stage and end stage.

3)   Ampere-Hour Integral Method
The most common technique for calculating battery SOC is ampere-hour method. The way of Ampere-hour Method to inference whether the battery is getting or releasing power is by testing current continuously and integral to get the value of SOC. Calculated as follows:

$$SOC = SOC_0 - \frac{1}{C_N} \int_{t_0}^{t_1} \eta i dt \tag{1}$$

In this formula, t_0 is the initial velocity, t_1 is the end of test time, SOC_0 is the initial power of battery, C_N is battery's rated capacity, η is the charge-discharge efficiency, i is the charge-discharge current. According to the formula(1), we can see several existing issues as follow: (1)To calibrate the initial value of SOC; (2) we need to Accurate calculates the charge-discharge efficiency; (3)we need to measure current accurately, otherwise it will cause accumulative error; (4) High temperature state and current volatile magnify error.

4)   Neural Network

As an intelligent technology, artificial neural network is strong self-learning and adaptive ability makes it good at association, generalization, analogy and extension. It is under the premise of building a reliable network model; rely on a large number of sample data. Containing high nonlinearity, fault tolerance capability and robustness, be able to give corresponding according to the external incentives. In addition, neural network is capable of estimating SOC when the initial SOC is unknown. However, the accuracy of SOC cannot guarantee and the SCM performance requirements are too high, so it is not widely used.

5)   KalmanFiltering

Kalman filtering is method to estimate the inner states of any dynamic process in a way that minimizes the mean of the squared error. It is a kind of Recursive data processing algorithm, be able to achieve the best estimation for the state of the power system at the minimum variance sense and has a strong correction function for the initial value of the SOC. But has a high accuracy requirement on the battery modelling and the computing ability [4].

A summary of above discussion is presented in table 1.

Table 1. Summary of SOC estimation techniques

Technique	Field of application	Advantages	Disadvantages
Discharge test	All battery systems	Easy and accurate	Offline, time intensive, loss of energy, modifies the battery state.
Open-circuit Voltage	Lead, lithium, Zn	Easy	Long rest time, offline
Ampere-hour Integral	All battery systems	Online, easy	Needs re-calibration points, consider battery loss
Neural Network	All battery systems	Online	Need training data of a similar battery
Kalman Filtering	All battery systems	Online, dynamic	Computationally intensive, needs a suitable battery model

Above all algorithms, ampere-hour integral method and open-circuit voltage method are most commonly used. Literature [5] [6] analysed the influence of several parameters to the result of ampere-hour integral method respectively, including the initial SOC (SOC_0), rated capacityC_N, -dischargeefficiency η. Conclusion was made that the initial SOC has the biggest effect on the battery SOC estimation accuracy. This article proposed a method to estimate SOC based on open-circuit voltage method and combined it with the equivalent circuit battery model, which shorten the standing time of open-circuit voltage method and reduce the influence of cumulative error in ampere-hour integral method. By comparing the results with discharge test method, verify the feasibility of the algorithm [7].

3   An Improved Algorithm Based on Open Circuit Voltage-Ampere Hour

To eliminate the accumulative error during ampere-hour method, we shorten the standing time of open-circuit voltage method, and modify the initial value of ampere-hour integral method with open-circuit voltage method continuously.

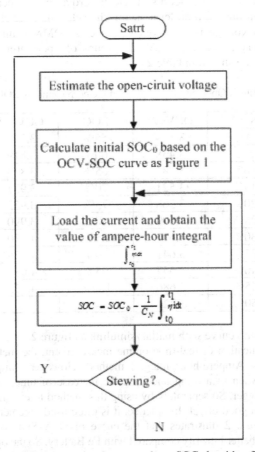

Fig. 1. A modified open-circuit-ampere-hour SOC algorithm flow chart

Combine the open-circuit voltage method with battery equivalent model, then estimate the open-circuit voltage with a voltage after stopping discharge, thus shorten the standing time of the open-circuit voltage method. Then estimate the initial SOC using the relationship between the open circuit voltage and SOC. Use ampere-hour integral method when charging and discharging and modify the SOC using the open-circuit voltage method to the resting batteries. Compensate for the accumulative error of ampere-hour method and achieve high estimate accuracy. Fig1 is the flow chart.

The implementation process of the whole algorithm includes two key parts: getting the relationship between OCV and SOC and getting the parameters of the equivalent circuit model of the battery. Both of them require a huge amount of experimental data collection.

3.1   Relationship between VOC and SOC

To acquire data to identify the parameters of the open-circuit voltage, a test was performed on the lithium-ion battery. Experiment was based on a single manganese acid lithium battery of 1300 mAH at room temperature.

First, discharge it to 2.6V and let it sit (SOC is zero at this moment) ; then charge it with 0.5C by the constant current, let it sit for 2h each time after charging 6 minutes, Battery open-circuit voltage is measured with precise DMMS until the open-circuit voltage of the battery is close to 4.2V. 21 groups of open circuit voltage and the corresponding SOC are shown in table 2.

Table 2. Open circuit voltage and the corresponding SOC

SOC	OCV(V)	SOC	OCV(V)
0%	3.194	55%	3.775
5%	3.458	60%	3.819
10%	3.488	65%	3.879
15%	3.521	70%	3.927
20%	3.568	75%	3.955
25%	3.609	80%	3.990
30%	3.625	85%	4.030
35%	3.641	90%	4.058
40%	3.657	95%	4.087
45%	3.686	100%	4.125
50%	3.725		

Map the relationship curve with matlab/simulink as figure 2.

The practical application of real-time online measurement, the method of estimating SOC is used mostly Ampere-hour integral method. However, Ampere-hour integral method exist error when measuring and with the increase of time, cumulative error is getting bigger and bigger. So separately by using this method to estimate the SOC of the battery doesn't obtain good effect. In practice, it is often used together with open circuit voltage method, Figure 2 illustrates that the curve of OCV-SOC of manganese acid lithium battery has a better linearity compared with Fe Battery. So the open circuit voltage fits more for estimating the SOC. Therefore we can modify the algorithm with figure 1.

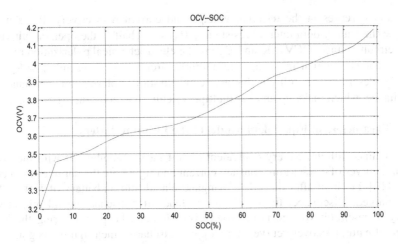

Fig. 2. Relationship curve of OCV-SOC of manganese acid lithium battery

3.2   Equivalent Model of the Lithium Battery

Equivalent circuit models use a combination of voltage and current sources, resistors and capacitors to model battery performance. It can be used to represent the electrical property of li-on batteries. The role of the battery model is to realize the conversion between the voltage of the battery and its open-circuit voltage in the running process of battery management system.

The common battery models include electrochemical model, equivalent circuitn model, neural network model and the specific factors model [8]. Among this, the equivalent circuit model is the based on the principle of battery power network theory to describe the performance of the battery. It has the advantages of having clear physical meaning and is convenient for mathematical analysis for parameter identification, so it is widely used in the simulation in the new energy car [9-11]. Considering in terms of the accuracy and complexity of the model, Choose the second-order RC equivalent circuit model as shown in figure 3.

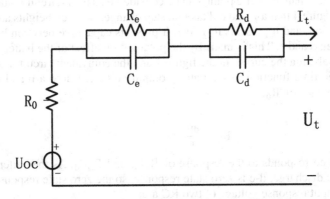

Fig. 3. Second-order RC equivalent circuit model

U_t and I_t represent the terminal voltage and current respectively, R_0 is used to describe the battery ohm internal resistance, U_{oc} on behalf of the open circuit voltage (open circuit voltage, OCV). R_e and C_e are the electrochemical polarization resistance and capacitance, which make up one multiple circuit, time constant $\tau_e = R_e C_e$; R_d and C_d describe the lithium battery concentration polarization resistance and capacitance, which make up another multiple circuit, time constant $\tau_d = R_d C_d$[12].

3.3   Parameters of Equivalent Model of the Lithium Battery

After obtained lithium battery equivalent circuit model, combine it with the voltage response curves of discharging constant current and figure out the parameters in circuit model [13]. Let it sit after the single manganese acid lithium battery of 1300 mAH is fully charged. Disconnect the dc electronic load after discharging 80 min at 0.5C. Battery open-circuit voltage is recorded every minute, keep monitoring the battery voltage of natural recovery curve after stopping discharge, then figure 4 is got.

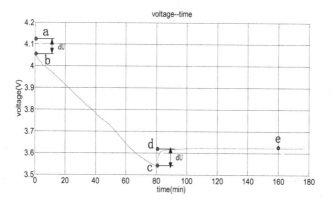

Fig. 4. Discharging voltage response curve

a-c is the curve when the battery is discharging at 0.5C; it was disconnected the load at c point; d-e is the battery's natural recovery curve after the discharge is ending; the voltage no longer change after e point, can be considered as open-circuit voltage. It can be seen from figure 4 that a-b and c-d take on step change, two step heights are equal in opposite direction, and it presents resistive. It presents capacitive between b and c, as well as between d and e. This is made by the polarization effect of the battery.

Through analyzing the curves in the figure 4 and the equivalent circuit model in the figure 3, fit the time function of the battery output voltage. Curve a-b and curve c-d reflect the influence of R_0.

As a result:

$$R_0 = \frac{dU}{I} \tag{2}$$

b-c and c-d corresponds to the response of $R_e C_e$ and $R_d C_d$. As the battery is fully stewing before discharge, d-e is zero state response, so the zero state response voltage and the zero input response voltage of two RC are:

$$u_{p1} = u_{ce} + u_{cd} = I(t)R_1\left(1 - e^{-t1/\tau e}\right) + I(t)R_2(1 - e^{-ti/\tau d}) \tag{3}$$

$$u_{p2} = u_{ce} + u_{cd} = U_1 e^{-t2/te} + U_2 e^{-t2/\tau d} \tag{4}$$

Assume that the open-circuit voltage is U_{ocv} and the voltage between d and e is $U(t)$, thus the terminal voltage of two capacitor U_{p2} is:

$$U_{p2} = U_{ocv} - U(t) \tag{5}$$

After combining formula (4) with formula (5), we are able to infer the formula of open-circuit voltage .

As a result:

$$U_{ocv} = U_{p2} + U(t) = U_1 e^{-t/\tau e} + U_2 e^{-t/\tau d} + U(t) \tag{6}$$

According to the experimental data that measured, combine with the formula (3) and formula (4), then fit the curve with MATLAB, we can get the value of U_1, U_2, τ_e, τ_d sepretely. Then we can estimate the open-circuit voltage after putting them into the formula (6). Thus greatly shorten the standing time.

4   Experimental Result Discussion

Figure 4 is the voltage response curve when discharged with constant-current, first, combine it with formula (3) and formula (4); then work out U_1 , U_2 , τ_e , τ_d use nonlinear least square curve fitting method; finally, fit the voltage curve of d-e through MATLAB, compared with the compared results, the result is shown in Figure 5.

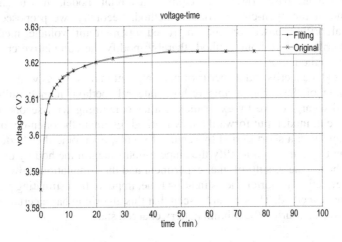

Fig. 5. The experiment and simulation comparison chart of d-e

As you can see in figure 5, the simulation of d – e fits well ; SEE is 2.78e-005, mean square error reach 9.36×10^{-4}. Discharge for 81 min by 0.65A before disconnecting the load. Take the measured data into the fitting curve, the results are shown in table 3. The measured open-circuit voltage is 3.623V, and the corresponding SOC is 29.38%.

Table 3. Different standing time and its corresponding estimation error rate

standing time/min	estimated voltage/V	Corresponding SOC	estimation error rate
5	3.624	29.49%	0.37%
10	3.624	29.49%	0.37%
20	3.623	29.38%	0
30	3.623	29.38%	0
60	3.623	29.38%	0

The traditional Open-circuit voltage method needs 1-2h before measuring Open-circuit voltage. While we can estimate the open-circuit voltage of the battery at the moment in a relatively short period of time using the proposed model parameters, thus can effectively shorten the estimate time, improve the estimate precision of the ampere-hour method.

5   Conclusion

In this paper, we made two contributions. First we presented a way to estimate open-circuit voltage based on battery equivalent circuit model, thus to shorten the standing time of the open-circuit voltage method. Secondly, we proposed an SOC estimation algorithm based on the improved open-circuit voltage method, and combined it with ampere-hour method, thereby modify the cumulative error for the ampere-hour method.

We analyzed the merits and demerits of main algorithm of SOC, as well as the major factors causing of the error in ampere-hour integral method, namely initial value. Getting the discharge curve by experiments, after combining with second order RC equivalent circuit model, put forward a new method. Estimate the open-circuit voltage of lithium battery in a short period by regular formula, overcome the shortcoming of long standing time, and then modify the ampere-hour integral method by open-circuit voltage method. The result of the experiment and the simulation shows that this method can effectively shorten the estimating time, improve the estimating precision of the ampere hour method. Thus it can be seen that this algorithm has a certain reference on the value for measuring the lithium battery charged state.

References

1. Zhang, Y.: LiFePO4 Dynamic Battery management system for Pure Electric Vehicle. Beijing Jiaotong University Electrical Engineering, Beijing (2009)
2. Cheng, K.W.E., Divakar, B.P., et al.: Battery-Management System and SOC Development for electrical vehicles. IEEE Transactions on Vehicular Technology 60(1), 76–88 (2011)
3. Liu, Y.Z., Zhang, Y.H., Wang, D.L.: VehiclePower Lithium Batteries. Power Electronics Technology 45(12), 48–50 (2011)
4. Wang, H., Zheng, Y.P., et al.: An Algorithm for predicting state of charge of Lithium Iron Phosphate Batteries. Chinese Journal of Power Sources 35(10), 1198–1207 (2011)
5. Li, Z., Lu, L.G., Ouyang, M.G.: Comparison of methods for improving SOC estimation accuracy throughampere-hour integral method. Journal of Tsinghua University 50(8), 1293–1301 (2010)
6. Lin, Z.Q., Teng, X.H.: An improved ampere-hour integral method for testing battery SOC. Journal of Guizhou University 26(6), 108–110 (2009)
7. Shi, W., Jiang, J.C., Li, S.Y., Jia, R.D.: Research on Estimationfor LiFePO4 Li-ion batteries. Journal of Electronic Measurement and Instrument 24(8), 769–774 (2010)
8. Wang, Z.G., Jin, X.M., et al.: A power battery model for power type based the analysis of the circuit transient. Journal of Beijing Jiaotong University 36(2), 91–94 (2012)
9. Li, S.H., Senior Member, et al.: Study of Battery Modeling using Mathematical and Circuit Oriented Approaches. In: Power and Energy Society General Meeting, pp. 1–8 (2011)
10. Hu, X.S., Li, S.B., Peng, H.: A comparative study ofequivalent circuit models for Li-ion batteries. Journal of Power Sources 198, 359–367 (2012)
11. Zhang, L., Zhu, Y.J., Liu, Z.Y.: Research on Joint Estimation for SOC and model parameters of Li-ion batteries. Journal of Electronic Measurement and Instrument 26(4), 320–324 (2012)
12. He, H.W., Xiong, R., Zhang, X.W., et al.: State-of-Charge Estimation of the Lithium-Ion Battery Using an Adaptive Extended Kalman Filter Based on an Improved Thevenin Model. IEEE Transactions on Vehicular Technology 60(4), 1461–1469 (2011)
13. Xiao, M., Choe, S.Y.: Dynamic modeling and analysis ofa pouch type LiMn2O4/Carbon high power Li-polymerbattery based on electrochemical-thermal principles. Journal of Power Sources 218, 357–36 (2012)

Application of the Real-Time EtherCAT in Steel Plate Loading and Unloading System[*]

Bin Jiao and Xiuming He

Shanghai Dianji University, Shanghai, China
jiaob@sdju.edu.cn, hejayming@126.com

Abstract. The paper firstly discusses the EtherCAT real-time Ethernet technology in detail, including operating principle, communication protocol and superior performance of EtherCAT Ethernet , synchronization, high-speed and so on. To show how to build up a master system based on configuration software TwinCAT and how to design a slave system considering the features of application, the methods of developing systems based on EtherCAT technology are proposed. Finally, a plate loading and unloading system based on EtherCAT technology is designed to obtain a faster response speed and higher synchronization accuracy. The system has realized high-speed remote data transmission and high-precision velocity control for Multi-axis which the traditional fieldbus can't achieve.

Keywords: EtherCAT, Real time Ethernet, Loading and unloading system, Synchronization.

1   Introduction

With the rapid development of industrial automation, in motion control system, requirements of real-time and network are also increasing, along with which a variety of real-time Ethernet was born. Traditional Fieldbus can not meet the requirements of the industrial field, such as, high-speed, high precision and multi-device. Development of real-time Ethernet, reduces the cost of automation equipment, improves the real-time response speed and conforms to the industrial automation trends[1-3].

The real-time Ethernet EtherCAT, is a real-time bus technology newly developed by German BECKHOFF. With superior performance, EtherCAT can not only make the real-time data exchanged between intelligent devices and efficient transmission of data sampling possible, but also meet the technical requirements of PC-based real-time control[4-5]; it satisfies the bus request of an integrated automation system proposed by the reference [6].

This paper describes the communication theory and superior performance of EtherCAT technology and introduces the design of an EtherCAT system, finally, puts forward an application of the EtherCAT technology in steel sheet loading and unloading systems.

---

[*] This work is supported by The Project of Shanghai Municipal Science and Technology Commission. (Grant No. 13DZ0511300).

K. Li et al. (Eds.): LSMS/ICSEE 2014, Part III, CCIS 463, pp. 268–275, 2014.
© Springer-Verlag Berlin Heidelberg 2014

2   A Real-Time Ethernet Technology Introduction-EtherCAT

EtherCAT (Ethernet for Control Automation Technology) is a real-time Ethernet fieldbus system. EtherCAT is fast, easy wiring; has a good compatibility and openness; is suitable for applications where need rapid control.

2.1   EtherCAT`s Operating Principle

Ethernet is a local area baseband bus network according to 802.3 using Carrier Sense Multiple Access/Collision Detection (CSMA/CD) media access control methods. Real-time Ethernet technology-EtherCAT uses a master-slave media access method. In an EtherCAT-based system, the master controls the slave to send or receive data from the station. Master sends a data frame, slave reads the output data in slave-related messages from the standing data when the frame goes through[7]. Meanwhile, the input data from the slave inserts into related packets of the same data frame. When all of the data frame from the station goes through all slave and finishes the exchange of data from the station, the data frame is returned by the end of EtherCAT slave system shown in Figure 1.

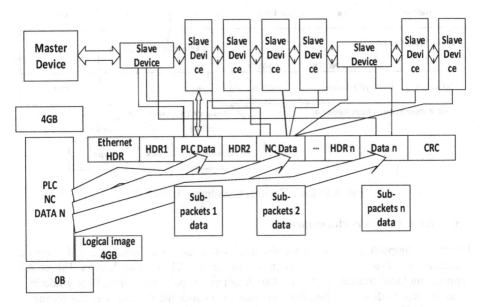

Fig. 1. Process data inserted in telegrams

2.2   EtherCAT Protocol

EtherCAT technology can fully take advantage of full-duplex Ethernet`s bandwidth. Using a master-slave media access technology, sent by the master Ethernet data frames to each slave, to communicate data. EtherCAT frame meets ISO/IEC802.3 standards, EtherCAT packet consists of a number of EtherCAT sub-packets which are located on a standard Ethernet frame structure of the data area. Each sub-packet servers a particular area of a logical process image region, which is up to 4GB bytes. Data does not depend on the order of the physical order of the Ethernet network terminals, and can be arbitrarily addressed. Thus, broadcasting and multicasting communication can be achieved between slave stations. If you want to maximize performance and EtherCAT components need to be on the same subnet as the controller to operate, you can be directly transmitted Ethernet frames.

Since the application is not limited to a single EtherCAT subnet, the EtherCAT protocol will be packaged into UDP/IP data packets by EtherCAT UDP, shown in Figure 2. Thus, any Ethernet protocol stack can be addressed to the control system into the EtherCAT. You can even make communication through a router connected across the other subnets[8].

According to the master-slave data exchange principle, EtherCAT is also very suitable for communication between the controller (master/slave). Addressable network variables can be used freely in process data and parameters, diagnostics, programming and various remote control services,which can meet a wide range of application requirements.

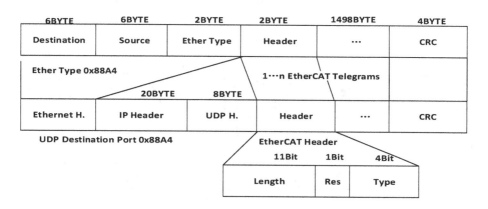

Fig. 2. EtherCAT frames according to IEEE 802.3

2.3   EtherCAT`s Performance

EtherCAT supports a variety of topologies, such as bus, star, ring, etc., and allows a combination of a variety of structures in EtherCAT system.Besides EtherCAT supports multiple transmission cables such as twisted pair, fiber optic light guide bus, so as to adapt to different situations, in order to enhance the flexibility of the wiring.

EtherCAT uses a precise clock synchronization system[9]. Data exchange system is completely based on pure hardware mechanisms, because communication utilizes a logical ring structure (physical layer by means of a full-duplex Fast Ethernet), the master clock can easily and accurately determine the individual from the station propagation delay skew . The distributed clocks are adjusted based on the master clock(typically, clock of the first slave station is choosed as the master clock), using precise and time-based synchronization errors identified within the network.

EtherCAT is direct access to integrated slave hardware and network controller and the processings of the protocol are implemented in hardware,so that it is completely independent of the real-time operating system protocol stack, CPU performance, or a software implementation, and the processing data transmission speed is faster than other protocols. For example, to transmits data of the distribution of 256 digital I / O takes only 11us, 1000 areal digital I/O takes only 30us, 200 analog I/O (16 bit) takes only 50us.

EtherCAT communication has strong diagnostic capabilities and can quickly shoot troubles. Meanwhile it also supports the master slave redundancy error detection to improve system reliability. EtherCAT achieves in integrating the safety-related communications and control communications in the same network as a whole, and follows the IEC61508 standard, meets the safety integrated level (SIL) 4 requirements[10].

3   Design of Systems Based on Real-Time EtherCAT Technology

EtherCAT system is a master-slave communication system, while the entire system consists of the master and slave. Slave station is connected to master via EtherCAT bus, and the entire system is controlled by the master. Design of EtherCAT system includes design of slave and master station.

3.1   Design of EtherCAT Master

Function of EtherCAT master can be achieved using TwinCAT- the configuration software provided by the company BECKHOFF. By configuring the configuration software to run on computers equipped with WinowsXP, you can realize the function of main station. TwinCAT communication system is shown in Figure 3. The system consists of a kernel mode and user-mode. In core mode system runs a real-time kernel (BECKHOFF real-time kernel), and the kernel is embedded in windows XP (or Windows NT) operating systems, to achieve real-time data communications. Various other logic devices of system utilize ADS (Automation Device Specification) interface to exchange information with ADS router. User programs are running in user mode. System executes the program by way of scanning cycles and time of each cycle can be set by system parameters. In a cycle BECKHOFF core owns CPU priority, so it can complete user control task first. When the task is completed, BECKHOFF core gives the right of using CPU back to operating system. Thus TwinCAT can achieve real-time control. TwinCAT also provides OCX, DLL, OPC interfaces to achieve human-machine interface functions.

In practical applications, user writes slave device configuration file (xml file) according to the characteristics of the slave device, so that the master is able to identify slave devices to complete related initialization and control.

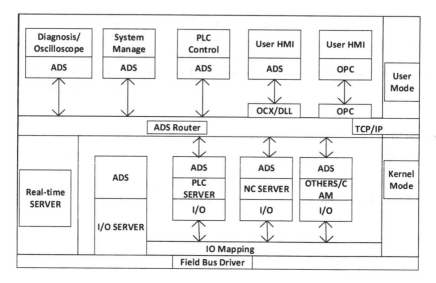

Fig. 3. TwinCAT communication system structure

3.2   Design of EtherCAT Slave

EtherCAT slave is the computer system on the bus, which consists of the Ethernet controller, microprocessors and sensors, actuators. Slave general structure is shown in Figure 4.

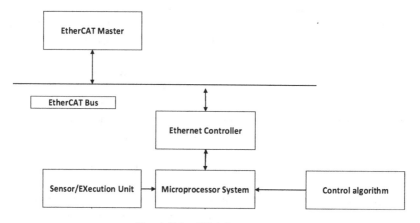

Fig. 4. EtherCAT slave system

EtherCAT controller, which is the most critical part of the EtherCAT slave, can achieve application layer processor of real-time Ethernet EtherCAT and mainly achieve application layer protocol. As to a logic control system, slave system can be designed into a simple digital IO, simply configured as digital IO interface without the need for application-layer Ethernet control processor. For a complex control system, based on application needs, the application layer processor should be designed into an embedded system. It mainly achieves application layer protocol and can also achieve emergency control algorithms to improve the reliability of the system.

4   Application of Real-Time EtherCAT in Steel Plate Loading and Unloading System

Loading and unloading system of the steel sheet, the single gripping mechanism to grab the steel sheet has length and weight restrictions. For length, weight larger steel grab institutions need multiple simultaneous operation to complete. This requires synchronization operation of multiple drive motor, and also high-precision motor-speed control. Real-time live performance of the network directly affects control accuracy and multi-motor synchronization performance, therefore, system puts forward higher requirements about the fieldbus or Etherne. Following is a brief introduction of a multi-axis steel loading and unloading control system based on real-time Ethernet EtherCAT.

4.1   Hardware Design of Multi-axis Loading and Unloading System Based on EtherCAT

Loading and unloading system(for example 4-axis system) is controlled by embedded PC that supports EtherCAT as a controller (such as Beckhoff CX1030), the multi-axis motion controller and servo amplifier as a drive (eg. SEW DHF41B and MXR), by control gear with encoder as the drive motor.   EtherCAT Slave Controller (eg. ESC20) is the connection between the embedded PC and the driver, which constitute a system with an embedded PC as the EtherCAT master and the driver as the EtherCAT slave. Figure 5 is the interface between EtherCAT Ethernet Controller ESC20-controller of multi-axis loading and unloading system based on EtherCAT, and application layer controller TMS320F2812 DSP. The Ethernet controller ESC20 is configured as a parallel port, through the DSP's external bus interface for data exchange.

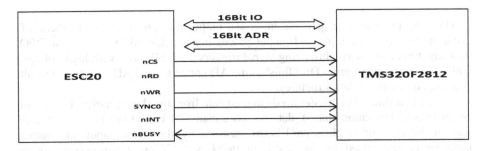

Fig. 5. Interface of ESC20 and TMS320F2812

ESC20 controller supports distributed clock signal to provide simultaneous sampling for each slave. As the controller of EtherCAT application layer communication, DSP mainly achieves application layer protocol of EtherCAT. By cyclic reading and writing on the application layer control registers of ESC20, DSP determines the control function between master and slave. Slave determines to complete the appropriate application layer functions according to the value of the controller. Finally, by writing the application layer status registers of ESC20 system responds to events both from controller and driver(such as location/constant speed).

4.2   Software Design

Software design process is shown in Figure 6.

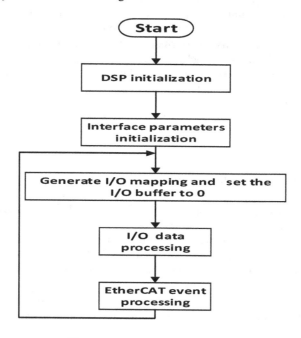

Fig. 6. Flow chart of software

The main program loop is about input and output data processing and EtherCAT event processing. Data input and output processing is achieved by defining a FIFO data structure. ESC20 synchronizing signal triggers the DSP INT1 with high voltage, within the interrupt program DSP finishes the AD converter and AD conversion result is written in the FIFO data structures.

Thus, when EtherCAT master needs to read data from the slave (master station will be set to Safe-Operation status), data(no more than 128 16-bit) in FIFO structure is sent to the relevant application ESC20 memory by the slave. Input and output updating is determined by the result of the EtherCAT event processing, which depends on state of the slaves.

After the master and the slave is built, you can achieve motion control as you want by configuring the TwinCAT parameters. TwinCAT provides virtual axis to control muli-axis. For example, you build a virtual axis as master axis and set the axis(motors) to follow the virtual axis with the same speed or different speed according to the application requirements. Besides, you can monitor speed of the actual axis by TwinCAT.

5   Conclusion

With advanced technology and superior performance EtherCAT will be widely used in multi-axis loading and unloading systems. It can solve the problem of high-speed real-time control signal transmission of multi-axis plate loading and unloading systems; improve multi-linkage system running synchronization; provide the conditions to improve the performance of the control algorithm; promote economic operation of multi-axis loading and unloading system.

References

1. Zhou, Y., Zhang, S., Huang, Q., Zhou, Y.: Synchronization of the Multi-axis Servo Motion Controller System. Science & Technology Review 28, 56–60 (2012)
2. Industrial Ethernet Technologies,
   `http://wenku.baidu.com/link?url=4vDE40W5ea2_UOE268EDg3-r6Q-`
   `kEjQXYvY27GfnwQSw3EHx2nz4ps8EM1_jS_IkYYBBScPUKAKFIlpwpVljbEf`
   `qfD3d6iSu-_ZVIOsY3eS`
3. Yu, S., Fan, L., Xia, B., Zhu, Z.: Study and application of ethercat distributed clock synchronization technology. Manufacturing Automation 6, 118–120 (2013)
4. Ma, D., Xiao, J., Wang, H.: Research on networked motion control system based on DSP and CAN bus. Manufacturing Automation 3, 5–7 (2013)
5. Ruan, Q., Wang, H., Shi, D., Liang, X.: Design of the High-performance AC Servo Control Systems Based on EtherCAT. Science & Technology Review 20, 60–61 (2010)
6. Huang, Q., Huang, S., Liu, Z., Wang, H., Pan, X., Chen, Z.: Design and Application of a High-Performance Servo System Based on EtherCAT. Small & Special Electrical Machines 4, 42–45 (2013)
7. Li, Z., Zheng, M., Wang, J., Zheng, A.: Design of Triaxial Motion Servo Control Systems Based on EtherCAT. Modular Machine Tool & Automatic Manufacturing Technique 2, 64–65 (2012)
8. Kang, C., Lin, Z., Ma, C., Huang, X., Fei, R.: Study on EtherCAT with TwinCAT master. Modern Manufacturing Engineering 11, 16–18 (2010)
9. Shi, D., Liu, J., Wang, H., Ruan, Q.: A Design of Real-Time Ethernet EtherCAT. Journal of Natural Science of Hunan Normal University 3, 37–40 (2011)
10. EtherCAT Technology Group.: EtherCAT technical introduction and overview. EtherCAT Technology Group, Germany (2007)

A Game Strategy for Power Flow Control of Distributed Generators in Smart Grids*

Jianliang Zhang, Donglian Qi**, Guoyue Zhang, and Guangzhou Zhao

Department of Electrical Engineering, Zhejiang Uinversity, Hangzhou 310027, China
{jlzhang,qidl,zhangguoyue,zhaogz}@zju.edu.cn

Abstract. We consider the distributed power control problem of distributed generators(DGs) in smart grid. In order to ensure the aggregated power output level to be desirable, a group of DGs with local and directed communications are expected to operate at the specified same ratio of their maximal available power output. To that end, the non-cooperative game is introduced and the DGs are modeled as self-interested game players. A new game model, termed state based weakly acyclic game, is developed to specify decision making architecture for each DGs, and at the point of the equilibrium of the game, the global objective of the power control problem can be achieved through autonomous DGs that are capable of making rational decisions to optimize their own payoff functions based on the local and directed information from other DGs. The validness of the proposed methodology is verified in simulation.

Keywords: Power flow control, distributed generators, non-cooperative game, distributed optimization.

1 Introduction

The growing threat of climate change and the depletion of non-renewable energy sources have led to the growth of distributed generators(DGs) in smart grid[1–3].However, with the increase of penetration of DGs in smart grid, many technical challenges appear in power systems. In a distribution network with a high level of penetration of DGs, any sudden disturbance may influence the information exchanges among DGs and trigger the reduction of their operation capability, even may result in a significant loss of active power support for power systems[8].

In order to control the total power output of DGs under a certain information exchange in a group, different control modes can be applied. In general, the centralized control modes such as the optimal power flow based approaches[4, 5],

---

* This work is jointly supported by National High Technology Research and Development Program of China(2014AA052501), National Nature Science Foundation of China(No.61371095 and 51177146) and the Fundamental Research Funds for the Central Universities(2014QNA4011).
** Corresponding author.

K. Li et al. (Eds.): LSMS/ICSEE 2014, Part III, CCIS 463, pp. 276–285, 2014.

are the standard approaches to dispatch traditional generators. However, centralized control need to collect the system-wide information and sending commend globally. Therefore, for a large number of DGs it would be too expensive to be implement and may not be robust to abnormal conditions in communication.

Instead, the decentralized control modes, such as the maximum photovoltaic power tracking[6] or the constant voltage and frequency with droop mode[7], are only efficient to control a small number of DGs by using its own information.

As the number of DGs increases, it is impossible for DGs to get the global information. However, it is practical that the DGs can regulate its power output level by using local and time varying communication with other DGs. Furthermore, the information flow among DGs may be directed because of the abnormal operation of a single communication channel. Either the centralized or decentralized control modes will be not practical to manage the total power output of DGs under local, directed and time varying communication conditions. Recently, the distributed control modes[8], which admit local communication and combine the positive features of both centralized and decentralized control modes, have been developed to control the DGs in a group.

In order to coordinate the DGs, local control law need to be designed for each DG such that the global objective can be realized based on the local, directed and time varying communication. However, designing local control laws for DGs with real-time adaption and robustness to dynamic uncertainties may come with several underlying challenges[9, 10]. Such challenges includes coordinating behaviors of selfish and rational DGs as well as dealing with overlapping and distributed information for a potentially large number of interacting DGs. Interestingly, these challenges are fitting into the category of non-cooperative game theory[11].

The main contribution of this paper is to develop a new game model, termed state based weakly acyclic game, to solve the power control problems of DGs under directed and local information exchanges. In the game model, the DGs in a group are modeled as self-interested decision makers, which interacts with each other to coordinate the individual power output to capture desired system level objective under minimal communication requirement, even though the corresponding communication topologies among DGs may be local, directed, and time-varying.

The rest of the paper is organized as follows: in Section 2, the power control problem of DGs is discussed. In Section 3, the communication among DGs is investigated. In section 4, the state based weakly acyclic game formulation for the power control problem is developed and the properties of the game model are analyzed. In Section 5, the simulation illustrates the validity of the proposed methodology and the conclusions are drawn in Section 6.

2   Problem Setup

2.1   Power Control Problem

Suppose there are $n \geq 2$ DGs which are geographically dispersed with directed and time varying information exchanges. The objective is to ensure the aggregated power output of all DGs in a group to be desirable by coordinating the power output of each individual DG. A simple solution is to specify the same power utilization profile for each DG i, which is defined by the ratio of real time power output p_i versus maximum available capability p_i^{max} as follows[8]:

$$\gamma^* = \frac{p_1}{p_1^{max}} = \frac{p_2}{p_2^{max}} = \cdots = \frac{p_n}{p_n^{max}}, \tag{1}$$

where γ^* is the desired utilization profile, which can be determined by the high level control in grid, such as the transmission and distribution control center[8, 12]. Then the objective is to establish a set of local control laws for DGs such that the power output ratios of all DGs seek to γ^* , even though the output capabilities of some individual DGs may have large swings.

Generally, the DGs in a group can be denoted by the set $N = \{1, 2, \ldots, n\}$. Each DG $i \subseteq N$ can make decision to select its ratio from its possible decisions set denoted by Ψ_i, which is a nonempty convex subset of $\mathbb{R}$. We denote a specific joint decision profile by the vector $\gamma \triangleq \{\gamma_1, \gamma_2, \ldots, \gamma_n\}$ and $\gamma \subseteq \Psi \triangleq \prod_{i \in N} \Psi_i$, where Ψ is the closed, convex and non-empty set consisting of all possible joint decisions. The global objective of the power control problem is to enable all DGs reach the same desirable ratios γ^*. So the global objective can be captured by a global objective function $\phi : \Psi \rightarrow \mathbb{R}$ that system designer seeks to minimize, where $\phi(\gamma) = \sum_{i \in N} (\gamma_i - \gamma^*)^2$. It is noticed that the global function ϕ is differentiable and convex, and the global objective can be achieved if and only if the global function ϕ acquires its minimum value 0. Consequently, the above DGs control problem can be formalized as the optimization problem as follows:

$$\begin{aligned} &\min_{\gamma_i} \phi(\gamma_1, \gamma_2, \ldots, \gamma_n) \\ &\text{s.t.} \gamma_i \in (0, 1), \forall i \in N. \end{aligned} \tag{2}$$

2.2   Problems to be Solved

Suppose the time varying and directed communication among DGs can be described mathematically by a time-varying and piecewise-constant matrix, whose dimension is equal to the number of DGs and its elements assume binary values. Without loss of any generality, the matrix can be defined as follows [18]:

$$\mathbf{S}(t) = \begin{pmatrix} s_{11}(t) & s_{12}(t) & \cdots & s_{1n}(t) \\ s_{21}(t) & s_{22}(t) & \cdots & s_{2n}(t) \\ \cdots\cdots\cdots\cdots\cdots\cdots \\ s_{n1}(t) & s_{n2}(t) & \cdots & s_{nn}(t) \end{pmatrix}$$

where $s_{ii}(t) = 1$ because DG can always acquire its own information. In general, $s_{ij}(t) = 1$ if the DG i can get the information of DG j for any $j \neq i$ at time t; $s_{ij}(t) = 0$ if otherwise. Over time, binary changes of $S(t)$ occur at an infinite sequence of time instants, denoted by $\{t_k : k \in \Omega\}$, where $\Omega \triangleq \{1, 2, \ldots, \infty\}$,and $S(t)$ is piecewise constant as $S(t) = S(t_k)$ for all $t \in [t_k, t_{k+1})$. Because the information flows among DGs are directed, the matrix is asymmetric. Next, there are two problems that will be solved to control the power of DGs.

Problem 1 (Local control laws design): Design the local control law for DG i to designate how the DG processes available information at time $t - 1$ in order to formulate a decision γ_i at time t as $\gamma_i(t) = U_i\big(s_{i1}(t)\gamma_1(t - 1), s_{i2}(t)\gamma_2(t - 1), \ldots, s_{in}(t)\gamma_n(t - 1)\big)$, where $i = \{1, 2, \ldots, n\}$, $U_i(\cdot)$ is the local control law for DG i at time t.

Intuitively, it would be sufficient for all the DGs to be controlled properly if each of them can receive enough information from its neighboring units. However, it is not practical. So we need to decide the minimum requirement on communication properly to ensure the system level objective is satisfied. So it will be the problem as below.

Problem 2 (Communication topology design): Determine the communication among DGs in order to ensure the global objective is desirable while minimizing the communication costs.

3   Communication Topology Design

In this section, the *Problem 2* will be solved first by using matrix theory analysis[21].

In order to guarantee the validness of the proposed control strategy with the minimal information requirement, the rule of the communication topology will be introduced.

Rule[21]:   The   sequence   of   sensing/communication   matrices $S_{\infty:0}$ = $\{S(t_0), S(t_1), \ldots\}$ should be sequentially complete.

The sequentially completeness condition [8, 12, 18, 21] is a very precise method to schedule local communication among DGs. Especially, it gives the cumulated effects in an interval of time and shows that the cumulated communication network can be connected even if the network may not be connected at some time instants.

4   Power Control of DGs Based on State Based Weakly Acyclic Game

In this section, the state based weakly acyclic game is developed to solve the *Problem 1* in directed and time varying information networks, and the local payoff function $U_{i,t}(\cdot)$ is designed to specify the local control laws for any DG i. The system level objective can be distributed optimized through individual DGs that are capable of optimizing their payoff functions[13–16] to make rational decisions γ_i.

4.1 State Based Weakly Acyclic Game

Definition 1(State Based Weakly Acyclic Game): Consider a game denoted by $G = \{N, \{A_i\}_{i \in N}, \{U_i\}_{i \in N}, X, f, \varphi\}$. It consists of a player set N and an underlying finite state space X. Each player i has a state invariant action set A_i, and a state dependent payoff function $U_i : X \times A \to \mathbb{R}$. The state transition function $f : X \times A \to X$ depends on both the state and action. At any time t_0, given any state action pair $(x(t_0), a(t_0))$ that is not a state action pair, there exists a differential and convex potential function $\varphi : X \times A \to X$ and a time $t \in [t_0, t_0 + T]$, where T is a non-negative constant, and there exists **a player** $i \in N$ with an action $a_i(t) \in A_i$ such that the following two conditions are satisfied provided that $a(t_0) = a(t_0 + 1) = \cdots = a(t - 1)$ and $x(t_0) = x(t_0 + 1) = \cdots = x(t - 1)$:

(1) $U_i(x(t), a_i(t_0), a_{-i}(t_0)) - U_i(x(t), a_i(t), a_{-i}(t_0)) > 0$ and
$\varphi(x(t), a_i(t_0), a_{-i}(t_0)) - \varphi(x(t), a_i(t), a_{-i}(t_0)) > 0$.

(2) For every action $a \in A$ and the ensuring state $\widetilde{x} = f(x, a)$, for a null action 0, the potential function satisfies $\varphi(x, a) = \varphi(\widetilde{x}, 0)$.

Then G is called state based weakly acyclic game with potential function φ.

4.2 State Based Weakly Acyclic Game Design

In this subsection, we will give the details of state based weakly game formulation for the power control problem. The game formulation design process is based on the work of [9, 10, 19, 20]. In contrary to these works, the designed game will admit the directed, time varying and not always connected communications.

(1) *Game player*: The DGs can be modeled as the game players in the state based weakly acyclic game. It is noticed that the interactions among DGs influence the total power and finally impact each other's benefits. Therefore, the DGs can be modeled as the game players to consider the influence of interactions and make rational decisions.

(2) *State Space*: The state space is denoted by X. Each state $x \in X$ is defined by the tuple $x = (\gamma, e)$, where $\gamma = (\gamma_1, \gamma_2, \cdots, \gamma_n)$ is the value of ratios and $e = (e_1, e_2, \cdots, e_n)$ is the estimation terms for DGs' ratios. Here $e_i = \{e_i^1, e_i^2, \cdots, e_i^n\}$ is DG i's estimations for the value of joint ratio γ.

(3) *Action Space*: Each DG i selects its action a_i from its action space A_i, which will change its ratio and change its estimation through the directed information exchange among each other. Here, $a_i = (\hat{\gamma}_i, \hat{e}_i), \hat{\gamma}_i \in \mathbb{R}$ indicates the change in the ratio γ_i and $\hat{e}_i = (\hat{e}_i^1, \hat{e}_i^2, \cdots, \hat{e}_i^n)$ indicates the change in the DG i's estimation term e_i. For any DG i, the changes in the estimation terms for some specific DG k is denoted by the set as $\hat{e}_i^k \triangleq \{s_{ji}\hat{e}_{i \to j}^k\}_{j \in N}$ which indicates the estimations that DG i pass to its neighbors with regard to DG k in the directed information manner.

(4) *State Transition Function*: Suppose the states evolve in a deterministic way. For any specific DG $k \in N$, define the initial ratios $\gamma(0)$ and initial estimations $e(0)$ satisfies

$$\sum_{i \in N} e_i^k(0) = n \cdot \gamma_k(0). \tag{3}$$

Note that satisfying above condition is trivial as $e_i^i(0) = n \cdot \gamma_i(0)$ and $e_i^j(0) = 0$ for any DG $i, j \in N, i \neq j$. The state transition function $\{f_i(x, a)\}_{i \in N}$ consists of two parts. The first part, denoted by $\{f_i^\gamma(x, a)\}_{i \in N}$, relates to the transition of ratio for any agent $i \in N$, and the second part, denoted by $\{f_{i,k}^e(x, a)\}_{i,k \in N}$, relates to the transition of estimations for any DG i with regard to any DG k. For any DG $i, k \in N$, given state as $x_i = (\gamma_i, e_i)$, then the two parts in the state transition function can be described as follows[20]:

$$f_i^\gamma(x, a) = \gamma_i + \hat{\gamma}_i. \tag{4}$$

$$f_{i,i}^e = e_i^i + n \cdot \hat{\gamma}_i + \sum_{j=1}^{n} s_{ij} \hat{e}_{j \to i}^k - \sum_{j=1}^{n} s_{ji} \hat{e}_{i \to j}^k. \tag{5}$$

$$f_{i,k \neq i}^e = e_i^k + \sum_{j=1}^{n} s_{ij} \hat{e}_{j \to i}^k - \sum_{j=1}^{n} s_{ji} \hat{e}_{i \to j}^k. \tag{6}$$

It is easily to verified that for all times $t \geq 1$ and all DGs $k \in N$, we have

$$\sum_{i=1}^{n} e_i^k(t) = n \cdot \gamma_k(t). \tag{7}$$

(5)*Local payoff functions*: Given $\gamma = (\gamma_1, \gamma_2, \ldots, \gamma_n)$ as the joint ratio profile for n DGs. The average of local payoff functions of DG $i's$ neighboring DGs are chosen to act as a new kind of local payoff function for DG i, which is

$$U_i(\gamma) = \frac{\sum_{j=1}^{n} s_{ij} U_j(\gamma)}{\sum_{j=1}^{n} s_{ij}}. \tag{8}$$

Because the DG may not have the complete knowledge about the true value of its neighboring DG' ratios as quickly as possible, especially in systems with intermittent communication or time delays. So we will make use of the estimation term $e = (e_1, e_2, \ldots, e_n)$ in the state space to estimate the true value of ratios. Accordingly, the DG $i's$ local payoff function can be rewritten as

$$U_i(e_j|_{s_{ij}=1}) = \frac{\sum_{j=1}^{n} s_{ij} \phi(e_j^1, e_j^2, \ldots, e_j^n)}{\sum_{j=1}^{n} s_{ij}}, \tag{9}$$

where ϕ is the optimization function in (2) for the power control problem. Meanwhile, considering the error caused by the introduction of the estimation items, thus we will have to minimize the errors so that the global objective can be achieved. Accordingly, the local payoff function can be written as

$$U_i(x, a) = U_i(e_j|_{s_{ij}=1}) + \alpha U_i^e(x, a), \tag{10}$$

where α is a positive tradeoff parameter and

$$U_i^e(x, a) = \frac{\sum_{j=1}^{n} \sum_{k=1}^{n} s_{ij}(e_i^k - \gamma_k)^2}{\sum_{j=1}^{n} s_{ij}}. \tag{11}$$

4.3   Analytical Properties of State Based Ordinal Potential Games

In this subsection, we will further analyze the analytical properties of the designed game and verify whether it is capable of solving the power control problem in (2) or not. First, we will verify whether the designed game model is a state based weakly acyclic game in Theorem 1p.

Theorem 1. *Model the power control problem in (2) as the game model in Section 4.2 with any positive constant α. Given the potential function as $\varphi(x,a) = \varphi^{\phi}(x,a) + \alpha\varphi^{e}(x,a)$ where $\varphi^{\phi}(x,a) = \sum_{j=1}^{n}\phi(e_{j}^{1},\ldots,e_{j}^{n})/n$ and $\varphi^{e}(x,a) = \sum_{i=1}^{n}\sum_{j=1}^{n}\sum_{k=1}^{n}s_{ij}(e_{i}^{k}-\gamma_{k})^{2}/n$. Then the game model in Section 4.2 is a state based weakly acyclic game.*

Proof. It is straightforward to verify the state based ordinal game designed in Section 4.2 is a state based weakly acyclic game.

The potential function $\varphi(x,a)$ captures the global objective of power control problem. As we know in [17], the designed game will guarantee the existence of an equilibrium, at the same time, the game model allows the existing learning algorithms to be directly used to update decisions of DGs[15, 16, 19]. However, whether the equilibriums are solutions to the power control problem will be verified in the Theorems 2 and 3.

Theorem 2. *Model the power control problem in (2) as the state based weakly acyclic game proposed in section 4.2 with any positive constant α. Suppose the communicaiton topology is directed, time-varying, and the sequence of communication matrixes is sequentially complete, then at the equilibrium point of the game model, we have $\forall i, k \in N, e_{i}^{k} = \gamma_{k}$.*

Proof. The proof process is based on the concept of equilibriums of the game model[17]. Considering the communication topology is directed, the proof process will recursively verify the equivalence of estimation term of any connected DGs in the directed communication. Combining with the equation (7), the proof will be completed easily. We will omit the details here.

Next, we will need to examine the relationship between the equilibriums and the optimal solutions to the power control problem in Theorem 3.

Theorem 3. *: Model the power control problem in (2) as the state based weakly acyclic game proposed in section (4.2) with any positive constant α. Suppose the communication topology is directed, time-varying, and the sequence of sensing/communication matrixes is sequentially complete, then the resulting equilibrium $(x,a) = ((v,e)(\hat{\gamma},\hat{e}))$ is optimal solution to the power control problem.*

Proof. Because of space limit, we will omit the proof.

5   Simulation Results

In this section, the effectiveness of the proposed game theoretical framework will be illustrated. Suppose there are 20 DGs with directed and time varying communications, and the desired power output ratio is $\gamma^* = 0.5$, which is determined by the high level control in grid.

First, the communication topologies for the power control problem will be defined as follows: at any time t_0, suppose there is only one different DG denoted by i, sends its information to all the other DGs, that is, $s_{ji}(t_0) = 1$ where $j \in N\backslash\{i\}$. It is noticed that information flow among DGs is directed and some entries in matrix may switch from 1 to 0 intermittently. Therefore, the communication matrix is asymmetric and time-varying. In order to ensure the total power output of DGs converges to the expected operational point as well as the system is robust to time varying and intermittent information conditions, the matrix sequence $S_{t_0}, S_{t_1}, \cdots, S_{t_l}$ corresponding to the topologies should be sequentially complete. That is to say, in certain time interval $[t_0, t_l]$, the composite graph over each of the intervals has at least one globally reachable DG [8, 18]. It is obvious that the sequentially complete condition can be satisfied in our communication design and the composite graph has at least one globally reachable DG .

Then, the game framework for the power control problem is established as follows. For any DG i, its state is defined as $x_i = (\gamma_i, e_i)$, where $e_i = (e_i^1, \cdots, e_i^{20})$, and e_i^k is DG $i's$ estimation for DG $k's$ ratio γ_k. The action of DG i is defined as $a_i = (\hat{\gamma}_i, \hat{e}_i)$ where $\hat{\gamma}_i$ is the change in the ratio of DG i, and $\hat{e}_i = (\hat{e}_i^1, \cdots, \hat{e}_i^{20})$ is the change in the estimation term of DG i. The payoff function for DG i is defined as follows.

$$U_i(x, a) = \frac{\sum_{j=1}^{20} s_{ij} \phi(e_j^1, e_j^2, \ldots, e_j^{20})}{\sum_{j=1}^{20} s_{ij}} + \alpha \frac{\sum_{j=1}^{20} \sum_{k=1}^{20} s_{ij}(e_i^k - \gamma_k)^2}{\sum_{j=1}^{20} s_{ij}}, \quad (12)$$

where α is a positive tradeoff parameter.

In order to solve the power control problem in the framework of state based weakly acyclic game, the potential function is defined as follows.

$$\varphi(x, a) = \frac{\sum_{j=1}^{20} \phi(e_j^1, e_j^2, \ldots, e_j^{20})}{20} + \alpha \frac{\sum_{i=1}^{20} \sum_{j=1}^{20} \sum_{k=1}^{20} s_{ij}(e_i^k - \gamma_k)^2}{20}, \quad (13)$$

where α is the positive tradeoff parameter.

The dynamics of the power control problem in the game formulation is described as follows:suppose at time t, the high level control command DG i to increase its output power ratio to the desired utilization profile γ^*. Then DG i distributes the profile γ^* to its neighboring DGs in the directed communication topology and commands them to increase their ratios. All the DGs update their actions by using gradient descent algorithm, and finally asymptotically reach the same desired utilization profile γ^* to satisfy the desired total power output.

The power control problem is simulated by applying the framework of state based weakly acyclic game with the parameter $\alpha = 0.2$. The results are presented in Fig. 1. The left figure in Fig. 1 shows the evolution of ratios of the 20 DGs, and the right figure in Fig. 1 plots the dynamics of the optimization function ϕ for the power control problem. Compared with the traditional gradient descent method, our proposed method will asymptotically converges to the desired value of 0 by using limited information.

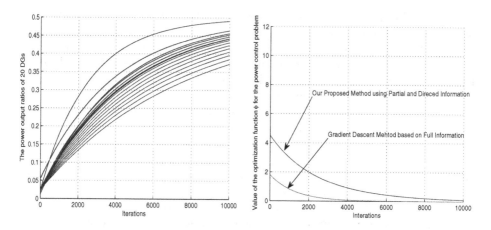

Fig. 1. Left figure shows the dynamics of the output power ratios of 20 DGs, and the right figure shows the dynamics of the value of the optimization function ϕ

6 Conclusions

In this paper, a new theoretical framework for analysis and design of power control problem for DGs is developed. This framework helps the system designer to implement the local control laws for DGs with the locality of information and guarantees the efficiency of the resulting equilibrium with regard to the global objective of the power control problem. There are two further research directions: (i) developing systematic procedures for designing the local payoff functions in the framework of state based games and (ii) exploring the influence of the parameter α to the dynamics of the game.

References

1. Katiraei, F., Iravani, M.: Power Management Strategies for a Microgrid with Multiple Distributed Generation Units. IEEE Transactions on Power Systems 21(4), 1821–1831 (2006)
2. Lopes, J., Hatziargyriou, N., Mutale, J., et al.: Integrating Distributed Generation into Electric Power Systems: A Review of Drivers, Challenges and Opportunities. Electric Power Systems Research 77(9), 1189–1203 (2007)
3. Kashem, A., Ledwich, G.: Multiple Distributed Generators for Distribution Feeder Voltage Support. IEEE Transactions on Energy Conversation 20(3), 676–684 (2005)
4. Gan, D., Thomas, R., Zimmerman, R.: Stability-constrained optimal power flow. IEEE Transactions on Power Systerms 15(2), 535–540 (2000)
5. Dent, C., Ochoa, L., Harrison, G., et al.: Efficient secure AC OPF for network generation capacity assessment. IEEE Transactions on Power Systerms 25(1), 575–583 (2010)

6. Kwon, J.M., Kwon, B.H., Nam, K.H.: Three-phase photovoltaic system with three-level boosting MPPT control. IEEE Transactions on Power Electronics 23(5), 2319–2327 (2008)

7. Diaz, G., Gonzaalez-Moran, C., Gomez-Aleixandre, J., et al.: Complex-valued state matrices for simple representation of large autonomous microgrids supplied by PQ and V f generation. IEEE Transactions on Power Systerms 24(4), 1720–1730 (2009)

8. Xin, H.H., Qu, Z.H., Seuss, J., Maknouninejad, A.: A Self-Organizing Strategy for Power Flow Control of Photovoltaic Generators in a Distribution Network. IEEE Transactions on Power Systems 26(3), 1462–1473 (2011)

9. Li, N., et al.: Designing Games to Handle Coupled Constraints. In: Proceedings of the 49th IEEE Conference on Decision and Control, pp. 250–255 (2010)

10. Marden, J.R., Wierman, A.: Distributed Welfare Games with Applications to Sensor Coverage. In: Proceedings of the 47th IEEE Conference on Decision and Control, pp. 1708–1713 (2008)

11. Fudenberg, D., Levine, D.: The Theory of Learning in Games. MIT Press, Cambridge (1998)

12. Xin, H.H., Gan, D.Q., Li, N., Dai, C.: A virtual power plant based distributed control strategy for multiple distributed generators. IET Control Theory and Applications 7(1), 90–98 (2013)

13. Marden, J.R., et al.: Cooperative Control and Potential Games. IEEE Transactions on Systems Man and Cybernetics Part B-Cybernetics 39, 1393–1407 (2009)

14. Arslan, G., et al.: Autonomous Vehicle Target Assignment: A Game Theoretical Formulation. Journal of Dynamic Systems, Measurement and Control-Transactions of the ASME 129, 584–596 (2007)

15. Marden, J.R., et al.: Joint strategy Fictitious Play with Inertia for Potential Games. IEEE Transactions on Automatic Control 54, 208–220 (2009)

16. Marden, J.R., et al.: Payoff based Dynamics for Multi-player Weakly Acyclic Games. SIAM Journal on Control and Optimization 48, 373–396 (2009)

17. Zhang, J.L., Qi, D.L., Zhao, G.Z.: A New Game Model For Distributed Optimization Problems With Directed Communication Topologies. Neurocomputing (to be appeared)

18. Qu, Z.H., et al.: Cooperative Control of Dynamical Systems with Application to Autonomous Vehicles. IEEE Transactions on Automatic Control 53, 894–911 (2008)

19. Young, H.P.: Learning by Trial and Error, Games and Economic Behavior 65, 626–643 (2009)

20. Li, N., et al.: Designing Games for Distributed Optimization. In: Proceedings of the 50th IEEE Conference on Decision and Control and European Control Conference, pp. 2434–2440 (2011)

21. Qu, Z.: Cooperative Control of Dynamical Systems: Applications to Autonomous Vehicles. Springer, London (2009)

Backstepping DC Voltage Control in a Multi-terminal HVDC System Connecting Offshore Wind Farms

Xiaodong Zhao[1], Kang Li[1], and Yusheng Xue[2]

[1] School of Electronics, Electrical Engineering and Computer Science,
Queens University Belfast, Belfast, BT9 5AH, Northern Ireland, United Kingdom
[2] State Grid Electric Power Research Institute, 210003, Jiangsu, China
{xzhao06,k.li}@qub.ac.uk

Abstract. Wind power is projected to play an important role in the current and future power systems. To integrate offshore wind farms to the existing onshore grid, voltage source converters (VSCs) based high voltage direct current (HVDC) transmission have drawn considerable interests from researchers and industries. As the most important variable in a DC system, DC voltage indicates the power balance between different terminals thus must be maintained in a safe range. To distribute transmitted wind power among distinct onshore AC grids, different operation modes need to be considered. This paper proposes to interoperate the backstepping DC voltage control method between various operation modes. DC cable dynamics are included in the DC voltage controller design to eliminate the effect of DC transmission line. Simulation is carried out in Matlab/Simulink to verify the backstepping enhanced DC voltage control method. It is shown that the transient stability can be improved with the backstepping method.

Keywords: backstepping control, MTDC, VSC, HVDC.

1 Introduction

Wind power is considered to be the most developed renewable energy in the world. By 2013, the wind energy has reached 117 GW installation capacity in Europe [1]. Due to the space requirement of large wind turbines and limited wind conditions onshore, offshore wind farms have become the key pillar of future energy for Europe [2]. Integrating a potential large number of offshore wind farms into the existing onshore AC grid has brought about the application of high voltage direct current (HVDC) transmission system. HVDC is considered to be more suitable for submarine transmission than traditional high voltage alternating current (HVAC) due to the intrinsic reactive current and power loss in an AC cable.

To connect the HVDC cable with AC network, voltage source converter (VSC) works as an interface to convert power from one side to the other. In an offshore HVDC transmission grid, more than two VSC terminals can connect to the same network. This will form a multi-terminal HVDC (MTDC) system. For the control of VSC in HVDC system, two separate control loops are often implemented using decoupled vector control. For steady and transient analysis, [3] studied the models of VSC-HVDC. Parameter uncertainties of a VSC-HVDC system have been studied in [4]. Inner current loop uncertainties have been considered in [5] and an adaptive backstepping control method

K. Li et al. (Eds.): LSMS/ICSEE 2014, Part III, CCIS 463, pp. 286–295, 2014.
© Springer-Verlag Berlin Heidelberg 2014

is proposed. A backstepping power controller is proposed in [6] integrating wind farms, without considering the DC voltage controller dynamics. For a two terminal HVDC system, [7] considers the DC cable dynamics based on backstepping method, however, the effect of inner current loop is not included. An adaptive backstepping design procedure with both cable dynamics and current loop is presented in [8]. However, these studies are limited to two terminal systems.

To control the DC voltage in a MTDC system, typical DC voltage control can be applied on one terminal with droop control on other terminals [9]. Another way is to share the DC voltage control burden among different terminals using droop control [10]. When droop control is deployed, the impact of transmission impedance on the DC power flow has been analysed in [11]. When outrage is occurred in one VSC, the effect of droop setting on the steady state voltage deviations is studied in [12]. A scheme for adapting the droop gains to share the burden according to the available headroom of each terminal is proposed in [13]. [14] proposed three different control strategies and operation modes for grid side converters in a four terminal HVDC system.

Based on the operation modes in [14], this paper propose a backstepping DC voltage controller under different modes. This paper is organized as follows. Section II introduces the system model. Three different operation modes for a four terminal HVDC grid are also outlined. Considering cable dynamics, operation and control of grid side converters are enhanced with backstepping controllers in Section III. Simulation based on Matlab/Simulink is carried out to verify the performance of the backstepping method on DC voltage control in Section IV. Finally, Section V is the conclusion.

2   System Modelling

In this paper, we will work on the system layout structure proposed in [14]. It is a four terminal HVDC system as shown in Fig. 1. Offshore wind turbines are connected using local AC cables to form one wind farm. All the active power collected by wind farms can be transmitted to DC grid if the wind farm AC voltage is controlled steady and forms a constant AC voltage source. For a long distance under sea transmission, the two wind farm converters are connected at a conjunction point. From this offshore conjunction point, power from the two wind farms are transmitted together to the onshore conjunction point where the power is then delivered to onshore grid 1 and 2 respectively. Two onshore grid side converters (GSVSC) are used to invert DC voltage back to AC for the AC connection.

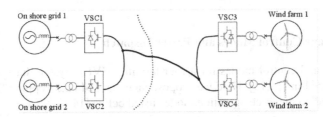

Fig. 1. Diagram of a four terminal HVDC system integrating two offshore wind farms

2.1 DC Cable Model

Each DC cable can be modelled using π sections. From each GSVSC's view, the connection to the onshore conjunction point is illustrated in Fig. 2. Here the DC cable is modelled as one π section for simplicity. In Fig. 2, R_1, L_1 are lumped cable resistance

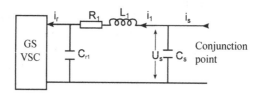

Fig. 2. DC cable connection from GSVSC to conjunction point

and inductance respectively. C_{r1} represents the DC side filter and cable capacitor. C_s is the capacitor at conjunction point. The two grid side converters will share the same capacitor at the conjunction point. Each converter can be modelled as a controlled current source if vector control is applied and the inner current loop dynamics can be ignored. After the above equivalence, the dynamic model for a four terminal system is shown in Fig. 3. VSC 1 and 2 represent the grid side converters and VSC 3 and 4 are wind farm side converters.

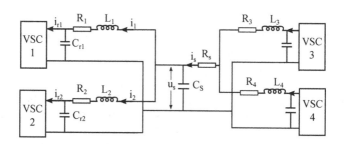

Fig. 3. Dynamic model for a four terminal HVDC system

2.2 DC Voltage Control Modes and Power Dispatch

The operation and control modes for a four terminal HVDC system have been studied and presented in [14,15]. Here we propose an enhanced backstepping DC voltage controller based on the three operation modes listed below [14,15].

Mode 1. In the first operation mode, GSVSC 1 has priority over GSVSC 2, which means unless the power need of GSVSC 1 has been satisfied, there will be no power transmitted to AC grid 2. Once the wind power exceeds certain level, DC voltage is no longer controlled by VSC 1 and will start to rise. VSC 2 then works as a DC voltage droop controller to deliver the extra power.

Mode 2. In operation Mode 2, the two grid side converters share the wind power generated according to a pre-determined ratio. This mode can be accomplished using the DC voltage droop control. Current delivered at each grid terminal is proportional to the voltage error from a reference value.

Mode 3. Mode 3 is a combination of the above two modes. GSVSC1 has some priority over GSVSC2 for certain amount of power. After the required amount of power has been delivered to AC grid 1, the two grid side converters will share the generated power thereafter.

3 DC Voltage Backstepping Control Design

Here we present a DC voltage backstepping control method, considering cable dynamics. It is based on adaptive backstepping method and works as an enhancement to the traditional DC voltage controller. For the convenience of a better understanding, the following state variable errors are defined first.

$$e_{us} = u_s^* - u_s, \ e_{i1} = i_1^* - i_1, \ e_{ur} = u_r^* - u_r \tag{1}$$

3.1 DC Voltage Control

Constant Voltage Control. In order to stabilize the capacitor voltage u_s, the Lyapunov function can be chosen as the energy on storage by the capacitor plus the wind farm power change.

$$V_s = \frac{1}{2}C_s e_{us}^2 + \frac{1}{k_{si}}\frac{1}{2}(i_s - \hat{i}_s)^2 \tag{2}$$

k_{si} is a weighing parameter and can also be shown to be the integration gain. i_s is the current from wind farm and $\hat{i}_s$ is the estimated current, for i_s can not be measured directly. The first order derivative of V_s needs to be negative in order to satisfy system stability requirement.

$$\dot{V}_s = e_{us}(-\hat{i}_s + i_1) + (i_s - \hat{i}_s)(-\frac{1}{k_{si}}\dot{\hat{i}}_s - e_{us}) \tag{3}$$

assuming that $\dot{i}_s = \dot{u}_s^* = 0$ in a short time interval. If the updating law is chosen as,

$$\hat{i}_s = -k_{si}\int e_{us}dt, \quad -\hat{i}_s + i_1 = -k_{sp}e_{us} \tag{4}$$

where k_{sp} is the proportional gain and $k_{sp} > 0$. The first order of V_s can be guaranteed that $\dot{V}_s < 0$ with a desired control law for i_1 as:

$$i_1^* = -k_{sp}e_{us} - k_{si}\int e_{us}dt \tag{5}$$

Voltage Droop Control. If more than one VSCs are regulating the DC voltage on C_s, the Lyapunov function can be chosen to ignore the current disturbance from the wind farm side. Following the same step with a constant voltage controller, this is a voltage droop controller and k_{sp} is the droop gain.

$$i_1^* = -k_{sp}e_{us} \tag{6}$$

3.2   DC Cable Backstepping Control

From the DC cable model in Section II, we have the DC cable dynamic equation as,

$$
\begin{aligned}
\dot{u}_s &= \frac{1}{C_s}i_s - \frac{1}{C_s}i_1 \\
\dot{i}_1 &= \frac{1}{L_1}(u_s - u_{r1} - R_1i_1 + \theta_1) \\
\dot{u}_{r1} &= \frac{1}{C_{r1}}(i_1 - i_{r1} + \theta_2)
\end{aligned}
\tag{7}
$$

Parameter change in inductor L_1, resistor R_1 and capacitor C_{r1} will cause inequality in (7), these uncertainties are denoted as θ_1 and θ_2.

Inductor Control. A new Lyapunov function can be chosen to include this new state variable and also to include parameter uncertainties θ_1.

$$V_{l1} = \frac{1}{2}L_1e_{i1}^2 + \frac{1}{2k_{l1i}}(\hat{\theta}_1 - \theta_1)^2 \tag{8}$$

where k_{l1i} is an adaptive gain for the parameter estimation $\hat{\theta}_1$. i_1^* represents the desired current and i_1 is the actual current. In order to guarantee i_1 can follow i_1^*, the control law for u_{r1} can be deduced from first order derivative of the new Lyapunov function. In order to make $\dot{V}_{l1} < 0$, we choose an updating law as,

$$
\begin{aligned}
\dot{\hat{\theta}}_1 &= -k_{l1i}e_{i1} \\
L_1\dot{i}_1^* - (u_s - u_{r1} - R_1i_1 + \hat{\theta}_1) &= -k_{l1p}e_{i1}
\end{aligned}
\tag{9}
$$

$k_{l1p} > 0$ is a positive gain. Then by choosing the control law of inductor voltage u_{r1} as,

$$u_{r1}^* = -L_1\dot{i}_1^* - k_{l1p}e_{i1} - k_{l1i}\int e_{i1}dt + u_s - R_1i_1 \tag{10}$$

we can make the first order of V_{l1} to be $\dot{V}_{l1} = -k_{l1p}e_{i1}^2 - e_{i1}e_{ur}$ If $e_{i1}e_{ur}$ can be eliminated at the next stage, then $\dot{V}_{l1} = -k_{l1p}e_{i1}^2 < 0$, thus the overall system is stable.

Capacitor Control. In order to eliminate voltage remaining error $e_{i1}e_{ur}$ from the previous stage, the new Lyapunov function considering capacitor C_{r1} can be constructed from previous stage.

$$V_{cr} = k_{cre}V_{l1} + \frac{1}{2}C_{r1}e_{ur}^2 + \frac{1}{2k_{cri}}(\hat{\theta}_2 - \theta_2) \tag{11}$$

where k_{cre} is the weighting parameter for the energy function from the last step, k_{cri} is the adaptive gain for parameter estimation $\hat{\theta}_2$. Similar to the derivative function in inductor voltage control, we have an updating law as:

$$\dot{\hat{\theta}}_2 = -k_{cri}e_{ur}$$

$$C_{r1}\dot{u}_{r1}^* - i_1 + i_{r1} - \hat{\theta}_2 - k_{cre}e_{i1} = -k_{crp}e_{ur} \tag{12}$$

The control reference for the input current to capacitor C_{r1} is

$$i_{r1}^* = k_{cre}e_{i1} - k_{crp}e_{ur} - k_{cri}\int e_{ur}dt + i_1 - C_{r1}\dot{u}_{r1}^* \tag{13}$$

i_{r1}^* is the reference current for VSC 1 DC side. Based on the power balance of converter, if vector control is used in the AC side current control, the reference current for the inner current loop can be easily obtained from i_{r1}^*.

Cable State Estimation. For the backstepping voltage controller, internal cable states will be used in some steps. However, not all the cable states can be measured directly in a usually long distance cable. A state observer can be used to estimate the internal cable states if the cable parameters are known. The DC voltage u_{r1} and current i_1 can be measured from the grid side terminals. For this one π section cable, only one state variable u_s needs to be estimated, hence a linear state observer should satisfy the requirement. A linear observe can be then built to estimate the remote side DC voltage u_s.

$$\dot{\hat{\mathbf{X}}} = (\mathbf{A} - \mathbf{LC})\hat{\mathbf{X}} + \mathbf{L}i_1 + \mathbf{B}u_{r1} \tag{14}$$

where $\mathbf{L}$ is observer gain.

4   Simulation Study

The proposed adaptive backstepping DC voltage control method is demonstrated in Simulink. Three operation modes as stated in Section II are investigated through simulation. The physical parameters of the MTDC system as shown in Fig. 3 are listed in Table 1. All the four VSC terminals, e.g. two grid side VSCs and two wind farm side

Table 1. Four terminal HVDC transmission line parameters

R_1	L_1	C_{r1}	R_2	L_2	C_{r2}	R_3	L_3	C_s	R_4	L_4
0.5Ω	$2.5mH$	$150uF$	1Ω	$5mH$	$150uF$	0.8Ω	$4mH$	$150uF$	1.5Ω	$7.5mH$

VSCs, have the same power and voltage ratings, at 500 MW, 400 kV respectively. The two grid side VSCs will work as DC voltage regulators. Both the cable states are estimated using state observers. Observer gains are chosen to be $[1.67 \times 10^5, 1 \times 10^4]$

based on the cable parameters to give an acceptable response. Parameters for the backstepping controller are tuned at each step and are listed as $K_{l1p} = 200$, $k_{l1i} = 3000$, $K_{cre} = 0.001$, $k_{crp} = 0.002$, $k_{cri} = 0.1$.

In the following simulation, both wind farms are equivalenced with lumped current source. A power step change is incurred in each wind farm at time $t = 0.05$ and $t = 0.15$ receptively. Power from wind farm 1 is increased from 50 MW to 200 MW at $t = 0.05$. Power from wind farm 2 is increased from 100 MW to 250 MW at $t = 0.15$. As a step input has an infinite spectrum, thus it can be used to inspect the potential stability capability of the controller. The two grid side converters are controlled using decoupled vector control in the current loop. Three different operation modes are modelled to investigate the voltage and current response of grid side converters.

4.1   Response of Mode 1

In operation mode 1, VSC 1 works as a DC voltage regulator. The reference voltage is set to be 400 kV. Any power change from the wind farm side is considered to be disturbance and VSCs are responsible to maintain the DC voltage at a fixed level. If the injected current exceeds the current limit of converter, VSC 1 can no longer regulate the DC voltage. Then VSC 2 will work as a DC voltage droop controller to deliver the

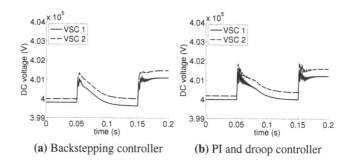

(a) Backstepping controller   (b) PI and droop controller

Fig. 4. DC voltage response with input power change in operation mode 1

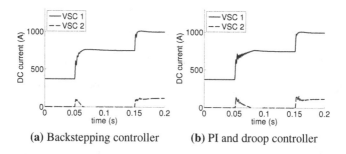

(a) Backstepping controller   (b) PI and droop controller

Fig. 5. DC current response with input power change in operation mode 1

extra power. Reference voltage for VSC 2 is set to be 401 kV considering any possible voltage drop on the transmission cable. Fig. 4 shows the voltage response following an input power change. Fig. 4 (a) is the voltage response using an adaptive voltage controller and Fig. 4 (b) is the result from a traditional PI controller. The current limit for VSC 1 is 1000 A. At time $t = 0.05$, delivered power is increased from 150 MW to 300 MW. However, current in VSC 1 is approximately 750 A after the change and is still under the current limit. Thus no power is delivered to VSC 2. This can be seen from DC current response in Fig. 5. It can be observed that the backstepping DC voltage controller can reduce the oscillations caused by power change and the effect on VSC 2. After $t = 0.15$, power is increased to 450 MW and current in VSC 1 is limited at 1000 A. The DC voltage will start to rise. When the DC voltage is above 401 kV, VSC 2 will work as a droop controller to deliver the extra power. Without considering the DC cable, the voltage response shows large oscillations. With the help of backstepping controller, this oscillation can be significantly reduced.

4.2 Response of Mode 2

In mode 2, both grid side converters work as droop controller and deliver power at the same time. Droop gain is 0.15 for VSC 1 and 0.2 for VSC 2. Reference voltages are all 400 kV. The voltage responses are shown in Fig. 6. In this mode, injected power are shared between these two grid side converters. When the DC cable is considered in the backstepping controller design, the current sharing ratio would be linear with droop gains around $0.15/0.2 = 0.75$.

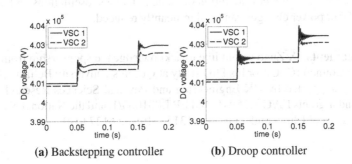

(a) Backstepping controller   (b) Droop controller

Fig. 6. DC voltage response with input power change in operation mode 2

4.3 Response of Mode 3

Mode 3 is the combination of mode 1 and 2. When the current in VSC 1 is below 1000 A, DC voltage is controlled by VSC using PI controller. At time $t = 0.15$, power from wind farm side is increased to 450 MW. The maximum power VSC 1 can deliver solely is around 400 MW according to the current limit. Extra power is shared between these two terminals. Fig. 7 shows the voltage trajectory when input power changes. It can be seen that the voltage overshoot can be significantly reduced when using the backstepping voltage controller.

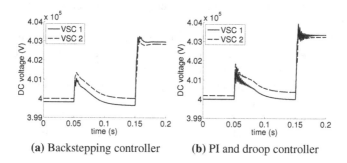

(a) Backstepping controller   (b) PI and droop controller

Fig. 7. DC voltage response with input power change in operation mode 3

5   Conclusion

In this paper, the DC cable dynamics in a multi-terminal HVDC system has been considered in the design of DC voltage controller. Based on different operation modes on the grid side converters, PI and droop based backstepping voltage controllers are constructed by including cable states to enhance the dynamic performance. A backstepping voltage controller is designed by considering one cable element at each step, and a state observer is used to provide necessary state estimation needed by the controller. Simulations using Matlab/Simulink have been carried out to compare the performance of the backstepping controller with traditional controller under different operation modes. It has shown that with the consideration of cable dynamics, oscillations and overshoot caused by input power changes can be significantly reduced.

Acknowledgments. X Zhao would like to acknowledge the sponsorship from Chinese Scholarship Council (CSC) for his PhD study at Queen's University Belfast. This work was partially supported by UK Engineering and Physical Sciences Research Council (EPSRC) under grants EP/G042594/1 and EP/L001063/1, and the National Natural Science Foundation of China under Grants 513111019 and 61273040.

References

1. E.W.E.T. Platform, Strategic research agenda market deployment strategy, European Wind Energy Association, Tech. Rep. (2014)
2. Gordon, S.: Supergrid to the rescue [electricity supply security]. Power Engineer 20(5), 30–33 (2006)
3. Xu, L., Li, S.: Analysis of HVDC light control using conventional decoupled vector control technology. In: 2010 IEEE Power and Energy Society General Meeting, pp. 1–8 (2010)
4. Milasi, R., Lynch, A.F., Li, Y.: Adaptive control of a voltage source converter. In: 2010 23rd Canadian Conference on Electrical and Computer Engineering (CCECE), pp. 1–4. IEEE (2010)
5. Ruan, S., Li, G., Jiao, X., Sun, Y., Lie, T.: Adaptive control design for VSC-HVDC systems based on backstepping method. Electric Power Systems Research 77(5), 559–565 (2007)

6. Wang, G.-D., Wai, R.-J., Liao, Y.: Design of backstepping power control for grid-side converter of voltage source converter-based high-voltage DC wind power generation system. IET Renewable Power Generation 7(2), 118–133 (2013)
7. Bregeon, S., Benchaib, A., Poullain, S., Thomas, J.: Robust DC bus voltage control based on backstepping and lyapunov methods for long distance VSC transmission scheme. In: 10th European Conference on Power Electronics and Applications, Toulouse, France (2003)
8. Zhao, X., Li, K.: Control of VSC-HVDC for wind farm integration based on adaptive backstepping method. In: 2013 IEEE International Workshop on Intelligent Energy Systems (IWIES), pp. 64–69. IEEE (2013)
9. Dierckxsens, C., Srivastava, K., Reza, M., Cole, S., Beerten, J., Belmans, R.: A distributed DC voltage control method for VSC MTDC systems. Electric Power Systems Research 82(1), 54–58 (2012)
10. Liang, J., Gomis-Bellmunt, O.: Control of multi-terminal VSC-HVDC transmission for offshore wind power. Power Electronics, 1–10 (2009)
11. Haileselassie, T.M., Uhlen, K.: Impact of DC line voltage drops on power flow of MTDC using droop control. IEEE Transactions on Power Systems 27(3), 1441–1449 (2012)
12. Beerten, J., Belmans, R.: Analysis of power sharing and voltage deviations in droop-controlled DC grids. IEEE Transactions on Power Systems 28(4), 4588–4597 (2013)
13. Chaudhuri, N.R., Chaudhuri, B.: Adaptive droop control for effective power sharing in multi-terminal DC (MTDC) grids. IEEE Transactions on Power Systems 28(1), 21–29 (2013)
14. Xu, L., Yao, L.: DC voltage control and power dispatch of a multi-terminal HVDC system for integrating large offshore wind farms. IET Renewable Power Generation 5(3), 223 (2011)
15. Xu, L., Yao, L., Bazargan, M., Yao, L.: DC grid management of a multi-terminal HVDC transmission system for large offshore wind farms. In: 2009 International Conference on Sustainable Power Generation and Supply, pp. 1–7 (April 2009)

APS Simulation System for 600MW Supercritical Unit Based on Virtual DCS Stimulative Simulator

Yue Wu[1], Daogang Peng[1,2], and Hao Zhang[1,2]

[1] College of Automation Engineering,
Shanghai University of Electric Power, Shanghai 200090, China
[2] Shanghai Key Laboratory of Power Station Automation Technology,
Shanghai 200090, China
jypdg@163.com

Abstract. Large capacity supercritical unit automatic power plant startup and shutdown (APS) control technology has been a research hotspot in the field of plant automation. According to characteristics of general supercritical unit start-up process, this paper introduces development process of supercritical unit APS stimulative simulation based on virtual DCS, including structure design, breakpoint design, interface configuration, development of functional groups and establishment of database, etc., which realizes auto startup process ranging from boiler ignition to full capacity, effectively improves production efficiency and makes significant reference value for practical application.

Keywords: Supercritical unit, APS, stimulative simulation, virtual DCS.

1 Introduction

As the rapid development of social economy, supercritical unit power generation technology is becoming an inevitable direction of coal-fired power generation in both domestic and foreign country while it is also the mainstream of clean energy generation technology in last decades [1]. This kind of unit is characterized by its multiple inputs and outputs, strong coupling between parameters and serious nonlinearity which sets high requirements for both manual operation and management [2]. During start-up and shutdown process, if system simply relies on conventional control system, unit running safety may be affected by disoperation on the one hand, let alone efficiency limit on the other [3].

Generally speaking, APS represents automatic power plant start-up and shutdown system which automatically completes whole process with little manual intervention. It effectively reduces labor intensity, standardizes normal process of start-up and shutdown, and avoids various unstable factors, therefore, increases reliability of the whole system [4].

This paper mainly introduces development process of mega supercritical unit based on simulative simulator including structure design, breakpoint design, interface configuration, functional groups development, database and interface establishment, etc. Cold start-up process is proved in simulation with high efficiency and safety.

K. Li et al. (Eds.): LSMS/ICSEE 2014, Part III, CCIS 463, pp. 296–303, 2014.

2   Overall Scheme Design

2.1   Framework Design

From the perspective of plant simulation, the whole APS can be divided into three layers which are unit logic management layer, functional group stepping procedure layer and subsystem step order layer [5] [6]. When it comes to stimulative simulator, as the absence of controlled object, virtual DCS is adopted in order to replace controlled object model. V-DPU (distributed processing unit) is almost the same as real DPU in hardware configuration.The only difference between them is the software configuration: V-DPU is not directly connected to IO card in the field which will send orders to regular control system and receive feedback from it. Meanwhile MMI (man machine interface) will be used to show orders, alarms and feedbacks, etc [7]. Detailed framework structure is shown in figure 1.

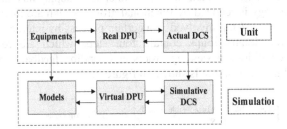

Fig. 1. System Framework Structure

2.2   Breakpoint Design

Currently, APS cannot realize its auto start-up and shutdown without manual intervention which is so-called 'one-click start-stop', mainly because the general control technology of power plant in domestic cannot meet the requirement [8].On the other hand, as the application of 'one-click start-stop' is few in domestic, relevant experience and parameters are extremely insufficient. So, breakpoint is adopted because it is relatively more common used and reliable. By cutting the whole process into small procedure and keep these precedure related, at the same time, relatively independent, breakpoint will ensure APS' efficiency and safety. Breakpoint details are listed below:

There are six breakpoints in the starting process which are preparation, cold washing, turbine run-up, synchronization and load up. The latter breakpoint will be started by manual confirm after the former one is finished. Before APS is started, related auxiliary systems must be checked [9]. Every breakpoint contains several steps which are specifically shown below:

Preparation: water supply system in, circulating water system in, water up in condenser, oil station of feed pump in, turbine oil system in, furnace bottom and cinder system in, etc.

Cold washing: condensate system in, condensate Fe<800ppb, condensate water PH qualified, water filled in furnace pump, auxiliary steam system in, water load in pipeline, shaft seal system in, deaerator heated, furnace opening wash and furnace pump dynamic wash, etc.

Ignition: specific system in, plasma system recovery, fuel check, plasma ignition, turbine bypass system in, separator water Fe<100pbb and main steam temperature>271°C, etc.

Impulse start: lube cooler in, DEH alarm reset, turbine latch-on, and ATC selection, speed up tp 3000 r/min, etc.

Synchronization: Auto-synchronization, HP heater in, coal pulverizing system in, fuel mode selection, circulating water pump in, etc.

Load up: DEH in, steam feed pump in, load up to 300MW, plasma system out, load up to full capacity, etc.

2.3   Control Strategy Design

Based on parameters offered by simulator, V-DPU in APS needs control strategy to realize its logical operation. The whole design process is modularized in order to be modified without chain reaction and structured to clearly display its control logic [10].

Unit control strategy is realized by functional groups which are intermediate link of APS. It can achieve start/stop involved in a certain process by interact with simulation model. Typical functional groups contain four steps which are receiving orders, checking launch condition, processing orders and sending feedback. Specific system startup functional group structure is shown in figure 2.

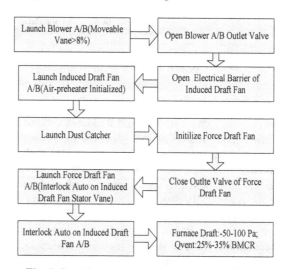

Fig. 2. Specific System Startup Functional Group

2.4   Stimulative Simulator Introduction

SPCS-3000 is a distributed control system which is used in this paper as stimulative simulator. Its open architecture, good hardware compatibility as well as large software scalability do a great job in realizing process automation and information integration.

Stimulative simulation technology based on virtual DCS appeared in 1960s. It is a combination between simulation technology and DCS technology which provides huge help in training operators and optimizing control strategy [11]. With the booming of power plant automation, more demands are asked in qualities of operators and control logic optimization, therefore, stimulative simulation technology will become more and more important.

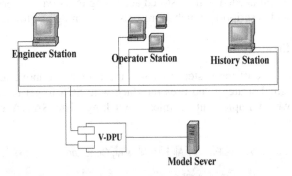

Fig. 3. Stimulative Simulation System Structure Chart

Traditional full scope simulator only considers DCS operator station of power plant without engineer station, so, simulation model and control strategy modification cannot be realized. While full-exciting simulator, on the other hand, reserves all functions in real DCS which is convenient for logic modification and optimization [12]. Its software and hardware configuration is also the same as practical operation. However, its high cost, complicated structure and long period development prevent its application from general power plant.

Integrated advantages and disadvantages of the above two kinds of simulator, stimulative simulator based on virtual DCS reveres all functions in real DCS while uses V-DPU instead of DPU to greatly reduce cost.

3   Implementation of APS Simulation Based on Stimulative Simulator

3.1   Database Establishment

Database is brain of the whole simulation system which is based on simulation model. The parameters of 600MW supercritical unit are listed below:

Table 1. Virtual Equipment Parameters

Item	Parameter
Boiler	*Supercritical once-through Benson boiler*
Turbine	*Supercritical once reheat HP-IP combined condensing steam turbine with single shaft, three overcasing, four condenser*
Generator	*Three phrase non-salient pole synchronization generator*
Main Transformer	*Double-winding oil immersion copper core single-phrase transformer group*

Independent control station will be constructed for data storage. And by using OPC server, system will realize communication between data points in SAMA graphic.

Based on the parameters above, control station of simulator offers four kinds of IO card where data point needed will be stored according to its type. Every data point has its own number and description which represents its meaning and limitation.

3.2   Control Strategy

Software control logic of the system transfers the flow chart above to SAMA graph, optimizes its original strategy and uses interface with SCS to get back to its original control circuit which implement seamless switching. The SAMA graph is shown below:

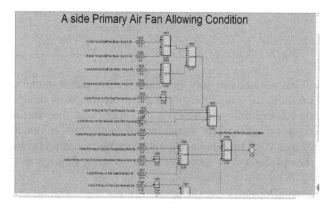

Fig. 4. SAMA Graph of Spec ific System Startup Allowing Conditions

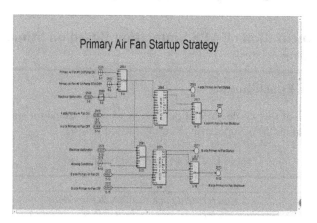

Fig. 5. SAMA Graph of Primary Air Fan Startup Strategy

Specific startup functional group contains following details:

(1) Launch air-preheater A/B;
(2) Launch induced draft fan A; Interlock auto on;

(3) Launch force draft fan A;

(4) If force draft fan B is not running, set 15% rotor blades of force draft fan A; Interlock auto on;

(5) Launch induced draft fan B; Interlock auto on;

(6) Launch force draft fan B;

(7) If force draft fan A is not running, set 15% rotor blades of force draft fan B; Interlock auto on;

3.3   Interface Configuration

Man-machine interface configuration is necessary to production process management and control where operators can operate all kinds of equipment by DCS and using keyboard and mouse to set values [13]. Flexible parameter interface shows operation status, processing information even malfunction reason on the screen for the sake of surveillance, which provides great help to standard operation in practice. What is more, the APS interface is close related to its structure which contains launch operation menu and shutdown operation menu. Put all operation and control information in detailed breakpoint images as well as functional groups and subfuctional groups. In the perspective of operators, dynamic operation information will be watched on the screen in a long time on one hand, which requires a clean and clear image, on the other hand, unit running process must be reflected faithfully which needs sufficient amount of information and data.

APS can be divided into two processes which are start and stop, as a result, the images contains launch and shut picture. The general start image is shown in figure 6.

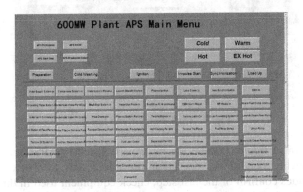

Fig. 6. APS Main Menu Image

(1) The general start menu of APS has for operational button on the top left corner which are allowing conditions: mainly contains auxiliary system running state, especially those do not belong to APS; Investment and withdraw: This is the master switch of the whole system without which other operation is invalid; Pattern choose: the start process ranges from water supply in coagulation filling tank to operation at full load and the stop process ranges from load down to unit shut down; Reset; Break point choose: choose one break point in six during start process which are preparation, cold rinse, firing up, rolling, paralleling in and load-up; choose one break point in three

during stop process which are load-down, generator phrased out and generator shutdown. Only one break point can be chosen at one time, the next one will reset the last.

(2) Unit mode selection is on the top right corner which is cold start, warm start, hot start and extremely hot start.

(3) Break point: as it is shown in figure 3, six break points have 58 specific steps which have two parts: sending orders and receiving feedback. When the step is chosen, system will automatically judge allowing conditions and operate virtual equipment if permitted. After that, virtual equipment will feed back to APS and system will use it as the starting signal of the next step.

Break Point Information:

Currently, APS cannot realize its auto start-up and shutdown without manual intervention which is called 'one-click start-stop' mainly because the general control technology of power plant in domestic cannot fit the requirement. On the other hand, as the application of 'one-click start-stop' is few in domestic, relevant experience and parameters are extremely insufficient. So, the system chooses to use breakpoint which is relatively more common and reliable. By cutting whole process into small pieces and keep these pieces related, at the same time, relatively independent which will ensure APS' efficiency and safety. Break point picture of APS is shown in figure 7.

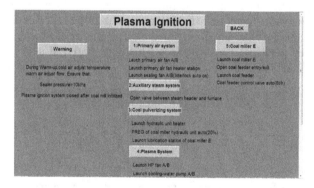

Fig. 7. Break Point Menu Image

As it is shown in figure 7, there are three parts listed in APS break point picture which are allowing conditions, operation procedure and feedback. The allowing conditions includes confirm of the last step, check equipment used in this step. System will only operate virtual equipment as described on screen after allowing conditions end its task. BACK button will get operator back to main menu.

4   Conclusion

With rapid development of supercritical unit, 600MW even 1000MW thermal power generating unit will be main force in power production in domestic. However, when compared with advanced countries such as Japan and USA, both research and application of APS in domestic lag far behind theirs [14]. Under such circumstances, this paper introduces development process of mega supercritical unit based on simulative platform including structure design, interface configuration, development of functional groups, establishment of database and conventional system interface processing, etc. By debugging and simulation, the system is proven effectively increase efficiency and makes valuable reference for practice.

Acknowledgment. This work was supported by the State Key Program of National Natural Science Foundation of China (Grant No. 61034004), Shanghai Science and Technology Commission Key Program (Grant No. 13111104300) and Shanghai Key Laboratory of Power Station Automation Technology (Grant No. 13DZ2273800).

References

1. Zheng, H.L.: The current situation and development of Automatic Plant Start-up and Shutdown system. Instrumentation Customer 19(6), 7–9 (2012)
2. Feng, T.Y.: Research on and Application of Automatic Plant Startup and Shutdown Technology for Million kW Supercritical APS Unit. Sci-Tech Information Development & Economy 20(30), 183–186 (2010)
3. Pan, F.P., Chen, S.H., Zhang, H.F., et al.: Design and application of 1 000 MW ultra-supercritical unit APS system. Electric Power 42(10), 15–18 (2009)
4. Lu, X.N., Duan, N.: Application of Automatic Start-up and Shutdown Technology in Sanhe Power Plant. Electric Power 34(3), 48–50 (2001)
5. Feng, L.G., Zhou, G.Q.: The Application of Automatic Plant Start-up and Shutdown System in Hejin Power Plant. Shanxi Electric Power 61(3), 54–57 (2005)
6. Gui, Y.S., Shen, C.Q., Hu, J., et al.: Application of APS to DCS retrofit for units. East China Electric Power 34(2), 51–53 (2006)
7. Pan, F.P., Chen, S.H.: Application of Automatic Plant Startup and Shutdown Control System to 600 MW Domestic Units. Guangdong Electric Power 21(12), 55–58 (2008)
8. Xiang, L., Guo, Q.S., Zhang, Y., et al.: Application of an Autometic Start and Stopping System in Baosteel Power Plant's 350 MW Generating Set. Power Equipment 19(5), 316–318 (2005)
9. Zhang, H.F., Pan, F.P., Luo, J., et al.: Design and Research of Human-machine Interface in Automatic Plant Startup and Shutdown System. Guangdong Electric Power 22(11), 66–69 (2009)
10. Zeng, W.N., Sang, Y.F.: Analysis of logic design for automatic procedure start - up/shut - down of units. Thermal Power Generation 41(6), 77–79 (2012)
11. Yu, Z.H.: Application Strategy and Experiment about Automatic Power Plant Start-up and Shutdown System. Guangxi Electric Power 30(5), 27–29 (2007)
12. Liu, K.X.: Analysis on Function and Design Scheme of APS in Thermal Power Unit. China Electric Power (Technology Edition) (2), 62–65 (2014)
13. Wang, L.D., Qin, L.: Application and Design of APS for Thermal Power Plant. Electric Power Technology 19(11), 18–26 (2010)
14. Li, F., Zhu, Y.Q., Pan, F.P., et al.: Design and application of APS function for air and flue gas system of 1000 MW ultra-supercritical unit. Electric Power Automation Equipment 30(8), 116–120 (2010)

A Multi-objective Chaotic Optimization Algorithm for Economic Emission Dispatch with Transmission Loss

Yijuan Di[1,*], Ling Wang[2], and Minrui Fei[2]

[1] School of Electronic Information and Electrical Engineering,
Changzhou Institute of Technology, Changzhou 213002, China
`poiuty@163.com`
[2] Shanghai Key Laboratory of Power Station Automation Technology,
School of Mechatronic Engineering and Automation,
Shanghai University, Shanghai 200072, China

Abstract. Economic emission dispatch (EED) in the power system is a non-linear constrained multi-objective optimization problem. In this paper, a new chaotic optimization algorithm for solving this complex problem is proposed. Two forms of logistic maps and marginal analysis for optimization are used in the proposed algorithm. The simulation results obtained by the proposed algorithm are compared with those of chaotic optimization algorithm and other approaches reported in recent literatures. The comparison results demonstrate the effectiveness of the proposed algorithm in solving the multi-objective EED problem.

Keywords: Economic emission dispatch, Chaotic optimization algorithm Logistic map, Marginal analysis method.

1 Introduction

With the increasing concern over the environmental pollution caused by thermal power plants, traditional economic power dispatch can not meet the need for environmental protections. Thus, minimum emissions have to be considered as one of the objectives along with cost economic in thermal power system, which is called as economic emission dispatch (EED) problem.

The EED problem is a complex non-linear constrained multi-objective optimization problem. Traditional methods [1-2] such as weight methods and mathematical programming, have limitations for solving multi-objectives EED problems, where the problem is converted to a single-objective problem. Recently, several Pareto-based multi-objective evolutionary algorithms [3-5] such as genetic algorithm (GA), particle swarm optimization (PSO) and differential evolution (DE) algorithm have been presented to solve the EED problems. However, these algorithms still have the problems of premature convergence and being trapped in the local optimum.

---

* Corresponding author.

K. Li et al. (Eds.): LSMS/ICSEE 2014, Part III, CCIS 463, pp. 304–312, 2014.

Due to the easy implementation and special ability to avoid search being trapped into local optima, chaos has been a novel optimization technique and chaos-based searching algorithms have aroused intense interests [6-7]. In this paper, a multi-objective chaotic optimization algorithm (MCOA) is proposed to solve the EED problem, where the logistic map in two forms is used to generate chaos sequence and a marginal analysis (MA) optimization is developed based on the marginal analysis method.

The rest of this paper is organized as follows. Section 2 gives the formulation of the EED problem. In Section 3, we describe the marginal model of EED problem. Section 4 proposes a multi-objective chaotic optimization algorithm. The efficiency of proposed approach is validated on test power systems in Section 5. This paper concludes with Section 6.

2   Problem Formulation

2.1   Objectives Functions

(i) The Minimum of Total Fuel Cost
Generally, the fuel cost function of each generating unit is a quadratic function of active power output. Thus, the total fuel cost of all generating units is expressed as:

$$\min F^{eco} = \sum_{i=1}^{N}[C^{eco}(P_i)] = \sum_{i=1}^{N}[a_iP_i^2 + b_iP_i + c_i] \tag{1}$$

where F^{eco} is the total fuel cost, P_i is the active power output of the ith unit, N is the number of generating units, $C^{eco}(P_i)$ is the cost function of the ith unit, and a_i, b_i and c_i are the cost coefficients of the ith unit.

(ii) The Minimum of Total Emission
The emission of atmospheric pollutants generated by each generating unit can be mathematically modeled as a quadratic function of active power output. Thus, the total emission of all generating units can be expressed as:

$$\min F^{emis} = \sum_{i=1}^{N}[C^{emis}(P_i)] = \sum_{i=1}^{N}[\alpha_iP_i^2 + \beta_iP_i + \gamma_i] \tag{2}$$

where F^{emis} is the total emission, $C^{emis}(P_i)$ is the emission function of the ith unit, and α_i, β_i and γ_i are the emission coefficients of the ith unit.

2.2   Constraints

(i) Power balance constraints

$$\sum_{i=1}^{N}P_i = P_D + P_L \tag{3}$$

where P_D is the load demand of the power system, and P_L is the transmission loss. The transmission loss P_L can be calculated as:

$$P_L = \sum_{i=1}^{N} \sum_{j=1}^{N} P_i B_{ij} P_j \qquad (4)$$

where B_{ij} is the element transmission of transmission loss coefficient matrix B.

(ii) Active power operating limits

$$P_i^{\min} \le P_{it} \le P_i^{\max} \qquad i \in N \qquad (5)$$

where $P_i^{\min}$ and $P_i^{\max}$ are the minimum and maximum active power outputs of the ith unit, respectively.

3   Marginal Model of EED Problem

Marginal analysis method (MA) [8-9] is a quantity analysis method widely used in the modern western economics and can help decision makers to determine the correct economic decision. Based on this concept, the marginal functions of fuel cost and emission of the ith generating unit based on Eq. (1)-(2) can be represented as follows.

(i) Marginal Fuel Cost

$$\begin{aligned} M_i^{eco} &= C^{eco}(P_i) - C^{eco}(P_i - 1) \\ &= a_i P_i^2 + b_i P_i + c_i - (a_i(P_i - 1)^2 + b_i(P_i - 1) + c_i) \\ &= 2a_i P_i - a_i + b_i, \qquad i \in N \end{aligned} \qquad (7)$$

(ii) Marginal Emission

$$\begin{aligned} M_i^{emis} &= C^{emis}(P_i) - C^{emis}(P_i - 1) \\ &= \alpha_i P_i^2 + \beta_i P_i + \gamma_i - (\alpha_i(P_i - 1)^2 + \beta_i(P_i - 1) + \gamma_i) \\ &= 2\alpha_i P_i - \alpha_i + \beta_i, \qquad i \in N \end{aligned} \qquad (8)$$

From Eq. (7) and Eq. (8), the marginal functions of the fuel cost and emission are linear and monotonic increasing.

4   Multi-objective Chaotic Optimization Algorithm

4.1   Initialization of the Population

The population of MCOA consists of N_P N-dimensions real-valued parameter vectors denoted as $P_i = [P_{i,1}, P_{i,2}, \ldots P_{i,j}, \ldots, P_{i,N}]$ $i \in N_P$ $j \in N$, where N_P is the size of population and N is the number of generating units.

The jth element P_j^0 of initial individual P^0 is randomly generated between the lower and upper bounds of the problem shown as:

$$P_j^0 = P_j^{\min} + r \bullet (P_j^{\max} - P_j^{\min}), \qquad j \in N \qquad (9)$$

where r is a random generated number between 0 and 1 uniformly distributed.

4.2 Marginal Analysis Optimization

The procedure of marginal analysis optimization operator is described as follows:

Step 1: Constraints handling for active power operating limits

If any element of new generated solution violates the active power operating limits, the following procedure will applied to modify its value to satisfy the inequality constraint.

$$P_i = \begin{cases} P_i^{min} & if \ P_i < P_i^{min} \\ P_i & if \ P_i^{min} \leq P_i \leq P_i^{max} \quad i \in N \\ P_i^{max} & if \ P_i > P_i^{max} \end{cases} \qquad (10)$$

Step 2: Calculate the difference P_{diff} of individual x_i

Calculate P_L by Eq. (4). Then, the difference P_{diff} of individual x_i can be calculated as:

$$P_{diff} = \sum_{i=1}^{N} P_i - P_D - P_L \qquad (11)$$

Step 3: Get new load increment

The new load increment ($step'$) is calculated as:

$$step' = \begin{cases} \min(step, P_{diff}) & if \ P_{diff} > 0 \\ \max(-step, P_{diff}) & if \ P_{diff} < 0 \end{cases} \qquad (12)$$

Step 4: Load dispatch

First, sort all generating units by the values of marginal fuel cost and marginal emission, which are calculated by Eq. (7) and Eq. (8). Get the number, denoted as j, of the unit with minimum marginal fuel cost and marginal emission. Then, update the value of the jth element of individual x_i as follows:

$$P_j' = P_j + step' \qquad (13)$$

Step 5: Stopping rule

If stopping criteria are met, the optimization process is stopped; otherwise go to step 1.

4.3 Chaotic Mutation Operation

The chaotic mutation operation generates chaos sequence by the well-known logistic map in two forms.

(i) Logistic1 [6]

$$x_{n+1} = u_1 x_n (1 - x_n) \qquad x_n \in (0,1) \qquad (14)$$

where x_n is the value of the chaotic variable x at the nth iteration, u_1 is the so-called bifurcation parameter ($u_1 \in (0,4]$), and the initial value x_0 is generated randomly between 0 and 1, with $x_0 \notin \{0.25, 0.5, 0.75\}$.

(ii) Logistic2 [10]

$$x_{n+1} = 1 - u_2 x_n^2 \qquad x_n \in (-1,1) \qquad (15)$$

where u_2 is the bifurcation parameter ($u_2 \in (0,2]$), and the initial value x_0 is generated randomly between -1 and 1, with $x_0 \notin \{-0.5, 0, 0.5\}$.

4.4   Implementation of MCOA for EED Problem

Generally, the key steps of the proposed MCOA for solving the EED problem can be described as follows:

Step 1: Initialize population.

Step 2: Marginal analysis optimization. When the new binary population generated violates system constraints, marginal analysis optimization is used to modify its value to satisfy system constraints.

Step 3: Chaotic mutation. To overcome the premature convergence and being trapped into local optimum in the optimization procedure, chaotic mutation operation generates new individuals by two logistic maps.

Step 4: Selection [11]. Evaluate the fitness of target individual and the corresponding trial individual and select the better one as offspring individual, which become the new ith individual of next generation.

Step 5: Stopping rule. If stopping criteria are met, the search process is stopped and return the best compromise solution [12]; otherwise go to step 3.

The flow chart of MCOA is shown in the Fig. 1.

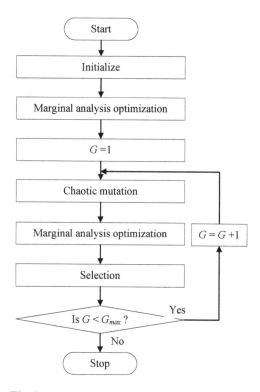

Fig. 1. The flow chart of MCOA for EED problem

5   Simulation Results

The test system consists of six generating units with transmission losses. The detailed data of generating units are derived from [13], which are given in Table A in Appendix as well as the transmission loss coefficients B. This system has three power demands, i.e. 500MW, 700MW and 900MW. The population size N_P, the maximum number of generations G_{max}, u_1 and u_2 are selected as 20, 100, 4 and 2, respectively. The different load increments (*step*) under three load demands are selected by trial and error as 29, 59 and 80, respectively. In this paper, the MCOA with logistic1 is denoted as MCOA (1) and that with logistic2 is denoted as MCOA (2).

To verify the advantages of the proposed MCOA for solving EED problem, the error is calculated as follows:

$$error = \sum_{i=1}^{N} P_i - P_D - \sum_{i=1}^{N}\sum_{j=1}^{N} P_i B_{ij} P_j \qquad (16)$$

The simulation results of the proposed approach are compared with those obtained by using COA [14], NR [15], FCGA [16], NSGA-II [13] and BBO [17], as shown in Table 1-3. In these tables, the transmission loss and the errors calculated by Eq. (16) with different algorithms are also illustrated.

Table 1. Simulation solutions of different algorithms (P_D =500MW)

Algorithm	P_1(MW)	P_2(MW)	P_3(MW)	P_4(MW)	P_5(MW)	P_6(MW)	Fuel cost($/h)	Emission(kg/h)	P_L	error
NR	59.8730	39.6510	35.0000	72.3970	185.2410	125.0000	28550.1500	312.5130	15.4505	1.7115
FCGA	65.2300	24.2900	40.4400	74.2200	187.7500	125.4800	28231.0600	304.9000	17.4107	-0.0007
NSGA-II	54.0480	34.2500	54.4970	80.4130	161.8740	135.4260	28291.1190	284.3620	20.5077	0.0003
BBO	49.8940	33.9080	62.9870	82.8810	153.4050	139.0324	28318.5060	279.3092	23.5364	-1.4290
COA	51.8960	23.2615	55.0767	56.4543	183.3342	152.0892	28250.3643	303.9951	22.1116	0.0003
MCOA(1)	49.4884	36.4505	40.1211	68.8950	179.7653	143.9110	28186.7739	300.9747	18.6313	0
MCOA(2)	48.4105	34.9770	35.3036	90.0968	185.2296	125.0000	28202.2589	303.4835	19.0177	-0.0002

Table 2. Simulation solutions of different algorithms (P_D =700MW)

Algorithm	P_1(MW)	P_2(MW)	P_3(MW)	P_4(MW)	P_5(MW)	P_6(MW)	Fuel cost($/h)	Emission(kg/h)	P_L	error
NR	85.9240	60.9630	53.9090	107.1240	250.5030	176.5040	39070.7400	528.4470	31.4573	3.4697
FCGA	80.1600	53.7100	40.9300	116.2300	251.2000	190.6200	38408.8200	527.4600	32.8547	-0.0047
NSGA-II	86.2860	60.2880	73.0640	109.0360	223.4480	184.1110	38671.8130	484.9310	36.2336	-0.0006
BBO	84.0180	61.9268	78.7284	110.9188	211.9852	187.8924	38828.2660	476.4080	38.7216	-3.2520
COA	77.9782	62.4097	35.7078	134.6418	246.5554	175.3184	38523.5052	527.7306	32.6119	-0.0006
MCOA(1)	78.9964	52.1021	49.6211	106.0974	254.0107	193.0799	38391.6773	524.0455	33.9078	-0.0002
MCOA(2)	75.5839	44.5263	56.8193	108.2143	255.0854	197.3753	38407.1234	525.9266	37.6045	0

Table 3. Simulation solutions of different algorithms (P_D =900MW)

Algorithm	P_1(MW)	P_2(MW)	P_3(MW)	P_4(MW)	P_5(MW)	P_6(MW)	Fuel cost($/h)	Emission(kg/h)	P_L	error
NR	122.0040	86.5230	59.9470	140.9590	325.0000	220.0630	50807.2400	864.0600	48.4269	6.0691
FCGA	111.4000	69.3300	59.4300	143.2600	319.4000	252.1100	49674.2800	850.2900	54.9258	0.0042
NSGA-II	120.0587	85.2020	89.5650	140.2780	288.6140	233.6870	50126.0590	784.6960	57.4041	0.0006
BBO	115.7087	100.7550	101.4450	145.0146	282.0580	212.4170	50297.2710	765.0870	62.4631	-5.0648
COA	110.8710	74.8440	57.4834	137.7023	316.9887	256.2932	49700.5409	848.4713	54.1838	-0.0012
MCOA(1)	103.2773	80.4243	68.0382	136.4611	325.0000	242.2829	49650.9139	840.8469	55.4838	0
MCOA(2)	95.3523	78.3287	77.5504	132.0115	325.0000	252.6683	49673.7254	846.4997	60.9112	0

Table 1-3 show that the proposed MCOA can find better fuel costs and emissions than NR, FCGA and COA. Compared with NSGA-II and BBO, MCOA achieves better fuel costs but poorer emissions and these solutions obtained by MCOA, NSGA-II and BBO are non-dominated solutions. From Table 1-3, it is also observed that the errors of MCOA under two logistic maps are least among seven algorithms. Meanwhile, the results of MCOA (1) are better than that of MCOA (2). Therefore, it is fair to claim that the proposed MCOA under two logistic maps can effectively solve the EED problem, and its performance is superior over other algorithms.

6   Conclusions

The EED problem is a non-linear multi-objective minimization problem with two conflicting objectives subject to various equality and inequality constraints. A multi-objective chaotic optimization algorithm is developed to solve the EED problem, where a marginal analysis optimization operator is developed and a chaotic mutation operation generates chaos sequence by the logistic map in two forms. Compared with other algorithms in recent literatures, the results demonstrate that the MCOA has good performance in solving the EED problem and outperform other algorithms.

References

1. Guo, C.X., Zhan, J.P., Wu, Q.H.: Dynamic economic emission dispatch based on group search optimizer with multiple producers. Electric Power Systems Research 86, 8–16 (2012)
2. Vahidinasab, V., Jadid, S.: Joint economic and emission dispatch in energy markets: A multiobjective mathematical programming approach. Energy 35(3), 1497–1504 (2010)
3. El-Sehiemy, R.A., El-Hosseini, M.A., Hassanien, A.E.: Multiobjective Real-Coded Genetic Algorithm for Economic/Environmental Dispatch Problem. Studies in Informatics and Control 22(2), 113–122 (2013)
4. Niknam, T., Doagou-Mojarrad, H.: Multiobjective economic/emission dispatch by multiobjective θ-particle swarm optimization. IET Generation, Transmission & Distribution 6(5), 363–377 (2012)

5. Lu, Y., Zhou, J., et al.: Environmental/economic dispatch problem of power system by using an enhanced multi-objective differential evolution algorithm. Energy Conversion and Management 52(2), 1175–1183 (2011)
6. Wang, Y., Zhou, J., Qin, H., Lu, Y.: Improved chaotic particle swarm optimization algorithm for dynamic economic dispatch problem with valve-point effects. Energy Conversion and Management 51(12), 2893–2900 (2010)
7. Cai, J., Ma, X., Li, Q., et al.: A multi-objective chaotic particle swarm optimization for environmental/economic dispatch. Energy Conversion and Management 50(5), 1318–1325 (2009)
8. Rusak, H.: The method of determining marginal costs of energy transmission in power network. International Journal of Electrical Power & Energy Systems 30(6-7), 428–433 (2008)
9. Mukherjee, A., Sarkar, S., Chakraborty, P.K.: Marginal analysis of water productivity function of tomato crop grown under different irrigation regimes and mulch managements. Agricultural Water Management 104, 121–127 (2012)
10. He, G.Y., He, G.W.: Synchronous chaos in the coupled system of two logistic maps. Chaos, Solitons and Fractals 23(3), 909–913 (2005)
11. Wang, L., Fu, X., Mao, Y., et al.: A novel modified binary differential evolution algorithm and its applications. Neurocomputing 98, 55–75 (2012)
12. Abido, M.A.: A niched Pareto genetic algorithm for multiobjective environmental/economic dispatch. International Journal of Electrical Power and Energy Systems 25(2), 97–105 (2003)
13. Rughooputh Harry, C.S., Ah King Robert, T.F.: Environmental/economic dispatch of thermal units using an elitist multiobjective evolutionary algorithm. In: Proceedings of the IEEE International Conference on Industrial Technology, vol. 1, pp. 48–53 (2003)
14. Shayeghi, H., Shayanfar, H.A., Jalilzadeh, S., Safari, A.: Multi-machine power system stabilizers design using chaotic optimization algorithm. Energy Conversion and Management 51(7), 1572–1580 (2010)
15. Dhillon, J.S., Parti, S.C., Khotari, D.P.: Stochastic economic emission load dispatch. Electric Power System Research 26, 179–186 (1993)
16. Song, H., Wang, G.S., Wang, P.Y., Johns, A.T.: Environmental/economic dispatch using fuzzy logic controller genetic algorithms. IEE Proceedings: Generation, Transmission and Distribution 144(4), 377–382 (1997)
17. Roy, P.K., Ghoshal, S.P., Thakur, S.S.: Combined Economic Emission dispatch biogeography based optimization. Electrical Engineering 92(4-5), 173–184 (2010)

Appendix

Table A. Fuel cost coefficients and emission coefficients

Unit	Fuel cost coefficients			Emission coefficients			P^{min}	P^{max}
	a($/hMW^2)	b($/hMW)	c($/h)	α(kg/hMW^2)	β(kg/hMW^2)	γ(kg/h)	(MW)	(MW)
1	0.15247	38.53973	756.79886	0.00419	0.32767	13.859	32 10	125
2	0.10587	46.15916	451.32513	0.00419	0.32767	13.859	32 10	150
3	0.02803	40.39655	1049.99770	0.00683	-0.54551	40.2669	03 5	225
4	0.03546	38.30553	1243.53110	0.00683	-0.54551	40.2669	03 5	210
5	0.02111	36.32782	1658.56960	0.00461	-0.51116	42.89553	1 30	325
6	0.01799	38.27041	1356.65920	0.00461	-0.51116	42.89553	1 25	315

The transmission loss coefficients B are:

$$B = \begin{bmatrix} 0.002022 & -0.000286 & -0.000534 & -0.000565 & -0.000454 & -0.000103 \\ -0.000286 & 0.003243 & 0.000016 & -0.000307 & -0.000422 & -0.000147 \\ -0.000533 & 0.000016 & 0.002085 & 0.000831 & 0.000023 & -0.000270 \\ -0.000565 & -0.000307 & 0.000831 & 0.001129 & 0.000113 & -0.000295 \\ -0.000454 & -0.000422 & 0.000023 & 0.000113 & 0.000460 & -0.000153 \\ 0.000103 & -0.000147 & -0.000270 & -0.000295 & -0.000153 & 0.000898 \end{bmatrix}$$

RSSI-Based Fingerprint Positioning System for Indoor Wireless Network[*]

Ruohan Yang and Hao Zhang

Department of Control Science and Engineering, Tongji University,
Shanghai 200092, P.R.China
xh_yrh@163.com, zhang_hao@tongji.edu.cn

Abstract. This paper presents a direct explicit method of the fingerprint positioning for indoor wireless network. In data collection, for the purpose of a reliable and stable signal, a feedback filter is added to the sampler. In positioning phase, the location clustering technique is used to exclude invalid reference points. Then a matching algorithm based on RSSI correlation coefficient is proposed, which can improve positioning accuracy. The example in the paper illustrates the effectiveness of the proposed positioning scheme.

Keywords: RSSI, fingerprint positioning, location clustering, matching algorithm.

1   Introduction

In recent years, the indoor positioning techniques have been well developed, due to its potential applications such as emergency rescue, library guidance and target tracking. In outdoor environment, global positioning system (GPS) can obtain accurate positioning results. However, because of the building block, GPS-based positioning methods cause large error when they are used in indoor environment. Moreover, it is difficult to get a suitable signal transmission when indoor environment is complicated and changeable.

There are many different indoor positioning methods for wireless network[1], for example, time-based positioning[2], angle-based positioning[3], received signal strength indicator(RSSI) based modeling positioning[4, 5, 6] and RSSI-based fingerprint positioning[7, 8, 9, 10, 11, 12, 13]. In current indoor positioning systems, the approaches of RSSI-based modeling positioning and RSSI-based fingerprint positioning are relatively mature. Though the approaches of RSSI-based modeling positioning are convenient and easy to implement, the signal strength will be largely affected by building blocks. By setting up a fingerprint database, the effects of complicated and changeable environment can be greatly reduced in RSSI-based fingerprint positioning schemes.

---

[*] This work is supported by the National Natural Science Foundation of China (61273026) and the Fundamental Research Funds for the Central Universities.

K. Li et al. (Eds.): LSMS/ICSEE 2014, Part III, CCIS 463, pp. 313–319, 2014.

The researches on fingerprint positioning were presented in the references [10, 11, 12], where k-nearest neighbor matching algorithm was used to search for K neighbor closest between classes of reference points and locate measured point based on Euclidean distance. However, the estimated values of measured points' coordinates were calculated by averaging the coordinates of valid reference points which may decrease the accuracy. A conception of RSSI-similarity degree was introduced in the references [13], which was used as weight factors to estimate the coordinates of measured points. However, the disturbances of RSSI values have not been excluded, so it is not very practical to be implemented in wireless network.

In this paper, we further develop the researches on RSSI-based fingerprint positioning schemes, which were presented in the references [10, 11, 12, 13]. In data collection, the feedback filter is used to obtain reliable and stable RSSI values. In positioning phase, with the aid of position clustering technology, we deal with the data on reference points, and reduce the amount of computation and exclude the invalid positioning data. Then a matching algorithm based on RSSI correlation coefficient is proposed, which can match the parameters of measuring point to that of reference point, and improve positioning accuracy. By testing and verifying the example, it shows that the results of the system are satisfactory.

2   Fundamental of Fingerprint Positioning Scheme

Due to the dependence of geographical environment, each site exhibits its uniqueness when signals transmit. This uniqueness is called fingerprint which means that there is only one RSSI vector for a certain site when the base stations have been set. Fingerprint positioning schemes behave well on the aspect of positioning accuracy since building blocks are considered. Thus, it seems suitable to apply fingerprint positioning schemes to indoor environment.

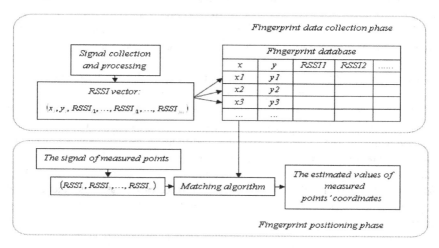

Fig. 1. The function diagram of fingerprint positioning scheme

Commonly, the fingerprint positioning scheme can be divided into two phases, including fingerprint data collection phase and fingerprint positioning phase. The function diagram is shown in Fig. 1.

3   Fingerprint Data Collection

3.1   Setup of Fingerprint Database

Assume that there are several base stations, and the locations of reference points are known. In data collection, each reference point collects RSSI values sent by every base station. Combining collected values with the reference point's location, one can get a vector, which is called fingerprint. For the reference point i, the fingerprint is shown as follow:

$$(x_i, y_i, RSSI_{i1}, \ldots, RSSI_{ij}, \ldots, RSSI_{im}), \quad i = 1, 2, \ldots, n; \quad j = 1, 2, \ldots, m$$

Where $RSSI_{ij}$ denotes RSSI value sent from base station j to the reference point i, (x_i, y_i) is the location of the reference point i. n is the amount of reference points and m is that of base stations.

Nowadays, most of fingerprint based positioning systems sample point by point when setting up fingerprint database. If there are few reference points, the positioning accuracy may be reduced. However, dense reference points may increase the computation so that computing time is longer and it is hard to obtain real time positioning. Thus, the setup of fingerprint database is essential to fingerprint based positioning, and it can directly affect positioning accuracy.

3.2   The Model of RSSI

Consider signal loss and the disturbance of indoor environment, the lognormal model[11] is used in the algorithm, which can be described by

$$RSSI(d) = RSSI(d_0) - 10\beta \lg(d / d_0) + X_\sigma \tag{1}$$

Where d_0 is the reference distance, β is attenuation coefficient, X_σ is disturbance of indoor environment.

According to Eq.(1), one can know that the distance between measured point and base station is longer when the RSSI value is larger, with constant attenuation coefficient β and transmitted signal strength value.

3.3   Signal Processing

In wireless network, RSSI value is highly vulnerable to the impact of external disturbance such as shade influence and multipath effect. In order to improve the accuracy of fingerprint positioning algorithm and obtain reliable and stable RSSI value, filter is used for processing primary signals.

There are many well-developed filtering schemes for excluding disturbing signal such as average filtering, feedback filtering and so on. Average filtering needs to sample a large amount of signals, which may lead a large time delay. Comparing with average filtering, feedback filtering updates data there is a new sample value. Thus it is practicable to add feedback filtering into real time positioning, and the filtering principle is shown in Eq.(2).

$$RSSI_{ij}(n) = \alpha \cdot RSSI_{ij}^{'}(n) + (1-\alpha) \cdot RSSI_{ij}(n-1) \tag{2}$$

Where $RSSI_{ij}^{'}(n)$ is the n-th sampling RSSI value, $RSSI_{ij}(n)$ is RSSI value through n times feedback filtering, and $\alpha \in [0,1]$ is the proportion of sampling values. Moreover, a larger proportion α results a faster data update. In general, α is no less than 0.75.

4 Fingerprint Positioning

4.1 Location Clustering

In positioning phase, if we match the parameters of measured point with that of all reference points, the computation will be really large. Hence, location clustering method is used to exclude invalid positioning data.

When there are numerous reference points in consider, the threshold is added to select the reference points which have similar RSSI value with measured point. Assume that threshold Δ_T is nonnegative, for base station j, one has

$$\left| RSSI_{Mj} - RSSI_{ij} \right| \leq \Delta_T \tag{3}$$

Where $RSSI_{Mj}$ is the RSSI value that measured point has received. The reference point i is regarded as valid reference point V_i when Eq.(3) is fulfilled for all $j = \{1, \ldots, m\}$.

4.2 Matching Algorithm

The value of RSSI could reflect the correlation between valid reference point and measured point. When the valid reference point is close to the measured point, the environment nearby is similar, that is, their RSSI values are similar. Thus, one can determine the distance from valid reference point to the measured point by RSSI-based correlation coefficient.

Assume that there are m base stations, the measured point is M, the valid reference point is V_i. The correlation coefficient between M and V_i is defined as

$$\rho(M,V_i) = \frac{\sum_{j=1}^{m}\left(RSSI_{Mj} - \overline{RSSI}_M\right)\left(RSSI_{V_ij} - \overline{RSSI}_{V_i}\right)}{\sqrt{\sum_{j=1}^{m}\left(RSSI_{Mj} - \overline{RSSI}_M\right)^2 \sum_{j=1}^{m}\left(RSSI_{V_ij} - \overline{RSSI}_{V_i}\right)^2}}$$

$$= \frac{m\sum_{j=1}^{m} RSSI_{Mj} RSSI_{V_ij} - \sum_{j=1}^{m} RSSI_{Mj} \cdot \sum_{j=1}^{m} RSSI_{V_ij}}{\sqrt{m\sum_{j=1}^{m} s_{Mj}^2 - \left(\sum_{j=1}^{m} s_{Mj}\right)^2} \cdot \sqrt{m\sum_{j=1}^{m} s_{V_ij}^2 - \left(\sum_{j=1}^{m} s_{V_ij}\right)^2}}$$

(4)

Where $RSSI_{Mj}$ and $RSSI_{V_ij}$ are the RSSI value measured point M and valid reference point V_i have received from base station j, respectively. From Eq.(4), one can obtain that the correlation coefficient is larger when the point M and V_i are nearer. Therefore, the measured point $(\hat{x}, \hat{y})$ can be estimated as

$$\hat{x} = \frac{\sum_{i=1}^{n} \rho(M,V_i)x_i}{\sum_{i=1}^{n} \rho(M,V_i)}, \quad \hat{y} = \frac{\sum_{i=1}^{k} \rho(M,V_i)y_i}{\sum_{i=1}^{k} \rho(M,V_i)}$$

5   Simulation Example

In this section, an example is provided to validate the effectiveness of the method proposed. Consider a room with 4 meters length and 4 meters width, where 7 base stations are set. Assume that the number of measured points is 25. The positioning results are shown in Fig.2. The average positioning error is 0.2714m. One can find that the accuracy of positioning is really low. Then the number of reference points is increased to 153, and the results are shown in Fig.3. The average positioning error is 0.2071m.

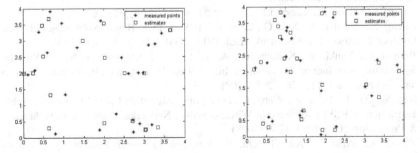

Fig. 2. The number of reference points is 25   **Fig. 3.** The number of reference points is 153

Noted that the more reference points are, the more accurate the positioning results are, when compare Fig.3 with Fig.2. Given that reference points' number is 285, the results are shown in Fig.4. The average positioning error is 0.1523m.

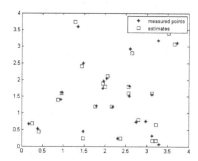

Fig. 4. The positioning results with 285 reference points

Fig.4 shows that the positioning accuracy is higher than that shown respectively in Fig.3 and Fig.2. However, when the number of reference points is considerably increased compared with Fig.3, it is obvious to know that the complexity of computation is larger, that is, the computing time is longer. So it can serve as the trade-off between accuracy and computing time.

6   Conclusion

In this paper, a direct explicit method of the fingerprint positioning for indoor wireless network is proposed. Location clustering technique is used to exclude invalid reference points, and a matching algorithm based on RSSI correlation coefficient is presented, which can improve positioning accuracy. The simulation example shows the effectiveness and simplicity of the proposed positioning scheme.

References

1. Wi-Fi Positioning Technology,
   http://labs.chinamobile.com/mblog/712208_82886
2. Cheng, X., Thaeler, A., Xue, G., Chen, D.: TPS: A Time-Based Positioning Schemes for Outdoor Wireless Sensor Network. IEEE INFOCOM 4(4), 2685–2696 (2004)
3. Belloni, F., Ranki, V., Kainulainen, A., Richter, A.: Angle-Based Indoor Positioning System for Open Indoor Environment. In: 2009 6th Workshop on Positioning, Navigation and Communication, pp. 261–265 (2009)
4. Ren, W., Xu, L., Deng, Z., Wang, C.: Positioning Algorithm Using Maximum Likelihood Estimation of RSSI Difference in Wireless Sensor Networks. Journal of Data Acquisition & Processing 21(7), 1247–1250 (2008)
5. Zhang, X., Zhao, P., Xu, G., Lin, R.: Research of Indoor Positioning Based on A Optimization KNN Algorithm. International Electronic Elements 21(7), 44–46 (2013)

6. Sakamoto, J., Miura, H., Matsuda, N., Taki, H., Abe, N., Hori, S.: Indoor Location Determination Using a Topological Model. In: Khosla, R., Howlett, R.J., Jain, L.C. (eds.) KES 2005. LNCS (LNAI), vol. 3684, pp. 143–149. Springer, Heidelberg (2005)
7. Brunato, M., Battiti, R.: Statistical Learning Theory for Location Fingerprinting in Wireless LANs. Computer Networks 47(6), 825–845 (2005)
8. Fang, S., Lin, T., Lin, P.: Location Fingerprinting In A Decorrelated Space. IEEE Transactions on Knowledge and Data Engineering 20(5), 685–691 (2008)
9. Milioris, D., Tzagkarakis, G., Papakonstantinou, A., Papadopouli, M., Tsakalides, P.: Low-Dimensional Signal-Strength Fingerprint-Based Positioning in Wireless. Lans Ad Hoc Networks 12, 100–114 (2014)
10. Liang, X., Gou, X., Liu, Y.: Fingerprint-Based Location Positioning Using Improved KNN. In: 2012 3rd IEEE International Conference on Network Infrastructure and Digital Content, pp. 57–61 (2012)
11. Peerapong, T., Xiu, C.: Indoor Positioning Based on Wi-Fi Fingerprint Technique Using Fuzzy K-Nearest Neighbor. In: Proceedings of 2014 11th International Bhurban Conference on Applied Sciences & Technology, pp. 461–465 (2014)
12. Ni, L.M., Liu, Y., Lau, Y.C., Patil, A.P.: LANDMARC: Indoor Location Sensing Using Active RFID. Wireless Networks 10(6), 701–710 (2004)
13. Tian, F., Dong, Y., Sun, E., Wang, C.: Nodes Localization Algorithm for Linear Wireless Sensor Networks in Underground Coal Mine Based on RSSI-Similarity Degree. In: 2011 7th International Conference on Wireless Communications, Networking and Mobile Computing, pp. 1–4 (2011)

Closed-Loop Test Method for Power Plant AVC System Based on Real Time Digit Simulation System

Hong Fan[1,*], Desheng Zhou[2], Aiqiang Pan[2], Chao Chen[3], Xinyu Ji[1], and Huiyan Gao[1]

[1] College of Electrical Engineering Shanghai University of Electric Power, Shanghai, China
fan_honghong@126.com
[2] Electric Power Research Institute Shanghai Municipal Electric Power Company, Shanghai, China
[3] Anhui Leadzone Smart Grid Technology Co., Ltd., Hefei Anhui, China

Abstract. A novel method of closed-loop test simulation for power plant AVC system on RTDS(Real Time Digital Simulator) is presented. The proposed closed-loop test simulation plat includes real time digital simulator models, power plant AVC system and RTU. System models include system plat model and central control system model which is built upon components library on RTDS. The proposed test method will test the AVC system in all system conditions and because of the use of RTDS,the test can be both accurate and credible.A real case is tested to verify the validity of the proposed method.

Keywords: Power system, Real time digital simulator (RTDS), Power plant AVC system, Closed-loop test.

1   Introduction

It is reactive power optimization and compensation of the power system that compose the important component of the safe and economic operation of power system. By rational allocation of reactive power and the reactive load compensation, it can not only maintain the voltage level and improve the stability of power system operation, but also reduce the loss of network which can make the power system operate safely and economically. In the traditional model which is controlled and managed by the voltage, the reactive power from generator and high voltage of bus are primarily adjusted manually by the operator in the power plant according to the requirement of the system. The method of adjustment can only guarantee the qualified voltage of some nodes in part time, while it is far from the economic operation and reliable power supply of power system.

Power plant automatic voltage control system utilizes the technologies of advanced electron, network communication and automatic control, receives bus voltage commands issued by dispatch center of provincial power online and tracking controls automatically reactive power of generator or high side real-timely. Power plant AVC

---

[*] Hong FAN, now being associate professor of Shanghai University of Electric Power.

K. Li et al. (Eds.): LSMS/ICSEE 2014, Part III, CCIS 463, pp. 320–330, 2014.

system can also control the rational flow of the reactive power area effectively and enhance the stability and security of the power system operation so that the quality of voltage has been ensured, the integral level of power network has been improved, the loss of active power has been decreased, the economic benefic has been exerted fully and at the same time the working intensity of the operators has been reduced.

The primary system of field operation imitated power plant in the real-time simulation system has been established in this paper, which sets up the closed-loop operation system between power plant AVC system and the real-time digital simulation system by making use of the characteristics and tests the functions by simulating the operational methods online of power plant AVC system.

2   The Introduction of Power Plant AVC System

2.1   System Structure

The main station and substation are involved in power plant AVC system and the former is installed in the dispatch center of regional network while the latter in the side of the power plant. The main station sends the voltage controlling commands of nodes to the substation according to the calculation of reactive power optimization and receives the feedback information of the substation. The substation is functionally divided into upper computer and lower computer. Upper computer, known as the central control unit, including the host and backup, is responding for receiving the commands of the main station and analyzing the current operation conditions, calculates and deals with tasks. Lower computer, also called executive termination, is the actuator of the regulating system, receiving the commands from the central unit

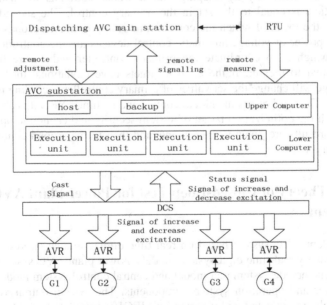

Fig. 1. AVC system structure scheme

and emitting adjusted instructions and auxiliary to the generators and the controlled commands from the DCS monitoring system of generators. The structure scheme of AVC system can be described in Figure 1.

Bus voltages are stabilized by the AVC substation adjusting reactive power from generators. The exciting current influences the reactive power and terminal voltage. The reactive power and terminal voltage will increase or decrease when the exciting current changes, which is achieved by adjusting the given voltage value of the excitation regulator AVR. The dispatch center of the main station issues the goal if voltage controlled and every once in a while sends bus voltage commands to some certain generator units and power plants and the commands and real-time data collected by RTU are received by the data communication processing platform of the power plant which sends the data to the substation by the communication network. Considering about the system, equipment failures and all limits of AVR and shutting, the substation calculates and provides adjusting schemes within the scope of the generator capacity in the current operation, then sends controlling signals to the excitation regulation or DCS or ECS, which can increase or decrease the given values of the excitation regulator to adjust reactive power of the generator, so that it can make the bus voltages maintain the range of dispatch center given.

2.2   Control Method

The measured AVC system adopts three-level voltage control modes in this paper. The primary voltage is controlled by the unit control. The generator excitation regulator is the controller, of which the control time is usually milliseconds to seconds. The control equipments are changed as close as possible to set compensation voltage fast and randomly by keeping output whose action is to ensure that the terminal voltage is equal to the set in the primary control. The second voltage is controlled by the local, of which the control time is approximate seconds to minutes. The reactive power of voltage automatic control devices are the controller, the main purpose of which is to coordinate the primary controller and ensure that the bus voltage is equal to the set. If the control targets have deviations, the controller if second voltage will change the set values of primary controller according to the preset control rules. The third one is the global control and its time constant is about minutes to hours. With the optimization target of system security and economic operation, the optimization results are given to each plants and then to the second controller as for the tracing target.

3   The Theory of Closed-Loop Test for Power Plant AVC System

The power plant AVC system based on real time digital simulation system includes three parts namely real-time digit system models, power plant AVC system and RTU. System models include system plat model and central control system model which are built by power and operation system components. The system output voltages and currents which are collected and inverted into IEC101 rule data by RTU real-timely, then are transmitted by the data buses to the substation. The control commands will return to AVC system in order to realize closed-loop test plat construction after AVC system calculating the control strategy. The diagram can be seen in the Figure 2.

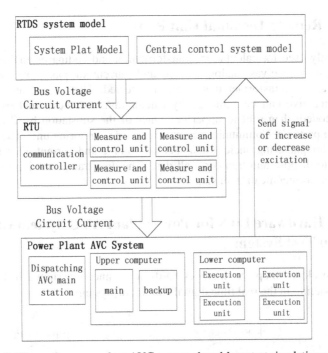

Fig. 2. Theory for power plant AVC system closed-loop test simulation plat

In the Figure 2, the real time digital simulation system which is used to simulate the characteristic electric changed of the controlled plants includes three parts: the first is the simulation of electrical system, that is to simulate electrical characteristic of plant real-timely based on the models and parameters of power plants; the second is simulating the monitor system of AVC system, namely simulating the monitor system voltage and reactive controlled of plant in DCS and the last is to provide communication interface between the system and RTU and AVC.

The models established in the real time digital simulation includes system plat model and central control system model, while system plat model includes primary system and excitation system. Central control system model is the integrated control system of the plant, which can display system data operating and receive input commands by the users. The real-time data of real time digital simulation system are collected by the RTU in time, then sent to the master control and inverted into IEC101 rule data which are sent to upper computer of the substation system. AVC main station of provincial dispatching gives control commands to AVC remote controls, then the upper computer of the substation accepts the control instructions of each unit by calculating the control strategy and sends to the lower computer, that is to say terminal which outputs increase or decrease excitation commands to real time digital simulation system to control reactive power generation to realize closed-loop plat construction.

4   The Remote Terminal Unit RTU

RTU is mainly used for calculation, transformation and coding of analogs in closed-loop test, which receives analog voltage and current of plant and switch values whether generator-transformer unit is grid-connected. The voltage and current rms, active and reactive can be achieved by calculating, and these analogs and switch values are coded, then sent to upper computer of the substation by IEC101 rule. It contains two parts i.e. Communication manager and monitoring unit. The latter one is responsible for gathering analog data such as voltage and current, then sends to the former one to dispose and transform all data. The data transformed are sent to AVC system by the communication manager.

5   The Hardware Link for Power Plant AVC System Closed-Loop Test System

AVC system closed-loop test includes hardware connection shown in Figure 3 and software modeling in the real time digital simulation system.

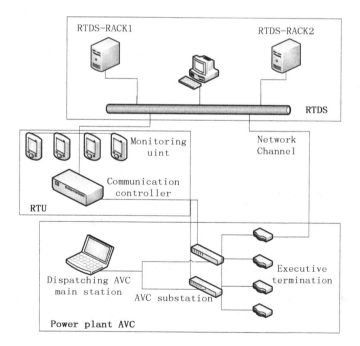

Fig. 3. Scheme of hardware link for power plant AVC system closed-loop test system

As Figure 3 shown, multi-channel analog outputs of real time digital simulation are connected to monitoring unit of RTU. The data collected by monitoring unit timely are sent to communication manager and the transformed data are sent to AVC system. Calculated by AVC system to get control strategies, terminal will send control demands to real time digital system. As IEC-104 rule test software of configuration of provincial dispatching AVC main station and Tsinghua AVC main station software in AVC system, portable computer is to simulate AVC main station, which sends target instructions to substation according to IEC-104 rule software and receives the status information of AVC substation.

6   System Modeling

In this paper the system model set in RTDS includes system platform model and the central control system model, while the system platform model contains the model of primary system and control system model.

6.1   Primary System Modeling

The primary system model simulates the primary system, which contains generators, two winds transformers, equivalent circuit, equivalent sources, equivalent load and secondary side measuring system a certain power plant. There are bus A and bus B in the simulating system. Each of them reserve two generator-transformer units, the outlets of the power plant are simplified as two equivalent circuit and the system side is simplified as an equivalent source. Secondary side measuring system includes telemeter spots and measured quantity.

The primary side models in RTDS contains generators, transformers, lines, sources, loads, PT and CT. The settable parameters in the models of generator are capacity of the generator, reactance parameters, time constant, voltage classes and so on. The settable parameters in the models of transformer are voltage classes, basic frequence, leakage reactance, no-load loss, etc. Loads are set as dynamic constant power models that can be changed according to the experiment.

6.2   Control System Model

Control system model contain the control of the units excitation system and the AVC excitation control system. To control the generator excitation system, including the reception of increasing and decreasing excitation as well as the control logic, on the one hand, it's doable to control by manual adjustment, on the other hand control by the outside AVC instruction is also available. Control system can also accomplish the control the active and reactive power output by the user commands sending from the central control system. After the above steps about platform models and corresponding sub models as well as the parameter setting, it's ready to compile the model and simulate the operation of the power plant.

AVC increase and decrease excitation system, on account of the AVC sending increase and decrease signal simultaneously, set up a control loop to distinguish increase or decrease.

6.3   Central Control System Model

Central control system, simulating the power plant centralized control center, is set up by the operation components in RTDS. Central control system model can display the running data of the system, which includes each bus voltage, the active/reactive power output of each generator, each winds' active/reactive power output of each transformer, the increase and decrease magnetic signal states of each generators, etc. Besides, by changing the generator phase angle manually, we can change the active/reactive power output of each generator, consequently changing the system operation condition. Excitation control model of the system can be tested by changing the increase or decrease signal manually.

6.4   System Modeling Process

System modeling steps are as follows:
1) Create a new construction system model file and save;
a) Build models of generator, steam turbine, transformer, breaker, load and connect them according to the electrical relationship;
b) Set various parameters of the models and circuit breaker signal;
c) Construct control models of each generator excitation;
d) Construct control model of AVC monitoring system;
e) Construct control model of dynamic loads;
    2) Complete the system platform construction of the model;
3) Create central control system and save the file;
a) Build the primary grid structure of the power plant;
b) Display the parameters of the system platform model which include bus voltage, generator active/ reactive output, breaker switching status, load switching status through meters or indicator lights;
c) Model each circuit breaker;
d) Model each dynamic load switcher;
4) Complete the construction of central control system model;
5) Debug the system to finish modeling the software of Close-loop test system.

7   Close-Loop Test Methods of the AVC System

Close-loop test methods of the AVC system are as follows:
1) In order to operate system, complete the system modeling and establish a hardware loop connection;

2) Test the fundamental function, safety performance as well as communication performance of the established model in accordance with the outline of tests which ruled in the power plant AVC operation manual;

3) Review test result.

In this study, we modeled the primary grid structure from a certain power plant. Build primary system model in RTDS software platform which contains generator model, two-winding transformer model, load model, equivalent source model and circuit breaker model. There are four generator, four two-winding transformer, two 500kV bus, two equivalent source and a dynamic load in this model. Figure 4 illustrates the power system network structure as follows.

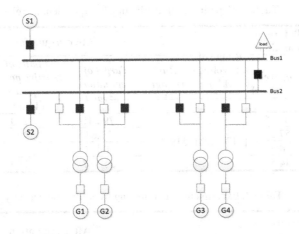

Fig. 4. Power network scheme of one power plant

The tests of AVC system contain three major parts which are parameter setting, fundamental function testing and safety performance testing. Here are some related test items.

1) Experiment of impulse width setting in units reactive power regulation device

This experiment is one of the tests about parameter setting. By adjusting AVC output impulse width, we can setup many parameters in "Reactive Power Output Setting" such as units signal output gap, maximal signal width, minimal signal width and slop, which gathered experience for field test and parameter setting.

2) Experiment of reactive power distribution performance between units

This experiment is one of the tests about basic function. Setup mode of AVC bus voltage regulation, test the soundness of reactive power distribution between each AVC substation, which refers to 4 distribution mode: distributing reactive power by equivalent power factor, distributing reactive power by equivalent margin, distributing reactive power by equivalent capacity and distributing reactive power equivalent. Then check the effect of above distribution modes.

Table 1. Experiment of distributing by equivalent power factor

| UNIT | Before regulation Reactive power (Mar) | Range of bus voltage (kV) | After regulation | | | |
			Bus voltage (kV)	Target of regulation (Mvar)	regulation result (Mvar)	Power factor
#1	62.37			134.373	135.36	0.979
#2	33.75	[514.6 ~520]	519.6	67.042	66.51	0.980
#3	24.75			53.74	53.64	0.979
#4	26.82			53.575	53.28	0.980

Table 2. Experiment of distributing by equivalent margin

| UNITS | Before regulation Reactive power (Mvar) | Range of bus voltage (kV) | After regulation | | | |
			Bus voltage (kV)	Target of regulation (Mvar)	Limits of reactive power capacity (Mvar)	Regulation factor
#1	63.45			79.241	220	0.100869
#2	57.24	[517~ 520]	519.7	71.64	200	0.100869
#3	48.6			60.846	170	0.100873
#4	44.55			55.187	150	0.100872

Table 3. Experiment of distributing by equivalent capacity

| UNITS | Before regulation Reactive power (Mvar) | Range of bus voltage(kV) | After regulation | | | |
			Bus voltage (kV)	Target of regulation (Mvar)	Actual reactive power result (Mvar)	Capacity (Mvar)
#1	64.98			106.699	102.06	-100-220
#2	63.27	[518.8~ 521]	520.8	89.862	88.11	-80-200
#3	63.54			70.606	68.04	-50-170
#4	63.00			70.606	67.95	-70-150

Table 4. Experiment of distributing equivalent.

| UNITS | Before regulation Reactive power (Mvar) | Range of bus voltage(kV) | After regulation | | |
			Bus voltage (kV)	Target of regulation (Mvar)	Actual reactive power result (Mvar)
#1	114.84			75.825	75.510
#2	113.04	[524.8~520]	520.2	75.825	73.800
#3	115.29			75.825	75.870
#4	114.75			75.825	78.120

3) Experiment of units reactive power reverse regulation

This experiment, one of the tests about basic function, tests the ability of reverse regulation and reasonable reactive power distribution when the reactive power is out of limits during the regulation. Setup the mode of AVC bus voltage regulation, then primary station issued reduction magnetic instruction till the units' reactive power reaches its bottom limits or primary station issued increasing magnetic instruction till the units' reactive power reaches its upper limits. Check the reverse regulation process of AVC.

4) Experiment of terminal voltage reverse regulation

This experiment, one of the tests about basic function, tests the ability of reverse regulation and reasonable reactive power distribution when the terminal voltage is out of limits during the regulation. Setup the mode of AVC bus voltage regulation, then primary station issued reduction magnetic instruction till the units' terminal voltage reaches its bottom limits or primary station issued increasing magnetic instruction till the units' terminal voltage reaches its upper limits. Check the reverse regulation process of AVC.

Table 5. Experiment of reverse regulation

UNITS	Reactive power （Mvar）	Top limit of reactive power （Mvar）	AVC output
#1	32.4	25	Lock, reduce magnetic and reverse regulate
#2	21.2	15	Lock, reduce magnetic and reverse regulate
#3	15	10	Lock, reduce magnetic and reverse regulate
#4	19.8	15	Lock, reduce magnetic and reverse regulate

5) Experiment of communication outage between AVC primary station and substation

This experiment, one of the tests about security performance, tests the action of the AVC substation devices facing the communication outage between AVC primary station and substation by disconnecting them and checking the movement of AVC substation.

6) Experiment of communication outage between AVC substation and RTU

This experiment, one of the tests about security performance, tests the action of the AVC substation devices facing the communication outage between AVC substation and RTU by disconnecting them and checking the movement of AVC substation.

7) Locking experiment

This experiment, one of the tests about security performance, tests the action of the AVC substation devices detecting the telemeter data overstepped the locking limits, which includes telemeter data locking such as bus voltage, units active/reactive power, terminal voltage, terminal current, station service, etc.

8   Conclusion

Power plant AVC system is a major method of optimal reactive power control as well as a necessary of stable operation in power system. Closed-loop test method for power plant AVC system based on real time digit simulation system laid foundation for AVC system off-line real time testing in laboratory. In this paper the test platform is able to give a comprehensive evaluation about various functions of the power plant AVC system. The research conclusion has a significant introduction for power plant AVC system.

Acknowledgments. This work is supported by National Natural Science Foundation of China (No.51307104) and Shanghai Green Energy Grid Connected Technology Engineering Research Center 13DZ2251900.

References

1. Yin, J.: Application of automation reactive power regulation and control devices to AVC system. East China Electric Power 33(9), 64–66 (2005)
2. Tang, J.H., Zhang, L.G., Zhao, X.L.: Application of the auto-voltage-control in power plants. Power System Protection and Control 37(4), 32–35 (2009)
3. Zhou, Y.H., Zhai, W.X., Ma, P.: Design of automatic voltage control (AVC) system for thermal power plant. Power System Protection and Control 40(9), 128–132 (2012)
4. Xie, F., Gong, X.M.: Security constraints control strategy optimization of AVC sub-station system in power plant. Power System Protection and Control 38(16), 147–149 (2010)
5. Wu, Z.W.: The development of the power plant side AVC substation system. Ph.D. Thesis, Hefei University of Technology, Hefei (2007)

Synchronous Current Harmonic Optimal Pulsewidth Modulation for Three-Level Inverter[*]

Jiuyi Feng, Wenxiang Song, Shuhao Jiang, and Haoyu Wang

School of Mechatronic Engineering and Automation, Shanghai University,
Shanghai 200072, China

Abstract. This paper introduces a method for three-level inverter operating at low switching frequency and presents a novel current harmonic optimum pulsewidth modulation strategy. Combining the selected harmonic elimination PWM (SHEPWM) with the current harmonic minimum PWM (CHMPWM), the switching angles can not only eliminate the selected harmonic but also reduce the current harmonic. The three-level inverter output voltage is analyzed using fourier series and the expression of the total harmonic current distortion(THD_i) rate is calculated. The THD_i is regarded as the objective function, with the voltage fundamental amplitude is equal to the reference voltage amplitude and the other harmonic amplitude is zero as nonlinear constraints. Then the switching angle is solved by calculating the THD_i. After researching its basic principle and the solution of the switching angle, results of the switching angle at 280Hz switching frequency are verified by using MATLAB software.

Keywords: three-level inverter, low switching frequency, selected harmonic elimination, current harmonic minimum.

1 Introduction

Three-level inverter has been widely used in high power motor such as mine hoisting, locomotive traction and so on with the development of the power electronics. However, it is really a severe test for the switching devices' life and heat dissipation, because the more the operating voltage raises, the more the output power of the inverter and the switching loss increase. So it is necessary to reduce the switching frequency of the power devices to about several hundred hertz to boost the inverter's output power[1]. Yet, in such a low switching frequency, SVPWM, the traditional method of PWM control, will produce a large number of low harmonics, which may cause serious distortion of the current harmonics[2], so that the optimized PWM strategies[3~5] are deadly need in this condition–the selected harmonic elimination PWM(SHEPWM) and the current harmonic minimum PWM(CHMPWM), for example.

---

[*] Project Supported by National Natural Science Foundation of China (51377102); Grants from the Power Electronics Science and Education Development Program of the Delta Environmental & Educational Foundation (DREG2013009).

K. Li et al. (Eds.): LSMS/ICSEE 2014, Part III, CCIS 463, pp. 331–340, 2014.

SHEPWM is put forward in 1973 by Hasmukh S.Patel and Richard G.Hoft[6]. It can eliminate the selected harmonics as well as create better output waveform, but the other harmonics will increase[7].

In 1977, Giuseppe S.Buja and Giovanni B.Indri propose the optimal pulse width modulation[8]. It shows the required performance index as the stator voltage Fourier series function so that the expected performance can be get after calculating the switching angles.

This paper combines the advantages of SHEPWM with the traditional CHMPWM, to make the switching angles can not only lower the total harmonic distortion of stator current but also eliminate the selected harmonics. Based on the Fourier analysis for the output of the three-level inverter's voltage wave, expression of the stator current's total harmonic distortion(THD$_i$) is easily got firstly. Then we make the THD$_i$ as the objective function, with the fundamental output voltage amplitude equals the set value and other harmonics' maintain zero as the nonlinear constraints. In this way, it will take a long time to calculate. So we just let the fundamental output voltage amplitude equals the set value as the nonlinear constraints, yet, applying the fundamental output voltage amplitude equals the set value and other harmonics' maintain zero as the fitness function of the Genetic Algorithm(GA) to calculate the initial value, which can eliminate the selected harmonics. Substituting the initial value into the objective function, then we get the switching angles for different modulation degrees. Finally, with the help of MATLAB, validity of the algorithms proposed is verified through the computer simulation.

2   Pinciples of CHMPWM

Fig. 1 shows the basic circuit structure of three-level voltage source PWM Inverter.

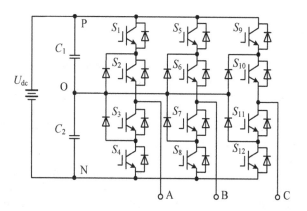

Fig. 1. Main circuit of three-level converter

Fig. 2 gives the waveform that the power electronic devices of phase A switching N times in a fundamental period.

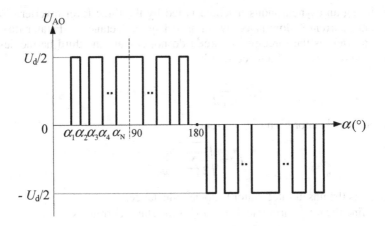

Fig. 2. Output phase voltage waveform

Based on the Dirichlet principle, phase A voltage can be presented by following Fourier series

$$U(t) = \frac{a_0}{2} + \sum_{n=1}^{\infty} (a_n \cos n\omega t + b_n \sin n\omega t) \quad n = 1, 2, 3 \cdots \tag{1}$$

where $a_n = \dfrac{2}{T} \displaystyle\int_{-T/2}^{T/2} U(t) \cos n\omega t \, dt$, $b_n = \displaystyle\int_{-T/2}^{T/2} U(t) \sin n\omega t \, dt$.

Usually, to simplify the solution of the nonlinear constraint equations, the phase voltage is made to be mirror symmetry to π and even symmetry to $\pi/2$ to eliminate even order harmonics and dc elements of inverter output voltage. (1) can be simplified as :

$$U(t) = \sum_{n=1,3,5\ldots}^{\infty} b_n \sin n\omega t \tag{2}$$

where $\omega t = \alpha$ and

$$\begin{cases} b_n = \dfrac{4}{n\pi} \cdot \dfrac{U_{dc}}{2} \sum_{k=1}^{N} (-1)^{k+1} \cos n\alpha_k & \text{for } n \text{ odd} \\ b_n = 0 & \text{for } n \text{ even} \end{cases}$$

is the amplitude of each order of harmonics; k is the equence number; αk satisfies

$$0 \le \alpha_1 < \alpha_2 < \cdots < \alpha_N \le \pi/2 \tag{3}$$

SHEPWM is to eliminate the selected harmonics to get the expected sinusoidal output voltage. however, CHMPWM just make the THD$_i$ as the performance index to calculate the switching angles. The calculation of this modulation mode depends on the nature of the load, but, for the AC motor, the circuit impedance is mainly determined by the leakage inductance of the stator [9].

Considering an asynchronous machine is fed by the three-level inverter. Suppose that the stator current is dominated by stator leakage inductance and stator flux is not saturated, as well as the three-phase circuit do not contain the third harmonics, then the amplitude of stator harmonic current is

$$I_{hn} = \frac{b_n}{n\omega L} \qquad n = 5, 7, 11, \cdots \tag{4}$$

Now the rms of stator harmonic current is

$$I_h = \sqrt{\frac{1}{2} \sum_{n=5}^{\infty} (\frac{b_n}{n\omega L})^2} \tag{5}$$

Where, L is the unsaturated stator leakage inductance.
Then define the total harmonic distortion of the stator current as

$$THD_i = \frac{I_h}{I_1} \times 100\% \tag{6}$$

By (2)~(5), the expression of THD_i is

$$THD_i = \frac{\sqrt{\frac{1}{2} \sum_{n=5}^{\infty} \left(\sum_{k=1}^{N} \frac{(-1)^{k+1} \cos n\alpha_k}{n^2} \right)^2}}{\sum_{k=1}^{N} (-1)^{k+1} \cos \alpha_k} \times 100\% \tag{7}$$

So (6) is the objective function to calculate the switching angles. Define modulation index as

$$m = \frac{b_1}{U_{dc}/2} \tag{8}$$

Where b_1 is the fundamental amplitude of stator voltage, U_{dc} is the DC bus voltage. By (8), the nonlinear constraint equations to solve the switching angles is

$$\sum_{k=1}^{N} (-1)^{k+1} \cos \alpha_k = \frac{\pi}{4} m \tag{9}$$

In the variable voltage variable frequency(VVVF) control of asynchronous motor, the modulation index often equals to

$$m = \frac{f}{f_s} \tag{10}$$

Where f is the stator current frequency, f_s is the rated frequency.
In the pactical motor driving systerm, the number of the switching angles always varies according to the change of the modulation index to get better optimal performance[10]. In this paper, the proposed subsection synchronization method is showed in Table 1.

Table 1. Subsection synchronization method

f/(Hz)	N	m	f_{sw}/(Hz)
45-50	5	0.9-1	225-250
35-45	7	0.7-0.9	245-315
25-35	9	0.5-0.7	225-315
20-25	11	0.4-0.5	220-275
15-20	15	0.3-0.4	225-300

When the stator frequency is reduced to several Hertz, CHMPWM is no longer superior to SVPWM, thus a handover between the two methods is necessary. Generally SVPWM is used when the modulation index is less than 0.3 [11].

3   Calculation of Switching Angles

(7) is a challenge with nonlinear constraints. Many solutions have been put forward to tackle this problem, such as Gradient Method , Ant Colony Algorithm[12], Genetic Algorithm(GA) [13], Particle Swarm Optimization (PSO)[14] and so on. However, Gradient Method has a really quickly iteration speed, it need a very precise iterative initial value. Genetic Algorithm is on the contrary. Thus, firstly, to get full use of these algorithms' advantages, a feasible initial value is obtained according to Genetic Algorithm. The *fmincon* function is then used to calculate all of the switching angles.

This paper makes the fundamental output voltage amplitude equals the set value and other harmonics' maintain zero as the fitness function of the Genetic Algorithm to get the expected initial value, which can eliminate the selected harmonics. Nonlinear equations are

$$
\begin{cases}
\sum_{k=1}^{N} (-1)^{k+1} \cos \alpha_k - \dfrac{\pi}{4} m = \varepsilon_1 \\[2mm]
\sum_{k=1}^{N} (-1)^{k+1} \cos 5\alpha_k = \varepsilon_2 \\[2mm]
\sum_{k=1}^{N} (-1)^{k+1} \cos 7\alpha_k = \varepsilon_3 \\[2mm]
\sum_{k=1}^{N} (-1)^{k+1} \cos 11\alpha_k = \varepsilon_4 \\[2mm]
\cdots\cdots
\end{cases}
\tag{11}
$$

Then the fitness function is

$$
\Gamma = 1 + \varepsilon_1^{\,2} + \varepsilon_2^{\,2} + \varepsilon_3^{\,2} + \varepsilon_4^{\,2} + \cdots\cdots
\tag{12}
$$

Generally, the GA function need to be performed only once in each subsection. The initial values got are shown in Table 2. Put the initial value and the objective function to the *fmincon* function with the help of MATLAB to calculate the switching angles in different modulation index. Figure 3 shows the switching angle curve.

The parameters of GA are set as follows:

Population size	3000
Selection function	Roulette
Crossover fraction	0.88
Mutation rate	0.12
Migration fraction	0.25

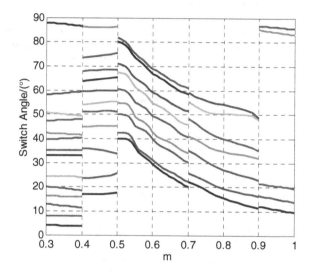

Fig. 3. Switching angle curve

Table 2. Initial value by GA

N	x_0 (rad)
5	[0.212,0.289,0.374,1.486,1.514]
7	[0.388,0.448,0.609,0.715,0.843,0.966,1.031]
9	[0.699,0.741,0.876,0.959,1.055,1.177,1.238,1.399,1.429]
11	[0.139,0.231,0.657,0.674,0.784,0.955,1.041,1.110,1.340,1.453,1.502]
15	[0.071,0.135,0.222,0.282,0.348,0.420,0.573,0.573,0.613,0.688,0.737,0.824,0.883,1.013,1.531]

4   Simulation

Take $N=7$, $m=0.8$ and $N=9$, $m=0.5$ for example. Then the stator current frequency—f is 40Hz or 25Hz, so that the switching frequency is about 300Hz. In the simulation, the DC bus voltage is 540V and the load is symmetrical three-phase inductive load, whose resistance values 12.6 and inductance values 40mH. Fig. 4, 5 shows the simulation result of CHMPWM.

Reference [15] gives the Formula to calculate the initial switching angles for SHEPWM. This paper uses that result to iterate a series of switching angles for different m . Fig. 6, 7 shows the spectrum analysis diagram of the SHEPWM , the traditional CHMPWM and the proposed CHMPWM. Obviously, the proposed CHMPWM can not only reduce the THD_i but also elimate the selected harmonics.

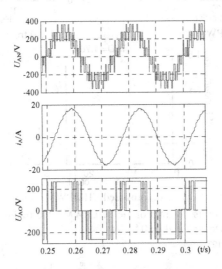

Fig. 4. Simulation of CHMPWM when $N=7$ and $m=0.8$

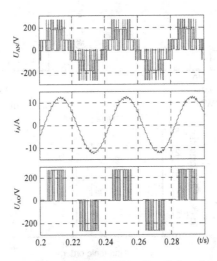

Fig. 5. Simulation of CHMPWM when $N=9$ and $m=0.5$

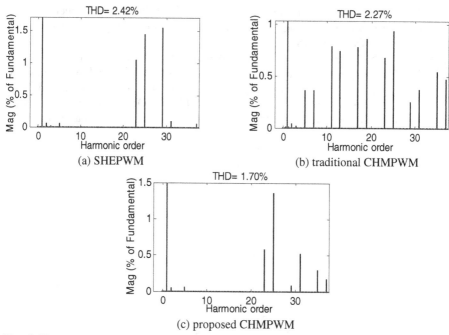

Fig. 6. The spectrum analysis diagram of the SHEPWM , the traditional CHMPWM and the proposed CHMPWM when $N=7$ and $m=0.8$

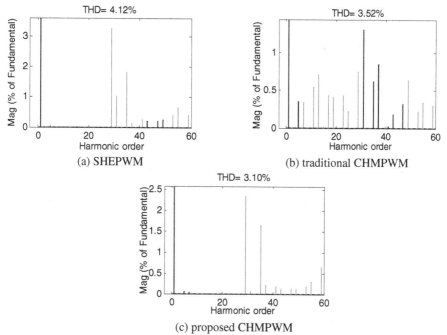

Fig. 7. The spectrum analysis diagram of the SHEPWM , the traditional CHMPWM and the proposed CHMPWM when $N=9$ and $m=0.5$

The simulation is carried out in different m between SHEPWM and CHMPWM. Table 3 shows their THD_i, from which we can see that the CHMPWM is better than SHEPWM.

Table 3. THDi of SHEPWM and CHMPWM

m	$THD_i(CHM)\%$	$THD_i(SHE)\%$
0.1	19.82	20.81
0.2	6.61	8.82
0.3	7.00	7.98
0.4	3.97	6.30
0.5	3.11	4.12
0.6	2.33	3.18
0.7	2.16	2.66
0.8	1.70	2.42
0.9	1.15	1.99
1.0	1.02	1.93

5   Conclution

This paper introduces the principle of CHMPWM and put forward a novel method to calculate the switching angles after analyzing the traditional CHMPWM. Apply the fundamental output voltage amplitude equals the set value and other harmonics' maintain zero as the fitness function of the Genetic Algorithm to get the expected initial value, which can eliminate the selected harmonics. Then put the initial value and the objective function to the *fmincon* function to calculate the switching angles in different modulation index. Finally, the simulation verifies the correction of the switching angles and the feasibility of this method.

However, the simulation is just for the open-loop control with resistance load. The closed-loop control hasn't been verified. Also, the phase's jump due to the pattern switchover hasn't been considered. Moreover, the influence of the dead band on the stator current harmonics isn't reflected in this simulation. All these are the emphases for the future research.

References

1. Franquelo, L.G., Leon, J.I., Dominguez, E.: New Trends and Topologies for High Power Industrial Applications: The Multilevel Converters Solution. Power Engineering. Energy and Electrical Drives, 1–6 (2009)
2. Holtz, J.: Fast Dynamic Cont rol of Medium Voltage Drives Operating at Very Low Switching Frequency— an Over-view. IEEE Trans. on Ind. Electron 55(3), 1005–1013 (2008)
3. Rathore, A.K., Holtz, J., Boller, T.: Synchronous Optimal Pulsewidth Modulation for Low-Switching- Frequency Control of Medium-Voltage Multilevel Inverters. IEEE Transactions on Industry Electronics 57(7), 2374–2381 (2010)
4. Xiao, F., Peng, D., Bingjie, Z., et al.: Research on subsection synchronization harmonic optimum pulsewidth modulation for inverters. In: Asia-Pacific Power and Energy Engineering Conference (APPEEC) (2010)
5. Xiao, F.: Control of Electrical Excited Synchronous Machine at Low Switching Frequency. China University of Mining and Technology, Xuzhou (2011)
6. Patel, H.S., Hoft, R.G.: Generalized technique of harmonic elimination and voltage control in thyristor inverter: part I—harmonic elimination. IEEE Trans. on Industry Application 9(3), 310–317 (1973)
7. Rodriguez, J.R., Pontt, J., Huerta, R., et al.: Resonances in a high-power active-front-end rectifier system. IEEE Transactions on Industrial Electronics 52(2), 482–488 (2005)
8. Buja, G.S., Indri, G.B.: Optimal pulsewidth modulation for feeding AC Motors. IEEE Trans. Ind. Appl. A1-13(1), 38–44 (1977)
9. Wang, R.: Research on Induction Motor Drive Technology for Locomotive Traction. Huazhong University of Science and Technology, Wuhan (2012)
10. Yah, S., Aflaki, M.A., Rezazade, A.R.: Optimal PWM for minimization of total harmonic current distortion in high-power induction motors using genetic algorithms. In: SICE-ICASE, pp. 5494–5499 (2006)
11. Zhang, Y.C., Zhao, Z.M., et al.: Study on a Hybrid Method of SVPWM and SHEPWM Applied to Three-level Adjustable Speed Drive System. Proceedings of the CSEE 27(16), 72–77 (2007)
12. Dorigo, M., Birattari, M., Stutzle, T.: Ant Colony Optimization. IEEE (S1556-603X) Computational Intelligence Magazine 1, 28–39 (2006)
13. Ozpineci, B., Tolbert, L.M., Chiasson, J.N.: Harmonic Optimization of Multilevel Converters Using Genetic Algorithm. In: 35 Annual IEEE Power Electronics Specialists Conference, Germany (2004)
14. Shi, Y., Eberhart, R.: Parameter selection in PSO optimization. In: Porto, V.W., Waagen, D. (eds.) EP 1998. LNCS, vol. 1447, pp. 591–600. Springer, Heidelberg (1998)
15. Zhang, Y.C., Zhao, Z.M.: Multiple Solutions for Selective Harmonic Eliminated PWM Applied to Three-Level Inverter. Transactions of China Electrotechnical Society 22(1), 74–78 (2007)

Game Theory Based Profit Allocation Method for Users within A Regional Energy Interchanging System

Hongbo Ren[1], Qiong Wu[1,2], Yinyin Ban[1], and Jian Yang[1]

[1] College of Energy and Mechanical Engineering, Shanghai University of Electric Power,
200090 Shanghai, China,
[2] Faculty of Environmental Engineering, The University of Kitakyushu,
808-0135 Kitakyushu, Japan
tjrhb@163.com

Abstract. As one of the most important for the energy-saving and emission-reduction measures, regional energy is expected to contribute much to the development a low carbon society in the local area due its benefits including the rational use of energy, using energy scientifically, comprehensive energy and integrated energy. The so-called regional energy interchanging system, is established by forming some energy communities which can generate and consume energy simultaneously, and are connected with each other through local micro electricity and heat grid. In the interchanging process, along with the improvement of energy use efficiency, additional issue will appear especially the fairness between end-users. Therefore, in this study, as a classic theory for dealing with the profit allocation problems, cooperative game has been employed for the analysis of profit allocation among the users within a regional energy interchanging system. According to the simulation results of a case study, the proposed method has been approved to be good solution for the profit allocation for the users in the energy interchanging system, realizing a perfect combination of efficiency and fairness between different users.

Keywords: Local energy, interchanging system, cooperative game, efficiency, fairness.

1 Introduction

As the world largest CO_2 emitter, China is facing pressure from both domestic and aboard. In addition, the fossil fuel based energy structure has led to not only the global environmental problems, but also local environmental issues (e.g., acid rain, heat island, etc.). In recent years, along with the urgent needs to save energy and reduce environmental emissions, the state has proposed a series of regulation polices towards the regulation of energy structure. In the local area, the energy development and utilization is also experiencing continuous innovation, and the means of rational and perfect use of energy is becoming more and more mature. Under these backgrounds, district energy is recognized to be one of the most effective measures to deal with the energy and environmental problems in the local area.

K. Li et al. (Eds.): LSMS/ICSEE 2014, Part III, CCIS 463, pp. 341–348, 2014.

Like other utilities including city water supply and power supply, district energy system is one of the city's infrastructures. To satisfy the electricity, cooling and heating demands for a certain area, it produces electricity, cold water and hot water based on a local energy centre, and then transmit them to various end-users through local energy transmission systems (e.g., water pipelines) [1-2].

In China, district energy appeared firstly in the northern area in about 60 years ago and mainly for heat supply from centralized heat plants. In recent 30 years, due to the development of air conditioning technologies, the transformation from decentralized cooling to centralized cooling and district cooling has become more and more popular. Especially, in the south of the Yangtze River, there have been some district cooling projects with an area of more than thousands square meters. Moreover, in recent ten years, along with the more and more serious energy and environmental issues, the combined cooling and heating system has been developed and the district energy development comes towards a comprehensive direction. Furthermore, in recent years, with the application of heat pump technologies, the combined cooling, heating and power (CCHP) system is proposed, which makes the energy use be more scientific and reasonable.

In this study, different from conventional district energy systems, a new form of district energy system, namely, regional energy interchanging system has been proposed. It forms some energy communities which can generate and consume energy simultaneously, and they are connected with each other through local micro electricity and heat grid. According to the concept of regional energy interchanging system, through the formation of local micro grids based compounded energy network, the electrical and thermal energy within the local area may be shared and interchanged. Compared with independent energy supply systems, regional energy interchanging system can increase system efficiency and reduce energy supply cost in the local area. However, because the end-users within the energy system may belong to different stake holders, the profits earned by the system cannot guarantee the profits of all independent users. In the local energy system, some end-users may partly act as an energy centre, which supply the energy demands of other users by producing more energy, and results in increased energy cost. In other words, the optimal decision for the whole system is not necessarily the optimal one for the single user. As to a single user, if the excess energy cost due to the surplus energy production cannot be compensated rationally, the economic incentives to join in the system will be lost and it will escape from the system. The economic benefits due to the establishment of the regional energy interchanging system should be allocated among the users according to their contribution to the whole system. Therefore, the fair and rational profit allocation mechanism is an important foundation for the existence of regional energy interchanging system, and also is the key point for the stable development of the regional energy interchanging system.

To solve this problem, in this study, the cooperative game theory which is recognized to the classical theory to study the system profit and its allocation, has been employed for the profit allocation problem among the users within the regional energy interchanging system. The execution of this study is expected to give a good solution for the optimal allocation of energy and resources with the principles of efficiency, equality and sustainability.

2   Concept of Regional Energy Interchanging System

As shown in Fig. 1, taking the heat interchanging between two buildings as an example, the heat equipments of the two buildings are connected by the heat pipelines, which can transmit the heat and cold water.

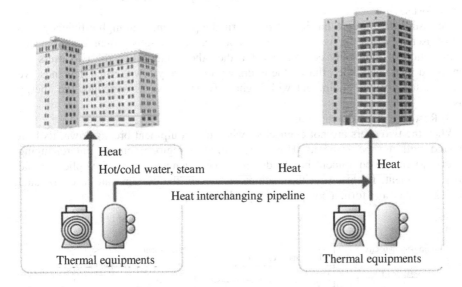

Fig. 1. Image of the regional energy interchanging system

Usually, the thermal equipment in a building is selected according to its peak demand which only has a small share in a total year; thus, the equipment operated under 50% of the rated capacity in most hours. In addition, various buildings may install different kinds thermal equipment with different efficiencies. Of course, the equipment with high efficiency may result in less energy consumption and CO_2 emissions. Thorough heat interchanging, the equipments with higher efficiency will enjoy the priority to run first and the running hours of the equipments with lower efficiency will be shorten. In this way, the energy consumption for the heat supply of the total building complex will be reduced and the cost will be reduced correspondingly. Furthermore, in order to operated the thermal equipments, some ancillary equipments (e.g., pumps) are necessary but has less relationship with the energy loads but dependent on the numbers of the thermal equipments. Under the situation of heat interchanging, when the heat load is relatively low, only a part of the thermal equipments should be on operation, which can reduce the energy consumption for ancillary equipments.

Fig. 2 illustrates the benefits of the regional energy interchanging system in a graphic way. Generally, they can be concluded as cost saving, size reduction and resource complementation, which are illustrated as follows in a detailed way.

(1) Size reduction.

Obviously, if the two users use their own equipments to satisfy their heat demands, the equipments should satisfy their peak demands while considering certain margins, separately. On the other hand, if the two users are connected within a regional energy interchanging system, because the peak hours of the two users do not happen simultaneously; thus, the overall size of the two equipments can be reduced.

(2) Cost saving

We assume that user A has installed a thermal equipment with high efficiency than that of user B. When the two users are not connected, they satisfy their heat demands by operation of their only equipments. On the other hand, when the two users are connected with each other, during the hours with relatively low thermal demands, we can operate only one equipment with higher efficiency, so as to reduce the energy costs.

(3) Resource complementation

When the two users are not connected, when one equipment breaks down, its heat demand will not be satisfied. However, within a regional energy interchanging system, when one equipment breaks down, its heat demands can be supplied by the other equipment. In this way, the equipments within the system can be cooperated with each other and form a complementary relationship.

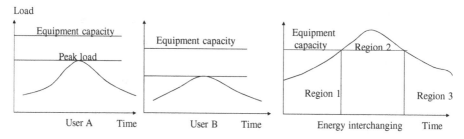

Fig. 2. Image of the benefits of regional energy interchanging system

According to the discussions of the benefits of regional energy interchanging system, the most important benefit is the improvement of overall efficiency which save the energy costs correspondingly. Therefore, how to allocation the saved energy cost is the main issued faced by the decision makers.

3   Theoretical Basis of Cooperative Game Theory

In a cooperative game, when the users cooperate with each other to form some coalitions, it is sure that different users may obtain different profits in different coalitions. Though one user can obtain its maximum profit in some coalition, it does not mean that the other ones can obtain their maximum profits in the same coalition. Therefore, it is important to decide how to obtain an equilibrium strategy for all users within the coalition. We consider the cooperation between any players in the games is allowed. In this study, cooperative game theory is introduced for the following analysis [3].

Analysis in cooperative game theory is centered on two major issues: coalition formation and distribution of wealth gained through cooperation. If the participants can obtain more profits through collaborating together than before, they will try their best to form a coalition rather than to participate the game individually. Every participant wants to obtain its maximum profit in the coalition, therefore, the satisfactory and reasonable scheme of allocation of profits in the coalition for each one becomes very important.

In this study, among the various solution forms of a cooperative game problem, Shapely value is employed for the following analysis.

When the players try to participate in the game, they will forecast that how much gain they can obtain in advance. Evaluating beforehand is important for all players on deciding whether joining the game or not. The Shapley value is the expected marginal amount which player contributes to coalition [4]. It is based on a particular concept of 'fairness' in distributing the total gain the grand coalition is capable of achieving which is expressed by the following equation

$$\phi_i = \frac{(m-1)!(n-m)!}{n!}\{v(S) - v(S - \{i\})\} \ . \tag{1}$$

where, m is the number of coalition S, n is the number of all the members in grand coalition N, $N - \{i\}$ is the coalition of not including member i.

Suppose the players (the element of n) agree to cooperate each other, because of random combination orders of cooperation, if a player i cooperates with the coalition which consists of members $S - \{i\}$, it receives the amount $v(S) - v(S - \{i\})$, the marginal amount which it contributes to the coalition, as payoff. Then the Shapley value ϕ_i is the expected payoff to player i. Under this randomization scheme, $(m-1)!(n-m)!/n!$ is the probability of that player i joins the coalition $S - \{i\}$. It can be seen that the sum of the coefficients $(m-1)!(n-m)!/n!$ is equal to 1. For the numerator is equal to the number of permutations of S in which i is preceded exactly by the elements of S, while the numerator is the total number of permutations.

4   Numerical Study

4.1   Research Object

In this study, in order to validate the method proposed above, a regional energy interchanging system composed of three energy users have assumed for analysis. For simplicity, only thermal energy demands (heating and cooling) are considered. Figs. 3-4 show the cooling and heating loads for the assumed three users, respectively.

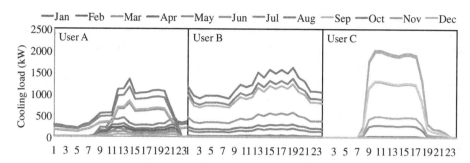

Fig. 3. Hourly cooling loads of the assumed three users

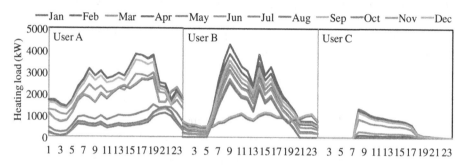

Fig. 4. Hourly heating loads of the assumed three users

In addition, we assume three users have introduced the thermal equipments with different efficiencies as shown in Table 1. The electricity tariff is 0.617 Yuan/kWh, the price of natural gas is 2.5 Yuan/m³.

Table 1. Efficiencies of thermal equipments for three users

Item	User A	User B	User C
COP of Cooling equipment	5.0	4.5	4.0
Efficiency of heating equipment	90%	80%	75%

4.2   Results and Discussions

In order to execute the analysis based on the cooperative game model, the energy cost (heating and cooling) for the situations with and without cooperation should be calculated. Table 2 shows the energy costs for various scenarios. If the three users supply their energy demands all by themselves, the total energy cost will be 9.17 million Yuan.

Table 2. Energy costs for various scenarios

Cooperative scenario	A	B	C	A+B	A+C	B+C	A+B+C
Cost (Yuan)	4.18	4.24	8.04	8.04	4.83	4.96	8.74

According the Shapley value method illustrated in section 3, the profit allocation methods for three users are detailed in Tables 3-5.

Table 3. Profit allocation results for user A

S	A	A+B	A+C	A+B+C	
v(S)	0	368053	107062	433181	
v(S-{i})	0	0	0	41159	
v(S)-v(S-{i})	0	368053	107062	392022	
|S|	1	2	2	3	
ϕ_i	0	61342	17843	130673	$\sum$ =209860

Table 4. Profit allocation results for user B

S	A	A+B	B+C	A+B+C	
v(S)	0	368053	41159.54	433181	
v(S-{i})	0	0	0	107062	
v(S)-v(S-{i})	0	368053	41159.54	326119	
|S|	1	2	2	3	
ϕ_i	0	61342	6860	108706	$\sum$ =176908

Table 5. Profit allocation results for user C.

S	A	A+C	B+C	A+B+C	
v(S)	0	107062	41160	433181	
v(S-{i})	0	0	0	368053	
v(S)-v(S-{i})	0	107062	41160	65128	
|S|	1	2	2	3	
ϕ_i	0	17844	6860	21709	$\sum$ =46413

Therefore, Table 6 summarized the profits for three users and the resulted energy costs for them.

Table 6. Profits and final costs of three users

User	Original cost (Yuan)	Profit allocation (Yuan)	Final cost (Yuan)
A	4175757	209859.8	3965898
B	4237114	176908.4	4060205
C	759415.5	46413.11	713002.4
Total	9172287	433181.3	8739105

In order to understand the benefits of develop a regional energy interchanging system, Table 7 shows the costs for the systems with and without cooperation. Generally, the scenario without cooperation encounters the largest annual total energy cost. The scenario with all three users under a cooperative framework results in the least annual total cost.

Table 7. Total costs for various scenarios.

Scenario	A+B+C	A+(B+C)	B+(A+C)	C+(A+B)	(A+B+C)
Cost (Yuan)	9172287	9131127	9065224	8804234	8739105

5   Conclusions

In this study, a new type of district energy system has been proposed, aiming at develop a cooperative framework among the users with their own equipments. In order to reach a rational and fair solution for the profit allocation among the users within the proposed regional energy interchanging system, the cooperative game theory has been employed for analysis. Especially, Shapley value method has been employed for the detailed analysis of the profit allocation. According to the calculation results of the case study, the regional energy interchanging system has results in reduced total energy cost. In addition, the Shapley value method proves to be a effective method for the profit allocation among the users within the energy system.

Acknowledgments. This work was supported in part by The Key Fund of Shanghai Science Technology Committee (No. 13160501000), as well as The Program for Professor of Special Appointment (Eastern Scholar) at Shanghai Institutions of Higher Learning.

References

1. Miao, L., Hailin, M., Nan, L., Huanan, L., Shusen, G., Xin, C.: Optimal option of natural-gas district distributed energy systems for various buildings. Energy and Buildings 75, 70–83 (2014)
2. Jean, D., Peter, W., Andrew, R.: The potential benefits of widespread combined heat and power based district energy networks in the province of Ontario. Energy 67, 41–51 (2014)
3. Forgo, F., Szep, J., Szidarovszky, F.: Introduction to the theory of games. Kluwer Academic Publishers, Dordecht (1999)
4. Owen, G.: Game theory. Academic Press, New York (1995)

Multi-agents Based Modelling for Distribution Network Operation with Electric Vehicle Integration

Junjie Hu[1], Hugo Morais[1, 2], Yi Zong[1], Shi You[1],
Henrik W. Bindner[1], Lei Wang[3], and Qidi Wu[3]

[1] Department of Electrical Engineering, Technical University of Denmark (DTU)
Elektrovej, Bld 325, 2800 Lyngby, Denmark
{junhu,morais,yizo,sy,hwbi}@elektro.dtu.dk
[2] GECAD - Knowledge Engineering and Decision Support Research Center
IPP - Polytechnic Institute of Porto, Porto, Portugal
[3] School of Electronics and Information, TongJi University, Shanghai, China
{wanglei,qidi}@tongji.edu.cn

Abstract. Electric vehicles (EV) can become integral part of a smart grid because instead of just consuming power they are capable of providing valuable services to power systems. To integrate EVs smoothly into the power systems, a multi-agents system (MAS) with hierarchical organization structure is proposed in this paper. The proposed MAS system consists of three types of agents: distribution system operator agent (DSO agent), electric vehicle fleet operator agent (EV FO agent or alternatively called virtual power plant agent) and EV agent. A DSO agent belongs to the top level of the hierarchy and its role is to manage the distribution network safely by avoiding grid congestions and using congestion prices to coordinate the energy schedule of VPPs. VPP agents belong to the middle level and their roles are to manage the charge periods of the EVs. EV agents sit in the bottom level and they represent EV owners and operate the charging behaviour of EVs. To simulate this collaborative (all agents contribute to achieving an optimized global performance) but also competitive environment (each agent will try to increase its utilities or reduce its costs), a multi-agent platform was developed to demonstrate the coordination between the interacting agents.

Keywords: Congestion Management, Electric Vehicles, Multi-agent Systems, Smart Grids, Virtual Power Players.

1 Introduction

To achieve the European energy roadmap 2050 [1] *"The EU is committed to reducing greenhouse gas (GHG) emissions to 80-95% below 1990 levels by 2050 in the context of necessary reductions by developed countries as a group"*, the decarbonisation of two main activities including power systems and transportations is necessary. Electric vehicles are important means to assure the GHG emission reduction goals in the transport sector. Furthermore, electric vehicles can be used to balance the intermittent

K. Li et al. (Eds.): LSMS/ICSEE 2014, Part III, CCIS 463, pp. 349–358, 2014.

renewable energy resources [2] [3]. The double benefits make the growing interests and the wide advocate on electric vehicles.

However, the integration of a high number of EVs into the electric system will require a significant grid capacity increase [1]. Uncontrolled charging of EV can create new load peaks during the day, increasing power losses, voltage deviations and network congestion [4]. Typically, the challenges in the distribution grid caused by the increasing electricity consumption from EVs are solved by expanding the grid to match the size and the pattern of demand. Alternatively, in a smart grid context, the grid capacity problem can also be solved smartly using advanced control strategies supported by an increased use of information and communication technology.

As discussed in [5], [6], an agent-based control system is very efficient to manage the complexities of a large systems like as the case in a future power distribution system. In [7], [8], the multi-agent concept is proposed for distribution system operation and control, especially considering the capacity management with a large penetration of electric vehicles [9]–[11] and more general loads [12].

In this paper, a multi-agent platform is implemented to simulate the interactions between the EV VPPs, the distribution system operator and the EVs owners. The main goal is to explicitly show the negotiation between the agents in order to avoid the congestion of the distribution network lines as well as the high voltage/medium voltage (HV/MV) power transformer. Besides, agents' operations are also described. In the developed agent-based control system, each EV VPP has the capability to manage the EVs' charge and discharge considering the energy prices and the EVs requirements (schedule trips, batteries technical limits, etc.). Before biding the energy schedule into the electricity spot market, a prior interaction between VPPs and DSO is required. The energy schedule of VPPs is sent to DSO and DSO evaluates the overall network performance considering the network technical constraints, namely the bus voltage magnitude and the lines thermal limits. If congestion exists, DSO uses market based control method to coordinate the energy schedule of VPPs.

This paper is organized as follows: after this introductory section, section 2 presents a description about the proposed multi-agent system, section 3 presents the operation of the agents in the multi-agent systems, section 4 describes the implemented multi-agents platform. In section 5, simulations are shown to illustrate the understanding. The main conclusions of the paper are drawn in section 6.

2   Multi-agents System Description

In Fig.1, a hierarchical agent-based control system is presented. Three types of agents are included in the system which enables the intelligent operation of a future power distribution system, especially considering a large scale integration of distributed renewable energy resources. The three types of agent consist of DSO agent, VPP agent and DER agent (Mainly EV agent is considered in this study). Each agent's role defined in this study is discussed in the following.

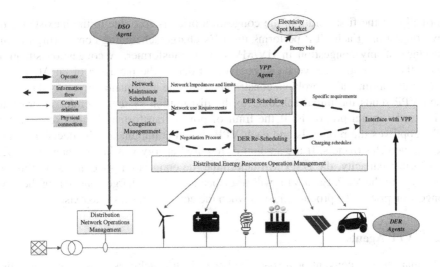

Fig. 1. An illustration of an agent-based control system with hierarchical architecture for future power distribution network management

DSO Agent: The DSO agent is responsible for technical operation of the distribution network such as preventing thermal overloading of power transformers and lines as well as voltage drops. Normal operation is mainly considered for this study, i.e., DSO can coordinate with VPPs to schedule the network utilization ahead.

VPP Agents: The VPP agents are responsible for managing the EV charging process by providing electricity to EVs. In addition, they aggregate a large scale of EVs and provide ancillary services to system operators such as distribution system operator by complying with the distribution network capacity limits.

EV Agents: The EV agent represents the EV owner or EV charging controller who can choose their interacting method with the VPP agent. For example, the EV agent can specify the desired state of charge of the battery at the time of departure, the desired departing time, etc. In a more advanced control environment, such as if the vehicle to grid (V2G) technology is feasible, the EV agent can define whether V2G is allowed in a planning period.

3   Operation of Agents in the MAS

3.1   DSO Agents

A much advanced control strategy to prevent the grid congestions is introduced for DSO agent. In the first step, DSO provides the VPPs information regarding the network characteristics, namely the impedances matrix, the technical limits for each line and for each bus and the congestion information in the power transformers.

Normally, in the first iteration, the congestion price is zero due to the inexistence of any congestion. Each VPP performs the EVs charge scheduling considering the inexistence of any congestion in HV/MV power transformer. A congestion situation inside the distribution network can be avoided due to the inclusion of an AC power flow in the optimization problem constraints using the information given by the DSO. Each VPP sends the initial schedule to the distribution system operator in order to validate the initial proposals. If the limit of power transformer is not excessed, the DSO approves the proposals and each VPP should communicate the decision to the electric vehicles owners. If the amount of required energy was higher than the power transformer capacity, the DSO will determine the congestion price and send this information to the VPP. Each VPP will re-dispatch the EVs charge considering the new congestion price. The process finishes when the congestion ceases to exist.

3.2   VPP Agents

VPP manage the conventional loads and the electric vehicles charge process. Discharge capability is not considered. Each VPP will try to schedule the EVs charge according to the spot market price and taking into account the network constraints defined by the DSO. The scheduling is formulated as a mixed-integer non-linear problem (MINLP) aiming to minimize the costs (1). The energy cost depends on the market price (external suppliers) as well as the HV/MV power transformer capacity use. The power losses in the distribution network are also considered in the problem. Expression (1) represents the implemented objective function.

$$
\min Z = \\
\sum_{t=1}^{T} \left(\left(\sum_{EV=1}^{N_{EV}} \left(P_{Ch(EV,t)} \right) + \sum_{l=1}^{N_l} \left(P_{Load(l,t)} \right) \right) \times \left(c_{Market(t)} - c_{Cong(t)} \right) + \\
\sum_{l=1}^{N_l} \left(P_{NSD(l,t)} \right) \times c_{NSD(l,t)} \right) \tag{1}
$$

In expression (1), the variable $P_{Ch(EV,t)}$ represents the charge power of electric vehicles. The loads are characterized by the consumption forecast $P_{Load(l,t)}$ and by the non-supplied demand $P_{NSD(l,t)}$. This parameter will be important to avoid the congestion situations in extreme situations, considering the cost $c_{NSD(l,t)}$. The $c_{NSD(l,t)}$ represents the cost with a penalization to load curtailment without any coordination or contract. The parameter $c_{Cong(t)}$ represents the HV/MV power transformer congestion cost. This parameter is different in each negotiation. In the first iteration $c_{Cong(t)}$ is zero, representing the situation without congestion. The variable $c_{Market(t)}$ represents the cost of energy supplied by the external suppliers. This value is based on the market prices during a day.

The problem constraints consider the first Kirchhoff law relating to the active and reactive power balance considering that the energy is supplied by external suppliers. The implemented AC power flow model considers the distribution network technical constraints regarding to the lines thermal limits and the bus voltage magnitude limits. The EVs constraints relating to the batteries energy capacity limits, the batteries internal energy balance and the charge rates are also considered.

3.3   EV Agents

In this case, the EVs agents only need to provide information regarding the trips requirements (distance and departure time) and the batteries status to their subscribed VPP agents. The VPP will use this information in the scheduling process, trying to estimate the required energy and the periods that each EV will be connected to the network. The EV's owners can also define some minimum limits regarding the energy stored in the EVs batteries. The minimum energy in the batteries avoids their fast degradation. However, the user can define higher values as minimum to prevent some unexpected situation.

4   Multi-agent Simulation Platform

A multi-agent system is implemented using the integrations of JACK software environment[1], MATLAB and GAMS[2]. JACK is an agent-oriented development environment built on top of Java programming language. GAMS is a high-level modeling system for mathematical programming and optimization. JACK is used to demonstrate the negotiation process between the agents. MATLAB and GAMS are used to calculate the optimization problem and support the decision making/operation of the agents. To simulate the derived problem, we further divide DSO agent into two agents who have distinctive roles and functions:

- **DSO Tech Agent**: A DSO Tech agent is responsible for congestion verifying after obtaining the power schedules of VPPs. The DSO agent communicates with the VPPs agent and the DSO market agent.
- **DSO Market Agent**: A DSO market agent is responsible for making of the shadow price. The market agent communicates with the DSO Tech agent and the VPPs agent.

Fig. 2 shows the whole design diagram for the desired multi-agent systems at the JACK platform, which is built according to the proposed solutions in this study. The diagram is explained according to the sequences which are essential to fulfil the distribution grid congestion management with electric vehicle integration and is divided into three parts: 1) the interaction between the EV agent and the VPP agent; 2) the interaction between the VPP agent and the DSO Tech agent; 3) the interaction between the VPP agent and DSO market agent.

---

[1] http://aosgrp.com/
[2] http://www.gams.com/

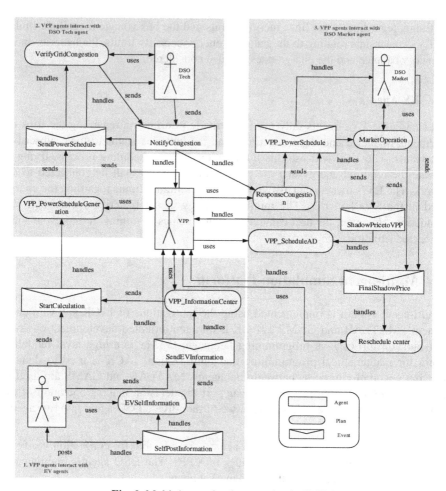

Fig. 2. Multi-Agents implementation in JACK

Due to the similarities, the content inside the boxes of the first interaction diagram, i.e., the interaction between the EV agent and the VPP agent is explained.

- Event **SelfPostInformation**: Posted by the EV agent, the purpose is to trigger the plan *EVSelfInformation*.
- Plan **EVSelfInformation**: The EV agent reads the information and sends an event named *SendEVInformation* to the VPP agent, and the event will be handled by the plan *VPP_InformationCenter*.
- Plan **VPP_InformationCenter**: The EV information will be collected here and prepared for schedule generation.
- Plan **VPP_PowerScheduleGeneration**: A MATLAB based program will be used to generate the charging power schedule. Inside the MATLAB program, GAMS is called. The power schedule will be sent to the DSO agent by the event *ChargingSchedule*.

5   Simulation Results

5.1   Simulation Results in MATLAB

This section presents the use of the developed platform considering four Virtual Power Players (VPPs), each one operating the resources in one single feeder of one substation. The considered network was proposed in [13] consisting in a 11 kV distribution network with 37 buses and 1908 consumers. The connection of this distribution network with transmission system is made through two HV/MV power transformers with a total capacity of 20 MVA. Each VPP manages a set of consumers, with different characteristics (domestic consumers (DM); industrial; government and services buildings; and commerce), and electric vehicles.

The initial scheduling is performed considering the loads demand forecast in each branch of substation and also the network constraints indicated by the DSO. The movements of electric vehicles were determined based on the UK transport report considering an average daily travel distance of 38 km and the trip energy requirements were calculated based on the normal consumption of the electric vehicles and also in the trips duration considering the range of average speeds [14].

The forecast load is near by the power transformers capacity in periods 75 to 82 (periods of 15 minutes). In these periods, the load consumption is higher than 18 MW. The considered market price is lower in the periods of high consumption. This type of prices can occur in situation with high renewable generation base. In the present case study, these prices were used to originate EVs charge in these periods allowing the use the negotiation mechanism. In Fig. 2 a) the initial scheduling for each VPP is presented. The scheduling for each VPP included the power to supply the load demand, the electric vehicles charge and the losses in the network. Fig. 2 b) presents the information regarding the EVs batteries states and the charge process for the VPP1. The other VPPs have a similar behavior.

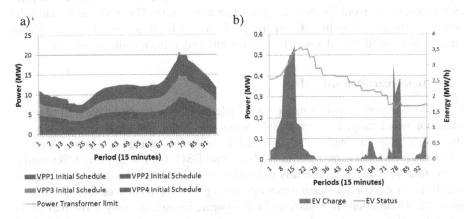

Fig. 3. a) Initial scheduled power; b) Initial EV state and charge schedule by VPP1

Analyzing Fig. 3 a) and b) it is possible to see the power transformers congestion in periods 76 to 80 but mainly in period 76 due to the lowest energy price in this period. VPPs try to schedule the EVs charge during the night (off-peak hours.). However, it is also necessary charge some EVs during the day to guarantee the enough energy to the required trips.

To avoid the power transformer congestion, the DSO should determine a new congestion price based on the congestion severity. The congestion cost increases with the congestion severity. In Fig. 4 a) it is possible to see the congestion price, determined by the DSO and the new scheduling by VPP resulting from the application of these new prices. Fig. 4 b) shows the final EVs scheduling to VPP1.

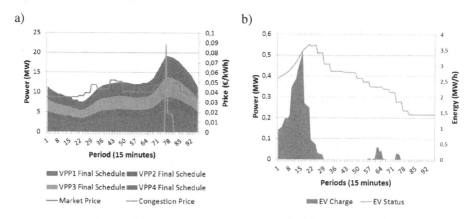

Fig. 4. a) Final VPPs scheduling and Market/Congestion Prices; b) Final Electric vehicles state and charge schedule by VPP1

As seen in Fig. 4 a) and b), the congestion price is very high in period 76 due to the high congestion level in this period in the initial scheduling. The new congestion prices are enough to avoid the power transformer congestion. This happens because the new EVs charge scheduling. Comparing 4 b) and 3 b), it is possible to see higher charge until period 29 and a very few energy charged in the remaining periods.

5.2   Demonstration with MAS

In JACK, there are a number of tools available to assist a detailed trace of the system execution which range from graphical tracing tools to logging tools. In this study, we run the program with the interaction diagram using a Java Args: - *Djack.tracing.idisplay.type=id*. As we have one DSO Tech agent, one DSO market agent, four VPP agents, and nearly one thousand EVs agents, the interaction diagram which shows the communication message among them is quite large. It is not wise to show the whole interaction diagram in this paper, instead, we only show part of the interaction diagram where the message sequence happens between the DSO Tech agent, the DSO market agent, the VPP agent VPP1 and VPP2, and two EV agents EV1 (subscribed to VPP1), EV2 (subscribed to VPP2), this is shown in Fig. 4. The sequence diagram starts from agent EV1 (holds for other EV agents) with an information sending and requirements specification. Then the VPP agent calculates the optimal charging schedule and sends the aggregated charging schedule to the DSO Tech agent. DSO Tech agent verifies whether congestions exist and sends the result to the VPP agents. If the congestion exists, the VPP agents coordinate with the DSO market agents until the congestion is solved. The rectangular box marked in the line of each agent represents their internal operations such as information preparation and calculations.

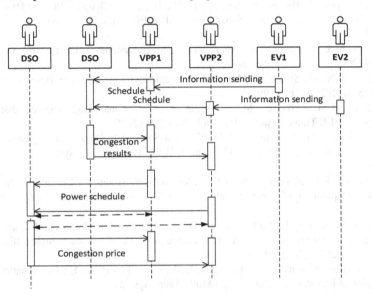

Fig. 5. Sequence Diagram between the chosen agents

With the interaction diagram facilitated by the external argument of displaying the 'ID' of the agents, it well emulates the information flows as well as the internal operations of the agents in the proposed control system.

6   Conclusions and Discussions

The growing integration of electric vehicles in power systems introduces new challenges in the distribution networks. In many situations some congestion problems can occur in different points of distribution networks. The most critical ones will be the lines thermal limits and also in the HV/MV power transformers. In this paper a hierarchical management structure is presented to integrate electric vehicles into the power distribution systems. Three types of agents are included in the system, i.e., DSO, VPP and EV agent. Internally, VPP agent centrally controls the charge period of all the EV agents and comply with the grid constraints imposed by DSO agent. The negotiation between VPP and DSO is discussed considering a market-based negotiation in congestion situations. To demonstrate the negotiations explicitly, a multi-agents system is built and explained in detail in the paper.

References

1. European Commission: Energy Roadmap 2050 (2011)
2. Sousa, T., Soares, J., Vale, Z., Morais, H.: Day-ahead resource scheduling in smart grids considering Vehicle-to-Grid and network constraints. Applied Energy 96, 183–193 (2012)
3. Tomić, J., Kempton, W.: Using fleets of electric-drive vehicles for grid support. Power Sources 168(2), 459–468 (2007)
4. Clement-Nyns, K., Haesen, E., Driesen, J.: The Impact of Charging Plug-In Hybrid Electric Vehicles on a Residential Distribution Grid. IEEE Trans. Power Syst. 25(1), 371–380 (2010)
5. Jennings, N.R., Bussmann, S.: Agent-based control systems. IEEE Control Syst. 23(3), 61–74 (2003)
6. Talukdar, S.N., De Souza, P., Murthy, S.: Organizations for computer-based agents. Eng. Intell. Syst. 1(2), 56–69 (1993)
7. Nordman, M.M., Lehtonen, M.: An agent concept for managing electrical distribution networks. IEEE Trans. Power Deliv. 20(2), 696–703 (2005)
8. Ren, F., Zhang, M., Sutanto, D.: A Multi-Agent Solution to Distribution System Management by Considering Distributed Generators. IEEE Trans. Power Syst. 28(2), 1442–1451 (2013)
9. Karfopoulos, E.L., Hatziargyriou, N.D.: A Multi-Agent System for Controlled Charging of a Large Population of Electric Vehicles. IEEE Trans. Power Syst. 28(2), 1196–1204 (2013)
10. Miranda, J., Borges, J., Valério, D., Mendes, M.J.G.C.: Development of a multi-agent management system for an intelligent charging network of electric vehicles. IFAC Proc. 18(pt. 1), 12267–12272 (2011)
11. Papadopoulos, P., Jenkins, N., Cipcigan, L.M., Grau, I., Zabala, E.: Coordination of the Charging of Electric Vehicles Using a Multi-Agent System
12. Greunsven, J.A.W., Veldman, E., Nguyen, P.H., Slootweg, J.G., Kamphuis, I.G.: Capacity management within a multi-agent market-based active distribution network. In: 2012 3rd IEEE PES Innovative Smart Grid Technologies, pp. 1–8 (2012)
13. Allan, R.N., Billinton, R., Sjarief, I., Goel, L., So, K.S.: A reliability test system for educational purposes-basic distribution system data and results. IEEE Trans. Power Syst. 6(2), 813–820 (1991)
14. Druitt, J., Früh, W.-G.: Simulation of demand management and grid balancing with electric vehicles. Power Sources 216, 104–116 (2012)

A Novel Method of Fault Section Locating Based on Prony Relative Entropy Theory

Ranyue Li [1], Chaoli Wang [1], and Xiaowei Wang [2]

[1] Department of Control Science and Engineering, University of Shanghai for Science and Technology, Shanghai 200093, P.R. China
[2] School of Electrical Engineering and Automation, Henan Polytechnic University, Jiaozuo, 454000, P.R. China
liranyue@163.com, clclwang@126.com

Abstract. A novel fault line section locating method of non-solidly grounded system based on Prony relative entropy theory was proposed in this paper. Firstly, piecewise Prony algorithm was used to fit the transient zero-sequence current signal of each detection point in the first T/20 cycle after fault occurs; secondly, transient zero-sequence dominant components were extracted and the relative entropy values of adjacent detection points were computed; lastly, the fault section was located by use of the feature that the transient zero-sequence currents from the same side of fault point possess high similarity while the opposite side of fault point possess low similarity. Simulation results verify the validity and accuracy of the method in this paper.

Keywords: Fault section locating, piecewise Prony, dominant component, relative entropy.

1   Introduction

Non-solidly grounded system is widely used in power distribution network with 3-35k voltage in china, but the transient characteristics of single phase to ground fault occurred is not clear, resulting in fault locating problem has not been well solved yet. Existing methods for fault locating can be divided into fault distance locating [1-3] and the fault section locating [4-7].

Reference [1-2] proposed a method of fault locating which based on 'S' injecting. This method injects AC measuring signal into the system by using voltage transformer and the fault point can be found according to the vanish point of measuring signal. But this method is not durable because it is restricted by the capacity of voltage transformer. Reference [3] locates the fault section based on the maximum related time of two transmitted waves generated by fault point. But this traveling wave method needs to accurately identify the traveling wave of the fault point and it is greatly influenced by the wave resistance of the line. Although fault distance locating law can calculate the distance between fault point and detection point, the investment are large for they are all required to add additional equipment. Reference [4-7] are belonging to fault section locating law, the transient

K. Li et al. (Eds.): LSMS/ICSEE 2014, Part III, CCIS 463, pp. 359–367, 2014.
© Springer-Verlag Berlin Heidelberg 2014

zero-sequence currents are extracted by using the feeder terminal units (FTU) which are installed on the lines. The master station will locate the fault section based on the sampling data upload from these FTU. Compare to fault distance locating law, fault section locating can lower the investment and the engineering applicability is stronger. Reference [4] locates the fault section according to the feature that the power direction polar from the opposite sides of fault point is opposite, which needs to install PT to measure the voltage. Reference [5-7] take transient zero-sequence current directly as the analytical data, then locate the fault section by using correlation [5-6] or approximate entropy [7], these methods are called zero sequence current law or related law.

A method of fault selection is proposed in reference [9] based on Prony relative entropy theory. Following this approach, a novel method of fault section locating based on Prony relative entropy theory was proposed in this paper [9].

2   Distribution Characteristics of Fault Transient Zero-Sequence Current

The zero module equivalent network diagram of a small current to ground system is showed in Fig.1. L is arc-suppression coil, S is generator winding. A, B, C and D are four current detection points on the line S_1. i_{0S}, i_{02}, i_{03}, i_{0A}, i_{0B}, i_{0C} and i_{0D} are respectively the zero-sequence current flowing through the generator, the line S_2, S_3, the detection points A, B, C and D. i_{C1}、i_{C2} are respectively the zero-sequence capacitance current to ground. Then:

$$i_{0A} = i_{0S} + i_{02} + i_{03} \tag{1}$$

$$i_{0A} = i_{C1} + i_{0B} \tag{2}$$

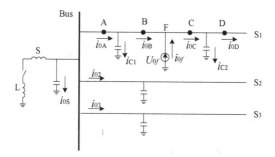

Fig. 1. Equivalent circuit of zero module network

The distribution characteristics of fault transient zero-sequence current in small current to ground system will be analyzed for Fig.1. The fault occurs at the point F in line S_1. At the moment of fault occurring, it is equivalent to add a virtual zero-sequence voltage U_{0f} at the fault point. The actual current flows between bus side and fault point is F→B→A, while the actual current flows between fault point and load side is F→C→D, so the differentiation between the waveform of i_{0B} and i_{0C} is large because they are at the opposite polarity. In formula (1), i_{0A} is the capacitive current to ground sum for non-fault lines. The distance between two detection points is normally short, so i_{C1} is very small even to neglect relative to i_{0A}. Then formula (2) can be simplified as $i_{0A} \approx i_{0B}$, that is to say, their waveforms are essentially the same. Similarly, i_{0C} and i_{0D} are basically the same. The comparison of transient zero sequence current at detection points AB, BC and CD are showed in Fig.2 when the initial phase angle is 90° and the earth resistance is 5 ohms. From Fig.2 we can see that the transient zero-sequence current waveforms at the same side of fault point (A, B, or C, D) possess high similarity, but the current waveforms at opposite side of fault point (B, C) possess low similarity. This characteristic can be used on fault section locating.

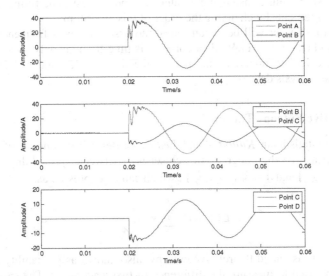

Fig. 2. Comparison of transient zero-sequence current

3   Prony Relative Entropy Theory

3.1   Prony Algorithm

Prony algorithm is used to fit a time signal $y(t)$ by the linear superposition of a series of (p) damped sinusoidal signals. Each component $q(t)$ has its own amplitude, phase, frequency and damping coefficient, and this paper call it the Prony basis function. Represented by the formula:

$$y(t) = \sum_{i=1}^{p} q(t) = \sum_{i=1}^{p} A_i e^{\alpha_i t} \cos\left(2\pi f_i t + \theta_i\right) \qquad (3)$$

where,

A_i : Amplitude of i-th frequency component

θ_i : Phase of i-th frequency component

α_i : Damping coefficient of i-th frequency component

f_i : Frequency of i-th frequency component

Prony algorithm can process real-time signal speedy, which is suitable for analysis of an exponential decay signal. It can also accurately reveals the main characteristics of the signal. But it's sensitive to noise. It works not very well when fiting a non-persistent or variable burst signal. In this regard, reference [10] introduced a modified Prony algorithm with adaptive, transient signal was adaptively divided into more than one field and separately fitted for each field. This paper calls the method as piecewise Prony algorithm. The simulation results in [9] verify the feasibility of Prony algorithm for power system fault transient signal analysis and process. The fitting accuracy can meet the requirements by changing the order of this algorithm.

Current transformer will be exhausted in the first 1/4 Power frequency cycle after fault occurs [11], so the sample data during this time frame is only valid. Taking into account the calculate quantity, this paper take the sample data in T/20 after fault occurs as the analysis data.

3.2   Relative Entropy Theory

In 1951, the statistician *Kullback* and *Leibler* presented an amount of information used to measure the degree of closeness between probability distribution $\chi = \{\chi_1, \chi_2 \cdots \chi_m\}$ and $\lambda = \{\lambda_1, \lambda_2 \cdots \lambda_m\}$, which can be expressed as:

$$L(\chi, \lambda) = \sum_{\eta=1}^{m} \chi_\eta \ln \frac{\chi_\eta}{\lambda_\eta} \tag{4}$$

This quantity is called the relative entropy, also known as probability distribution, which can be used to measure the difference of two waveforms. The smaller relative entropy is, the higher the similarity of the two signals is. Conversely, the similarity of the two signals is lower.

Prony relative entropy theory combines piecewise Prony algorithm with relative entropy theory. A series of Prony base function will be acquired by using piecewise Prony algorithm when fitting a transient zero-sequence current. The Prony base function which has the maximum energy instead of transient zero-sequence current signal is studied to eliminate the effect of random interference signal or high frequency signal [9]. Then the relative entropy values of adjacent detection points will be computed. The transient zero-sequence current waveforms at the same side of fault point possess high similarity, so the relative entropy between them is small, conversely, the relative entropy between the opposite side of fault point is high for they possess low similarity.

4   Fault Locating Mechanism

Assumed there are m current detection points in a line totally, FTU will upload the transient zero-sequence current of these current detection points to the master station in real time. Then the master station will locate the fault section based on these sample data. Concrete steps are as follow. The flow chart of fault section locating is shown in Fig.3.

Step 1: Sampling period is Δt, the number of sample data is N in T/20 after fault occurs, then: $N\Delta t = T/20$.

Step 2: Fit transient zero sequence currents in each detection points using piecewise Prony algorithm. Select the transient zero-sequence current dominant component I_k ($k=1,2,...m$) according to the maximum amplitude principle [9]. I_k instead of transient zero-sequence current is studied.

Step 3: Consider a total system $\sum I = \sum_{k=1}^{m} I_k$, where $k=1,2,...m$.

Step 4: At n-th sampling time point, i.e. $t = n\Delta t$ ($n=1,2,...N$), the proportion of I_k in the total system $\sum I$ at k-th detection point is $I_{k,t}$

$$I_{k,t} = \frac{I_k}{\sum I} \ (k=1,2,...m) \tag{5}$$

Step 5: Compute the Prony relative entropy value of adjacent detection points.

$$M_{k,k+1} = \sum_{n=1}^{N} \left| I_{k,t} \ln \frac{I_{k,t}}{I_{k+1,t}} \right| \ (k=1,2,...m-1) \tag{6}$$

where, $t = n\Delta t$ ($n=1,2,...N$).

Step 6: Find the maximum one among $M_{k,k+1}$ ($k=1,2,...m-1$), then the fault occur between k-th detection point and $k+1$.

5   MATLAB Simulation Example

Build a Simulation model of small current to ground system in MATLAB/Simulink, which is shown in Fig.4. The length of overhead lines $S_1=20km$, $S_2=15km$, $S_3=24km$, $S_4=30km$, $S_5=16km$, $S_6=30km$. Positive sequence parameters $R_1=0.17\Omega/km$, $L_1=1.2mH/km$, $C_1=9.697nF/km$. Zero sequence parameters $R_0=0.23\Omega/km$, $L_0=5.48mH/km$, $C_0=6nF/km$. Sampling frequency $fs=1000kHz$. The connection type of transformer is Y/Δ, transform ratio is 220kV/35kV. Single-phase fault to ground occur at 10km of line S_1, A, B, C and D are four current detection points on the line S_1, which are respectively installed at 8.5km, 9.5km, 10.5km and 11.5km of line S_1.

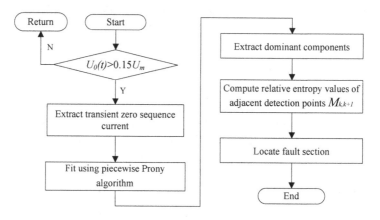

Fig. 3. Flow chart of fault area locating

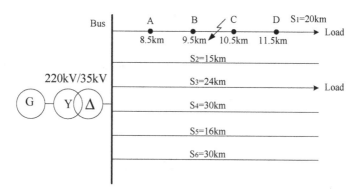

Fig. 4. Simulation model of small current to ground system

Set the initial phase angle is 90°, the earth resistance of fault is 5 ohms. Fitting of transient zero-sequence current of detection point A in T/20 after fault occurs is shown in Fig.5. It indicates that the fitting works very well.

There are many Prony basis functions to fit an actual current signal, but only the dominant component is selected. The dominant components at each detection point when the initial phase angle is 90° and the earth resistance is 5 ohms are showed in Tab. 1. Form Tab. 1, we can see that the four fitting parameters at the same sides of fault point (A, B, or C, D) are close similar, while the opposite sides of fault point (B, C) are vary widely.

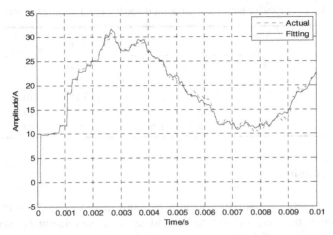

Fig. 5. Fitting of transient zero-sequence current of detection point A

Table 1. Fitting parameters of each line on dominant component

Parameter	A	B	C	D
A_i	15.8321	15.2082	0.8082	0.7517
α_i	0.5244	0.5358	0.2295	0.2156
f_i	52.6696	52.7701	190.7678	194.5218
θ_i	-1.7678	-1.3541	-1.1145	-1.1584

When the initial phase angle consists on 90° and the earth resistance changes, the Prony relative entropy values of the dominant component among adjacent detection points are shown in Tab. 2. While the earth resistance consists on 5Ω and the initial phase angle changes, the Prony relative entropy values are shown in Table 3. The simulation results in Tab. 2-3 indicate that the Prony relative entropy between B and C ($M_{B,C}$) is much larger than $M_{A,B}$ and $M_{C,D}$ no matter what fault condition is. Then the fault section locating result is obtained that the fault occurs at the section between detection point B and C, which is corresponds to the fact.

Table 2. Prony relative entropy of adjacent detection point in different earth resistance

Earth Resistance (Ω)	$M_{A,B}$	$M_{B,C}$	$M_{C,D}$	Fault Section
5	17.9314	839.4912	12.1106	BC
50	8.9205	780.3482	6.3924	BC
100	3.1223	875.8765	3.4113	BC
200	0.7732	980.4901	1.2248	BC
500	7.5965	736.7724	20.4869	BC

Table 3. Prony relative entropy of adjacent detection point in different initial angle

Initial Angle (degree)	$M_{A,B}$	$M_{B,C}$	$M_{C,D}$	Fault Section
0	3.1948	1134.1306	5.0275	BC
5	4.0777	1057.4364	3.8918	BC
30	29.4907	2649.3304	115.1581	BC
60	4.3673	3382.1199	3.6532	BC
90	17.9314	839.4912	12.1106	BC

6   Conclusion

This paper presents a fault line section locating method based on Prony relative entropy theory in small current to ground system. Simulation results verify the validity and accuracy of the method in this article.

This method uses FTU to extract transient zero-sequence current in each detection points. The dominant components instead of transient zero-sequence currents are used to eliminate the effect of random interference signal or high frequency signal. Compare to the method of power, this method needn't install PT to detect voltage, which can lower the investment. The shortcoming of this article is that the module of small current to ground system is a single hub network system, it requires further study for complex localization of topological grid system issues.

Acknowledgments. This paper was partially supported by The National Natural Science Foundation (61374040), Key Discipline of Shanghai (S30501), Scientific Innovation program (13ZZ115), Graduate Innovation program of Shanghai (54-13-302-102).

References

1. Zhang, H.F., Pan, Z.C., Sang, Z.Z.: A new fault locating method for power system with floating neutral based on signal injection. Automation of Electric Power Systems 28, 64–66 (2004)
2. Pan, Z.C., Zhang, H.F., Zhang, F., et al.: Analysis and modification of signal injection based fault line selection protection. Automation of Electric Power Systems 31, 71–75 (2007)
3. Yan, F., Yang, Q.X., Qi, Z., et al.: Study on Fault Locating Scheme for Distribution Network Besaed on Travelling Wave Theory. Proceedings of the CSEE 24, 37–43 (2004)
4. Hang, L.L., Xu, B.Y., Xue, Y.G., et al.: Transient Fault Locating Method Based on Line Voltage and Zero-mode Current in Non-solidly Earthed Network. Proceedings of the CSEE 32, 110–115 (2012)
5. Ma, S.C., Xu, B.G., Gao, H.L.: An Earth Fault Locating Method in Feeder Automation System by Examining Correlation of Transient Zero Mode Currents. Automation of Electric Power System 32, 48–51 (2008)

6. Tian, S., Wang, X.W., Wang, J.J.: Comparative Research on Fault Locating by Transient Zero-Module Current and Transient Zero-Module Power Based on Correlation Analysis. Power System Technology 35, 206–221 (2011)
7. Sun, B., Xu, B.Y., Sun, T.J.: New Fault Locating Method Based on Approximate Entropy of Transient Zero-module Current in Non-solidly Earthed Network. Automation of Electric Power System 33, 83–87 (2009)
8. Ma, S.C., Gao, H.L., Xu, B.G., et al.: A survey of fault locating methods in distribution network. Power System Protection and Control 37, 119–124 (2009)
9. Wang, X.W., Wu, J.W., Li, R.Y., et al.: A Novel Method of Fault Line Selection Based on Voting Mechanism of Prony Relative Entropy Theory. Electric Power 46, 59–64 (2013)
10. Sun, X.M., Gao, M.P., Liu, D.C., et al.: Analysis and processing of electrical fault signals with modified adaptive prony method. Proceedings of the CSEE 30, 80–87 (2010)
11. Shu, H.C., Peng, S.X.: Distribution network fault line detection using the full waveband complex relative entropy of wavelet energy. High Voltage Engineering 35, 1559–1564 (2009)

PV Fouling Detecting System Based on Neural Network and Fuzzy Logic

Xuejuan Chen, Chunhua Wu, Hongfa Li, Xiayun Feng, and Zhihua Li

Shanghai University, Shanghai Key Laboratory of Power Automation
Shanghai, Yanchang street 149, 200072 Shanghai, China
{Angelaventana,wuchunhua,lihongfa}@shu.edu.cn,
715643564@qq.com, lzh_sh@staff.shu.edu.cn

Abstract. PV fouling detecting system based on neural network and fuzzy logic is proposed. Comparing with traditional methods, the proposed method is rapid adaptive and universal to all PV power station. Neural network is used to predict the maximum power point (MPP) of a PV module under any lighting conditions. Then fuzzy logic rule is used to identify the fouling condition according to the result from neural network prediction. The experiment shows that the neural network can precisely predict the MPP under any lighting environment and the fuzzy logic rules can precisely identify the fouling condition of PV modules.

Keywords: PV, fouling detecting, neural network, fuzzy logic.

1   Introduction

Energy plays an important part in our daily life. Solar energy is clean and environmental-friendly. Besides, it is inexhaustible. In recent years, with the prosperous of PV industry, several model PV power stations have been built domestically, thus promoting the efficient use of new energy [1]. However, more and more people focus on solar energy. Many domestic scholars mainly focus on the maximum power point tracking (MPPT) control methods, so few research works are related to PV module fouling detecting.

However, in the operation of PV power stations, the PV modules are widely affected by environmental factors, especially fouling. Fouling blocks light transmittance thought the PV module surface, then reducing the solar irradiance received by the cell, thus reducing the output power of PV modules [2,3]. If the fouling condition is serious, partial shade may form, thus bringing many other problems. In severe situations, hot spot may occur and irreversible damage may occur to PV modules. So fouling effect of PV modules cannot be ignored.

2   Dust

2.1   Basic Feature of Dust

Dust is the solid particles with a diameter less than 0.92mm [4]. Its concentration in different urban areas is different. Due to small particles and light weight, dust is easily

K. Li et al. (Eds.): LSMS/ICSEE 2014, Part III, CCIS 463, pp. 368–377, 2014.

carried by wind. A large quantity of dust goes up and down with the help of wind and atmosphere. Gravity, Van der Waals forces, electrostatic force also contribute to the formation of dust [5]. Dust lies on the PV module and fouling is then formed.

The sources of dust are from nature and human works. When rock weathered, small dust particles are formed. When dust dries, it is carried by wind and flying in the atmosphere. Anthropogenic source is much more complex. There are 3 main sources: particles from industrial fuel after burning, dust from construction and ground dust caused by traffic [6].

According to the chemical nature, dust can be divided into 3 categories: acidic dust, alkaline dust and neutral dust; according to attached patterns, dust can be divided into 2 categories: dry pine fouling and adhesive fouling. The common pattern of dust is dry pine fouling. Fluffy dust attaches to the PV module surface evenly. Strong wind or simply a wipe can clean up. Adhesive fouling is formed by chemically active dust. When in wet situation, hydrolysis happens, then gelatinous substance forms and attaches to PV modules. When it dries, hard shell crystals form [7]. Its shape in the PV modules' surface is uneven state.

2.2   Effect of Fouling

When dust covers evenly on PV modules, there are mainly 2 kinds of effects: blocking the incident light, thus reducing the solar radiation intensity the PV module actually receives; block the radiating of PV module and break the thermal equilibrium, and then the temperature will rise as the PV module can't radiate timely and the heat accumulates. The two effects ultimately result in decreasing the output power of PV module, affecting the power efficiency.

When active chemically dust accumulates on PV modules, affected by rain, it may form binding substance. Some are washed away by the rain; the others form small plaques when dried, resulting in partial shade. In more severe case, the small plaque can lead to hot spot, causing irreversible damage.

3   MPP Prediction Based on Neural Network

3.1   Calculation of MPP

Solar cell takes the advantage of the photovoltaic PN junction effect to convert light energy into electrical energy. It consists of a current source, a diode and resistors. Its common DC equivalent model is shown below [8].

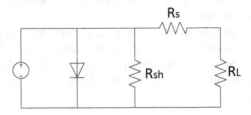

Fig. 1. Actual solar cell equivalent circuit diagram

MPP can be calculated by (1) according to the mathematic model of PV module [9].

$$P = [I_{sc}(R_s + R_p)U - U^2 - PR_s]/R_p + \frac{U_{oc}U - I_{sc}(R_s + R_p)U}{R_p} e^{\frac{q(U^2 + PR_s - U_{oc}U)}{AKTU}} \quad (1)$$

I_{sc} represent short circuit; U represent output voltage; U_{oc} represent open-circuit voltage; R_s represent series resistor; R_p represent shunt resistor; A is the ideal diode constant; q is the amount of charge; K is Boltzmann constant; T is the environment temperature.

By (1), MPP of PV module can be calculated precisely under any lighting conditions. However, the parameters given by manufactures exclude series resistor and shunt resistor. The ideal diode constant is also unknown. It is difficult to estimate the value of series resistor, shunt resistor and ideal diode constant with only an I-V curve under standard test condition (STC) given by the manufacturer. So this model is applied mostly in simulation but not in real situation. Many parameters can't be got in actual use. And the equation is an implicit equation. It needs complex computing to solve. So in field experiment MPP value can't be calculated by the equation.

Since it is difficult to calculate MPP under any lighting condition only with a simple equation, it is necessary to find other method to estimate MPP in a simple way.

3.2   MPP Prediction

There are 2 categories of prediction method according to physical parameters: one is to predict the solar irradiance, and then calculated the output power according to the prediction; the other one is to directly predict the output power. For the prediction of solar irradiance, it often combines with weather forecast or it depends on a small-scale weather station. So when it depends on the prediction of solar irradiance, it is the average value of the day, not the specific moment.

Building a PV system model is the main prediction method in nowadays. The usual way is to utilize the parameters under (STC) of short-circuit and open-circuit provided by the manufacture and the parameters of MPP to establish equations and solving the model parameters [10].

For instants, Zai Zhaiteng uses LambertW Function to get the current and voltage functions [11]. And he shows the relationships of different models and their parameters under different conditions and predicts the output feature of PV arrays according to the relationships [12].

Establishing PV output model and solving the parameters need a lot of computing. However, it can't describe the output feature under any solar irradiance and environmental temperature. Using simply models, such as the rough model of numerical fitting, the accuracy may be not very high, but it is enough for estimating the fouling condition.

4 Prediction

4.1 Neural Network

Since the 80th century, artificial neural network has caused widespread concern due to its unique advantages. The basic idea is to simulate the nervous system of the human brain, so that the machine has abilities of perception, learning and reasoning that the human brain has.

BP(Back Propagation) neural network is a multi-layer feedback leaning method. It has a strong arithmetic reasoning ability and is one of the widely used neural networks today. The basic method of the algorithm is least squares learning algorithm, using gradient search technique, adjusting the weights to reach the minimum total error of the network. The learning process is a process which is backward error propagation, and during which process the weights are corrected. This algorithm is especially suitable for the fitting nonlinear characteristics. Through training, the expected output can be precisely fitted.

4.2 Prediction of MPP Value Based on Neural Network

Considering the advantage of neural network in predicting the output of a nonlinear system, a BP neural network is established to predict the MPP value. Large numbers of experimental data were collected to train the network and finally got the MPP prediction value under any lighting condition.

A neural network was established referred to fig.3. The input layer is responsible for receiving the external information. The input parameters are solar irradiance measured by solar meter, the average temperature of the back board of the PV module, open-circuit voltage, short-circuit current, voltage and current of MPP; in the middle are the hidden layers, which are responsible for the exchange and processing of the information.

For output node k, we use (2) to update its weight. For hidden node I, we use (3) to update its weight. In the two equations, η is the learning speed; d_k is the expecting output; o_{pk} is the output of output node k; o_{pi} is the output of hidden node i; o_{pj} is the output or input of node j. δ_k is defined as $\delta_k = o_k(1-o_k)(d_k - o_k)$.

$$\Delta w_{ki} = \eta o_{pk}(1-o_{pk})(d_{pk} - o_{pk})o_{pi} \tag{2}$$

$$\Delta w_{ij} = \eta o_{pi}(1-o_{pi})(\sum_{k=1}^{L} \delta_{pk} w_{ki})o_{pj} \tag{3}$$

There are 6 hidden layers in the system. The number of hidden layers is very important in BP neural network. Generally, if there are too few hidden layers, complex mappings can't be created and the training error will be big; if there are too many hidden layers, the training process will take a lot of time, yet the error is not necessarily a minimum. So, by several experiments, a six-hidden layers are selected as the best choice.

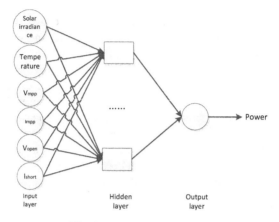

Fig. 2. Neural network module

The training data are more than 400 groups of data got by experiment (using clean PV module). They are collected by a data collection software system especially designed for the experiment (It will be introduced later). Then a neural network training system is set up using MATLAB. By the training, with large information forward propagation and error backward propagation, the weights of layers adjust continuously and the output error is reduced and achieves the requirement. The simulation result is shown in fig.3.

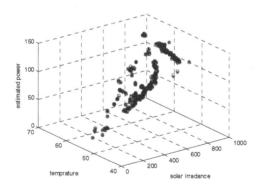

Fig. 3. Training result

'*' in the figure represent the original data and 'o' represent the data predicted by the BP neural network. We can see that most of the two groups of data can overlap. It means that the prediction value can match the original data at most of the time. So the neural network we established can perform well. So the neural network can achieve the result we expected.

In order to ensure the training quality, 500 times of training is set. In fig.4 we can see fast decline in error in the beginning. The result is good within 100 times. If we set fewer training times, the training can be completed in a short time.

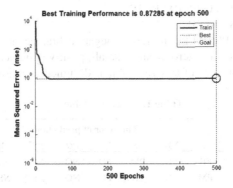

Fig. 4. Training error

So when predicting the PV output power, using a neural network would be a better choice. And the result is good.

5   Fuzzy Logic Rules

5.1   Introduction of Fuzzy Logic Rules

Comparing with classical control method, fuzzy logic does not need an accurate mathematical model. So for some object which cannot get a mathematical model or hard to establish an accurate mathematical model, it is better to use fuzzy logic method. Especially for nonlinear problems, fuzzy logic can provide a good solution. At present, the application of fuzzy logic is rapidly developing. It can be applied in the field of judgment, reasoning, prediction, identification, planning, decision making and problem solving, etc. Fuzzy logic has become an important part in the intelligent control field.

5.2   Design of Fuzzy Logic Rules

Fuzzy logic rules are used to identify the fouling condition in the system. The input parameters are the MPP value predicted by neural network and the actual measured MPP value.

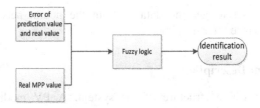

Fig. 5. Fuzzy decision flowchart

In the fuzzy logic design, the predicted MPP value and the actual MPP value are mapping to fuzzy set theory domain Ee and Er.

One fuzzy set has 8 fuzzy subsets and the other one has 6 fuzzy subsets. They are:

Ee={NB,NM,NS,NO,PO,PS,PM,PB};

Er={NB,NM,NS,PS,PM,PB};

NB,NM,NS,NO,PO,PS,PM,PB represent negative big, negative middle, negative small, negative zero, positive zero, positive small, positive middle, positive big.

Thought repeatedly practice, a fuzzy rule table is designed as follows.

Table 1. Fuzzy rule table

Real measurement	The error of prediction							
	NB	*NM*	*NS*	*NO*	*PO*	*PS*	*PM*	*PB*
NB	PB	PB	PM	PO	NO	NS	NM	NB
NM	PM	PS	PS	PO	NO	NS	NS	NB
NS	PB	PM	PS	PO	NO	NO	NM	NB
PS	PM	PS	PS	PO	NO	NS	NS	NM
PM	PS	PS	PO	PO	NS	NM	NM	NB
PB	PS	PO	PO	PO	NS	NM	NB	NB

5.3   Fuzzy Identification

Through the above fuzzy identification, we know that under the condition of PO and NO, there is little difference between the predicted MPP value and the actual MPP value. It is within the tolerance. When NS appears, the actual MPP value is a little smaller than predicted MPP value. There may be little fouling. At this time, one can considering the cleaning cost and the power loss cost, and then decide whether it needs cleaning. When NM or NB appears, there is a big difference between the predicted MPP value and the actual MPP value. In such case, the fouling condition may be serious and needs an immediate cleaning. When PS, PM or PB appears, it indicate that predicted MPP value is bigger than the actual MPP value. In such case, there is exception. It may be due to a sudden change in solar irradiance (e.g. a piece of cloud has just covered the sun suddenly). Or it is just because of neural network prediction failure. So we can't identify the condition by the invalid information. In such case, another identification needs to be made.

6   Experiment

Experiments were set to get the data to train the neural network and to check prediction performance after training.

6.1   Experiment Description

Fig. 6 shows the whole structure of the system. A PV module (HQ-190W), a controller, a solar meter, a temperature meter and a PC host were used to establish the system. The PV module is working at maximum power point tracking (MPPT) model most of the time, expect when open-circuit voltage, short-circuit current need to be get. The controller is used to get the open-circuit voltage, short-circuit current occasionally. Fig.6 shows the real equipment of the experiment.

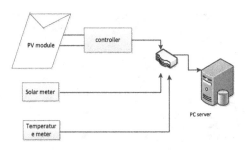

Fig. 6. Structure of the system

Solar meter (TES1333R) is placed in the same direction and the same angel of PV module so as to get the correct solar irradiance the module receives. The sensors of the temperature meter are placed at the back of the PV module. Their average temperature is the temperature of the module. The solar meter, the temperature meter and the controller communicate with the host by RS232 communication lines. Then a hub is used to collect the information and sends the information to the PC server.

When collecting normal data (using clean PV module), all the data were restored in a database for a later use. When testing the prediction performance, the data were stored temporarily and the prediction was made online.

6.2   Experiment Performance

A clean PV module was used in the first part of the experiment. Though online detecting software, we can see that the PV module is working normally. There were no warnings of serious fouling. The whole MPPT figure was show in Fig.7. The points which were much lower were due to thick cloud.

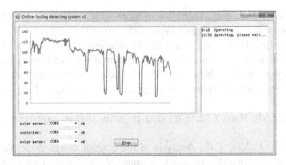

Fig. 7. Operation of system (clean PV module)

Then, in the second part of the experiment, dust was placed on the PV pale artificially. The system ran again. This time, the system detected the big difference between the predicted MPP value and the real MPP value. A warning was given in Fig. 8.

Fig. 8. Operation of system (fouling PV module)

7   Summary

In this paper, neural network and fuzzy logic identification are used to detect the fouling condition of PV modules. Take the advantage of prediction in nonlinear field of neural network, the maximum power point value of PV modules under any lighting environment is predicted. By training and simulation on the MATLAB platform, it shows that neural network can precisely predict the generating power of PV power station. Then ,the fuzzy logic rules is used to value the difference of predicted MPP value and the actual MPP value, so the fouling condition of PV module can be identified. This method proposed a good method to identify the fouling condition. Compare with traditional methods, it does not require modeling of the system. And it is a universal method. By experiments, this method can precisely identify the fouling condition of PV module.

References

1. Zhao, Z.M., Lei, Y., He, F.B.: Overview of large-scale grid-connected photovoltaic power plant. Automation of Electric Power System 35, 101–107 (2011)
2. Zhang, F., Bai, J.B., Hao, Y.Z.: Effect of airborne dust deposition on PV module surface on its power generation performance. Power System and Clean Energy 28, 82–86 (2012)
3. Hou, J.S., Wang, K.H., Shi, Y.L.: The influence of temperature and dust to the independent system of solar photovoltaic generation in Handan. Energy Conservation 28, 82–86 (2012)
4. Du, P.X., Ma, Z.M., Han, Y.M.: City dust pollution and management. Urban Problem 2, 46–49 (2004)
5. Gong, H.X., Wu, J., Xie, Y.K.: Analysis of the general design principles of pptimal dust removal plan of photovoltaic modules. Journal of Chongqing University of Technology (Natural Science) 27, 58–61 (2013)

6. Wang, X.L.: Pollution analyse of street dust in core district of Chongqing. Southwest University, Chongqing (2008)
7. Fali, J.U.: Study on The Effect of Photovoltaic Power Generation Project By Dust. Chongqing University (2010)
8. Zhai, Z.T., Cheng, X.F., Yang, Z.J.: Analytic solutions of solar cell model parameters. Acta Energiae Solaris Sinica 30, 1078–1083 (2009)
9. Wei, D., Lou, H., Xiao, C.Y.: Calculation of Maximum Power Point and Solution of Model Parameters for Solar Photovoltaic Output Characteristics. Proceedings of the CSEE 33, 1–7 (2013)
10. Engin, K., Boztepe, M., Colak, M.: Development of suitable model for characterizing photovoltaic arrays with shaded solar cells. Solar Energy 81, 329–340 (2007)
11. Chen, C.S., Duan, S.X., Yin, J.J.: Design of Photovoltaic Array Power Forecasting Model Based on Neutral Network. Transactions of China Electrotechnical Society 24, 153–158 (2009)
12. Zha, Z.T.: The output characteristic predieting of PV array in arbitrary condition. University of Science and Technology of China (2008)

Optimal DG Integration in Active Distribution Network Based on S-OPF

Yang Fu[1], Chunfeng Wei[1], Zhenkun Li[1], and Yiliu Jiang[2]

[1] Shanghai University of Electric Power, Yangpu District, Shanghai 200090, China
[2] Changzhou Power Supply Company, Changzhou, Jiangsu 213003, China
mfudong@126.com

Abstract. Active distribution network (ADN) is an indispensable content of smart distribution network under smart grid framework. On the basis of integration modes study, an economy optimal model of DG integration from a DSO perspective is proposed and two DG integration indexes are defined, respectively, integration ratio and GC ratio. The hourly sequential model is adopted to simulate load and DG uncertainties and a method of DG optimal integration based on stochastic optimal power flow(S-OPF) is presented. Monte Carlo Simulation aiming to get expected value and variance of network parameter precisely is used in stochastic power flow. The influence of integration indexes on economy in two scenarios is analyzed and some useful conclusions are achieved.

Keywords: active distribution network (ADN), stochastic optimal power flow(S-OPF), integration mode, integration index, Monte Carlo Simulation, hourly sequential model.

1   Introduction

With the exhaustion of traditional energy and growth of electricity demand, new energy power generation technologies will continue to mature, Distributed Generation (DG) will be integrated in all levels of distribution network and the increase of DG penetration is the development trend of future distribution network. Due to the uncertainties of demand and DG, a series of issues will come out in the operation of distribution network with high penetration of DG. Active Distribution Network (ADN) [1] and Microgrid [2] are two feasible technology routines. ADN is a public distribution network managed by Distribution System Operators (DSOs) to make easier integration of DG and optimize the operation state of the network.

At present, a lot of scholars at home and abroad have started research in this field and achieved some achievements. For example, multi-objective planning models of ADN are established in literature [3, 4], whereby the literature [3] uses multi-scenario analysis to deal with uncertainty of DG output and demand. Literatures [5, 6] take advantage of Stochastic Power Flow (SPF) to deal with uncertainties. Traditional distribution network planning meets all kinds of operation condition of extreme cases at the expense of a larger margin [7], but ADN planning approach should take into

K. Li et al. (Eds.): LSMS/ICSEE 2014, Part III, CCIS 463, pp. 378–387, 2014.

account possible uncertainty of operation condition at planning stage [8]. Literatures [9] establish a probabilistic model of the wind turbine and photovoltaic generator and consider that wind, solar and load are independent of each other, but in fact there is a time-based coupling relationship between them. Most of ADN related literatures are too conservative about the penetration of DG integration and less of them involve control strategy of DG operation and reverse power flow transport [3, 9].

Based on the above analysis, this paper presents an approach of optimal DG integration in ADN based on Stochastic Optimal Power Flow (S-OPF), which is conductive to future studies of energy storage allocation in ADN. Simulation results demonstrate the effectiveness of the proposed method, verify the necessity of Active Management (AM) and draw some useful conclusions about DG integration indexes.

2   Mathematical Model

The installation of DG units in current distribution network still follows the "fit and forget" policy that the DGs operate in prefixed range under local control without considering the global state of distribution network. This kind of DG integration mode is defined as the FF Mode. With the increasing penetration of DG in distribution network, Active Management (AM) is needed for each DG. By Generation Curtailment (GC) [5, 6] of DG, adjusting the DG output power factor and other measures to make distribution network operate in a safe state, it is defined as the AM Mode. It is not necessary and economic to monitor and control each DG, especially decentralized DG with small capacity. Therefore a part of DGs with large capacity are needed AM technology which is defined as Mixed Mode.

2.1   Objective Function

Mathematical model sets the economic optimal as objective function and the model takes into account the whole life cycle of the equipment, expressed as follows:

$$\min obf = C_{\text{inv}} + C_{\text{O\&M}} + C_{\text{fine}} - C_{\text{cs}} \tag{1}$$

where C_{inv} is investment costs, $C_{\text{O\&M}}$ is operation and maintenance costs, C_{fine} is fine costs of risk, C_{cs} is financial subsidy of renewable energy from civil social.

2.1.1   Investment Costs

Investment costs include the new installation costs of DG, DG connection costs, construction costs of ADN, expressed as follows:

$$C_{\text{inv}} = C_{\text{DG}} + C_{\text{con}} + C_{\text{ADN}} = [\sum_{i=1}^{N_{\text{DG}}} (c_{\text{DG},i} w_{\text{DG},i} + c_{i,\text{con}} l_i) + c_{\text{ADN}}^{\text{fix}} + N_{\text{DG}}^{\text{AM}} \cdot c_{\text{ADN}}^{\text{var}}] \cdot c_{\text{AP}} \tag{2}$$

where N_{DG} is the number of DG installed in the distribution network, $N_{\text{DG}}^{\text{AM}}$ is the number of DG adopting AM technology, $w_{\text{DG},i}$ is capacity of i-th DG, $c_{\text{DG},i}$ is the investment costs of i-th DG per unit capacity, l_i is the distance between i-th DG and the connection node, $c_{i,\text{con}}$ is the investment costs of connection line of i-th DG per unit length. c_{AP} is capital recovery factor. The construction cost of ADN is simplified as the sum of the fixed costs $c_{\text{ADN}}^{\text{fix}}$ and variable costs $N_{\text{DG}}^{\text{AM}} \cdot c_{\text{ADN}}^{\text{var}} \cdot c_{\text{ADN}}^{\text{var}}$ is the construction costs of communication and control equipment required for a DG integration in ADN.

2.1.2 Operation and Maintenance Cost, Fine Cost of Risk, Financial Subsidy of Renewable Energy

The sum of the three part can be expressed as f, as follows:

$$
\begin{aligned}
f &= C_{\text{O\&M}} + C_{\text{fine}} - C_{\text{cs}} \\
&= \sum_{i=1}^{N_{\text{DG}}} c_{\text{om},i} w_{\text{DG},i} + 8760 \cdot (c_{\text{loss}} P_{\text{loss}} + c_{\text{trans}} P_{\text{trans}} + \sum_{i=1}^{N_{\text{load}}} c_{\text{fine}} P_{li} - \sum_{i=1}^{N_{\text{DG}}} c_{\text{cs}} P_{\text{DG},i})
\end{aligned} \tag{3}
$$

where operation and maintenance costs include operation and maintenance costs of DG, annual network loss costs, annual costs of energy imported from transmission system. $c_{\text{om},i}$ is operation and maintenance costs of i-th DG per unit capacity, c_{loss} is energy loss costs per unit power, c_{trans} is the costs of energy imported from transmission system per unit power, P_{loss} is the network loss, P_{trans} is the energy imported from transmission system, if the value of P_{trans} is negative, it means that the distribution network transfers power to the network with higher voltage level and generate revenue. N_{load} is the number of load nodes, P_{li} is active power of i-th load, c_{fine} is fine cost of risk per unit load. $P_{\text{DG},i}$ is active power output of i-th DG, c_{cs} is the active power subsidy of DG per unit power, including environmental benefits. If the optimal power flow calculation has no solution which means constraints are violated, so fine cost of risk is considered in objective function.

2.2 Constraint Condition

Constraints are expressed as follows:

$$P_{gi} - P_{li} - \sum_{j \in \Omega_i} p_{ij} = 0 \tag{4}$$

$$Q_{gi} - Q_{li} - \sum_{j \in \Omega_i} q_{ij} = 0 \tag{5}$$

$$V_i^{\min} \le V_i \le V_i^{\max} \tag{6}$$

$$p_{ij} \le p_{ij}^{\max} \tag{7}$$

$$\left| P_{\text{T},i} \right| \le e_i P_{\text{T},i}^{\text{rate}} \tag{8}$$

$$P_{\text{GC},i} \le \lambda P_{\text{DG},i} \tag{9}$$

$$\left| \cos \varphi_{\text{DG},i} \right| > \psi \tag{10}$$

where, P_{gi}, Q_{gi} is the active and reactive generation at bus i, respectively; Q_{li} the reactive load at load i; p_{ij}, q_{ij} the active and reactive power flow between node i and node j, respectively; Ω_i the nodes set connected to bus i, V_i, $V_i^{\min}$ and $V_i^{\max}$ the voltage, minimum voltage and maximum voltage at bus i, respectively; $p_{ij}^{\max}$ the maximum power flow between node i and j; $P_{T,i}$ the active power of i-th main transformer, its value is positive if the power flow transmit from high voltage level to low level otherwise negative. $P_{T,i}^{rate}$ is the rated power of i-th main transformer; e_i the load-capacity-ratio of i-th main transformer. $P_{DG,i}$ is rated power of i-th DG, $P_{GC,i}$ the reduction power due to GC operation of i-th DG which will be described in detail in Section 2.3. $\cos\varphi_{DG,i}$ is power factor of i-th DG, ψ adjustable power factor of DG.

2.3   DG Integration Indexes

1) Integration ratio γ

Integration ratio is proposed to descript the proportion of DG with and without AM in ADN, namely:

$$\gamma = \sum_{i=1}^{N_{DG}^{AM}} w_{DG,i} \left/ \sum_{i=1}^{N_{DG}} w_{DG,i} \right. \tag{11}$$

In the situation that certain capacity constraints are satisfied, "fit and forget" policy can be adopted for small-capacity DGs that means they operate under local control according to prefixed value rather than under global real-time control.

2) GC ratio λ

Due to the high penetration of DG accessing to the end of the feeder which will result in violations of node voltages, branch capacities and other constraints, large volatility of DG output makes the distribution network may cannot consume all the DG output power, so appropriate Generation Curtailment measure is needed to meet the operation constraints. GC ratio of DG adopting AM is defined as the maximum ratio of the allowable power reduced by GC operation and the rated power, defined as follows:

$$\lambda = P_{GC}^{max} \left/ P_{DG} \right. \tag{12}$$

where, P_{GC}^{max} is maximum allowed power reduction by GC operation of DG; P_{DG} the rated power of DG. GC ratio of DG is an important performance index of control equipment of DG in ADN, the value of which determines the maximum allowable DG penetration in ADN value to some extent. The higher the value is, the higher the allowable penetration of DG.

3   Uncertainties Simulation Model in ADN

In order to simulate the operation of actual distribution network, DG output and load demand use hourly sequential model. Meanwhile considering a certain error between the predicted and actual values, a disturbance variable is added on the basis of the predicted value in a simulation.

3.1   Hourly Sequential Model of Wind Turbine

The output of wind turbine (WT) can be approximately expressed as a function of wind speed[5]. Forecast wind speed data of 8760 hours in a year can be obtained by HOMER software through history monthly average wind speed from meteorological sites.

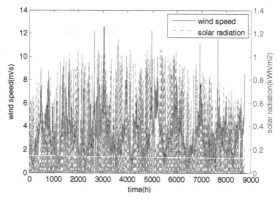

Fig. 1. Yearly wind speed and solar radiation of Shanghai

Take Shanghai (longitude 121.45°E, latitude 31.4°N) as an example, Weibull shape factor is 1.98, autocorrelation is 0.9, diurnal pattern is 0.2, hour of peak in a day is 14h, the wind speed is presented as a solid line in figure 1.

3.2   Hourly Sequential Model of Photovoltaic

The power output of photovoltaic (PV) is related to daily radiation, conversion efficiency, installation angle of photovoltaic panel, temperature and so on [9]. The generated solar radiation of Shanghai is presented as a dotted line in figure 1.

3.3   Hourly Sequential Model of Load

Daily fluctuation characteristics of load have a great relationship with load type and load patterns of the same type are also slightly different in four seasons throughout the year. In this paper, loads are categorized into the agricultural load, commercial load, industry load and residential load and a total of 16 load patterns are needed considering different seasons. Load data of 8760 hours in a year can be simulated by various types of load pattern in four seasons.

4   Solution of DG Optimal Integration Model

4.1   OPF

OPF itself is a kind of mathematical optimization problem. Under the situation that network structure A, its parameter p, load demand, maximum active and reactive power of DG are known, OPF minimizes the costs of energy imported from transmission system through adjusting the control variable u (actual output of DG, energy imported from transmission, etc.) and meeting the constraints of load demand and operation requirement. This is similar with control measures and operational goals in ADN, so OPF calculation is used to determine the runtime status of ADN.

During power flow calculation, if DGs connect the network using FF Mode, they are set as PQ buses and their output only contains active power; if using AM Mode, they are initialized as PV buses and upper limit of DG's active power is determined by wind speed and solar radiation, lower limit of active power is determined by equation (9), upper and lower limits of reactive power is determined by equation (10).

4.2   Stochastic Power Flow Based on Monte Carlo

Monte Carlo simulation randomly generates any time of a year, samples the corresponding load distribution and DG output based on load and DG output hourly sequential models built, calculates network operating parameters through determinate power flow, then get the statistical analysis data. In sampling process, if convergence criterion is not satisfied, sampling continues and sample sequence of the network operating parameters is obtained by calculating; if satisfied, probability statistics of the above sample sequence is done, mathematical expectation and variance of the sample sequence is considered as the mathematical expectation and variance of the research problem.

In the above process, the convergence criterion is judged at each sample in order to improve the calculation speed. As a result of relative error is used as the convergence criterion, the function that used to judge convergence excludes investment costs, only contains the subfunction f expressed in section 2.1.2. The subfunction of i-th sample simulation is defined as f_i, the estimated value of subfunction of i-th sample $\hat{f}_i$ and its practical estimation formula of relative error is defined as follows:

$$\hat{f}_i = \sum_{i=1}^{N} f_i \bigg/ N \tag{13}$$

$$\varepsilon_i \leq u_{1-\alpha/2} \left(\sqrt{\sum_{i=1}^{N} (f_i - \hat{f})^2} \bigg/ \sqrt{N(N-1)} \hat{f} \right) \tag{14}$$

where, $1-\alpha$ is a given probability (α is a given level of significance), $u_{1-\alpha/2}$ is the lower side of the median of the standard normal distribution N (0,1) at a given probability. In this article $1-\alpha$ is taken as 0.95, corresponding $u_{1-\alpha/2}$ is 1.959 963 984. When the error estimated by equation (18) is smaller than the prefixed value of relative error ε_{ref}, the algorithm converges.

4.3   Optimization Algorithm of DG Integration

Each DG integration solution is economically assessed in accordance with the following steps.

1) Initialize parameter. Input network parameters and DG connection solution, and calculate investment cost by equation (2). Set the relative error of Monte Carlo simulation, $N=1$.

2) Build hourly sequential model of load and DG output, respectively.

3) Sample $t \in [1, 8760]$ according to hourly sequential model and generate corresponding random variable of load and DG output.

4) Select the slack bus according section 4.1, set PV and PQ buses and calculate upper and lower limits of active and reactive power of slack and PV buses.

5) Minimize the cost of energy imported from transmission which is set as the objective function of OPF and complete deterministic OPF calculation.

6) Reset slack bus or PV bus if the power constraint is violated and return to step 5; Save sample sequence of subfunction f_i, power flow, network losses, voltage of buses, real output of DGs and reduction power due to GC operation and so on if not.

7) Calculate the estimated value of subfunction $\hat{f}$ by equation (13) and relative error by equation (14). If the relative error meets accuracy requirements compared with ε_{ref}, return to step 8; if not, $N=N+1$, return to step 3.

8) Carry out probability statistics. Calculate objective function *obf* of DG integration by equation (1), set the mathematical expectation and variance of the sample sequence as the mathematical expectation and variance of the solution.

5   Simulation Results and Discussion

5.1   Test Case of ADN

The test case in this paper is a 48 bus test system to simulate the real medium voltage distribution network of Shanghai China. The system construction is shown in figure 2.

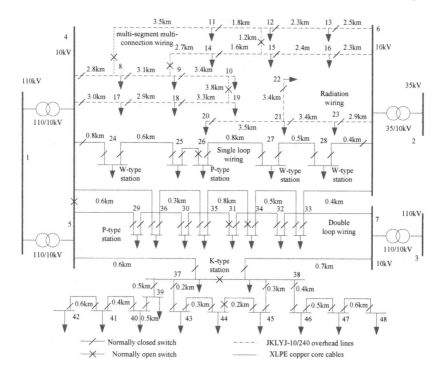

Fig. 2. A Shanghai 48 bus test system for ADN research

The wiring patterns of overhead line of Shanghai distribution network commonly are radiation and multi-segment multi-connection, mainly agricultural load. The wiring patterns of cable are mainly single loop, double loop and switching stations. Assume that the load supplied by a single loop wiring is residential load, load supplied by a double loop wiring is industry load and load supplied by switching station is commercial load.

5.2   Basic Simulation Data

It is assumed that GC ratio is taken as 50%, allowable power factor range of DG is taken as 0.9 lead or lag. To facilitate the study of the meaning of AM to DG in ADN, the following two scenarios of DG connection are defined.

Scenario 1: DGs orderly connect to ADN. The total capacity of DGs connected to each 10kV feeder is not greater than the feeder capacity, the rest of DGs are connected to 10kV buses in 35/10kV or 110/10kV substation.

Scenario 2: DGs randomly connect to ADN. At different penetration condition, the probability of DG connected to 10kV buses and load nodes is the same and DG connection is independent of each other.

5.3   Analysis of Simulation Results

The economics of DG integration in distribution network at the different penetration is shown in figure 3 in the FF Mode and the AM Mode. It is observed that, the economics of photovoltaic connected to the distribution network in Shanghai alone is much better than the wind turbine, mainly because solar radiation resource in Shanghai is relatively rich then wind resources. The transverse line in the figure indicates the objective function is 127.49 million Yuan when no DG is connected to the network.

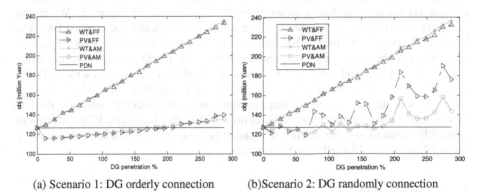

(a) Scenario 1: DG orderly connection   (b)Scenario 2: DG randomly connection

Fig. 3. Economic comparison of FF Mode and AM Mode

As shown in figure 3(a), the DGs orderly connecting to network, economy of ADN is better than distribution network without DG when the PV penetration is 0~200%, in which the economy is the best and objective function is 119.06 million Yuan when the PV penetration is 20%. Economy of FF Mode is slightly better than AM Mode when the penetration is 0~220% and is worse than FF Mode when the penetration is above 220%.The overall economy of scenario 2 in which the PVs randomly connect to network is worse than scenario 1 in which the PVs orderly connect as shown in figure 3 (b). When the penetration is greater than 65%, the economy of AM Mode is far superior to FF Mode and the overall difference between the two increases with the penetration grows. As can be seen from comparison of the two pictures in figure 3, the potential of AM is greater when DG randomly access.

In Mixed Mode when DGs randomly connect to network, the impact of DG integration ratio on economy at different penetration (PR) is shown in figure 4. The small range of fluctuation in the figure is due to the randomness of selecting DG to be under active control. According to the comparison of the three curves, the DG integration ratio should increase in order to achieve optimal economy when the penetration continuously grows, for example, DG integration ratio should be taken as 50% and 70% when penetration is 150% and 200% respectively.

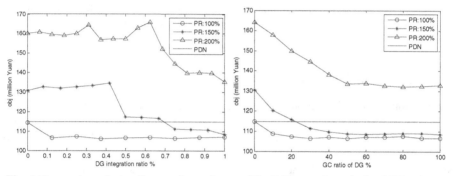

Fig. 4. Economic analysis of integration ratio   **Fig. 5.** Economic analysis of GC ratio

In AM Mode when DGs randomly connect to network, the impact of GC ratio of DG on economy at different penetration (PR) is shown in figure 5. When the penetration is 100%, the objective function achieves the optimal value when GC ratio is taken as about 20%, the further increase of GC ratio of DG has litter impact on economy, but the requirements of control technology for DG is also higher; when penetration is 150% and 200%, the GC ratio of DG should be taken as 40% and 50% respectively to make the objective function achieve optimal value. According to comparison of the three curves, when the penetration continuously improves, the GC ratio to make objective function reach optimal value also increases. If the maximum allowable penetration is determined in a planning area, the recommended value of GC ratio of DG can be obtained from the curves.

6   Conclusion

1) The test case shows that in the trend DG penetration increases continually, especially when DGs randomly connect to network, AM Mode has a greater advantage compared with FF Mode. 2) The proposed concept and recommended

value of DG integration ratio at Mixed Mode provide a feasible idea for the operation and management of DG in future distribution network. 3) Meanwhile, the recommended value of GC ratio of DG in AM Mode at different penetration can be obtained through the experimental data.

Acknowledgments. This work was supported by Shanghai Green Energy Grid Connected Technology Engineering Research Center, Project Number: 13DZ2251900.

References

1. D'Adamo, C., Jupe, S., Abbey, C.: Global survey on planning and operation of active distribution networks; update of CIGRE C6.11 working group activities. In: CIRED 2009 (20th International Conference on Electricity Distribution), Prague, CZ (2009)
2. Chrn, J., Yang, X., Zhu, L., et al.: Microgrid Multi-objective Economic Dispatch Optimization. Proceedings of the CSEE 33(19), 57–66 (2013) (in Chinese)
3. Borges, C.L.T., Martins, V.F.: Multistage expansion planning for active distribution networks under demand and Distributed Generation uncertainties. Electrical Power and Energy Systems 36, 107–116 (2012)
4. Cell, G., Pilo, F., Soma, G.G.: Active distribution network cost/benefit analysis with multi-objective programming. In: CIRED 2009 (20th International Conference on Electricity Distribution), Prague, CZ, pp. 1–5. IET Services Ltd. (2009)
5. Cheng, H.Z., Zhang, J.T., Wu, O., et al.: Siting and sizing of distributed wind generation under active management mode. Journal of Electric Power Science and Technology 24(4), 12–18 (2009) (in Chinese)
6. Pilo, F., Celli, G., Mocci, S.: Multi-objective programming for optimal DG integration in active distribution systems. In: 2010 IEEE Power and Energy Society General Meeting, Minneapolis, MN, pp. 1–6 (2010)
7. Fu, Y., Wei, C.F., Li, Z.K., et al.: Optimal Partitioning of Substation Service Areas Considering Impacts of Geographic Information and Administrative Boundaries. Power System Technology 37(1), 126–131 (2014) (in Chinese)
8. Fan, M.-T., Zhang, Z.-P., Su, A.-X., et al.: An Investigation of Enabling Technologies for Active Distribution System. Proceedings of the CSEE 33(22), 12–18 (2013) (in Chinese)
9. Zeng, B., Liu, N., Zhang, Y.-Y., et al.: Bi-level Scenario Programming of Active Distribution Network for Promoting Intermittent Distributed Generation Utilization. Transactions of China Electrotechnical Society, 28(9), 154—163 (2013) (in Chinese)

Impact of Wind Power Penetration on Unit Commitment

Qun Niu, Letian Zhang, and Hongyun Zhang

School of Mechatronic Engineering and Automation,
Shanghai Key Laboratory of Power Station Automation Technology, Shanghai University,
Shanghai 200072, China
comelycc@hotmail.com

Abstract. Wind farm outputs have the features of intermittence and variability which impose a significant impact on the operation of power systems. In this paper, Latin hypercube sampling (LHS) and reduction technique is used to simulate the 24h power output of a wind farm. Then a model of unit commitment (UC) with predicted wind power (UCW) is established, and a harmony search (HS) with arithmetic crossover operation (ACHS) is employed for solving this problem. The results are analyzed in detail, which assess the impact of wind power on UC and demonstrate that ACHS is practicable for UCW problem with comparison with other proposed HS methods.

Keywords: Wind power, Unit commitment, Harmony search, Crossover.

1 Introduction

Wind power as an environmental friendly green energy has attracted more and more attention in recent years. However, the uncertainty and fluctuation of the wind energy has huge impact on the unit commitment of conventional power systems. Thus, it is vital to make more accurate prediction of wind farm power output and schedule the thermal generations reasonably.

Unit commitment (UC) as a large-scale, mixed-integer and non-linear optimization problem subject to various constraints including unit characteristics and power grid demands, is always a key issue in operation and control of conventional power systems [1]. To solve this problem, many conventional methods and heuristic optimization algorithms have been applied to the UC problem such as Lagrange relaxation method [2], dynamic programming [3], genetic algorithm (GA) [4], particle swarm algorithm (PSO) [5], and differential evolution algorithm (DE) [6].

To evaluate the impact of wind power on UC, a proper UC model must be established first. Based on conventional UC model, some updated models have been proposed which consider the thermal UC with wind farm power output (UCW). In [7], Wang presented a security-constrained UC model with forecasted intermittent wind power generation, and assessed the effectiveness of wind power generation on the security of power system operation. In [8], Aiden investigated high levels of wind power penetration in UC and power system dispatch, and compared two different modes of optimization. In [9], Roy proposed an auto-regressive moving average time series model to generate the wind power for UC risk analysis, and suggested that the

K. Li et al. (Eds.): LSMS/ICSEE 2014, Part III, CCIS 463, pp. 388–397, 2014.

operation capacity could be further increased when wind power was considered. In [10], Colm proposed a stochastic UC with rolling planning wind forecasts, and analyzed how the error of wind forecast changed the utilization of UC. By considering the stochastic characteristic of wind power to the UC, more robust schedules should be investigated.

In this paper, an efficient method which utilizes Latin hypercube sampling (LHS) and scenarios reduction technique is proposed for forecasting wind power generation at each time interval. With LHS, based on the expected wind power generation over a 24h period, 1000 different scenarios are generated, and each scenario has a given probability. Then, scenarios reduction technique is applied to these samples for eliminating the one with low probability and close distanced scenarios. With this method, the number of wind power generation scenarios will be strictly limited under normal predicted bound and will be used for UCW system suitably.

With the preparation of wind power generation profile, a UCW model is presented which take into account the wind power prediction to the spinning reserve constraints of UC. In order to solve this problem, a new variant of harmony search (HS) [11] with arithmetic crossover operation (AC) [12], namely ACHS, is employed which is first proposed by the authors in [13]. ACHS combines the searching ability of HS and exploitation of crossover operation, and is an improved solver for non-linear, non-convex optimization problems. This paper sequentially investigates ACHS algorithm and apply to the large-scale and mix-integer UCW problems. The simulation results are presented and the impact of wind power on conventional UC problem is analyzed, and it is shown that ACHS offers a competitive alternative for UCW optimization problems in comparison with a few other methods.

2   Wind Power Forecasting

Wind power forecasting is an important issue in analyzing the influence of wind power fluctuation on the power system. In this paper, each hourly random wind power generation is taken into account which is based on the expected wind farm outputs. It is assumed that the wind power is subject to a normal distribution $N(\mu, \sigma^2)$ where μ stands for the mean value of the forecasted wind power and σ is standard deviation of the forecasted wind power. Monte Carlo simulation is used to generate wind power scenarios subject to $N(\mu, \sigma^2)$. To guarantee the accuracy of wind power prediction, Latin hypercube sampling (LHS) technique is utilized to generate 1000 samples with expected wind power given in Table 1 [7] with 10% percentage for volatility.

Table 1. Expected wind power

Hour	Wind(MW)	Hour	Wind(MW)	Hour	Wind(MW)	Hour	Wind(MW)
1	44	7	100	13	84	19	10
2	70.2	8	100	14	80	20	5
3	76	9	78	15	78	21	6
4	82	10	64	16	32	22	56
5	84	11	100	17	4	23	82
6	84	12	92	18	8	24	52

In order to use representative scenario with a high probability of forecasted wind power, a scenario reduction technique is employed which decreases the 1000 wind scenarios to just one representation. The detailed procedure is shown as follows [14].

Step 1: Set S is the 1000 wind scenarios, and DL is the scenarios to be deleted. The initial DL is null. Calculate all distance between each scenario pairs:

$$D_{S1,S2}=D(X_{S1},X_{S2})=\sqrt{\sum_{i=1}^{24}\left(x_{S1}^{i}-x_{S2}^{i}\right)^{2}} , \quad S1, S2\in S \quad \text{and} \quad S1\neq S2 . \quad X_{S1} \text{ and } X_{S2}$$

are the scenarios of wind power. x_{S1}^{i} and x_{S2}^{i} are the wind power of time i in scenarios $S1$ and $S2$.

Step 2: For each scenario k, choose the minimum distance with scenario k $D_{k}(r) = \min D_{k,s}$, $k, s\in S$ and $s\neq k$. r is the index that has minimum distance with scenario k.

Step 3: Calculate each $FD_{k}(r) = P_{k}\cdot D_{k}(r)$, $k\in S$. P_{k} is the probability of scenario k. Select scenario m which has minimum $FD_{m} = \min FD_{k}$, $k\in S$.

Step 4: $S = S -\{m\}$, $DL = DL+\{m\}$, $P_{r} = P_{r}+P_{m}$.

Step 5: Repeat steps 2-4 until the number to be decrease to 1.

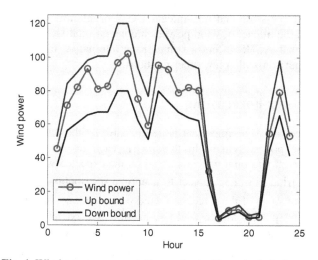

Fig. 1. Wind power representation and actual bound of wind power

The final wind power representation is shown as Fig. 1. The actual wind power bound of $\pm 20\%$ expected wind fluctuation is also given in Fig. 1. It can be seen that the wind power representation fits well with the bound of possible wind power error tolerance, and therefore it is appropriate for simulating wind power fluctuation.

3   UCW Model

With a 24h period wind power output, the UCW model is to determine the unit state and power output to minimize the operation cost for the forecasted schedule. Thus, the objective function of UCW can be expressed below.

$$Min \ F = \sum_{t=1}^{T}\sum_{n=1}^{N}\left\{f_{fc,n} + f_{sc,n}\left[1-u_{n}\left(t-1\right)\right]\right\}\cdot u_{n}\left(t\right) \tag{1}$$

where T is the total number of scheduling periods, N is the number of units and $u_{n}(t)$ is ON/OFF (1/0) status of unit n. $f_{fc,n}$ and $f_{fc,n}$ are the fuel cost and start-up cost which are expressed as (2) and (3), respectively.

$$f_{fc,n} = a_{n}\left(P_{n}\left(t\right)\right)^{2} + b_{n}P_{n}\left(t\right) + c_{n}u_{n}\left(t\right) \tag{2}$$

$$f_{sc,n} = \begin{cases} C_{n}^{hot} & M_{n}^{down} \leq T_{n}^{off} \leq M_{n}^{down} + CH_{n} \\ C_{n}^{cold} & T_{n}^{off} > M_{n}^{down} + CH_{n} \end{cases} \tag{3}$$

where $P_{n}(t)$ means the output power of unit n at time t. a_{n}, b_{n} and c_{n} are the fuel cost coefficients of unit n. C_{n}^{hot} and C_{n}^{cold} are the hot/cold start-up cost of unit n, respectively. T_{n}^{off} is the duration for which unit n is OFF. M_{n}^{down} is the minimum down time of unit n, and CH_{n} means the cold start time of unit n. Further, some constraints which should be met are given as follows.

1) *Power balance*: Wind energy fluctuation is transmitted to the power system, and these uncertain wind power and output from the thermal units must satisfy the load demand, which is defined as (4).

$$\sum_{n=1}^{N}P_{n}\left(t\right)u_{n}\left(t\right) + P_{w}\left(t\right) = Pd\left(t\right) \tag{4}$$

where $P_{w}(t)$ is the output power of wind farm at time t, and $Pd(t)$ is the power load demand at time t.

2) *Generation range*: The output of thermal unit must satisfy the limit range, which is represented as (5).

$$P_{n}^{min} \leq P_{n}\left(t\right) \leq P_{n}^{max} \tag{5}$$

where P_{n}^{min} and P_{n}^{max} are the minimum/maximum power limits of unit n.

3) *Spinning reserve*: The spinning reserve contributes to incidental sudden decrease in wind power and unpredictable generator outrages and forecast error in load demands. To guarantee a reliable system, the spinning reserve in this paper is shown as (6).

$$\sum_{n=1}^{N}P_{n}^{max}u_{n}\left(t\right) + \eta_{r}P_{w}\left(t\right) \geq Pd\left(t\right)\cdot\left(1+SR\right) \tag{6}$$

where η_{r} is the prediction error of wind power, and SR is the requirement of spinning reserve.

4) *Minimum up/down time*: When a unit is ON/OFF, there is a limited minimum time after it can be set to OFF/ON respectively.

$$\begin{cases} T_n^{off} > M_n^{down} \\ T_n^{on} > M_n^{up} \end{cases} \tag{7}$$

where T_n^{on} is the duration for which unit n is switched ON, and M_n^{up} is the minimum up time of unit n.

4 Method

UCW problem can be regarded as a non-linear, mixed-integer, large-scale combination optimization problem. In this paper, a new variant of harmony search (HS) [11] with arithmetic crossover operation (AC) [12], namely ACHS, is employed to solve this problem. ACHS is first proposed by authors in [13], and applied to dynamic economic dispatch (DED). In HS, the global optimized information, which can guide the solutions to high quality searching field and improve the convergence speed, is not fully utilized. Therefore, the crossover operation is employed which plays a key role in genetic algorithm (GA), and it recombines the new generated solution obtained by HS and the current global best one. This recombined solution is compared with the current best solution, and if the recombined solution wins, it will replace the current best solution in the population. The detail procedure can be found in [13]. Although the ACHS has good efficiency in DED, DED is only applicable to optimization problems with continuous searching space. This paper expands the application of ACHS to mixed-integer UCW problem, and the implementation of ACHS to UCW is given as follows.

Step 1: *Preparation.* Three steps will be taken in this step. First, collect wind power data. Second, the priority order of units for scheduling is listed based on the maximum limit of units P_n^{max}. A unit with higher P_n^{max} will be given a higher priority to be committed. Finally, determine the parameters in ACHS and the UCW model.

Step 2: *Initialize.* Initialize a population of unit states and unit power output randomly as follows.

$$u_n(t) = \begin{cases} 1 & rand \le 0.5 \\ 0 & rand > 0.5 \end{cases}, \quad n \in N, t \in T$$

$$P_n(t) = \left(P_n^{min} + rand \cdot \left(P_n^{max} - P_n^{min} \right) \right) \cdot u_n(t), \quad n \in N, t \in T$$

Step 3: *Repair constraints.* In this process, the population of unit states must satisfy the spinning reserve (6) and minimum up/down time constraints (7). If the committed units do not conform to the constraints, adjust the units one by one according to the priority order given in Step 1. Then, with the states of units, adjust the power outputs of thermal units, and make them satisfied the power balance (4) and generation range (5).

Step 4: *Calculate the fitness.* Based on the unit states and power output, calculate the objective function according to (1).

Step 5: *Update.* Use the ACHS to update the solution of UCW system, and the detailed procedure is shown below. It is assumed that *State* and *Power* constitutes a solution in the populations of unit states and power output which gives *HMS* individuals.

If $rand \leq HMCR$

    $r = ceil(rand \cdot HMS)$

    $u_{n,new}(t) = State(r, n+(t-1) \cdot N)$

    $P_{n,new}(t) = Power(r, n+(t-1) \cdot N)$

    If $rand \leq PAR$

        If $u_{n,new}(t) == 1$

            If $rand < 0.5$

                $u_{n,new}(t) = 0$

                $P_{n,new}(t) = 0$

            Else

                $P_{n,new}(t) = P_n^{\min} + rand \cdot (P_n^{\max} - P_n^{\min})$

            End if

        Else if $u_{n,new}(t) == 0$

            If $rand < 0.5$

                $u_{n,new}(t) = 1$

                $P_{n,new}(t) = P_n^{\min} + rand \cdot (P_n^{\max} - P_n^{\min})$

            Else

                $u_{n,new}(t) = 1$

            End if

        End if

    End if

Else

    If $rand < 0.5$

        If $rand < 0.5$

            $u_{n,new}(t) = 1$

            $P_{n,new}(t) = P_n^{\min} + rand \cdot (P_n^{\max} - P_n^{\min})$

        Else

            $u_{n,new}(t) = 0$

            $P_{n,new}(t) = 0$

        End if

    Else

        If $u_{n,new}(t) == 1$

            $P_{n,new}(t) = P_n^{\min} + rand \cdot (P_n^{\max} - P_n^{\min})$

        End if

    End if

End if

$rr = rand$

$$P_{n,new}(t) = P_{n,new}(t) \cdot rr + P_{n,best}(t) \cdot (1 - rr)$$

If $rand < 0.5$

$$u_{n,new}(t) = u_{n,best}(t)$$

End if

Step 6: *Select.* Repair constraints and calculate the fitness of new individual. If the new fitness is better than current best fitness, choose the new solution to take place the worst in the population.

Step 7: Repeat. Step 5 – Step 6 until the limited iteration is reached.

5   Simulation and Results

An IEEE 10-unit system is used to evaluate the wind power impact on unit commitment. The 24h period wind farm outputs are generated as described in section 2, and the parameters of the 10-unit system are collected from [15]. The spinning reserve for this system is 10%, and the prediction error of wind power is 80%.

Table 2. 10-unit system without considering wind power

Hour	U1	U2	U3	U4	U5	U6	U7	U8	U9	U10	$f_{sc,n}$
1	455	245	0	0	0	0	0	0	0	0	0
2	455	295	0	0	0	0	0	0	0	0	0
3	455	370	0	0	25	0	0	0	0	0	900
4	455	455	0	0	40	0	0	0	0	0	0
5	455	390	0	130	25	0	0	0	0	0	560
6	455	360	130	130	25	0	0	0	0	0	1100
7	455	410	130	130	25	0	0	0	0	0	0
8	455	455	130	130	30	0	0	0	0	0	0
9	455	455	130	130	85	20	25	0	0	0	860
10	455	455	130	130	162	33	25	10	0	0	60
11	455	455	130	130	162	73	25	10	0	0	60
12	455	455	130	130	162	80	25	43	10	10	60
13	455	455	130	130	162	33	25	10	0	0	0
14	455	455	130	130	85	20	25	0	0	0	0
15	455	455	130	130	30	0	0	0	0	0	0
16	455	310	130	130	25	0	0	0	0	0	0
17	455	260	130	130	25	0	0	0	0	0	0
18	455	360	130	130	25	0	0	0	0	0	0
19	455	455	130	130	30	0	0	0	0	0	0
20	455	455	130	130	162	33	25	10	0	0	490
21	455	455	130	130	85	20	25	0	0	0	0
22	455	455	0	0	145	20	25	0	0	0	0
23	455	420	0	0	25	0	0	0	0	0	0
24	455	345	0	0	0	0	0	0	0	0	0

Table 3. 10-unit system with wind power

Hour	U1	U2	U3	U4	U5	U6	U7	U8	U9	U10	P_w	$f_{sc,n}$
1	455	199.4	0	0	0	0	0	0	0	0	45.6	0
2	455	223.8	0	0	0	0	0	0	0	0	71.2	0
3	455	313	0	0	0	0	0	0	0	0	82.0	0
4	455	376.8	0	0	25	0	0	0	0	0	93.2	900
5	455	455	0	0	9	0	0	0	0	0	81.0	0
6	455	407.1	130	0	25	0	0	0	0	0	82.9	1100
7	455	443.3	130	0	25	0	0	0	0	0	96.8	0
8	455	397.5	130	90.4	25	0	0	0	0	0	102.1	1120
9	455	455	130	130	29.8	0	25	0	0	0	75.2	520
10	455	455	130	119	104	40	37.6	0	0	0	59.4	340
11	455	455	130	128.2	129.9	20	25	11.6	0	0	95.3	60
12	455	455	130	130	162	20	25	20.1	10	0	92.9	60
13	455	455	124.2	103	139	20	25	0	0	0	78.8	0
14	455	455	130	104.5	25	0	48.3	0	0	0	82.2	0
15	455	379.9	130	130	25	0	0	0	0	0	80.1	0
16	455	277.7	128.9	128.9	29.6	0	0	0	0	0	32.3	0
17	455	334.5	130	51.3	25	0	0	0	0	0	4.3	0
18	455	351.1	123.7	130	31.3	0	0	0	0	0	8.9	0
19	455	450.6	130	130	25	0	0	0	0	0	9.4	0
20	455	455	130	101.7	162	42.3	39.2	10	0	0	4.8	660
21	455	455	120.7	128.5	31.3	79.6	25	0	0	0	4.9	0
22	455	455	0	0	77	33.2	25	0	0	0	54.8	0
23	455	340.7	0	0	25	0	0	0	0	0	79.3	0
24	455	291.5	0	0	0	0	0	0	0	0	53.5	0

Table 4. Comparison of different methods for UCW system

Type	Without wind power			With wind power		
	Best	Mean	Worst	Best	Mean	Worst
HS	567424.3	568526.7	569021.4	534002.8	534596.0	535023.7
GHS	565491.6	566947.5	567809.2	532877.5	533515.4	534068.6
IHS	566474.4	567082.1	567963.4	532793.4	533499.0	533826.2
ACHS	563977.0	563987.2	564018.3	531836.2	532518.4	532836.1

To investigate the effect of the wind power, ACHS is applied to two simulation cases. 1) 10-unit system without considering wind energy. 2) 10-unit system with wind power. For these two cases, the control parameters of ACHS are the same with [13], except the number of maximum fitness evaluations FES = 20000. In order to show the searching ability of ACHS, some other HS algorithms are also utilized to solve the UCW problem, including basic harmony search (HS) [11], global-best harmony search (GHS) [16] and improved harmony search (HIS) [17]. Their control parameters are the same with references [11] [16] and [17], and the FES is set to 40000. All simulations are independently run 30 times in Matlab 7.10 and carried out on an Intel(R) core(TM) i5-3470 CPU and 8 GB memory PC.

Table 2 shows the best results of 10-unit system without considering wind power. From the Table 2, it is clear that U1 and U2 are always committed all day, especially, U1 is under its maximum power output. The operation cost is $563977.0 ($559887 + $4090), and 10 units all have been committed for satisfying the spinning reserve and power balance constraints.

Table 3 shows the best results of 10-unit system when wind power is connected to the power grid. Comparing Table 2 and Table 3, it can be seen that the states of U1, U2, U7 and U9 are not changed, but the working hours of the other six units decrease at different degrees when wind power is connected, and especially, U10 is turned off all day. From Table 3, when wind energy is considered, the operation cost is $531836.2 ($527076.2 + $4760). The fuel cost decreases by $32810.8 ($559887 - $527076.2), but the start-up cost increases $670 ($4760 - $4090). This is caused by the fluctuation of wind power. The wind power increases the spinning reserve of UCW system which makes it unnecessary for some thermal units to be turned on in order to satisfy the spinning reserve constraints, however, the closing duration for these units may be under the hot start-up time condition. Thus, it can be concluded that the uncertainty and fluctuation of wind energy may affect the operation of conventional thermal generation units, and the units will adjust themselves for adapting the variations. In order to avoid significant negative impact of wind power on grid operation, accuracy of wind prediction is necessary in system scheduling.

To investigate the searching ability of ACHS, the HS, GHS and IHS are also used to solve the UCW problem. Table 4 shows the results of these methods. From the Table 4, ACHS can find better solutions than HS, GHS, and HIS with less FES. For mean results of these methods, ACHS has high robustness for UCW problem.

6   Conclusion

This paper has evaluated the impact of wind power penetration on unit commitment by introducing a UCW model. For accurate forecasting of wind power, scenario generation and reduction techniques are applied to account for the uncertainty of wind farm output, and simulation results show the conformation of ±20 % error limited bounds using the proposed technique. Then, the mixed-integer, large-scale and non-linear UCW problem has been modeled and solved by ACHS methods. The simulations consider two different cases for a 10-unit system with and without wind energy. The results show that wind power penetration can decrease the operation cost of conventional thermal generation significantly, and improving the accuracy of wind forecast can be further beneficial to power scheduling. Comparing ACHS with other HS methods basic HS, GHS and IHS demonstrates that ACHS has shown its potential effectiveness and searching ability for UCW optimization problems.

Acknowledgments. This work is supported by the National Natural Science Foundation of China (61273040), Shanghai Rising-Star Program (12QA1401100), and the project of Shanghai Municipal Education Commission (12YZ020).

References

1. Padhy, N.P.: Unit Commitment – A Bibliographical Survey. IEEE Trans. Power Syst. 19, 1196–1205 (2004)
2. Ongsakul, W., Petcharaks, N.: Unit Commitment by Enhanced Adaptive Lagrangian Relaxation. IEEE Trans. Power Syst. 19, 620–628 (2004)
3. Prateek, K.S., Sharma, R.N.: Dnamic Programming Approach for Large Scale Unit Commitment Problem. In: 2011 International Conference on Communication Systems and Network Technologies, pp. 714–717. IEEE press (2011)
4. Damousis, I.G., Bakirtzis, A.G., Dokopoulos, P.S.: A Solution to the Unit-commitment Problem Using Integer-code Genetic Algorithm. IEEE Trans. Power Syst. 19, 577–585 (2004)
5. Yuan, X., Su, A., Nie, H., Yuan, Y., Wang, L.: Unit Commitment Problem Using Enhanced Particle Swarm Optimization Algorithm. Soft Comput. 15, 139–148 (2011)
6. Yuan, X., Su, A., Nie, H., Yuan, Y., Wang, L.: Application of Enhanced Discrete Differential Evolution Approach to Unit Commitment Problem. Energy Convers Manage. 50, 2449–2456 (2009)
7. Wang, J.H., Shahidehpour, M., Li, Z.Y.: Security-Constrained Unit Commitment With Volatile Wind Power Generation. IEEE Trans. Power Syst. 23, 1319–1327 (2008)
8. Tuohy, A., Meibom, P., Denny, E., Mark, O.M.: Unit Commitment for Systems With Significant Wind Penetration. IEEE Trans. Power Syst. 24, 592–601 (2009)
9. Roy, B., Bipul, K., Rajesh, K., Ramakrishna, G.: Unit Commitment Risk Analysis of Wind Integrated Power Systems. IEEE Trans. Power Syst. 24, 930–939 (2009)
10. Colm, L., Mark, O.M.: Impact of Wind Forecast Error Statistics Upon Unit Commitment. IEEE Trans. Sustain. Energy 3, 760–768 (2012)
11. Geem, Z.W., Kim, J.H., Loganathan, G.V.: A New Heuristic Optimization Algorithm: Harmony Search. Simulation 76, 60–68 (2001)
12. Amjady, N., Nasiri-Rad, H.: Nonconvex Economic Dispatch with AC Constraints by A New Real Coded Genetic Algorithm. IEEE Trans. Power Syst. 24, 1489–1502 (2009)
13. Niu, Q., Zhang, H.Y., Wang, X.H., Li, K., George, W.I.: A Hybrid Harmony Search with Arithmetic Crossover Operation for Economic Dispatch. Electrical Power and Energy Systems 62, 237–257 (2014)
14. Wu, L., Shahidehpour, M., Li, T.: A Computationally Efficient Mixed-Integer Linear Formulation for the Thermal Unit Commitment Problem. IEEE Trans. Power Syst. 22, 800–811 (2007)
15. Carrion, M., Arroyo, J.M.: Stochastic Security-Constrained Unit Commitment. IEEE Trans. Power Syst. 21, 1371–1378 (2006)
16. Mahdavi, M., Fesanghary, M., Damangir, E.: An Improved Harmony Search Algorithm for Sovling Optimization Problems. Appl. Math. Comput. J. 188, 1567–1579 (2007)
17. Omran, M.G.H., Mahdavi, M.: Global-best harmony search. Appl. Math. Comput. J. 198, 643–656 (2008)

Application of Information-Gap Decision Theory to Generation Asset Allocation

Yanan Zhao and Shaohua Zhang

Key Laboratory of Power Station Automation Technology,
School of Mechatronic Engineering and Automation,
Shanghai University, Shanghai 200072, China
eeshzhan@126.com

Abstract. In the deregulated electricity market, the generation company (GenCo) can sell electricity power through several trading choices such as bilateral contracts and the spot market. These trading choices have different risk characteristics. Especially, the risk faced by the GenCo in the spot market trading is extremely large. To seek the maximum profits and the minimum risk simultaneously, the GenCo should allocate its generation capacity among these trading choices reasonably. A risk management method based on the information-gap decision theory (IGDT) is proposed to evaluate different generation asset allocation strategies under serious uncertainty of spot market prices. An information-gap model is used to describe the volatility of spot market prices around the forecasted prices. Robustness of the decisions against low spot prices is evaluated using a robustness model and windfall higher profit due to unpredicted higher prices is modeled using an opportunity function. Numerical simulation is used to illustrate the proposed method.

Keywords: Electricity market, generation asset allocation, risk management, information-gap decision theory.

1   Introduction

With the restructuring and deregulation of electric power industry, generation companies (GenCos) become the main players in power markets, and they are provided with different trading choices to supply electricity, such as bilateral contracts and wholesale spot markets. These choices have different risk characteristics, especially, the risk faced by the GenCo in the spot market trading is extremely large due to the serious uncertainty in spot prices. As such, the GenCo should allocate its generation asset among these trading choices reasonably [1]. In addition, how to obtain the maximum profit while controlling the risk becomes the most important problem that a GenCo has to solve [2].

In general, price uncertainties are formulated by mathematical abstractions such as probability density functions or fuzzy logic membership [3]. Based on these formulations, risk management methods in electricity markets are extensively investigated. The scenario-based method is one of the risk management techniques and it focuses on finding optimal strategies based on a limited number of possible prices.

K. Li et al. (Eds.): LSMS/ICSEE 2014, Part III, CCIS 463, pp. 398–408, 2014.

The mean-variance method is also widely used for risk management, but it depends on the normal distribution of prices, which is not satisfied when the electricity supply is tight [4]. The conditional value at risk (CVaR) is a more conservative index compared to the value at risk (VaR), since it measures the average loss exceeding the VaR value for a given confidence level. However, the CVaR approach also requires some assumptions about the uncertainty, such as its probability density function [5], [6]. Lack of information brings many challenges to the generation assets allocation.

The information-gap decision theory (IGDT), which has been developed by Ben-Haim (2001) [7], is a newly developed alternative for decision-making under uncertainty that involves no measure functions—neither probabilistic density nor fuzzy membership functions. The IGDT method focuses on the disparity between what is known and what could be known. The choice of uncertainty parameters is based on a maximizing robustness or a minimizing opportunity rule. The IGDT-based models are recently applied to various decision-making processes under uncertainty. Hybrid IGDT-probability method is used for consumer demand construction in [8]. Electricity procurements of large consumers using the IGDT method are discussed in [9], [10], where bilateral contracts, pool markets, and self-generating facilities are considered as the resources of electricity. In [11], IGDT based robust decision making tool for distribution network operators in load procurement is proposed. Application of IGDT to self-scheduling of GenCos is addressed in [12], but the generation assets allocation between forward and spot markets is not considered.

This paper develops a non-probabilistic IGDT-based generation assets allocation method for a GenCo. Bilateral contracts and the spot market are considered as trading choices for the GenCo. The proposed method does not maximize the profit but assesses the risk aversion or risk-seeking nature of some allocation strategies with regard to the maximum profit. Numerical simulation is presented to illustrate the reasonableness and effectiveness of the method.

2   Information Gap Decision Theory

The Information Gap Decision Theory (IGDT) is a method to describe the uncertainties which can not be described using PDF (Probability Distribution Function) or MF (Membership Function) due to the lack of sufficient information. IGDT models the errors between the actual and forecasted parameters [13]. It is based on quantitative models and provides numerical decision-support assessments. Using this method, the decision maker can recognize priorities, evaluate risks and opportunities, and make more informed decisions ultimately [14].

Consider a typical optimization function as follows:

$$y = \max_{d} B(X,d) \ . \tag{1}$$

$$H(X,d) = 0 \ . \tag{2}$$

$$G(X,d) \geq 0 \ . \tag{3}$$

where X is the vector of input parameters (which are subject to severe uncertainty) and d is the vector of decision variables. H and G are the equality and inequality constraints respectively. $B(X,d)$ describes the relations between the decision variables (d) and input uncertain parameters (X).

The uncertainty of parameters in IGDT method is usually defined using the envelope bound model, as follows:

$$X \in U(\alpha, \tilde{X}) \ . \tag{4}$$

$$U(\alpha, \tilde{X}) = \left| \frac{X - \tilde{X}}{\tilde{X}} \right| \le \alpha \ . \tag{5}$$

where α is the uncertainty level of parameter X, $\tilde{X}$ is the forecasted value of X and $U(\alpha, \tilde{X})$ is the set of all values of X whose deviation from $\tilde{X}$ will never be more than $\alpha\tilde{X}$. The decision maker does not know the values of X and α. This model describes that the actual value X fluctuates around its forecasted value $\tilde{X}$.

For better clarification, assume that the function B describes the system model (e.g. set of constraints describing generation allocation to different markets), X is a vector of input uncertain parameters to the system which are subject to severe uncertainty (e.g. electricity price without any historic data) and y is the output variable (e.g. total profit for selling generation). d denotes the set of decision variables (e.g. amount of allocated energy to different trading markets like spot market and bilateral contracts).

In case the uncertain input parameters X are equal to their predicted values ($X = \tilde{X}$) then solving the (1) to (3) gives the predicted value of $y = \tilde{y}$. However, if the value of X is unknown then the IGDT method tries to find a solution for the problem which is robust or opportunistic against the error in predicting the value of X. Two different performance functions could be defined based on the risk management strategy of the decision maker.

A risk-averse decision maker desires to allocate in a way to be immune against effects of unfavorable deviations of the uncertain parameter from the corresponding forecast value. The robustness function $\hat{\alpha}(B_c)$ expresses the greatest level of uncertainty at which the minimum generation profit cannot be lower than a given cost level B_c, where $B_c = (1 - \sigma) \times \tilde{y}$, σ is the degree that decision maker tolerates the deterioration of objective function due to forecasting error of input parameter X. In other words, this function describes the risk-aversion potential of a bidding strategy. Therefore, we can define it through an optimization problem:

$$\hat{\alpha}(B_c) = \max_d \alpha \ . \tag{6}$$

s.t.

$$\min B(X,d) \ge B_c \ . \tag{7}$$

$$\forall X \in U(\alpha, \tilde{X}) \ . \tag{8}$$

$$H(X,d) = 0 \ . \tag{9}$$

$$G(X,d) \ge 0 \ . \tag{10}$$

An opportunity function models the possible high profits for the risk-seeker decision maker. The opportunity function $\hat{\beta}(B_w)$ is the lowest value of α such that the maximum profit could potentially be as great as a given cost level B_w, $B_w = (1+\sigma) \times \tilde{y}$, as follows:

$$\hat{\beta}(B_w) = \min_d \alpha \ . \tag{11}$$

s.t.

$$\max B(X,d) \geq B_w \ . \tag{12}$$

$$(8)\text{-}(10) \ . \tag{13}$$

Compared with other traditional risk management methods, the advantages of the IGDT are: ① the risk factor is not initially determined. ② the proposed method provides a robust strategy to avoid low profits or to take advantage of high profits. ③ the method has an actual advantage to find the optimum strategy for any profit levels, not for a given price scenario. Thus the IGDT method is especially applicable for the decision making in the condition that the information is seriously lacking.

3   Theoretical Model

We propose an IGDT-based generation assets allocation formulation for a GenCo. The GenCo can sell electricity through bilateral contracts and the spot market. The decision variable q is the amount of electricity allocation in each market, $q = [p_{s,t}, p_{l,t}]$. The specifications of bilateral contracts are known. The forecast values of future spot prices, denoted by $\tilde{\lambda}_{s,t}$, are assumed to be available. The spot price $\lambda_{s,t}$ is uncertain and it can be expressed as an information-gap model:

$$U(\alpha, \tilde{\lambda}_{s,t}) = \{\lambda_{s,t} : \left| \frac{\lambda_{s,t} - \tilde{\lambda}_{s,t}}{\tilde{\lambda}_{s,t}} \right| \leq \alpha\}, \alpha \geq 0 \ . \tag{14}$$

According to the above description, the revenue of the GenCo is calculated as:

$$B(\lambda_s, q) = \sum_{t=1}^{T} \left\{ \sum_{l=1}^{L} \lambda_{l,t} \cdot p_{l,t} + \lambda_{s,t} \cdot p_{s,t} - C(p_t) \right\} \ . \tag{15}$$

where T is the number of time intervals. L is the number of bilateral contracts. $p_{s,t}$, $\lambda_{l,t}$ are the spot trading generation quantity and prices respectively. $p_{l,t}$, $\lambda_{l,t}$ are the trading quantity and price of contract l. p_t is the total generation quantity at time t, which is stated as follows:

$$\sum_{l=1}^{L} p_{l,t} + p_{s,t} = p_t \ , t = 1, 2, \cdots, T \ . \tag{16}$$

$C(p_t)$ is the fuel cost, which is defined as follows:

$$C(p_t) = a p_t^2 + b p_t + c \ . \tag{17}$$

where a, b, c are cost parameters for the GenCo.

The capacity limits of generation units can be stated as follows:

$$p^{\min} \le p_t \le p^{\max} \ . \tag{18}$$

where $p^{\min}, p^{\max}$ are the minimum and maximum outputs of the GenCo, respectively.

The amount of electricity allocation to the bilateral contracts must be between the allowable bounds, which is defined as follows:

$$p_{l,t}^{\min} s_l \le p_{l,t} \le p_{l,t}^{\max} s_l \ . \tag{19}$$

where $p_{l,t}^{\min}, p_{l,t}^{\max}$ are the minimum and maximum power pertaining to contract l at time t. s_l is a binary variable, which is equal to 1 if bilateral contract l is selected, and 0 otherwise.

3.1   Robust Generation Assets Allocation Model

The risk-averse GenCo desires to allocate in a way to be immune against losses or low profit due to unfavorable deviations of spot prices from the forecast values. In this condition the IGDT model's objective is to maximize the uncertainty parameter while the required performance is satisfied:

$$\hat{\alpha}(B_c) = \max_q \alpha \ . \tag{20}$$

s.t.

$$\min \sum_{t=1}^{T} \left[\sum_{l=1}^{L} \lambda_{l,t} \cdot p_{l,t} + \lambda_{s,t} \cdot p_{s,t} - C(p_t) \right] \ge B_c \ . \tag{21}$$

$$\lambda_{s,t} \ge (1 - \alpha)\tilde{\lambda}_{s,t} \ . \tag{22}$$

$$\lambda_{s,t} \le (1 + \alpha)\tilde{\lambda}_{s,t} \ . \tag{23}$$

$$(16) - (19) \ . \tag{24}$$

Given the uncertainty α, it is readily seen that the minimum profit in Eq.(15) occurs in the lowest price allowed by the IGDT, which is equal to $(1 - \alpha)\tilde{\lambda}_{s,t}$. Based on this analysis, the optimization problem can be simplified.

The solution of the above optimization problem will give the robust generation allocation based on the defined value of B_c. In other words, the solution guarantees a minimum profit of B_c if all hourly absolute relative forecast errors are less than $\hat{\alpha}(B_c)$. Note that $\hat{\alpha}(B_c)$ is the maximized tolerable fluctuation.

3.2   Opportunistic Generation Assets Allocation Model

A risk-taking GenCo desires to benefit from prices favorable variations using an opportunity function. Following Eq.(15), the mathematical formulation of the opportunity function in generation allocation problem can be expressed as follows:

$$\hat{\beta}(B_w) = \min_q \alpha \ . \tag{25}$$

s.t.

$$\max \sum_{t=1}^{T}\left[\sum_{l=1}^{L}\lambda_{l,\mathfrak{t}}\cdot p_{l,\mathfrak{t}} + \lambda_{s,\mathfrak{t}}\cdot p_{s,\mathfrak{t}} - C(p_t)\right] \geq B_w \ . \tag{26}$$

$$\lambda_{s,t} \geq (1-\alpha)\tilde{\lambda}_{s,t} \ . \tag{27}$$

$$\lambda_{s,t} \leq (1+\alpha)\tilde{\lambda}_{s,t} \ . \tag{28}$$

$$(16) - (19) \ . \tag{29}$$

Given the uncertainty α, it is readily seen that the maximum profit occurs in the highest spot prices allowed by the IGDT model, which is equal to $(1+\alpha)\tilde{\lambda}_{s,t}$. Hence, the opportunity model can be simplified to single objective optimization problem. Thus if future prices deviate from the forecasted prices by $\hat{\beta}(B_w)$, a greater profit of B_w may be achieved. Note that $\hat{\beta}(B_w)$ is the minimum required favorable price variations that make B_w achievable.

The generation asset allocation problem is formulated as a mixed-inter nonlinear problem and can be solved using SBB/CONOPT under GAMS [15].

4   Numerical Simulation

Numerical testing results are presented in this section. The historical locational marginal prices (LMP) for the PJM Western Hub depicted in Fig.1 are used as the expectation of the forecast spot market prices. The time horizon contains 24 periods (one day). A GenCo is assumed to have cost parameters of $a=\$0.00048/(MW)^2h$, $b=\$16.19/MWh$, $c=\$1000.0$. Its generation output is from 150MW to 455MW. In this study, 5 bilateral contracts are considered. Table 1 shows the bilateral contracts' specifications.

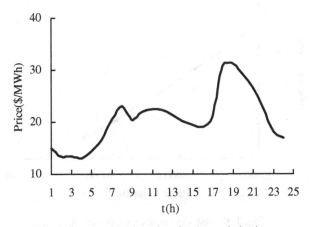

Fig. 1. Forecasted prices data for the study horizon

Table 1. Bilateral contracts' specification

Contract number	Min(MW)	Max(MW)	Price($/MWh)
1	25	40	23.1
2	20	45	22.4
3	35	65	21.7
4	40	60	20.8
5	45	80	19.6

We first solve the deterministic original next-day generation allocation problem based on the forecasted spot prices. The expected maximum profit is equal to $33982, which is the result of maximizing Eq.(15) with the estimate of spot price values. In this case, the GenCo should allocate 45.1% of its generation to the bilateral contracts, and 54.9% to the spot market. From an uncertainty perspective, we can say that $\hat{\alpha}(\$33982) = \hat{\beta}(\$33982) = 0$. The results related to the robustness function will be useful if the GenCo wants a risk-averse strategy and the opportunity function-related strategies can be used when the GenCo chooses a risk-taking decision making model.

4.1   Robust Generation Assets Allocation

We then solve the optimization problem Eq. (20)–Eq. (24) for different critical profits of B_c. The robustness function is depicted in Fig.2, where it is clear that robustness decreases with B_c. If the consumer desires high robustness it will obtain a low profit. Inversely, if the GenCo obtains a low profit this indicates its allocation strategy is highly robust and more risk-averse. Fig.3 shows that the generation percentage of bilateral contracts decreases with B_c, while percentage of spot market increase with B_c. It indicates when the GenCo chooses a more robust allocation strategy, it is better to allocate more generation to the market whose prices have no uncertainty.

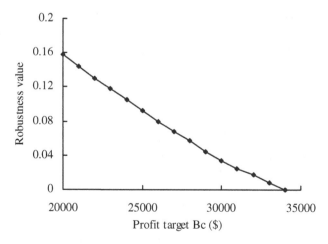

Fig. 2. Robustness curve as a function of B_c

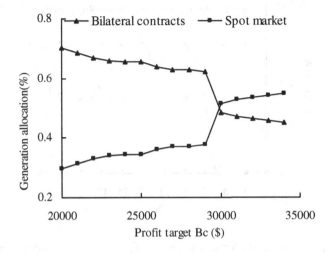

Fig. 3. Robust allocation as a function of B_c

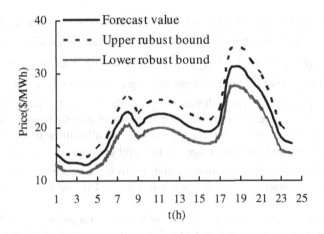

Fig. 4. The robustness region for B_c =$23000 for a 24-hour period

For example, a critical profit of B_c =$23000 is guaranteed if none of the hourly errors are more than $\hat{\alpha}(B_c)$ =0.118 or 11.8%. The robustness region for this particular example is depicted in Fig.4. In other words, if after-the-fact actual prices fall within this range, the profit gained by the GenCo will be at least $23000. The generation allocation strategy for $\hat{\alpha}(B_c)$ =11.8% is shown in Fig.5. We can see that bilateral contracts undertake the main task of guaranteeing the robustness of generation allocation strategies.

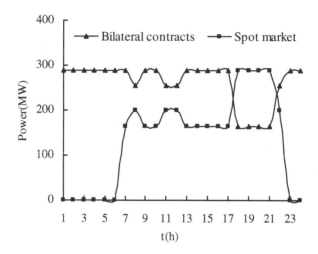

Fig. 5. Generation allocation strategy for B_c =$23000 for a 24-hour period

4.2   Opportunistic Generation Allocation

Variations of the resulting opportunity parameter versus the desired target profits are presented in Fig.6. The opportunity value $\hat{\beta}(B_w)$ indicates that if the spot prices are equal to or more than $100*(1+\hat{\beta}(B_w))$ % of the best estimated values, a greater profit of B_w may be achieved. Observe from this figure that higher desired target profit requires higher favorable price deviations from the forecast values, i.e., higher or more frequently positive price spikes. The increase in benefit means corresponds to increased risk. Fig.7 show that the percentage of allocation to the spot market increases with B_w and the percentages of bilateral contracts decrease with B_w. This is a logical behavior because choosing a high profit level means deciding to take high risk. Thus, a comparatively higher portion is allocated to the market with uncertain prices.

For example, we assume the GenCo decides to reach a profit of $42000, which is 23.5% higher than the expected profit ($33982), the observed prices must be at least 5.9% higher than the forecast prices. This means that if spot prices increase by 5.9% or more, the benefit may achieve $42000. At this moment, the GenCo should sell30.1% of the generation through bilateral contracts, 69.9% of generation through spot market.

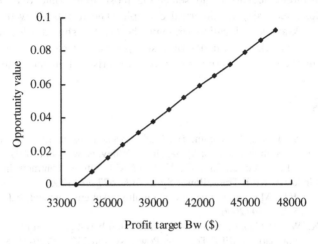

Fig. 6. Opportunity function as a function of B_w

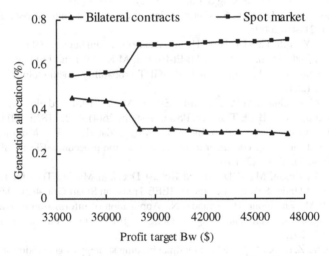

Fig. 7. Opportunistic allocation as a function of B_w

5   Conclusions

In this paper, a novel IGDT-based method for generation asset allocation of generation companies under uncertain spot market prices is proposed. For a risk-averse GenCo, the proposed robust model guarantees a minimum critical profit if the spot prices fall within a maximized robustness region. For a risk-seeker GenCo, the proposed opportunistic model enables the GenCo to benefit from unpredictable high price spikes and potentially gain a target profit. This method provides a wide range of information regarding to different decisions. Accordingly, the GenCo can evaluate a number of different decisions and select the most appropriate one. The proposed method is illustrated using a numerical example. The results show that in order to have a more risk-averse allocation strategy, the GenCo should allocate most of its generation to the markets with deterministic prices. Inversely, more allocation of electricity to markets with uncertain prices results in risk-taking strategies.

References

1. Shahidehpour, M., Li, Z., Yamin, H.: Market Operations in Electric Power Systems: Forecasting, Scheduling, and Risk Management. Wiley, New York (2002)
2. Tanlapco, E., Lawarrée, J., Liu, C.: Hedging with futures contracts in a deregulated electricity industry. IEEE Trans. Power Syst. 17(3), 577–582 (2002)
3. Liu, M., Wu, F.F.: Managing price risk in a multimarket environment. IEEE Trans. Power Syst. 21(4), 1512–1519 (2006)
4. Guan, X.X., Wu, J., Gao, F., Sun, G.J.: Optimization-based generation asset allocation for forward and spot markets. IEEE Trans. on Power System 23(4), 1796–1818 (2008)
5. Alexander, S., Coleman, T.F., Li, Y.: Minimizing CVaR and VaR for a portfolio of derivatives. Journal of Banking & Finance 30(2), 583–605 (2006)
6. Rockafellar, R.T., Uryasev, S.: Optimization of conditional value-at-risk. The Journal of Risk 2(3), 21–41 (2000)
7. Ben-Haim, Y.: Info-Gap Decision Theory. Academic, SanDiego (2001)
8. Zare, K., Moghaddam, M.P., Sheikh-El-Eslami, M.K.: Demand bidding construction for a large consumer through a hybrid IGDT-probability methodology. Energy 35(7), 2999–3007 (2010)
9. Zare, K., Moghaddam, M.P., Sheikh-El-Eslami, M.K.: Risk-based electricity procurement for large consumers. IEEE Trans. on Power Systems 26(4), 1826–1835 (2011)
10. Zare, K., Conejo, A.J., Carriónc, M., Moghaddamd, M.P.: Multi-market energy procurement for a large consumer using a risk-aversion procedure. Electric Power Systems Research 80(1), 63–67 (2010)
11. Soroudi, A., Ehsan, M.: IGDT Based Robust Decision Making Tool for DNOs in Load Procurement Under Severe Uncertainty. IEEE Trans. on Smart Grid 4(2), 886–895 (2013)
12. Ivatloo, B.M., Zareipour, H., Amjady, N.: Application of information gap decision theory to risk-constrained self-scheduling of GenCos. IEEE Trans. on Power Systems 28(2), 1093–1102 (2013)
13. Nojavan, S., Zare, K., Feyzi, M.R.: Optimal bidding strategy of generation station in power market using information gap decision theory (IGDT). Electric Power Systems Research 96, 56–63 (2013)
14. Soroudi, A., Amraee, T.: Decision making under uncertainty in energy systems: state of the art. Renewable and Sustainable Energy Reviews 28, 376–384 (2013)
15. Brooke, A., Kendrick, D., Meeraus, A., Raman, R.: GAMS: A User's Guide. GAMS Development Corp., Washington, DC (1998)

Vision-Based Lane Detection Algorithm
in Urban Traffic Scenes

Feng Ran[1], Zhoulong Jiang[2], and Meihua Xu[2]

[1] Microelectronic Research and Development Center, Shanghai University, Shanghai, China
ranfeng@shu.edu.cn
[2] School of Mechatronics Engineering and Automation, Shanghai University, Shanghai, China
{jzl,mhxu}@shu.edu.cn

Abstract. Lane departure warning system plays an important role in driver assistance systems. The proposed algorithm assumes that lanes are always the straight lines and whole algorithm is based on Hough transform. Due to the complexity of urban traffic scenes, false lane detections are highly caused by warning lines and signs whose shapes and colors are similar to the lane boundary. In this study, we improve the accuracy of the lane detection base on Hough, a score function based on the width between left and right lanes is proposed to obtain reliable lane detect results on urban traffic scene. Meanwhile, a list of candidate lanes is constructed at the least of execution time. Experiments under various scenes showed that the proposed lane detection method can work robustly in the real-time.

Keywords: Lane detection, Intelligent vehicle, Machine vision, driving assistance.

1   Introduction

Driving is a very common activity in our daily life. It is extremely enjoyable until we face a nasty situation, such as traffic-violation, flat-tire, accidents etc. Accidents are the most vital situations and cause a great loss to human lives and assets. Therefore, it is necessary to investigate a driver assistance which can remind the driver of danger when needed. Advanced driver assistance system (ADAS) can play a positive role in improving driver's awareness and hence performance by providing relevant information when needed.

Lane Departure Warning (LDW) system is a typical application in the field of ADAS. LDWS helps prevent the driver from unintended lane departure which is caused by driver's fatigue, drowsiness and improper driving maneuver. Most of LDWS are based on the vision system, processing images captured by a camera attached in the front of a car. Since erroneous findings will generate wrong steering commands which may jeopardize vehicle safety, a robust and reliable algorithm is a minimum requirement. Many approaches have been applied to lane detection which can be categorized in three main classes: feature-based[1-3], region-based[4-5] and model-based[6]; Yifei Wang[7] presented a lane-detection and tracking system based

K. Li et al. (Eds.): LSMS/ICSEE 2014, Part III, CCIS 463, pp. 409–419, 2014.

on a novel feature extraction approach and Gaussian Sum Particle filter(GSPF). It is able to improve most of existing feature maps by removing the irrelevant feature points produced by unwanted objects in the scene. The GOLD[8] (Generic Obstacle and Lane Detection system) developed by Massimo Bertozzi remapped each pixel of the image towards a different position. The resulting image represented a top view of the road region in front of the vehicle, as it was observed from sky. With this image, it could detect the parallel lane boundaries. Chris Kreucher[9] proposed a method which is based on a novel set of frequency domain features that capture relevant information concerning the strength and orientation of spatial edges.

In previous studies, few studies focused on the urban traffic road scenes due to its complexity.Complicated lane marks in the urban traffic scenes easily cause the lane fault detections with the studies focused only on the highway scenes. In muti-lane marked roads, lane mark detection based on Hough Transform alone can be easily mistaken. Traffic signs and warning lines are hard to be distinguished from the lane boundaries.

To reduce the misrecognize rate in these cases, we proposed a discrimination algorithm based on the width between left and right lanes. This algorithm achieves a high reliablity with small computation complextity.

2   Lane Marks Detection

2.1   Image Pre-processing

CMOS camera is fixed on the front-view mirror to capture the road scene. In this paper, it was assumed that the input to the algorithm was a 720×480 RGB color image. Therefore the first thing the algorithm to do is to convert the image to a grayscale image, in order to minimize the processing time.

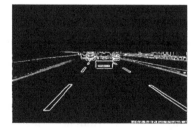

(a) Image after dividing   (b) Edge detection result

Fig. 1. Image Pre-processing result

A.   Setting ROI
Road always located in the low part of the image. To speed up the image processing, we take the low part as region of interest, results can be found in Fig.1.(a)
B.   Edge Extraction
Due to the sharp contrast between the road surface and painted lanes, edge detection can be very useful to determine the location of boundaries.

The edge of image $f(x, y)$ at the point $d(x, y)$ can be represented as the follows:

$$\nabla f = [G_x \ G_y] = [\partial f / \partial x \ \ \partial f / \partial y]^T \tag{1}$$

Where G_x and G_y are approximated by 3x3 Sobel mask:

$$G_x = f(x+1, y-1) + 2f(x+1, y) + f(x+1, y+1)$$
$$- f(x-1, y-1) - 2f(x-1, y) - f(x-1, y+1) \tag{2}$$

$$G_y = f(x-1, y+1) + 2f(x, y+1) + f(x+1, y+1)$$
$$- f(x-1, y-1) - 2f(x, y-1) - f(x+1, y-1) \tag{3}$$

Magnitude $\nabla f(x, y)$ is represented in (4).

$$\nabla f(x, y) = \sqrt{G_x^2 + G_y^2} \approx |G_x| + |G_y| \tag{4}$$

Then, an edge image can be extracted by magnitude estimation described in (5).where T is aquired from the average of $\nabla f(x, y)$.

$$g(x, y) = \begin{cases} 1, & \nabla f(x, y) > T \\ 0, & \nabla f(x, y) < T \end{cases} \tag{5}$$

Edge detection result is shown in Fig.1.(b).

2.2   Hough Transform

Lane detector used is a standard Hough transform with a restrict search space. Hough transform is used to get slopes and intercepts of candidate boundaries. If there is a straight line y=kx+b, its parameter space polar coordinate lines can be expressed as the following equation:

$$\rho = x \cos \theta + y \sin \theta \tag{6}$$

Where x and y are the coordinate value of a pixel in image, ρ is the distance from the origin to the fitted line, θ is the angle between the normal line and x-axis. As illustrated in Fig.2 (a), parameters ρ and θ are showed by (ρ_L, θ_L) and (ρ_R, θ_R). Equation (6) is applied to all pixels of edge points obtained from edge detection. As a result, each point (x, y) in the image are mapped into a sinusoid in parameter space and we can get two accumulate arrays, $H_L(\rho_L, \theta_L)$ and $H_R(\rho_R, \theta_R)$. All elements in accumulators are initialized to zero and we set the increments of ρ and θ to 1 and 1°, respectively. Considering the likelihood of locations of the road, we reduced the search range of θ and increased the search efficiently. Yu Tianhong[11] proposed that angles between the normal line and x-axis usually falls in the range of [15°,75°], so in this paper, we set the range of θ_L and θ_R to [15°, 75°] and [105°, 165°], respectively.

In order to improve the precision of ρ_L, ρ_R, θ_L and θ_R, these parameters are defined as the average of the corresponding parameters of inner and outer boundaries. They are shown by $(\rho_{L_{in}}, \theta_{L_{in}})$, $(\rho_{L_{out}}, \theta_{L_{out}})$, $(\rho_{R_{in}}, \theta_{R_{in}})$ and $(\rho_{R_{out}}, \theta_{R_{out}})$.

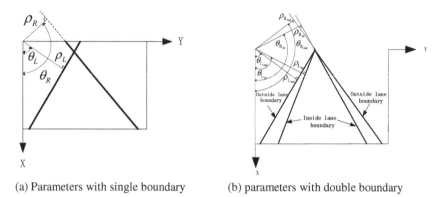

(a) Parameters with single boundary   (b) parameters with double boundary

Fig. 2. Image coordinates

If we find one θ of inner or outer boundary, we can find another boundary near the founded boundary. For example, if we find $\theta_{L_{in}}$, then we can seek $\theta_{L_{out}}$ in $[\theta_{L_{in}} - 5, \theta_{L_{in}} + 5]$ to reduce the searching time and improve the reliability. Parameters of inner and outer boundaries can be found in Fig.2 (b).

In general, we may often encounter with such scenarios: more than one boundary exist in a captured image and it may cause error identification of current lanes, as shown in Fig.3 (a). This image was captured when the vehicle was changing lanes. To solve this problem, we defined a score function to get the correct lanes.

2.3   Lane Boundary Matching

Lane boundary pairs from candidate boundaries, as shown in Fig.3 (b).

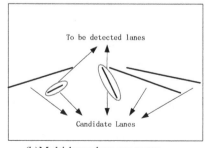

(a) Multi-boundary case   (b)Multi-boundary geometry

Fig. 3. Image coordinates

Firstly, if we want to get the correct lane boundary pair, a list of candidate boundaries should be constructed. In this paper, we introduced a method which can search with limited number of candidate boundaries at the least of time. Suppose we have got two accumulate arrays which contain the voting values of left and right boundaries. Voting values are acquired by Hough transform and the boundary with the largest voting value has a good possibility of the being correct lane. Based on prior knowledge and large number of experiments, we can know that if a lane has larger voting value and larger angle between the lane and y-axis, it will have a better possibility of being the correct lane. As shown in Fig.3 (a), it describes a left half-plane of an image and L1, L2, L3, L4 have the top four voting values. Based on prior knowledge and experiments, we came to a conclusion that if L2 has a less voting value and a smaller angel between normal line and x-axis than L1, then L2 is almost impossible to be the correct lane. Similarly, L3 cannot be the correct lane. Although L4 has the smallest voting value among these lines, it has the largest angle between normal line and x-axis and might be the correct lane, so it will be chosen as candidate boundaries. Result of candidate boundaries searched is showed in Fig.4 (b). It usually occurs at these scenarios: car is changing lanes; current lane is a dashed line or being severely occluded, and as a result, voting value of the current lane pair is not large enough.

In order to make a further reduction of the time of searching candidate lanes, we presented searching steps as follows. Firstly, we will find lines which have the maximum voting values in each array, taking them as the candidate boundaries of left and right lanes. Then we can obtain their parameters θ and ρ, named as (θ_{L1}, ρ_{L1}) and (θ_{R1}, ρ_{R1}), respectively. Secondly, we will find lines which have the second largest voting values in each array in the angle range from $(90^o - \theta_{L1})$ to 90° and to θ_{R1}. In considering that another boundary of the same lane lies in $[\theta - 3^o, \theta + 3^o]$, that is, in this paper, we consider the angle difference between inside boundary and outside boundary of a lane is within 3°. So, we search the second candidate boundaries in $[90^o - (\theta_{L1} + 3^o), 90^o]$ and $[90^o, \theta_{R1} - 3^o]$, respectively. Their parameters θ and ρ are named as (θ_{L2}, ρ_{L2}) and (θ_{R2}, ρ_{R2}). Then we will update the searching space like $[90^o - (\theta_{L2} + 3^o), 90^o]$ and $[90^o, \theta_{R2} - 3^o]$. Above steps will be repeated to obtain each candidate boundary's parameter until the voting value is lower than a certain value or their angle have exceeded the given range.

After obtaining candidate boundaries, we will choose the best boundary pair from candidate boundaries as the left and right lanes. Suppose we have got n+1 candidate boundaries of left and m+1 candidate boundaries of right, and Fig.5 defines the width between left and right candidate boundaries. Each left and right candidate lane boudary are matched as a candidate lane boundary pair. Equation (7) to (9) represent the lane boundary matching criterion, where W_{ij} is the sum of lane boundaries's width which obtained from (7),and i, j are the candidates of left and right lane boundaries

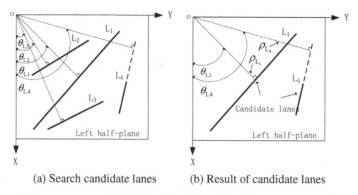

(a) Search candidate lanes   (b) Result of candidate lanes

Fig. 4. Candidate Lanes Search

respectively. d_{ij} is the lane width difference between current width w_{ij} and updated width $\hat{w}_{ij}$, and V_{ij} is the sum of voting values of left and right boundaries.

$$W_{ij} = w_{ij}[0] + w_{ij}[1] + w_{ij}[2] + w_{ij}[3], i = 1, 2, ...n, j = 1, 2, ...m \qquad (7)$$

$$d_{ij}[k] = \left| w_{ij}[k] - \hat{w}_{ij}[k] \right| \qquad (8)$$

$$V_{ij} = v_{Li} + v_{Rj} \qquad (9)$$

We get the correct lane pair based on the following rules:lane pair with the minimum lane width has a good possibility of being the correct pair. Lane width difference between updated and current width should be small for the correct lane boundary pair because of similar lane width in image squence.The voting value of a lane boundary pair obtained in (10) should be the largest one if this pair is the correct lane boundary. Yu-Chi Leng [12] proposed a score function ,as defined in (10) ,are to find small lane width ,small lane width difference and large sum of voting value.

$$S(i, j) = \sum_{k=0}^{3} d_{ij}[k] + \beta_1 \bullet W_{ij} - \beta_2 \bullet V_{ij} \qquad (10)$$

Where β_1 and β_2 are the weighting factors, and the lane boundary pair which has the smallest scores value would be selected as the detected lane boundary. Though it did improve the recognition rate to a certain extant, it has the problem of great order differences of magnitude among $d_{ij}[k]$, W_{ij} and V_{ij}. As a result, it will bring some problems. In a certain structured road, lane width changes little, but voting value will be changeful for different road conditions. For example, in a certain road with solid lanes, voting value of the correct lane ranges from 80 to 120 while of noise ranges from 0 to 20. In a road with dashed lanes, voting value of the correct lanes ranges 30 to 45 and noise still ranges 0 to 20. Since the weighting factors are constants, when voting value declines with road conditions, the percentage of width factor will be

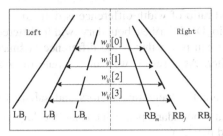

Fig. 5. Genometry of lane boundary position

increased. That is, it may cause error detection when the width of wrong lane boundary pair is smaller than the correct one, while the difference between the voting value of the correct pair and of the wrong is not large enough. As a result, lane width plays a decisive role, that is, whose lane width is smaller, it will be chosen as the correct pair. In this paper, we presented a method to compute the possibility of the candidate boundary pair with more efficiency and with smaller computing. We defined the ratio of boundary width between LB_i and RB_i as follows:

$$ratio\_W_{ij} = \frac{W_{ij}}{sum\_W} \tag{11}$$

Where W_{ij} is obtained from (8), and sum_W can be obtained as follows:

$$Max\_W = \max\{W_{00},W_{01},...,W_{ij},...,W_{nm}\} \tag{12}$$

$$Min\_W = \min\{W_{00},W_{01},...,W_{ij},...,W_{nm}\} \tag{13}$$

$$sum\_W = Min\_W + Max\_W \tag{14}$$

As we can see, ratio_W_{ij} is an increasing fuction and it indicates the less possibility of being the correct lane boundary pair from the width factor, that is ,whose width is larger, its possibility of being the correct pair is smaller, and the value is maintained at a range from 0 to 1.Then we set the ratio of voting values as follows:

$$ratio\_V_{ij} = \frac{V_{ij}}{sum\_V} \tag{15}$$

Where V_{ij} is obtained from (9), and sum_V can be obtained as the follows:

$$Min\_V = \min\{V_{00},V_{01},...,V_{ij},...,V_{nm}\} \tag{16}$$

$$Max\_V = \max\{V_{00},V_{01},...,V_{ij},...,V_{nm}\} \tag{17}$$

$$sum\_V = Min\_V + Max\_V \tag{18}$$

As we can see, ratio_V_{ij} is also an increasing fuction. When voting value declines, due to the current lanes are dashed lines and servely occluded or the change of viewing angle caused by changing lanes, ratio_V_{ij}still mantians a similar value while

MAX_V declines. And ratio of width difference is defined as follows. Where W_{ij} is obtained from(7). ratio_D_{ij} describes boundary width difference between updated and current width. We can prove that if the difference is smaller, then the value of ratio_D_{ij} will be smaller. As a result, we can compute the possibility of candidate boundaries as follows:

Where α, β and γ are the weighting factors and satisfy the identity(21).Lastly, (22) shows that the lane boundary pair which has the largest possibility would be selected as the detected boundary.

$$ratio\_D_{ij} = \frac{\sum_{k=0}^{3}\left|w_{ij}[k] - \hat{w}_{ij}[k]\right|}{\sum_{k=0}^{3}(w_{ij}[k] + \hat{w}_{ij}[k])} \approx \frac{\left|W_{ij} - \hat{W}_{ij}\right|}{W_{ij} + \hat{W}_{ij}} \tag{19}$$

$$P(i, j) = \alpha \bullet ratio\_V_{ij} - \beta \bullet ratio\_W_{ij} - r \bullet ratio\_D_{ij} \tag{20}$$

$$\alpha + \beta + \gamma = 1 \tag{21}$$

$$(\hat{i}, \hat{j}) = \arg \max_{i,j} P(i, j) \tag{22}$$

Where $\hat{i}$ is the extracted left lane boundary, and $\hat{j}$ is the extracted right lane boundary.As shown in Fig.6. (a), it shows the results after Hough transform and parts of fitting straight-lines are obvious wrong. Due to the use of prior knowledge, we can get the correct boundary pair from candidate lanes, as shown in Fig.6. (b).

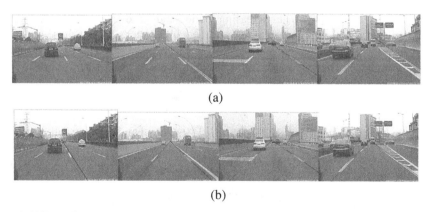

(a)

(b)

Fig. 6. Effect of proposed mechanism: (a) Results after Hough transform (b) Results of proposed method

3   Experiment Result

In this paper, our proposed algorithm is developed and tested on the PC-based experimental platform with an Intel Pentium IV 2.2GHz CPU inside, using the C programming language. Due to the pure C programming language, it can be easily transplanted to embedded systems. Size of captured images is 720×480 pixels and the frame frequency is 30fps. Experiment results showed that the average image processing time was about 30 ms/frame, which could meet real-time and stability request of the vision navigation system. The images in Figure 7show the results of our method under various conditions,such as (a) ~ (d) bright sunny day,(e) ~(h) cloudy day with printing on road surface, (i) ~(l) night with different intensity of light. Test results on these vedio clips are shown in Table 1. 33292 frames from these video clips are used for testing.Here "Correct Detection" means both left and right lane parameters are correctly detected. If one parameter is not correctly detected, the the result is reported in "False Detection".

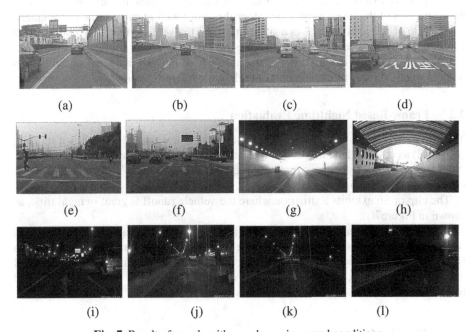

(a)   (b)   (c)   (d)

(e)   (f)   (g)   (h)

(i)   (j)   (k)   (l)

Fig. 7. Result of our algorithm under various road conditions

Table 1. Test Results on video clip

Day/Night	Scene	Frames	Correct Detection	False Detection	Correct Rate
Day	S1	4387	4074	313	92.87%
	S2	3657	3376	281	92.32%
	S3	10259	10063	196	98.09%
	Total	18303	17513	790	95.68%
Night	S4	5377	4834	543	89.90%
	S5	5368	5210	158	97.06%
	S6	4244	3891	353	91.68%
	Total	14989	13935	1054	92.97%
Day & Night	Total	33292	31448	1844	94.46%

3.1   Frame-Based Daytime Evaluation

The clip of S1 exhibits a situation which has a lot of continuous curve lanes. Due to the restriction of the linear model, our alogrithm can't well identify some curve lanes.

The clip of S2 was captured in a situation where the vehicle runoff is great. Current lanes are always severely occluded, making the lane identifications become difficult.

The clip of S3 was captured in the situation where most of lanes are straight roads and has relatively less of cars.It achives high rate of identification.

3.2   Frame-Based Nighttime Evaluation

The clip of S4 shows the situation when the car is in a weak light, as shown in Fig7. (i), and (l) as a result, lane marks is not clear very much and causes false detections.

The clip of S5 shows the situation which has relatively less of cars with enough light, as shown in Figure 7 (k).

The clip of S6 exhibits a situation where the vehicle runoff is great in night time, as shown in Figure7(j).

Note that in this paper, we fouced on improving the accuracy of lane detection based on Hough transform, and we didn't use any lane track method, like Kalman filters or other homologous methods. Of cource combined with these methods will make the system work better, however it is not the main point of this paper, these will be the work of next stage.

4   Conclusion

In this paper, we proposed a method which constructed a list of candidate boundaries. It is simple ,fast and accurate. Then we defined a score function to get the possibility of each candidate boundary pair and get the correct boundary pair as the current left and right lane.The experiment results showed that this algorithm has a good ability of real time and robustness. And in order to improve its mobility, portability and convenience, we need to investigate the hardware system to compute the proposed algorithm and improve its real timing and precision.

Acknowledgments. This work was financially supported by the financial support of Shanghai Economic and Information Technology Committee by the Annual Projects for Absorption and Innovation of Shanghai Imported Technology under Grant (No.11XI-15), the funded project of the national natural science foundation under Grant (No. 61376028) and the funded project of the Shanghai Science and Technology Commission under Grant (No. 13111104600).

References

1. Goldbeck, J., Huertgen, B., Ernst, S., Kelch, L.: Lane following combining vision and DGPS. Image and Vision Computing 18, 425–433 (2000)
2. Jung, C.R., Kelber, C.R.: Lane following and lane departure using a linear-parabolic model. Image and Vision Computing 23, 1192–1202 (2005)
3. Liu, W., Zhang, H.L., Duan, B.B., Yuan, H., Zhao, H.: Vision-Based Real-Time Lane Marking Detection and Tracking. In: Proc. of the IEEE Intelligent Transportation Systems (2008)
4. Lookingbill, A., Lieb, D., Thrun, S.: Optical Flow Approaches for Self-supervised Learning in Autonomous Mobile Robot Navigation. Autonomous Navigation in Dynamic Environments (2007)
5. lvarez, J.M.A., López, A.: Novel Index for Objective Evaluation of Road Detection Algorithms. In: Proc. of the IEEE Intelligent Transportation Systems (2008)
6. Wang, Y., Shen, D.G., Teoh, E.K.: Lane Detection Using Catmull-Rom Spline. In: Proc. of the IEEE Intelligent Vehicles (1998)
7. Wang, Y., Dahnoun, N., Achim, A.: A novel system for robust lane detection and tracking. Signal Processing 2, 319–334 (2012)
8. Bertozzi, M., Broggi, A.: GOLD: A parallel real-time stereo vision system for generic obstacle and lane detection. IEEE Transactions on Image Processing 1, 62–81 (1998)
9. Kreucher, C., Lakshmanan, S.: Lane: A lane extraction algorithm that uses frequency domain features. IEEE Transactions on Robotics and Automation 2, 343–350 (1999)
10. Yu, H.Y., Zhang, W.G.: Lane tracking and departure detection based on linear mode. Processing Automation Instrumentation 30(11), 1–3 (2009)
11. Yu, T.H.: Study on Vision based Lane Departure Warning System. Univ.of Jilin, Jilin (2006) (in Chinese)

Research on the Reliability of Narrow-Band Frequency-Sweep Electromagnetic Descaling Instrument

Wei Sun[1], Qian-Qian Gong[1], Wen-Jie Wang[2], Li-Hong Gai[3], and Li Li[1,*]

[1] Mechanical Engineering and Automation, Shanghai University, Shanghai, China 200072
lili.shu@shu.edu.cn
[2] Shengli Oil Field Electric Power Head Office, Dongying 257000
[3] Enviromental Engineering CO, LTD of Shandong Academy of Environmental Science,
Jinan 250013

Abstract. By producing some form of excitation signal, the new-style narrow-band frequency-sweep magnetic descaling instrument forms the electromagnetic field inside the pipeline, so as to achieve the effect of descaling, while the power amplification part of the instrument generally has the disadvantage of poor reliability. This paper puts forward two methods, shortening the distance between driving signal and the gate of transistor, and reducing the number of via holes, to reduce parasitic parameters. Meanwhile, measure the temperature of transistors with a thermistor directly, when the temperature exceeds the threshold, the adjustable rectifier is shut down automatically to protect the H bridge and enhance the reliability of the magnetic descaling instrument. Experiments show that the reliability of modified magnetic descaling instrument has been greatly improved by using the methods proposed.

Keywords: electromagnetic sweep, reliability, H-bridge, thermistors.

1 Introduction

The scaling of Oilfield water injection system has brought serious harm to production. Commonly used chemical scale removal method is mainly to add scale inhibitors, but this method has many shortcomings: the high cost of production, long periodic acid pickling, a narrow using scope, environmental pollution [1]. Physical cleaning methods appeared in recent years, damage into scale by creating certain conditions or changing external conditions [6]. Among them, the electromagnetic descaling method has achieved the rapid development and widely application, with the advantages of quick effect, simple installation, low operation cost, pollution-free, and so on [5]. By producing some form of excitation signal, magnetic descaling instrument forms the electromagnetic field within the pipeline. When the water flows through the pipe after magnetic treatment, it is not easy to form scale on pipe wall or scale is easy to fell off the wall, so as to achieve the descaling effect [2].

At present, the commonly used frequency-sweep electromagnetic descaling instrument has good effect when dealing with static water, but it doesn't fit with the flowing water in

---

* Corresponding author.

K. Li et al. (Eds.): LSMS/ICSEE 2014, Part III, CCIS 463, pp. 420–425, 2014.
© Springer-Verlag Berlin Heidelberg 2014

the pipeline, because the change of frequency from minimum to the maximum needs a scanning time, and there is only a short time is effective in this scanning period (i.e., action period of the best working frequency), therefore most of the scanning time is useless. Because of not considering the relationship between the cycle of frequency-sweep and the flow velocity, the descaling efficiency of the frequency-sweep magnetic descaling instrument is low, which is generally less than 40%. In this paper, the narrow frequency-sweep electromagnetic descaling instrument is improved on the basis of the frequency-sweep electromagnetic descaling instrument. Instead of outputting a wide range of frequencies, it outputs a narrow range of frequencies, which covers the natural frequencies of water quality changing in a certain range, and then the descaling efficiency of the frequency-sweep magnetic descaling instrument will be greatly improved.

The defect of the electromagnetic descaling instrument is its power amplification part is not stable. Especially under high voltage, and high frequency, the output voltage will produce oscillation with a larger amplitude, which results in the transistor overheating seriously, even burning. This defect severely reduces the service life of descaling apparatus, increases the maintenance costs and brings huge economic losses to industrial production. For the heat problem of the serious transistor fever, patent "electromagnetic resistance descaling instrument of frequency conversion and direction conversion with temperature and over-current protection" [10] has transistor temperature protection by using the temperature sensor measuring the temperature of the heat sink of transistor. If transistor temperature rises in a short period of time, the temperature sensor can't catch the change of the temperature for the short period of time because the temperature conduction takes time.

Aiming at the problems above, the paper enhances the stability of output of power amplification part of frequency sweep descaling instrument, mainly by reducing the parasitic parameters, and by measuring temperature of the transistor of H bridge directly with a thermistor. These methods have greatly enhanced the reliability of the power amplification part of the frequency-sweep magnetic descaling instrument.

2   The Structure of Magnetic Descaling Instrument

The excitation signal generator part of magnetic descaling instrument mentioned in this paper, can be divided into three parts, including the control unit module, adjustable rectifier module and H bridge module. Among them, the adjustable rectifier module can convert industrial alternating current (AC) to pulsating direct current (DC) signal which is proportional to the analog control signal of the control unit. Coil is winding on the pipeline, after the H bridge module converting the voltage on the bus to square wave, the square wave is loaded on the coil, therefore the instrument produces high-frequency oscillating magnetic field inside the pipe [3].

The structure of H bridge module is shown as figure 1. Four surge absorption modules (6, 7, 8 and 9), respectively connect with four transistors (2, 3, 4, and 5) in parallel, to absorb the voltage spike and current spike, which is caused by the shutdown and break-over of transistors, so as to protect the transistors [7]. Meanwhile, in order to guarantee the reliability of the instrument, a thermistor in the H bridge module is added, as shown in figure 1. The thermistor (10) connects with four transistors and the control unit (11), and transmits the temperature of the transistor to the control unit (11) in time.

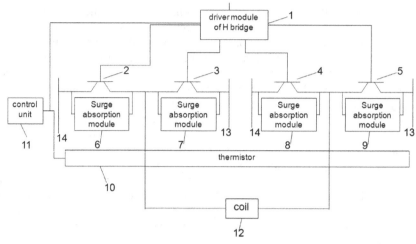

Fig. 1. Structure of H bridge module

When the temperature is higher than the threshold, the control unit (11) will shut down the adjustable rectifier automatically, to protect H bridge, so as to ensure that the H bridge can run reliably for a long time.

3   Experiment、 Results and Analysis

3.1   Experiment and Results

In this paper, the PCB of narrow frequency-sweep magnetic descaling instrument, which is the first generation of PCB of the H bridge module, is designed. Connect the instrument to power, set parameters, and then use the oscilloscope to measure the output voltage waveform of the first generation H bridge module. In figure 2 (a), there are the physical map of the first generation of H bridge, PCB graph of the distance between drive signal and the gate of transistor and the output waveform graph, respectively.

From the figure 2 (a), it is shown that the output voltage of H bridge is unstable with the phenomenon of damping vibration, and the transistors overheats seriously. After checking the circuit, it is found that these problems are caused by parasitic capacitance and parasitic inductance. Then the second generation of the H bridge module [9] , whose PCB layout is changed greatly to reduce the generation of parasitic capacitance and parasitic inductance as far as possible, is designed. In this experiment two methods as follows are used to reduce parasitic parameters: (1) reduce the distance between drive signal and the gate of transistor, (2) reduce the amount of the via holes as far as possible. Figure 2 is the contrast figure of the second generation PCB and the first generation PCB. Customize the second generation of the H bridge according to its PCB, finish welding then replace the first generation of H bridge module with the second generation of the H bridge module, and then connect the instrument to power, set the same parameters as the first experiment, and then use the oscilloscope to measure the output voltage waveform of the second generation of H bridge module. Figure 2 (b) are the physical map of the second generation of H bridge, PCB graph of the distance of drive signal and the gate of transistor and the output waveform graph of H bridge, respectively.

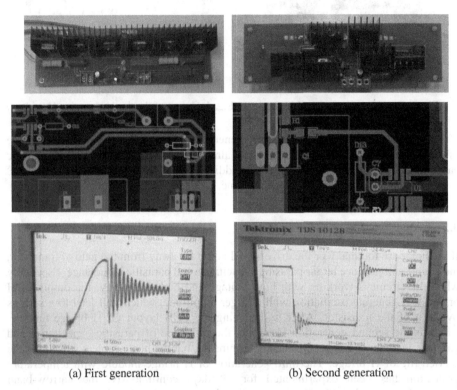

(a) First generation   (b) Second generation

Fig. 2. The physical map of H bridge, PCB diagram and H bridge output waveform

From figure 2(b), it can be seen that the second generation of the H bridge module has already reduced the influence of parasitic capacitance and parasitic inductance, and has better results. But when the instrument works for a long time, the transistors still overheat, especially when the instrument works for long without a break, the overheating of transistors is particularly serious [8]. In order to enhance further the reliability of the H bridge module, the third generation of the H bridge module is designed, which inherits the advantages of the second generation PCB layout, simultaneously, add a thermistor on the transistor [4]. For the third generation of H bridge module (shown as figure 3), do the experiment as the two experiments above, and then use the oscilloscope to measure the output voltage waveform, which is shown in figure 4.

Fig. 3. The third generation H bridge   **Fig. 4.** Output of the third generation H bridge

From figure 4, it is found that the third generation H bridge has inherited the advantage of the second generation H bridge, it has already reduced the influence of parasitic capacitance and parasitic inductance. In addition, the added thermistor can keep H bridge module working safely for a long time.

3.2   Analysis of Experimental Results

If via holes are too many or the drive signal is too far away from the gate of transistor, the circuit will produce larger parasitic capacitance and parasitic inductance. Especially when the circuit is working with high voltage and high frequency, the output of H bridge will generate oscillation with a larger amplitude, which will lead to a serious overheat of the transistor. After re-designing the PCB layout Of H bridge module, which is the second generation H bridge, the influence of parasitic capacitance and parasitic inductance has been basically eliminated, but the heat problem still limits the service life of H bridge. The third generation of H bridge module with temperature protection has been experimented for 20 days continuously, the narrow-band frequency-sweep electromagnetic descaling instrument has been working normally during this period, which proves that the third generation of H bridge has solved the problem of the transistor heat well.

4   Conclusion

At present, the narrow-band frequency-sweep electromagnetic descaling instrument is gradually replacing chemical cleaning methods in the industrial field, so the reliability of the narrow-band frequency-sweep electromagnetic descaling instrument becomes especially important. This paper mainly researches on the reliability of the power amplification part of the current narrow-band frequency-sweep electromagnetic descaling instrument. The experiment shows that if via holes are too many or drive signal is too far away from the gate of transistor, the circuit will produce larger parasitic capacitance and parasitic inductance, which results in instability of the output of the instrument. For the heat problems of the transistor, a thermistor is added directly on transistors to ensure that the H bridge module can work safely for a long time, and experiments show that the third generation of H bridge module with the temperature protection has worked safely for 20 days continuously. The methods of enhancing the stability of the instrument mentioned above have overcome the unstable faults of narrow- frequency sweep magnetic descaling instrument, and have reduced the failure probability of equipment, so that a lot of maintenance funds are saved, and a very good application prospect is expected.

References

1. Xiao, Z.L., Pu, C.S.: Review of technology and application on magnetized water treatment for anti-scaling. Fault-Block Oil & Gas Field 17, 121–125 (2010)
2. He, J., Zhao, Z.Z., Li, Y.H., Bai, G.C., Zhao, H.Y.: Research status and progress of the physical methods descaling technology. Industrial Water Treatment 30, 5–9 (2010)
3. Jiang, W.B., Chen, H.D., Li, Q., Li, X.: The design of a new type of electromagnetic field generator used for descaling. Journal of Institute of Metrology of China 22, 263–267 (2011)
4. Miao, X.F.: Experimental study on Circulating cooling water of thermal resistance monitoring and high frequency scaling. Chongqing university (2013)
5. Alimi, F., Tlili, M., Ben Amor, M., Maurin, G., Gabrielli, C.: Influence of magnetic field on calcium carbonate precipitation in the presence of foreign ions. Surface Engineering and Applied Electrochemistry 45, 56–62 (2009)
6. Gryta, M.: The influence of magnetic water treatment on CaCO3 scale formation in membrane distillation process. Separation and Purification Technology 80, 293–299 (2011)
7. Marzoughi, A., Iman-Eini, H.: Selective harmonic elimination for cascaded H-bridge rectifiers based on indirect control. In: 3rd Power Electronics and Drive Systems Technology, pp. 79–85. IEEE Computer Society, New Jersey (2012)
8. Das, M.K., Capell, C., Grider, D.E.: 10 kV, 120 a SiC half H-bridge power MOSFET modules suitable for high frequency, medium voltage applications. In: Energy Conversion Congress and Exposition, pp. 2689–2692. IEEE Computer Society, New Jersey (2011)
9. Reusch, D., Strydom, J.: Understanding the effect of PCB layout on circuit performance in a high-frequency gallium-nitride-based point of load converter. IEEE Transactions on Power Electronics 29, 2008–2015 (2014)
10. Gao, C.G., Ding, L.P., Song, J.E., Hou, X.J.: The electromagnetic resistance descaling device of frequency conversion and direction conversion with temperature and over-current protection: China, CN201120001943.8, pp. 10–15 (2011)

Torque Calculation for a Radial Field Permanent Magnet Coupler in a Pump by Analytical Technique

Shiqin Du

School of Electric, Shanghai Dianji University, No.690 Jiangchuan Road,
Shanghai 200240, China
dusq@sdju.edu.cn

Abstract. The magnetic force pumps are widely used in place of petro-chemical industry, drugs manufacture and treatments for heart disease for their compact structure and reliable working. The design of the electromagnetic devices requires accurate calculation of the field parameters. Taking the bigger airgap and regular airgap boundary condition of the radial field permanent magnet coupler into consideration, analytical method is employed to solve the governing partial differential equations in the article. The electromagnetic field distribution for the inner rotor and outer rotor are obtained respectively by the analytical method, thus the torque is calculated and torque-angle curve is drawn.

Keywords: radial field, permanent magnet coupler, analytical method, torque calculation.

1 Introduction

Magnetic force pump which outer magnetic rotor and inner magnetic rotor go together is a device driven by magnetic force. When outer magnetic rotor rotates, inner magnetic rotor and the pump rotation shaft keep pace with it, to complete task of conveying liquid. Due to the dividing cover arranged between the inner rotor and the outer rotor, inner magnetic rotor and pump output shaft are enclosed in a sealed cavity, preventing leakage of fluid to the outside and cooling magnetic driving device. One of the main part of magnetic force pump is magnetic coupler. In the paper a radial field magnetic coupler is studied, as shown in Figure 1.

The magnetic force pumps are widely used in place of petro-chemical industry, drugs manufacture and treatments for heart disease for their compact structure and reliable working⦿ Pumping acids, Lye, scarce valuable liquid, especially in cases of flammable, explosive liquid pumping [1-2].

To design a high performance magnetic force pump, the calculation of magnetic field and electromagnetic torque of the magnetic coupler is a key step[3-5].

Research method of permanent magnet magnetic field and magnetic force transmission problems, generally comes down to solving certain partial differential equations. Numerical analysis method and analytical analysis method of electromagnetic field are main methods.

K. Li et al. (Eds.): LSMS/ICSEE 2014, Part III, CCIS 463, pp. 426–433, 2014.

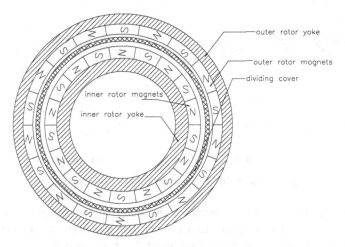

Fig. 1. Structure of radial field magnetic couple

However, the process of finite element method contains three steps, i.e., processing before, calculation and processing after. Modeling in the air gap of electric machines is often challenging. The air gap length may be several orders of magnitude smaller than the circumference. Corresponding commercial software is expensive, and operation personnel need to be specially trained. Electromagnetic field analytical method has a long history, and a lot of documents dealing with it in magnetic field of permanent magnet motor were published in the past. Qishan Gu etc. published articles about analytical method for calculation of electromagnetic field for permanent magnet motor, researched the air-gap magnetic field of permanent magnet motors, edge effect and slotting effect[6-8]. Z. Q. Zhu etc. published a series of articles for a systematic analysis of no-load magnetic field, the armature reaction and load magnetic field, slotting effect of permanent magnet brushless DC motor by way of analytical methods [9-12].

These methods have been used in the analysis and design of permanent magnet motor, the following will be used in the resolution of magnetic field distribution of the radial magnetization magnetic coupler and its torque calculation.

2   Analytical Model for Radial Field Magnetic Coupler

When the mathematical model of magnetic field is described by the Laplace's equation or Poisson's equation, the problem will be how to solve this type of boundary value problems. Method of separation of variables is a basic method for solving boundary value problems. In case of potential function changes with two or three coordinate variables and field boundaries consistent with the coordinate surfaces, the method of separation of variables can be used to solve the boundary value problem.

Using polar coordinates to analyze the radial magnetization magnetic Coupler, isolation materials is non-magnetic materials, equivalent to air-gap. Calculation models of inner and outer magnetic rotor magnetic field are drawn, as shown in Figure 2.

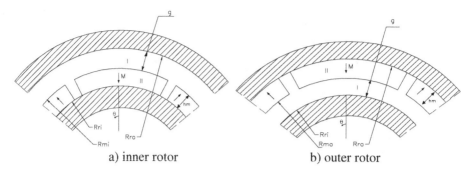

a) inner rotor   b) outer rotor

Fig. 2. Coupler topologies of inner rotor and outer rotor

In Figure 2, Rro is outer diameter of the outer rotor magnets, Rmo is inner diameter of the outer rotor magnets, Rmi is outer diameter of the inner rotor magnets, Rri is inner diameter of the inner rotor magnets, g is equivalent air gap length, Variable θ is the angle between radial direction of the point and the pole center line, M is magnetization. The assumptions used in the analysis is , (a)inner and outer rotor core's permeability is infinite, (b)uniform permanent magnet radial magnetization, magnets work in the second quadrant with linear demagnetization characteristics, and $\vec{M} = \vec{B}_r / \mu_0$, $\vec{B}_r$ is the residual magnetism density, μ_0 is air permeability. Radial magnetization distribution, as shown in Figure 3.

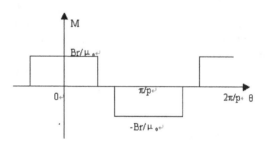

Fig. 3. Radial magnetization

Corresponding inner magnetic rotor model, using the scalar magnetic potential φ, the magnetic field strength can be expressed as,

$$\vec{H} = -\nabla \varphi \tag{1}$$

in air gap (region I),

$$\vec{B}_I = \mu_0 \vec{H}_I \tag{2}$$

in permanent magnets (region II),

$$\vec{B}_{II} = \mu_m \vec{H}_{II} + \mu_0 \vec{M} \tag{3}$$

and thus come to the scalar magnetic potential equation, in air gap,

$$\nabla^2 \varphi_I = 0 \tag{4}$$

in permanent magnets,

$$\nabla^2 \varphi_{II} = \frac{div\vec{M}}{\mu_r} \tag{5}$$

In polar coordinates, for a uniform radial magnetization, magnetization $\vec{M}$ can be expressed as

$$\vec{M} = M_r \vec{r} + M_\theta \vec{\theta} \tag{6}$$

among it , $M_r = \sum_{n=1,3,5...}^{\infty} M_n \cos np\theta$, $M_\theta = 0$, p is pole pairs, θ is the angle between radial direction of the point and the pole centerline, $M_n = 2(B_r/\mu_0)\alpha_p \dfrac{\sin(n\pi\alpha_p/2)}{(n\pi\alpha_p)/2}$, α_p is the effective pole-arc coefficient.

From $div\vec{M} = \dfrac{M_r}{r} + \dfrac{\partial M_r}{\partial r} + \dfrac{1}{r}\dfrac{\partial M_\theta}{\partial \theta} = \dfrac{M_r}{r}$, the scalar magnetic potential mathematical model in the air gap and in the permanent magnets is obtained respectively,

in the air gap

$$\frac{\partial^2 \varphi_I}{\partial r^2} + \frac{1}{r}\frac{\partial \varphi_I}{\partial r} + \frac{1}{r^2}\frac{\partial^2 \varphi_I}{\partial \theta^2} = 0 \tag{7}$$

in the permanent magnets

$$\frac{\partial^2 \varphi_{II}}{\partial r^2} + \frac{1}{r}\frac{\partial \varphi_{II}}{\partial r} + \frac{1}{r^2}\frac{\partial^2 \varphi_{II}}{\partial \theta^2} = \frac{M_r}{\mu_r r} \tag{8}$$

Two components of $\vec{H}$ is $H_r = -\dfrac{\partial \varphi}{\partial r}$, $H_\theta = -\dfrac{1}{r}\dfrac{\partial \varphi}{\partial \theta}$. The boundary conditions which satisfy formula (7) and (8) are as follows:

$$① \quad \varphi_I(r,\theta) = \varphi_I(r,-\theta) \qquad \varphi_{II}(r,\theta) = \varphi_{II}(r,-\theta) \tag{9}$$

$$② \quad \begin{cases} H_\theta(r,\theta)|_{r=R_{ro}} = 0 \\ H_{\theta II}(r,\theta)|_{r=R_{ri}} = 0 \\ B_{rI}(r,\theta)|_{r=R_{mi}} = B_{rII}(r,\theta)|_{r=R_{mi}} \\ H_\theta(r,\theta)|_{r=R_{mi}} = H_{\theta II}(r,\theta)|_{r=R_{mi}} \end{cases} \tag{10}$$

According to mathematical models and its boundary conditions, scalar magnetic potential expressions are obtained, and thus air-gap flux density expressions are obtained.

When $np \neq 1$ components of magnetic flux density in air gap expression are,

$$B_{rl\ i}(r,\theta) = \sum_{n=1,3,5,\cdots}^{\infty} \frac{\mu_0 M_n}{\mu_r} \frac{np}{(np)^2 - 1} R_{mi}^{-(np-1)}$$

$$\bullet \left\{ \frac{(np-1)R_{mi}^{2np} + 2R_{ri}^{np+1}R_{mi}^{np-1} - (np+1)R_{ri}^{2np}}{\frac{\mu_r+1}{\mu_r}\left[R_{ro}^{2np} - R_{ri}^{2np}\right] - \frac{\mu_r-1}{\mu_r}\left[R_{mi}^{2np} - R_{ro}^{2np}(R_{ri}/R_{mi})^{2np}\right]} \right\} \tag{11}$$

$$\bullet \left[r^{np-1} + R_{ro}^{2np} r^{-(np+1)}\right]\cos np\theta$$

$$B_{\theta\ i}(r,\theta) = \sum_{n=1,3,5,\cdots,}^{\infty} \frac{(-\mu_0 M_n)}{\mu_r} \frac{np}{(np)^2 - 1} R_{mi}^{-(np-1)}$$

$$\bullet \left\{ \frac{(np-1)R_{mi}^{2np} + 2R_{ri}^{np+1}R_{mi}^{np-1} - (np+1)R_{ri}^{2np}}{\frac{\mu_r+1}{\mu_r}\left[R_{ro}^{2np} - R_{ri}^{2np}\right] - \frac{\mu_r-1}{\mu_r}\left[R_{mi}^{2np} - R_{ro}^{2np}(R_{ri}/R_{mi})^{2np}\right]} \right\} \tag{12}$$

$$\bullet \left[r^{np-1} - R_{ro}^{2np} r^{-(np+1)}\right]\sin np\theta$$

Accordingly, for the outer magnetic rotor model, the same method get air-gap flux density expressions

When $np \neq 1$, components of magnetic flux density in air gap expression are [9]:

$$B_{rl\ o}(r,\theta) = \sum_{n=1,3,5,\cdots,}^{\infty} \frac{\mu_0 M_n}{\mu_r} \frac{np}{(np)^2 - 1} R_{mo}^{-(np-1)}$$

$$\bullet \left\{ \frac{(np-1)R_{mo}^{2np} + 2R_{ro}^{np+1}R_{mo}^{np-1} - (np+1)R_{ro}^{2np}}{\frac{\mu_r+1}{\mu_r}\left[R_{ri}^{2np} - R_{ro}^{2np}\right] - \frac{\mu_r-1}{\mu_r}\left[R_{mo}^{2np} - R_{ri}^{2np}(R_{ro}/R_{mo})^{2np}\right]} \right\} \tag{13}$$

$$\bullet \left[r^{np-1} + R_{ri}^{2np} r^{-(np+1)}\right]\cos np\theta$$

$$B_{\theta\ o}(r,\theta) = \sum_{n=1,3,5,\cdots,}^{\infty} \frac{(-\mu_0 M_n)}{\mu_r} \frac{np}{(np)^2 - 1} R_{mo}^{-(np-1)}$$

$$\bullet \left\{ \frac{(np-1)R_{mo}^{2np} + 2R_{ro}^{np+1}R_{mo}^{np-1} - (np+1)R_{ro}^{2np}}{\frac{\mu_r+1}{\mu_r}\left[R_{ri}^{2np} - R_{ro}^{2np}\right] - \frac{\mu_r-1}{\mu_r}\left[R_{mo}^{2np} - R_{ri}^{2np}(R_{ro}/R_{mo})^{2np}\right]} \right\} \tag{14}$$

$$\bullet \left[r^{np-1} - R_{ri}^{2np} r^{-(np+1)}\right]\sin np\theta$$

With flux density expressions of PM inner rotor and PM outer rotor in the air-gap, combined with the actual situation of air-gap length of magnetic couplers are relatively large, gap is a linear medium for magnetic field calculation, composition of airgap magnetic field excited by both PM inner rotor and PM outer rotor can be expressed by way of superposition, as below,

$$\begin{cases} B_{rl}(r,\theta) = B_{rl\ i}(r,\theta) + B_{rl\ o}(r,\theta) \\ B_{\theta l}(r,\theta) = B_{\theta l\ i}(r,\theta) + B_{\theta l\ o}(r,\theta) \end{cases} \tag{15}$$

3   Calculation of Analytical Method for Radial Magnetization Magnetic Coupler

Finite element method in dealing with complex boundary and the magnetic field calculation of nonlinear problems, effectiveness of its application and breadth of its application has been verified by many times practice, here used as the validation software.

Study on analytical method for magnetic field and torque calculation of radial magnetic field couplers, analyses the main characteristics of this method, as follow. (1)In theory, visual concepts of precise solution when analytical method applied to the rather special boundary conditions appear, understand the effects of parameter changes on the magnetic field is helpful, and you can take this parameter adjustment in the field. (2) In calculating electromagnetic torque of the magnetic coupling, finite element method for air-gap regional subdivision has certain requirements, and using analytical method is easy.

Taking a 18 poles, work torque of 200Nm, maximum torque of 400Nm of the radial magnetizing magnetic Coupler, as magnetic field calculation example, concreting these features of the analytical method, and the corresponding finite element calculation result is given to be compared with. Specific parameters of the example are shown in table 1

Table 1. Parameters of sample magnetic coupler

Name	Value	Unit
Outer diameter of outside magnetic rotor	220	mm
Outer diameter of outside magnetic rotor magnets	200	mm
Inner diameter of outside magnetic rotor magnets	186	mm
Outer air gap length	1	mm
Isolation cover length	2	mm
Inner air gap length	2	mm
Outer diameter of inner magnetic rotor magnets	176	mm
Inner diameter of inner magnetic rotor magnets	162	mm
Inner diameter of inside magnetic rotor	128	mm
Axial length	60	mm
Residual magnetization of magnets	1.1	T
Pole pairs	9	pair
Pole-arc coefficient	1	

Based on the above figures, in case of internal and external magnetic pole center line forming a 90°electric   angle, using analytical method and finite element method for calculation of radial magnetic flux density distribution in gap center, respectively, and the results shown in Figure 4.

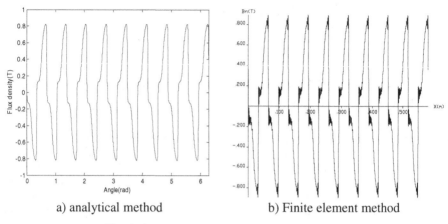

a) analytical method   b) Finite element method

Fig. 4. Comparison of resultant field distribution

Further, based on synthesis of the inside and outside magnetic rotor's radial and tangential component of the magnetic field in the middle air gap, calculate torque-angle characteristics of the magnetic coupler, curve as shown in Figure 5.

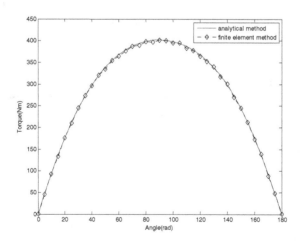

Fig. 5. Torque-angle curve

4   Conclusions

Due to radial field permanent magnet coupler's greater air radial length, and its inside and outside magnetic rotor with regular boundary conditions, better analytical method calculation condition is met, analytical method can be used. Analytical method results in the electromagnetic torque calculation are consistent with results obtained by finite element method in the example. The above-mentioned analytical method simplifies the design of radial field permanent magnet coupler, increases predicting capacity as parameters' adjustment to the effect of electromagnetic torque.

Acknowledgments. The article is supported by the doctoral scientific research foundation of shanghai dianji university(13C411).

References

1. Yang, J.L., Xu, W., Geng, D.: Characteristics and application of magnetic drive pump. Food Safety Guide 8, 68–69 (2010) (in Chinese)
2. Yan, X.L., Ren, Z.L., Han, A.: The design of magnetic-driving pump of high-carbon organic acid. Gansu Science Journal 17(3), 69–71 (2005) (in Chinese)
3. Kong, F.Y., Chen, G., Cao, W.: Numerical calculation of magnetic field in magnetic couplings of magnetic pump. Mechanical Engineering Journal 42(11), 213–217 (2006) (in Chinese)
4. Ren, Z.L., Ding, C.B., Xue, F.L.: Finite element analysis and optimal design of magnetic driver. Gansu Science Journal 18(4), 117–120 (2006) (in Chinese)
5. Wang, Y.: Calculation of magnetic field in permanent magnet device. Magnetic Materials and Devices 10, 49–60 (2007) (in Chinese)
6. Gu, Q.H., Gao, H.Z.: Effect of slotting in PM eletric Machines. Electric Machines and Power Systems 10(4), 273–284 (1985)
7. Gu, Q., Gao, H.: Airgap field for PM electric machines. Electric Machines and Power Systems 10(6), 459–470 (1985)
8. Gu, Q.S., Gao, H.Z.: Effect of fringing in PM eletric machines. Electric Machines and Power Systems 11(2), 159–169 (1986)
9. Zhu, Z.Q., David, H., Bolte, E.: et al: Instantaneous Magnetic Field Distribution in Brushless Permanent Magnet dc Motors, Part I : Open-Circuit Field. IEEE Trans. on Magn. 29(1), 124–135 (1993)
10. Zhu, Z.Q., David, H.: Instantaneous Magnetic Field Distribution in Brushless Permanent Magnet dc Motors, Part II : Armature-Reaction Field. IEEE Trans. on Magn. 29(1), 136–142 (1993)
11. Zhu, Z.Q., David, H.: Instantaneous Magnetic Field Distribution in Brushless Permanent Magnet dc Motors, Part III : Effect of Stator Slotting. IEEE Trans. on Magn. 29(1), 143–151 (1993)
12. Zhu, Z.Q., David, H.: Instantaneous Magnetic Field Distribution in Brushless Permanent Magnet dc Motors, Part IV : Magnetic Field on Load. IEEE Trans. on Magn. 29(1), 152–158 (1993)

Information Fusion for Intelligent EV Charging-Discharging-Storage Integrated Station

Guangning Su[1], Da Xie[2], Yusheng Xue[3], Chen Fang[1], Yu Zhang[1], and Kang Li[4]

[1] State Grid Shanghai Municipal Electric Power Company, 200122, Shanghai, China
[2] Shanghai Jiao Tong University, 200240, Shanghai, China
[3] State Grid Electric Power Research Institute, 210003, Jiangsu, China
[4] Queen's University, Belfast, Northern Ireland, UK

Abstract. According to the distribution of information flow in the integrated EV station, the detailed parameters of the station and car terminals can be obtained through the monitoring and control system. Radio frequency identification is used to collect data from battery system of electric vehicles, combined with IOT (internet of things) and GPS, the information of batteries can be acquired quickly and accurately. Moreover, the state of batteries can be diagnosed and faults be handled timely. Based on the collection of above parameters of the integrated station and the grid, the grid-connected control strategy is proposed according to the grid state and energy flow of the integrated station. The experimental result suggests that the grid-connected control strategy based on information fusion can help to implement peak load shifting effectively and timely, and provide energy flow support when the grid is in a heavy-loaded state so it can restore to its normal state.

Keywords: Electric vehicle, integrated station, information flow, radio frequency identification, information fusion, grid-connected control.

1   Introduction

The problems of primary fossil fuel resource consumption and environment pollution caused by traditional automobile industry are becoming increasingly drastic, the electric vehicles (EVs) have attracted attentions and have achieved rapid development due to their advantages in terms of low consumption, no pollution and no noise [1]-[4].

As important parts of the electric vehicle industry chain, the structure and operation mode of energy supply facilities have potential impact on the electric vehicle industry development and the grid they are connected to. People have research on them and proposed many effective charging strategies [5]-[6]. In [7] an agent-based decision support system is presented for identifying patterns in residential EV ownership and driving activities to enable strategic deployment of new charging infrastructure. In [8] the potential benefits of using control mechanisms that could be offered by a Home Energy control box are investigated in optimizing energy consumption from PHEV charging in a residential use case, and present smart energy control strategies based on the quadratic programming for charging PHEVs, aiming to minimize the peak load and

K. Li et al. (Eds.): LSMS/ICSEE 2014, Part III, CCIS 463, pp. 434–448, 2014.

flatten the overall load profile, then compare two strategies, and benchmark them against a business-as-usual scenario assuming full charging starting upon plugging in the PHEV.

Intelligent grid is the development direction and trends of the future grid. Along with the power expansion and plug-in of numerous electrical equipment, high-speed and accurate information processing technology has become a key factor in the development of intelligent grid. In [9] the differences between multi-sensor integration and multi-sensor fusion are given. Multi-sensor integration refers to the synergistic use of the information provided by multiple sensory devices to assist in the accomplishment of a task by a system. An additional distinction is made between multi-sensor integration and the more restricted notion of multi-sensor fusion. Multi-sensor fusion, refers to any stage in the integration process where there is an actual combination (or fusion) of different sources of sensory information into one representational format. Information fusion, just like multi-sensor fusion, refers to the analysis, processing and integration of information in different time and forms for further processing and response. The rapid development of multi-source information fusion technology provides a new way of thinking for the grid development direction of intelligent and fault-quick-diagnostic [10]-[11]. [12] presents a fault diagnosis system based on D-S evidence theory and neural networks. The method can determine the fault type accurately and has been validated in the fault diagnosis of turbine equipment in a power plant. [13] analyzes the sag vulnerability area using the concept of data fusion. And structure of multilayer information fusion of Common Information Model is built based on the SCADA system, Management Information system of Safe Production and Equipment Manage system, thus obtaining required data for calculating sag vulnerability area of on-line power system. Then, this method is applied to a large region network of China and results prove its availability. It can help with the operational judgment while in faulty state and give a decision support to select appropriate operation mode for dispatcher.

Although information fusion has made significant progress in terms of fault diagnosis with its advantages of fast and accurate results, the use of information fusion in electric vehicle charging-discharging-storage integrated station has been rarely researched. As an important power supply facility, the integrated station will generate a large amount of information along with the power flow. To ensure that its operation is reliable and batteries are charged and exchanged in a timely and efficient means, it is necessary to analyze and study its internal information flow. This paper studies the distribution of information flow in the intelligent charging-discharging-storage integrated station and study grid-connected control strategy based on information fusion processing, which is of great benefit to the charging of EVs and safety operation of the grid.

2   Principle of Charging-Discharging-Storage Integrated Station

As an energy supply system for EVs, the integrated station has the advantages of improving battery efficiency and coordinating with the grid and loads. The integrated

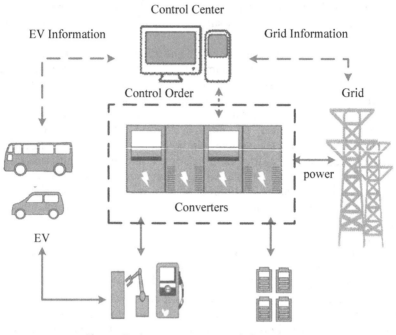

Control Center

EV Information

Grid Information

Control Order

Grid

power

Converters

EV

Charge Exchange System   Echelon Battery System

Fig. 1. Power flow and information flow inside the Integrated Station

station mainly contains five parts, namely multi-purpose converter device, charge exchange system, echelon battery system, information collection device and control center as shown in Figure 1. A multi-purpose converter device, which is composed of two groups of converters in parallel, connects charge exchange system and echelon battery system and serves for power exchanging. The integrated station can exchange power with the power grid through a set of multi-purpose converters, charge EVs with charger and implement battery replacement for EVs using changing-battery robot. Echelon battery system can exchange power with the grid through another set of multipurpose converters or charge the charge exchange system directly.

Information flow involves information exchange. Along with the energy flow, a large amount of information is produced, such as battery remaining power for EV, remaining mileage, location, destination, battery model, and the corresponding quantity and power of charge exchange system, battery quantity and power of echelon battery system, the operation mode and operating parameters of the multi-purpose converter device. If the above information are monitored and managed reasonably and effectively, we can control energy flow comprehensively. Combining grid information with the information of the integrated station, we can further optimize the operation and management of the integrated station. Figure 1 also shows the information flow distribution diagram of the integrated station. Information flow transmission and control situation are shown in table 1.

Table 1. Transmission and Receiving of Information Flow

Message sender	Message receiver	Context of the information flow	Information code
EV Terminals	Charge exchange system	Information of the battery replaced	I1
	GPS	Information of EV	I2
Charge exchange system	EV Terminals	Information of battery installed	I3
	Information aggregator	Information of charge exchange system	I4
Echelon battery system	Information aggregator	Information of cascaded battery system	I5
Multi-purpose converter device	Information aggregator	The current power flow information	I6
EMS of power grid	Information aggregator	The current load and predicted load	I7
GPS	Information aggregator	Location information of EVs	I8
Information aggregator	Control center	Multi-source information fusion	--
Control center	Charge exchange system	Energy flow-changing control	I9
	Echelon battery system	Energy flow-maintenance control	I10
	Multi-purpose converter device	Power-flow control	I11

3   Battery Management Mode of Charging-Discharging-Storage Integrated Station

As the direct power supply device for EVs, sufficient power and desirable performance guarantee for batteries is particularly important. To achieve the goal, an advanced efficient battery management system is needed.

The battery management system of the integrated station is mainly managed by people at present. When the EV pulls in, the staff records the code-message of the battery box. When changing the battery box is needed, the staff registers the information of battery box. There are a number of problems with changing battery manually and the present management mode, such as low efficiency, high error rate, long battery replacement cycle, lack of unified and efficient management, etc. Since there is no tracking and prediction system, it's hard for EVs to communicate with a repair center for timely maintenance when battery failure occurs. The problems mentioned above are the bottlenecks for rapid construction and development of the integrated station and they need to be addressed properly and promptly.

RFID (radio frequency identification) is a non-contact automatic identification technology. It makes use of radio signals to transmit data and information immediately in a specific space [14]. Based on the advantages of fast reading speed, safe and reliable information, large data storage and excellent security, RFID develops fast.

Its application scope increases and the industry scale expand constantly. At present, the technology has been used mostly in the fields of manufacturing, product security, library management, and warehouse management, etc. [15]-[16].

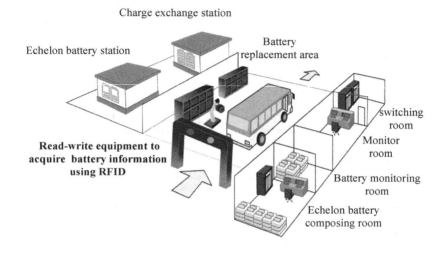

Fig. 2. Reading and writing of the battery RFID tags

Given the aforementioned advantages, RFID can be used to establish a battery resource information management database for the integrated station. We can store the type, charge and discharge times and power state of batteries in the database. If we update the real-time battery information to SCADA (supervisory control and data acquisition) system, IOT (Internet of things) and GPS, we can greatly improve battery information acquisition speed and battery replacement efficiency. At the same time, battery energy flow can be tracked in real-time and controlled effectively. The process of reading and writing of the battery RFID tags is shown in figure 2.

IOT is a sensing device for passing information. It is a network allowing the interconnection of different things according to the agreed protocol [17]. SCADA system in charge exchange system and echelon battery system can know the battery number, power state, working conditions and many other details in real-time.

GPS provides navigation services for EVs in real-time, all day and globally [18]. Using the received information from the EV battery terminals, GPS updates its internal battery information management database system in real-time, and sends information to the aggregator timely.

Making use of RFID, IOT and GPS, we can achieve advanced battery management for EVs. Battery management system based on IOT is shown in figure 3. When changing batteries for EVs, the remaining power, battery ID and other information should be sent to exchange station and information after replacement also need to be recorded. In the driving process, information of EVs is transmitted to GPS. Together with IOT, the information is transmitted to and handled by the information aggregator.

Through analyzing data received by IOT from the SCADA system of cascaded battery exchange and charge system and by GPS of EV battery information of EVs on road , advanced battery management of the integrated station can be achieved. Through the detection of the battery state of the charge, the faulty batteries are sent to repair centers and battery exchanging forecast is transmitted to the charge exchange station in advance. By analyzing the position and the route of EVs, which are monitored by GPS, suggestions can be provided to EV users while on road , information of EV faults is fed back to repair centers, and faulty batteries can be sent to repair centers.

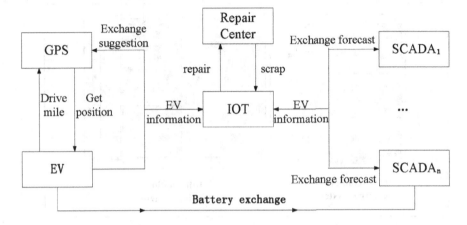

Fig. 3. Electric vehicle battery management system based on IOT system

With the above management system, we can not only significantly improve battery information acquisition speed and battery replacement efficiency, effectively track the battery energy flow control, but also can diagnose the state of the battery accurately, through which we can take timely diagnosis and treatment procedures for battery failures that may occur at any time.

4   The Information Processing System for Charging-Discharging-Storage Integrated Station

There are a lot of data to be collected and fused in the multi-purpose converter device, charge exchange system and echelon battery system of the integrated station. Given the complexity of the information flow, the information management system of the integrated station is divided into four parts, namely monitoring system, information fusion system, grid-connected control system and operation management system, as shown in figure 4.

4.1   Monitoring System

Monitoring system can collect and monitor data of smart EV terminals and all the equipment of the integrated station, including electricity distribution equipment, battery boxes, battery box replacement equipment and multi-purpose converter device. The data is transmitted to a higher dispatching system.

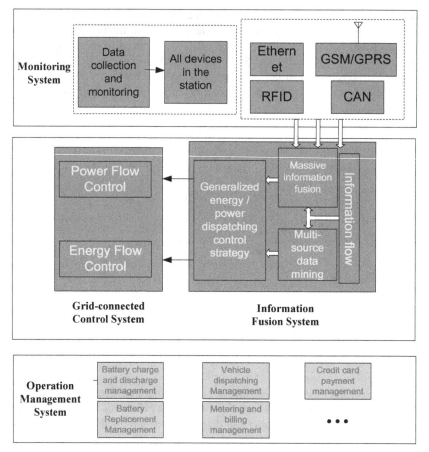

Fig. 4. Subsystems of integrated station supervisory system

When EV batteries are charged or replaced in the battery replacement station, the station can acquire the information of battery ID, power state and using times via EV terminals using RFID. Information of replaced batteries is recorded and transmitted to information aggregator. When an EV leaves the station, the information of route and location can be acquired by GPS through EV terminals, and the information is transmitted to information aggregator via IOT.

Equipment information of the integrated station is recorded and stored through collection and monitoring equipment and transmitted to information aggregator through CAN, such as the operation mode and operating parameter of the multi-purpose converter device, charge exchange system and the work state of its battery box replacement equipment and type and power state of battery box, the work state and power state of battery box of echelon battery system.

The integrated station exchanges active and reactive power with grid through two groups of multi-purpose converter devices. The EMS transmits collected information to the information aggregator through CAN, including grid voltage and current amplitude and phase, grid load and load forecasting, etc.

4.2   Information Fusion System

The information fusion system is the control center for all information received from the information aggregator. The information aggregator sends heterogeneous multi-source information to the control center. The higher level management platform of the control center combines with EMS, IOT, station monitoring and other information to achieve monitoring and analysis of energy and power flow. By making optimal decisions in real time though generalized power / energy scheduling control strategy, we can achieve the orderly management of energy information, and then obtain the energy flow and power flow adjustment strategy of the exchange station, echelon stations and multipurpose deflector, and perform associated control actions.

Since the monitoring equipment acquire and upload large numbers of information data, we use fusion technology to deal with different types of information data at different levels. According to the work characteristics and operating mode of integrated power station, the fusion processing can be divided into three types: reference values fusion, measured values fusion and decision fusion.

Reference fusion is to provide valid referenced data for data fusion by analyzing those data from different data acquisition systems and updating reference value based on the actual needs. For instance, the charging information of batteries in charge exchange system should contain two kinds of reference values: charging reference value and the reference value of return echelon battery system. The charging reference value is the threshold charging value of battery replacement. If the charging value is higher than charging reference value, the battery is ready to be used in battery replacement. Otherwise, it should be promptly charged. Return echelon battery system reference value refers to the power reference value need to be retreated, when charging value is below a certain limit the battery will be badly damaged. So, the exterior batteries in the integrated station (IS) have two kinds of reference value as well, that is, charging reference value and exchanging reference value. When the reserve power is short on the circumstance that power grid cannot supply the power of IS, we should decide the charging and exchanging reference value according to the battery information acquired inside and outside IS.

Reference fusion is used to get reference value. So it is necessary to make sure that all the values be used to evaluate the reference value is accurate so as to make rational analysis and processing. Therefore, reference fusion is often based on Kalman filter algorithm, the joint probabilistic data association, multi-hypothesis, interactive model method and sequential processing theory.

Fusion of measured values is to integrate actual data and operating parameters of a variety of devices, to compare with the reference value, and make a judgment and treatment. For instance, we take battery data information into consideration. Firstly, classify a series of different IDs. Secondly, fuse battery power information of the same series. Thirdly, compare the actual power data with the corresponding reference value data of batteries inside and outside IS. If the series of batteries is inside IS, we decide whether the batteries should be charged for replacement or be returned to the echelon charging station. Otherwise, we decide whether they should be charging or exchanged.

Measured values fusion is used to compare value measured with the reference value. Therefore Kalman filter algorithm is always used to make sure the value measured is reliable.

Decision fusion is to send data acquired by certain collector to the fusion processing center directly. And the fusion center make the final decision based on the identification of local collector of each part. Decision fusion is to fuse specific decision goals, and the outcomes directly affect decision-making. For instance, we take multi-purpose converter device into consideration. When the grid voltage is abnormal, the device should have LVRT (Low Voltage Ride Through) function, which requires decision fusion of the working mode of converter devices. And make quick decisions to ensure the operation safety and reliability of IS and power grid.

Decision fusion aims at specific decision target, it mainly based on Bayes estimates, expert systems D-S evidence theory, Fuzzy set theory.

Under normal operating conditions, the reference values of all kinds of data are determined, and for which there is no need for reference fusion, but there is a need to carry out fusion of a large amount of battery information, or to take decision fusion for relatively small number of multi-purpose converter devices. When the external environment changes, such as blackout after power failure and cataclysm of electric car trips, we need to regulate the data reference.

4.3   Operations Management System

Operations management system can independently achieve operation monitoring and equipment's management inside IS, can perform operations management commands from the higher level system to the charging-discharging-storage integrated station as well, and deal with the system in battery charge and discharge management, battery replacement management, vehicle dispatching management, metering and billing management, and so on.

Information fusion technology can always obtain information in regards to the remaining battery power and the remaining mileage of electric vehicles through GPS satellite navigation system. And give reasonable recommendations on whether there is a need to charge the battery or do battery replacement. Operations management system is based on the results of the information fusion analysis, and gives alert and suggestions to electric vehicles' drivers, and sends commands in advance for the possible or upcoming battery charging and exchanging work.

Operations management system is also responsible for statistical analysis, such as charge exchange system and echelon battery system, and corresponding quantity, power, lifespan. Then dispatch the batteries in the IS that has been long used or with low power to the echelon battery system and do statistical analysis and treatment of scrapped battery boxes in the echelon battery system.

Through statistical listing analysis of battery information inside and outside IS, operation management system can monitor and report the operating status of all battery boxes at any time, which is convenient for to make timely and effective treatment of any faulty battery boxes. On the one hand, this function of operation management system ensures the reliable operation of battery boxes. On the other hand, it simplifies the workflow of the staff, and improves efficiency greatly.

4.4   Grid-Connected Control System

The grid-connected control system is based on the charging and discharging command and generation demand after fusion treatment, to control the input and output of sub-devices in IS, such as energy storage and convention system, to achieve grid-connected control and a variety of operating response of IS, which is based on broad energy/power dispatch control strategy.

The SOC (State of charge) of charge exchange system (which is denoted as A) and echelon battery system (which is denoted as B) are changing over time. According to the current operating status of power grid (which is denoted as G), we will discuss the following three scenarios, namely normal peak load condition, the normal valley load condition and overloading condition. On the one hand, the integrated station need to reserve certain energy to maintain its operation off-grid, we define it as 20% of the total energy according to our experience. On the other hand, the batteries' charging cycle counts are limited, so we define 80% as the value to be charged to get optimal effect.

4.4.1   Normal Peak Load Condition

According to the SOC of exchanging system and echelon system, the normal peak load condition can be divided into seven possible operating modes.

(1) Both exchanging system and echelon system discharge power to the grid, i.e., A-> G, B-> G.

When the grid is in peak load condition, the IS takes advantage of integrated battery energy storage, combined with the station's battery charge, discharge and exchange programs, and select a part of suitable batteries to discharge to power grid and support the load peak clipping of the grid on the circumstance that the energy reserve is not sufficient inside the station. In this case, echelon system should take the maximum discharging power.

(2) Echelon system charge power to exchanging system, i.e., B-> A.

When the SOC of exchanging system is lower than 0.2 (SOC <0.2), and the SOC of echelon system is higher than 0.2 (SOC> 0.2), or the SOCs of the two station is between 0.2 and 0.8, we would rather charge power from the echelon system to the exchanging system than charging from the power grid, which can not only improve energy efficiency, but also reduce the burden of the power grid.

(3) Echelon system charge power to exchanging system and discharge power to power grid, i.e., B-> A, B-> G.

When the SOC of echelon system is sufficient (SOC>0.8) and the SOC of exchanging system is normal (0.8>SOC>0.2), the echelon system charge power to the exchanging system to ensure the normal operation, and discharge power to the power grid to play a positive role in clipping the peak load.

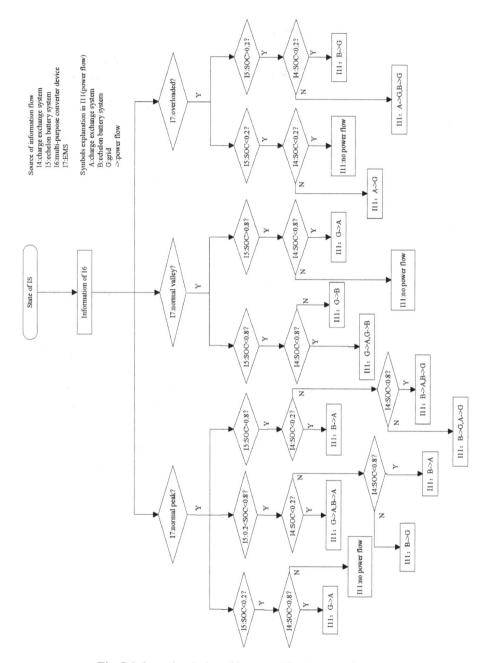

Fig. 5. Information fusion of Integrated Station operating state

(4)Power grid charges power to the exchanging system, and the echelon system discharges power to exchanging system, i.e., G-> A, B-> A.

When the SOC of exchanging system is insufficient (SOC <0.2), and the SOC of echelon system is normal (0.8>SOC>0.2), in order to ensure the normal operation of exchanging system, echelon system and the grid should charge the exchanging system to alleviate the load pressure of power grid.

(5) Echelon system discharge power to power grid, i.e., B-> G.

When the SOC of exchanging system is sufficient (SOC >0.8), and the SOC of echelon system is normal (0.8>SOC>0.2), in order to ensure the normal operation of exchanging system, only echelon system is available to charge to power grid to play a positive role in clipping the peak load.

(6) Power grid charge to exchanging system, i.e., G-> A.

When the SOC of exchanging system is normal (0.8>SOC>0.2), and the SOC of echelon system is insufficient (SOC <0.2), in order to ensure the normal operation of exchanging system, only power grid is available to charge the exchanging system.

(7) No energy exchange.

When the SOC of exchanging system is sufficient (SOC >0.8), and the SOC of echelon station is insufficient (SOC <0.2), in order to ensure the normal operation of the exchanging system, there is no any power exchange with power grid.

4.4.2   Normal Valley Load Condition

According to the SOC of exchanging system and echelon system, the normal valley load condition can be divided into four possible operating modes.

(1) Power grid charge power both to exchanging system and to echelon system discharge power to the grid, i.e., G-> A, G-> B.

When the SOCs of both exchanging system and echelon system are not full (SOC <0.8), power grid, on the valley state or even light load, should charge to the exchanging system and echelon system to play an active role in valley filling.

(2) Power grid charge to exchanging system, i.e., G-> A.

When the SOC of echelon system is sufficient (SOC>0.8), and the SOC of the exchanging system is not full (SOC <0.8), power grid charges the exchanging system to play an active role in valley filling.

(3) Power grid charges the echelon system, i.e., G-> B.

When the SOC of exchanging system is sufficient (SOC>0.8), and the SOC of echelon system are not full (SOC <0.8), power grid charges the echelon system to play an active role in valley filling.

(4) No energy exchange.

When the SOC of both exchanging system and echelon system is sufficient (SOC>0.8), there is no any power exchange with power grid.

4.4.3   Overloading Condition

(1) Both exchanging system and echelon system discharge power to the grid, i.e., A-> G, B-> G.

When the SOCs of the exchanging system and echelon system are higher than 0.2 (SOC>0.2), and the power grid is under abnormal overloaded condition, then priority should be given to support power to the grid. In the meantime, discharge to the power grid for auxiliary support to improve the overloaded condition, and regulate the power grid state into a normal operation state.

(2) Exchanging system discharge power to the grid, i.e., A-> G

When the SOC of echelon system is lower than 0.2 (SOC <0.2), and the SOC of exchanging system is higher than 0.2 (SOC> 0.2), only exchanging system is available to discharge power to grid to provide energy support.

(3) Echelon system discharge power to the grid, i.e., B-> G

When the SOC of exchanging system is lower than 0.2 (SOC <0.2), and the SOC of echelon system is higher than 0.2 (SOC> 0.2), only echelon system is available to discharge power to grid to provide energy support.

(4) No energy exchange.

When both the SOC of exchanging system and the SOC of echelon system are pretty low (SOC <0.2), and the power demand cannot be met, then there will be no energy exchange between the systems and the grid. In practice, we should make reasonable forecast and planning according to the operation state trends to avoid this kind of situation.

Based on the information data flow in Table 1, we can reasonably and effectively control the operation state of IS, through information fusion. According to the real-time status of power grid and the corresponding state information of the echelon battery system and charge exchange system inside IS, we can perform fast and reliable grid-connected control based on the results of information fusion. The control flow is shown in Figure 5.

Table 2. The example of grid-connected control based on information fusion

Time	State of grid	State and response of echelon station (B)	State and response of exchanging station (A)
9:00	peak load	SOC>0.8	SOC>0.8
		B—>G, A—>G	
10:30	peak load	SOC>0.8	0.8>SOC>0.2
		B—>G, B—>A	
11:00	peak load	0.8>SOC>0.2	0.8>SOC>0.2
		B—>A	
12:00	valley load	0.8>SOC>0.2	0.8>SOC>0.2
		G—>A, G—>B	
13:00	valley load	0.8>SOC>0.2	SOC>0.8
		G—>B	
14:00	peak load	SOC>0.8	SOC>0.8
		B—>G, A—>G	
14:30	overloading	SOC>0.8	SOC>0.8
		B—>G, A—>G	
14:45	peak load	0.8>SOC>0.2	0.8>SOC>0.2
		B—>A	
15:30	valley load	SOC<0.2	0.8>SOC>0.2
		G—>A, G—>B	
16:00	valley load	SOC<0.2	SOC>0.8
		G—>A, G—>B	

4.5   Grid Control Case

An IS contains charge exchange system and echelon battery system, which is shown in Figure 1. According to the grid-connected control strategy proposed above, its operation state is monitored in a certain period, which is shown in Table 2.

This case shows that, the grid-connected control strategy discussed above can play a role in clipping the peak load and filling the valley load, according to the current state of power grid. When power grid is operating in overloaded condition, IS can provide power support to the grid, be an auxiliary to power grid for improving the degree of overloading, and regulate power grid into normal state.

5   Conclusion

Information fusion strategy proposed in this paper has been implemented in an actual project. Analysis and study of the information flow generated along the energy flow in the integrated station, the following three aspects have been covered in this paper:

(1)Analyze the distribution and transmission of information flow in the integrated station;

(2)The battery management system is managed using RFID, IOT, GPS technology. As a result, battery information collecting speed and accuracy and battery replacement efficiency are all improved.

(3)Using information fusion technology, grid-connection control for the integrated is studied and a grid-connection strategy is proposed based on the technology.

The practical example shows that the grid-connection strategy based on information fusion technology achieves the role of load shifting for the grid. When the grid is overloading, the integrated station can support the grid and the grid will be adjusted to normal operation state.

Acknowledgement. This work is supported by NSFC-EPSRC Collaborative Project (NSFC-No.513111025-2013, EPSRC-EP/L001063/1), State Grid Corporation of China.

References

1. Su, W., Eichi, H.R., Zeng, W., Chow, M.: A survey on the electrification of transportation in a smart grid environment. IEEE Transactions on Industrial Informatics 8(1), 1–10 (2012)
2. Ipakchi, A., Albuyeh, F.: Grid of the future. IEEE Power and Energy Magazine 7(2), 52–62 (2009)
3. Smith, M.: Batteries versus biomass as a transport solution. Nature 457(7231), 785 (2009)
4. Sovacool, B.K., Hirsh, R.F.: Beyond batteries: An examination of the benefits and barriers to plug-in hybrid electric vehicles (PHEVs) and a vehicle-to-grid (V2G) transition. Energy Policy 37(3), 1095–1103 (2009)
5. Singh, M., Kar, I., Kumar, P.: Influence of EV on grid power quality and optimizing the charging schedule to mitigate voltage imbalance and reduce power loss. In: 2010 14th International Power Electronics and Motion Control Conference, T2, pp. 196–203. IEEE Press (2010)

6. Deilami, S., Masoum, A.S., Moses, P.S.: Real-time coordination of plug-in electric vehicle charging in smart grids to minimize power losses and improve voltage profile. IEEE Transactions on Smart Grid 2(3), 456–467 (2011)

7. Sweda, T., Klabjan, D.: An agent-based decision support system for electric vehicle charging infrastructure deployment. In: 2011 IEEE Vehicle Power and Propulsion Conference, pp. 1–5. IEEE Press (2011)

8. Mets, K., Verschueren, T., Haerick, W.: Optimizing smart energy control strategies for plug-in hybrid electric vehicle charging. In: 2010 IEEE/IFIP Network Operations and Management Symposium Workshops, pp. 293–299. IEEE Press (2010)

9. Luo, R.C., Kay, M.G.: Multisensor integration and fusion in intelligent systems. IEEE Transactions on Systems, Man and Cybernetics 19(5), 901–931 (1989)

10. Liu, Y., Wang, Y., Peng, M., Guo, C.: A fault diagnosis method for power system based on multilayer information fusion structure. In: 2010 IEEE Power and Energy Society General Meeting, pp. 1–5. IEEE Press (2010)

11. Hu, H., Qian, S., Wang, J., Shi, Z.: Application of Information Fusion Technology in the Remote State On-line Monitoring and Fault Diagnosing System for Power Transformer. In: 8th International Conference on Electronic Measurement and Instruments, pp. 550–555. IEEE press (2007)

12. Xia, F., Zhang, H., Huang, C.: Fault diagnosis on power plant with information fusion technology. In: IECON 2011-37th Annual Conference on IEEE Industrial Electronics Society, pp. 2370–2375. IEEE Press (2011)

13. Xue, J., Tao, S., Zhou, S., Xiao, X.: Analysis of sag vulerability area of a large power system based on information fusion of CIM. In: 2013 International Conference on Technological Advances in Electrical, Electronics and Computer Engineering, pp. 483–487. IEEE Press (2013)

14. Want, R.: An introduction to RFID technology. IEEE Pervasive Computing 5(1), 25–33 (2006)

15. Tu, M., Lin, J., Chen, R.: Agent-based control framework for mass customization Manufacturing with UHF RFID technology. IEEE Systems Journal 3(3), 343–359 (2009)

16. Ranky, P.G.: Engineering management-focused radio frequency identification (RFID) model solutions. IEEE Engineering Management Review 35(2), 20–30 (2007)

17. Zhou, J., Hu, L., Wang, F.: An efficient multidimensional fusion algorithm for IoT data based on partitioning. Tsinghua Science and Technology 18(4) (2013)

18. Moore, P., Crossley, P.: GPS applications in power systems. Part 1: Introduction to GPS. Power Engineering Journal 13(1), 33–39 (1999)

Experimental Study on EV Purchases Assisted by Multi-agents Representing a Set of Questionnaires

Yusheng Xue[1], Juai Wu[2,1], Dongliang Xie[1], Kang Li[3], Yu Zhang[4],
Fushuan Wen[5], Bin Cai[2,1], Qiuwei Wu[6], and Guangya Yang[6]

[1] State Grid Electric Power Research Institute, Nanjing 210003, China
[2] Nanjing University of Science & Technology, Nanjing 210094, China
[3] Queen's University, Belfast, Northern Ireland, UK
[4] State Grid Shanghai Municipal Electric Power Company, Shanghai 200122, China
[5] College of Electrical Engineering, Zhejiang University, Hangzhou 310027, China
[6] Technical University of Denmark, Lyngby 2800, Denmark

Abstract. An experimental economics (EE) method is used to analyze the influences of subjective willingness on the development of the electric vehicle (EV) industry. It is difficult to run large-scale EE-based simulations and to support decision optimizations due to the limited number of qualified human participants and the incomparability among repeated trials. Taking the customers' willingness to buy EVs as an example, this paper extracts multi-layer correlation information from a limited number of questionnaires and builds a multi-agent model to match the probabilistic distributions of multi-responder behaviors, for the purpose of reflecting the truly statistic information embedded from the questionnaires. The vraisemblance of both the model and the algorithm is validated by comparing the agent-based Monte Carlo simulation results with the questionnaire-based deduction results. Based on the work presented in this paper, the influence of a key factor on the EV development can therefore be analyzed by using a simulation platform with mixed inputs from agents modelled in this paper and human participants.

Keywords: Experimental economics, human experimenters, multi-agents, behavior analysis, knowledge extraction, willingness to buy EVs.

0   Introduction

Nowadays, China is facing an unprecedented challenge of serious air pollution. In 2013, large-scale smog influenced more than 100 large cities located in 30 provinces. The annual average number of smog days is 29.9 in 2013, reaching the peak value in the past 52 years. Drastic growth of motor vehicles partly contributes to the heavy smog in metropolis areas. There are about 0.14 billion cars on road in China, which is only second to the USA and is still increasing by 13.7% annually and 25% nitric oxide emissions are from vehicles. Intensive exhaust gas emission is a major source of urban air pollution [1].

The electric vehicle (EV), including battery EV and plug-in hybrid EV, is a very promising technical alternative to replace oil-powered vehicles. Governments of many

K. Li et al. (Eds.): LSMS/ICSEE 2014, Part III, CCIS 463, pp. 449–459, 2014.

cities have put forward incentive policies to support the development of EV industry. In order to avoid the blindness in this process, it is necessary to evaluate customers' response to policies quantitatively in advance, and to deal with challenges when integrating a huge number of EVs into power systems [2]. Mass data should be collected, modelled and analyzed to help understand actual operation of EV-involved systems, policy implementation and their consequences, customers' preferences on buying and using EVs, and what's more, the relations between the EV's prospect and participants' willingness.

Borrowing the simulation-based optimization and decision-making concepts in natural scientific researches, experimental economics (EE) analyzes the intrinsic laws guiding human behaviors in a given economic environment through controllable and repeatable experiments, in order to prove theoretical hypothesis in economics and validate the consequence of a given economic policy [3-4]. It offers a powerful support platform for mechanism research and decision optimization by considering the influences of subjective willingness of participants and game behaviors, which is absent in the traditional researches of market economy and social science. However, a classical EE-based simulation is not practical as it is time-consuming in training participants and conducting repetitive experiments. Besides that, the consistency of a participant's behaviors in the same scenario of different trials cannot be guaranteed. On the contrary, special subjective willingness and game behaviors in the minority could be ignored if all the human participants were replaced by well-programed multi-agents.

Combining multi-agents and human participants, and using the former to replace the majority of participants through matching the statistical distribution of a certain group of people and the latter to represent the behaviors of other participants, the EE-based simulation can be used to analyze social scientific problems quantitatively [5-7]. It is critical to ensure the consistency of statistical results between multi-agents and corresponding human participant group. Moreover, the difficulties in data mining, information extraction and multi-agent modeling have to be overcome before reaching the final goal.

The relevant data of car purchasing and driving behaviors [8-10] can be obtained from transaction records, videos, GPS trackers and ultra-sonic detections, etc. for fossil fuel-powered vehicles [11]. However, EV's data can hardly be obtained from the above-mentioned methods due to rare number of application cases. A questionnaire survey [12-13] sometimes is the only option. It is a simple, efficient, and repeatable method to investigate and understand subjective behavior preferences of human participants, which is popular in marketing, social science and economics studies.

Current researches of the agent algorithm mainly focus on the development of an agent's logic deduction ability [14-15], using decision optimization theories or expert-knowledge systems to describe internal mechanisms of a decision-making process mathematically. But the behavior of "well-developed" multi-agents is statistically far different from those of the experimenters.

This paper extracts multi-layer correlation information from different factors from questionnaires about customers' willingness to buy EVs. This can reflect fully the multi-dimensional correlation among uncertain behaviors of participants. Embedded with this information, a stochastic multi-agent model can be created by fitting with the probabilistic distributions of the behaviors. Then, any target vehicle described by a combination of factors can be used to evaluate the goodness of fit in the purchase ratio between the outputs from the multi-agents and questionnaires.

1   Probabilistic Information of Respondents' Willingness

1.1   Characteristic Factors of EV Purchasers' Willingness

According to the world-wide investigations of consumers' willingness to buy EVs, the Deloitte's Global Manufacturing Industry Group published a report [16] indicating that the top 5 factors an EV purchaser considers are the maximum range, minimum charging time, price difference compared with the same class of oil-powered vehicles, purchase price and fuel price. The report gives the statistical result for each factor indicating a psychological threshold people of a given percentage have willingness to buy EVs, but the result considering the correlative information among factors is not presented. Due to the lack of necessary information, the joint distribution among the 5 factors cannot be obtained, which is indispensable to a complete understanding of individual purchasing willingness. The reason is that an agent considers to buy an EV if the minimum requirements of all the factors are satisfied. The satisfaction of any single factor is not a sufficient condition, but a necessary one.

Table 1. Questions Involved in the Questionnaire

①What is the minimum range that an EV would need before you would consider buying it ?
A.80km or less B.160km or less C.320km or less D.480km or less E.640km or less
②Considering your expected vehicle use, what is the longest time to fully recharge the battery that you would consider acceptable when buying an EV ?
A. 8h or less   B. 4h or less   C. 2h or less   D. 1h or less   E. 0.5h or less
③How much more would you be willing to pay for an EV compared to a similar vehicle with a gasoline engine ?
A. Same price or less   B.10% more   C.20% more   D.30% more   E.40% more
④If you were considering buying an EV, in which of the following price ranges would you be shopping ?
A.≤￥120,000   B.≤￥180,000   C.≤￥240,000 D.≤￥300,000 E. >￥300,000
⑤At what price for gasoline would you be much more likely to consider buying an EV ?
A.￥6.75/lit   B.￥8.00/lit   C.￥9.50/lit   D.￥10.50/lit   E.>￥10.50/lit
⑥If you were considering buying an EV, what is your main concern ? (Please rank order)
A. Range   B. Charge time   C. Price difference   D. Purchase price   E. Fuel price

Therefore, a questionnaire survey mainly among Chinese young people between the age of 25 and 35 (see Table 1) was conducted by the authors of this paper. Respondents were asked to answer what are the minimum acceptable psychological thresholds of the given factors when he/she was choosing an EV. Consequently, the survey gathered 200 effective questionnaires. A simple counting work gives almost the same distributions over the psychological threshold of any single factor, which matches well with the

China part in the Deloitte's report [16]. The data therefore provides a credible source for mining deep information including the joint probability distributions hiding in different questionnaires.

1.2   Ranking the Importance of Characteristic Factors

Although each question in the questionnaire involves only 1 characteristic factor, the answer sheet reveals a deterministic opinion about the relations among all factors. Therefore, the joint probability information of the mentioned psychological thresholds of the 5 factors must have been embedded in the universal set of the questionnaires.

Table 2. Multi-layer distributions of factors' importance (No correlation considered)

Choosing frequency	Range	Charge time	Price difference	Purchase price	Fuel price
1st	73	23	21	74	9
2nd	57	75	25	24	19
3rd	38	52	32	48	30
4th	23	29	64	36	48
5th	9	21	58	18	94

If multi-agent model is adopted to replace a group of respondents, the threshold value of each factor for each agent must be tuned based on the high-dimensional joint probability distribution reflecting respondents' willingness. 200 samples are far from enough to create an accurate joint probability of the 5 factors by any conventional method. In order to overcome this difficulty, one more question is included in the questionnaire to request each respondent sorting the factors by their importance. The proportion of a factor being chosen as the first important one is listed in the following in the descending order: purchase price (37.0%), range (36.5%), charge time (11.5%), price difference (10.5%), and fuel price (4.5%). The proportions of these factors selected as the second important ones are: purchase price (12.0%), range (28.5%), charge time (37.5%), price difference (12.5%), and fuel price (9.5%), resulted from Table 2. However, it is impossible to know from these questionnaires the probability that respondents take the range as the second important factor while taking the purchase price as the first important one. This question is not answered by this table, as well as from the results shown in [16]. The actual conditional probability is 40.5%, and is not equal to the total probability 28.5%. In order to reproduce the joint probabilistic distribution of such group of factors after bulk samplings, the sampling agents have to be simulated based on the joint probabilistic distribution.

In order to store joint probabilistic data effectively, the data structure with multiple layers in the applications of conditional probability is used. The concept of "sorted importance layer" is defined because the ordering of factors in the sampling process may obviously affect the use of questionnaire information. In the data structure, layer i corresponds to the frequency distribution of the ith important factor in all n-i+1 possibilities, and the joint frequency distribution between layer i and n-i possibilities in layer i+1, here n is the total number of factors. The latter distribution depends not only on the sampling in the current layer, but also on previous sampling results. The ordering of the sampling process uses the importance data associated to sample factors, rather than the reverse. The joint probability of a layer with smaller i implies larger entropy, so it needs support from a sufficient number of samples. The entropy of layers with larger i is relatively smaller, so approximate statistical results could be used.

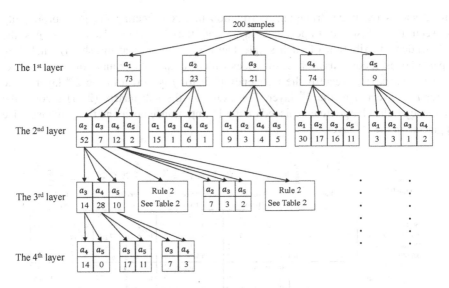

Fig. 1. Multi-layer frequency distributions

Given the above description, Fig. 1 shows a tree-like correlation data structure among adjacent layers for sorting factor importance, reflecting the correlation of frequency distributions. Since there is no uncertainty associated in the final layer, the "tree" contains 4 layers to present the correlation among 5 factors. Here, a_k (k =1,2,3,4,5) respectively represents the 5 features, namely range, charge time, price difference, purchase price and fuel price.

1.3   Conducting Rules for Insufficient Correlation Data of Importance

The joint frequencies emerged on the correlation "tree" are counted along with the layers from top to bottom. However, in the counting, the case of insufficient samples often emerges in the bottom layers of the "tree". If the frequency is too low at certain node, which means the information provided to next layers could be meaningless, approximate distributions should be used to replace the original ones for the purpose of filling up missing information. The designed rules to control the replacement are as following:

Rule 1: If the number of samples in the selected layer is sufficient, the frequency counting is strictly conducted on the corresponding data from questionnaires.

Rule 2: If the number of samples in the selected layer is insufficient, i.e. the number is less than a threshold value α (α is set as 8 in Fig. 1), the correlation between factors is ignored and the independent distribution of the corresponding factor is used directly.

Rule 2 will be adopted in the cases of quite low frequencies; therefore the influence on the accuracy is very limited, which is confirmed through a number of simulations.

1.4   Information Extraction about Factors' Psychological Thresholds

Fig. 2 shows the upper-triangular correlation matrix reflecting the joint probability between the 1st sorted importance layer and the 2nd layer (Table 3 shows the independent distribution of each single factor's psychological threshold). Including repeated ones, it contains $n \times n$ sub-matrices with n equal to the number of factors. Here, the sub-matrix (i, j) records the frequency if factor j is drawn in the 2nd layer while factor i is selected in the 1st layer. For example, the sub-matrix $(2, 4)$ records the frequency if the charge time is placed on the 1st importance layer and the purchase price is placed on the 2nd layer.

The choosing frequency		0-80	80-160	160-320	320-480	480-640	≤8	≤4	≤2	≤1	≤0.5	≥40%	≥30%	≥20%	≥10%	≤0%	□30	≤30	≤24	≤18	≤12	6.75	8.00	9.50	10.50	□10.50
		Range					Charge Time					Price Difference					Purchase Price					Fuel Price				
0-80		0	0	0	0	0	4	7	2	0	1	0	0	0	3	11	0	1	2	2	9	4	1	4	2	3
80-160		0	0	0	0	0	16	26	11	6	3	1	0	10	12	39	0	2	6	25	29	4	8	20	18	12
160-320	Range	0	0	0	0	0	18	29	24	7	4	0	0	12	32	38	0	6	11	37	28	3	12	34	20	13
320-480	(km)	0	0	0	0	0	7	7	6	3	1	0	4	1	2	17	2	2	3	6	11	2	3	5	9	5
480-640		0	0	0	0	0	7	2	6	1	2	0	0	0	1	17	0	0	2	6	10	2	3	3	4	6
≤8							0	0	0	0	0	0	2	6	9	35	0	1	11	15	25	5	5	19	13	10
≤4	Charge						0	0	0	0	0	1	0	7	25	38	0	7	4	33	27	5	10	26	15	15
≤2	Time						0	0	0	0	0	0	0	9	9	31	1	2	3	20	23	4	9	12	18	6
≤1	(h)						0	0	0	0	0	0	2	1	5	9	1	1	5	5	5	1	2	6	3	5
≤0.5							0	0	0	0	0	0	0	0	2	9	0	0	1	3	7	0	1	3	4	3
≥40%												0	0	0	0	0	0	0	0	1	0	0	0	0	1	0
≥30%	Price											0	0	0	0	0	0	2	1	1	0	0	1	2	1	0
≥20%	Difference											0	0	0	0	0	1	3	7	7	5	1	4	7	6	5
≥10%												0	0	0	0	0	0	6	8	28	8	2	8	17	14	9
≤0%												0	0	0	0	0	1	0	8	39	74	12	14	40	31	25
□30																	0	0	0	0	0	0	1	0	0	1
≤30	Purchase																0	0	0	0	0	1	1	4	2	3
≤24	Price																0	0	0	0	0	2	3	9	4	6
≤18	(¥10,000)																0	0	0	0	0	4	10	30	23	9
≤12																	0	0	0	0	0	8	12	23	24	20
6.75																						0	0	0	0	0
8.00	Fuel																					0	0	0	0	0
9.50	Price																					0	0	0	0	0
10.50	(¥/lit)																					0	0	0	0	0
□10.50																						0	0	0	0	0

Fig. 2. The joint distribution of 2-layer factors' psychological threshold

The dimension of each sub-matrix is decided by the number of the corresponding factor's thresholds, which is equal to 5 in Fig. 2. Every row represents a threshold value $d_{1.g_1}$ of the 1st important factor, and every column represents certain threshold value $d_{2.g_2}$ of the 2nd important factor. The value of a sub-matrix's element records the joint frequency in which thresholds $d_{1.g_1}$ and $d_{2.g_2}$ are both selected by the questionnaire's responders. Fig. 3 shows the flow chart of obtaining the complete joint frequency $d_{1.g_1}$ $d_{2.g_2}$ $d_{3.g_3}$ $d_{4.g_4}$ $d_{5.g_5}$ among all factors.

Table 3. The independent distribution of each single factor's psychological thresholds, $d_{1.g_1}$

Characteristic factor	Grade	Psychological threshold value of each grade	Choosing ratio %
Range (km)	1	0-80	7.0
	2	80-160	31.0
	3	160-320	41.0
	4	320-480	12.0
	5	480-640	9.0
Charge time (h)	1	≤ 8	26.0
	2	≤ 4	35.5
	3	≤ 2	24.5
	4	≤ 1	8.5
	5	≤ 0.5	5.5
Price difference	1	$\geq 40\%$	0.5
	2	$\geq 30\%$	2.0
	3	$\geq 20\%$	11.5
	4	$\geq 10\%$	25.0
	5	$\leq 0\%$	61.0
Purchase price (¥ 10,000)	1	>30	1.0
	2	≤ 30	5.5
	3	≤ 24	12.0
	4	≤ 18	38.0
	5	≤ 12	43.5
Fuel price (¥ /lit)	1	6.75	7.5
	2	8.00	13.5
	3	9.50	33.0
	4	10.50	26.5
	5	> 10.50	19.5

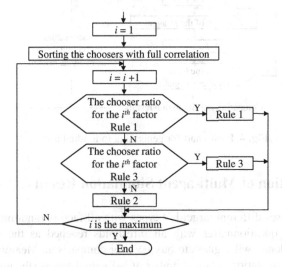

Fig. 3. The choosers' sorting algorithm based on factors' psychological thresholds

Since there are more choices for a factor's psychological threshold, Rule 3 is introduced to fully use information in answer sheets and reduce approximation error.

Rule 3: If the number of samples is less than a threshold value β in the group where respondents choose exactly the given psychological threshold, a new group will be used for counting where respondents choose a value equal or larger than the given psychological threshold.

Rule 3 will be considered prior to Rule 2 if the frequency in a sub-matrix's element is lower than β. Only if the number of samples is still insufficient after applying Rule 3, Rule 2 will be launched.

2   Construction of the Multi-agent Model

The first half of Fig. 4 shows the algorithm to build multi-agents reflecting customers' willingness to buy EVs based on multi-dimensional information inside the questionnaires. Its key part is the extraction of joint probabilistic distributions from factors' sorted importance data, and the distribution of factors' psychological thresholds. The second half of Fig. 4 uses these distributions to generate individual agents as many as needed.

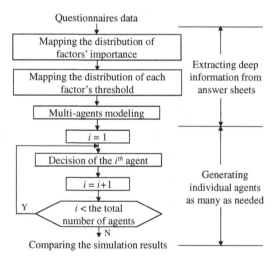

Fig. 4. Flow chart for generating individual agents

3   Verification of Multi-agent Simulation Results

The verification uses different target EV types with all factors randomly selected, and uses the ratio of questionnaires with all thresholds reached as the standard result reflecting respondents' willingness to buy EVs for comparison. Meanwhile, similar to the Monte-Carlo simulation, a large number of individual agents (the number of agents is 100,000 in the cases below) are generated by the multi-agent model to obtain the ratio of choosers who satisfy with this EV type (the purchase ratio below for short), as well as errors compared with standard results. The statistics of simulation errors from a lot of trials of variant EV types will show the effectiveness of the rules above and the multi-agent model.

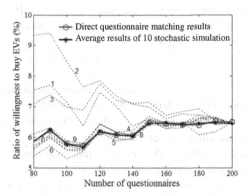

Fig. 5. The comparison between agent-based simulation results and direct questionnaire matching results

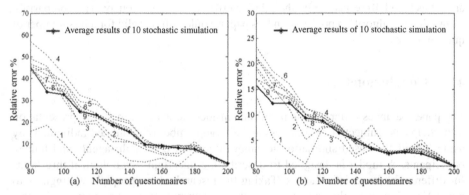

Fig. 6. The influences of the number of questionnaires on the simulation error

Table 4. Simulation scenarios

Vehicle parameters	Range (km)	Charge time (h)	Price difference	Purchase price (¥ 10,000)	Fuel price (¥ /lit)
Scenario	160	0.5	≥ 10%	12	9.5

In Fig. 5, the horizontal axis represents the number of questionnaires used for counting actual ratio of willingness to buy EVs; the vertical axis is the ratio. Taking a target EV shown in Table 4 as an example, the solid line with circles in this figure reflects the actual ratio from different number of questionnaires. As shown in this figure, if the number of questionnaires used is equal or greater than 150, the purchase ratio of all the respondents is better matched. The circle over 200 on the horizontal axis is the actual purchase ratio (6.5%) for the target EV, which is also the standard result used to evaluate the simulation accuracy.

Fig. 5 also includes a solid line marked with *, highly matched with the actual purchase ratio, which reflects the average purchase ratio of 10 stochastic simulation results. 9 dash lines represent the average ratios from 1-9 stochastic simulation results respectively. The numbers tagged beside dash lines are their simulation times. Obviously, dashed lines converge quickly to the solid line if the simulation times increase.

Fig. 6 shows the influence of the above rules on the simulation error. The horizontal axis represents the number of questionnaires taken for extracting the distributions of willingness to buy EVs. The vertical axis shows the relative error. Otherwise the symbols and legends are similar to corresponding ones in Fig. 5. Fig. 6(a) shows the result where threshold α and β that control the switches towards rule 2 and 3 respectively are both set as 0, indicating that these two rules will never be used. The figure shows that the relative error can be limited to a level below 10% only when the number of questionnaires is equal or greater than 150, which indicates that the degree of information loss influences the simulation precision seriously. In Fig. 6(b), α and β are set as 8 and 40 respectively, the relative error decrease obviously. The comparison shows that the algorithm proposed in this paper would still be valid for a less number of questionnaires.

4   Conclusions

The paper balances different computation considerations between strong presentation of participants' subjective willingness and a large number of simulated individuals by using hybrid EE based simulation techniques combined of multi-agents and human participants. A model is designed to describe the uncertain psychological thresholds for different characteristic factors. Taking the research about people's willingness to buy EVs as an example, the joint probability distribution can be extracted by this model from a limited number of relevant questionnaires. A probabilistic multi-agent model is then constructed to fully reproduce the distribution of respondents' willingness. Bulk tests on variant target EV types prove the vraisemblance of the model. Based on that, agents, instead of a lot of respondents, with a few of human participants can attend any times of EE based simulation. The method increases the scale of simulation, maintains comparability among repeated trials, holds the color of an EE method, and reflects the effect of online attendance of human participants.

Acknowledgement. This work is supported by NSFC-EPSRC Collaborative Project (NSFC-No.51361130153, EPSRC-EP/L001063/1), State Grid Corporation of China.

References

1. Zhang, B., Yang, Y., Zhong, Y.: This year the national average of 29.9 days of haze TICO most of the past 52 years,
   http://www.wantinews.com/news-6111485-This-year-the-national-a
   verage-of-299-days-of-haze-TICO-most-of-the-past-52-years.html
2. Xue, Y., Xiao, S.: Generalized congestion of power systems: insights from the massive blackouts in India. Mod. Power Syst. Clean Energy 2, 91–100 (2013)
3. Du, N.: Experimental Economics (in Chinese). Shanghai University of Finance and Economics Press, Shanghai (2008)
4. Xue, Y., Peng, H., Wu, Q.: Analysis and simulation of interaction between electricity market and power system dynamics (in Chinese). Automation of Electric Power Systems 23, 14–20 (2006)
5. Huang, J., Xue, Y., Xu, J., Xue, F., Zou, Y.: Dynamic simulation platform for power market and power system part one function design (in Chinese). Automation of Electric Power Systems 10, 16–23 (2011)
6. Xie, D., Xue, Y., Xue, F., Luo, J.: Dynamic simulation platform for power market and power system part two support layer design (in Chinese). Automation of Electric Power Systems 11, 1–7 (2011)
7. Xie, D., Xue, Y., Xue, F., Luo, J.: Dynamic simulation platform for power market and power system part three application layer design (in Chinese). Automation of Electric Power Systems 12, 7–14 (2011)
8. Wu, Q., Nielsen, A.H., Østergaard, J.: Driving pattern analysis for electric vehicle (EV) grid integration study. In: Proceedings of IEEE Conference on Innovative Smart Grid Technology Europe, Gothenburg, Sweden, pp. 1–6 (2010)
9. Wang, H., Wen, F., Xin, J.: Charging and discharging characteristics of electric vehicles as well as their impacts on distribution systems (in Chinese). Journal of North China Electric Power University 5, 17–24 (2011)
10. Qian, K., Zhou, C., Allan, M.: Modeling of load demand due to EV battery charging in distribution systems. IEEE Trans. on Power Systems 2, 802–810 (2011)
11. Shao, C., Zhao, Y., Wu, G.: Review of road traffic data collection technology (in Chinese). Modern Transportation Technology 6, 66–70 (2006)
12. Navigant Consulting Inc.: Electric vehicle consumer survey,
    http://www.navigantresearch.com/wp-assets/uploads/2013/11/WP
    -EVCS-13-Navigant-Research.pdf
13. Deloitte Touche Tohmatsu: 2014 Global automotive consumer study.,
    http://www.deloitte.com/assets/Dcom-UnitedStates/Local%20Asset
    s/Documents/us_auto_GlobalAutomotiveConsumerStudy_012314.pdf
14. Song, Y., Gao, Z.: Survey on the application of agent technology in electricity market (in Chinese). Proceedings of the CSU-EPSA 3, 111–117 (2008)
15. Sujil, A., Agarwal, S.K., Kumar, R.: Centralized multi-agent implementation for securing critical loads in PV based microgrid. Mod. Power Syst. Clean Energy 1, 369–378 (2014)
16. Deloitte Touche Tohmatsu: Unplugged: Electric vehicle realities versus consumer expectations,
    http://www.deloitte.com/view/en_US/us/Industries/Automotive-
    Manufacturing/f769ebb8bf4b2310VgnVCM1000001a56f00aRCRD.htm#

Temperature Characteristics Research on LiFePO$_4$ Cells Series Battery Pack in Electric Vehicles

Fei Feng, Rengui Lu, Shaojie Zhang, Chunbo Zhu, and Guo Wei

School of Electrical Engineering and Automation, Harbin Institute of Technology,
West Dazhi Street 92, 150001 Harbin, China
ffe423@126.com, {lurengui,zhuchunbo}@hit.edu.cn,
zhangshaojie.1990@163.com,

Abstract. Due to cell-to-cell variations in battery pack, it is hard to manage cells of the battery pack safely and effectively. As a result, the battery pack performance is rapidly degraded, which in turn spread the differences in individual cells. Ambient temperature is a significant factor that influences characteristics of lithium-ion battery and cells variations in the pack. This paper tries to put effort on researching the temperature characteristics of cells series battery pack. The battery model parameters identification tests are designed to analyze the inconsistency characteristics of cells (such as open circuit voltage (OCV), ohmic and polarization resistances, and polarization capacitance) under various ambient temperatures. The results indicate that ohmic and polarization resistances are most significantly increased as the temperature decreases, while the opposite is true for OCV. The variation of cells inconsistency characteristics is obvious along with the temperature change.

Keywords: LiFePO$_4$, Battery pack, Ambient temperature, Resistance, Capacity, Open circuit voltage (OCV).

1 Introduction

In recent years, rechargeable Li-Ion cells have become a more and more favorable choice for EVs applications because of the high power density, high energy density and long lifetime[1-3]. Due to the insufficient voltage and capacity of one single cell, battery cells are connected in series and parallel to meet the requirement for various EVs. Cell variations are not only a critical factor that strongly influences the pack performance, but also bring difficult to model the battery pack accurately. As a result, accurate parameters estimation for battery pack, the basis of the battery management, remains very challenging and problematic. Therefore, it is of great significance for researching the characteristics inconsistency on cells series battery pack.

Cell variations of the capacity, internal resistance, and voltage come principally from two aspects. One is the variation within the manufacturing process. We shall call it "intrinsic cell variations". For instance, due to inhomogeneity in ink mixing and calendaring or deviation in the electrode footprint during cell fabrication, origins of

K. Li et al. (Eds.): LSMS/ICSEE 2014, Part III, CCIS 463, pp. 460–468, 2014.

capacity variations are physical in nature[4]. The other is the different temperature gradient, charge/discharge rate, and degradation along the battery pack during use. We shall call it "extrinsic cell variations". In this paper, the "intrinsic cell variations" is the focus of research.

Ambient temperature is a significant factor that influences characteristics of lithium-ion battery. The hybrid power pulse characteristic (HPPC) tests at different temperatures are conducted to identify the model parameters of cells in battery pack. Therefore, we will able to analyze the characteristics inconsistency at various ambient temperatures.

The remainder of the paper is organized as follows. In section 2, the Thevenin battery model is selected as the object for parameters identification. Section 3 introduces the model parameters identification method. Section 4 describes the cells test, as a case, a total of 26 LiFePO₄ cells are used for carrying out the cells filtering process. In section 5, the results of cells parameters identification at various ambient temperatures is illustrated and the correlation analysis is given. In the final section, some conclusions and final remarks are drawn.

2   Modeling for LiFePO₄ Battery

A suitable battery model is essential to the development of the model-based BMS in real EVs, which requires less computation power and fast response to ever-changing road conditions[5]. To predict the performance of batteries, several different mathematical models exist [6]. Several existing battery models for characterizing battery electrochemical behavior have been constructed by using mathematical approaches from either macroscopic or microscopic perspectives. For microscopic models, partial differential equations are entailed for modeling the electrochemical interaction between two electrodes and electrolytes in order to determine the battery's effective capacity, current–voltage relationship, and heat generation[7, 8]. However, such mathematical models are generally complex and need a numerical method to solve the problems, which are subject to initial and boundary conditions. On the other hand, the electrical models that regard the battery cell as a lumped system have been widely introduced to capture the dynamic characteristics in terms of current and voltage in order to ease the computation[9, 10]. These have been demonstrated to agree with the dynamical response in discharge or charge operation within the order of 5% error, which is acceptable in terms of an engineering viewpoint. Due to the lower amount of required computation relative to the microscopic models, they are capable of determining the battery status online via the designed algorithm. In Refs.[11, 12], seven different battery models are introduced and evaluated. As the basis for model parameter identification, the selected model should not only provide a good fit to the dynamic characteristics of a battery but also have a suitable computational complexity[13]. The Thevenin battery model shown in Fig. 1 is selected as the battery model. The electrical behavior of the Thevenin model can be expressed by Eq.(1).

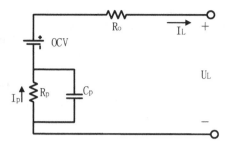

Fig. 1. Schematic diagram of the Thevenin model

$$\begin{cases} \dfrac{dI_P}{dt} = \dfrac{1}{R_P C_P}(I_L - I_P) \\ U_L = OCV - I_L \times R_o - I_p \times R_p \end{cases} \tag{1}$$

Where U_L and I_L indicate the battery terminal voltage and the load current; OCV is the internal voltage source of the battery; Ro and Rp denote the ohmic resistance and polarization resistance, respectively; Ip is polarization current through polarization resistance. The polarization capacitance Cp is used to describe the transient dynamic voltage response during charging and discharging.

3   The Parameters Identification Method

For the proposed equivalent circuit model, the commonly used parameters identification approach is multiple linear regression method with hybrid pulse power characterization data. To identify the parameters in the battery model, we should discretize the dynamic nonlinear model. Then a regression equation is built for the discretization system. We use the regression equation for the battery model which is shown in Eq. (2).

$$\begin{cases} U_{L,k} = OCV - I_{L,k} \times R_o - I_{p,k} \times R_p \\ I_{p,k} = (1 - \exp(\dfrac{-\Delta t}{\tau_P})) \times I_{L,k} + \exp(\dfrac{-\Delta t}{\tau_P}) \times I_{p,k-1} \end{cases} \tag{2}$$

where, $I_{p,k}$ is the out flow current of R_p at the kth sampling intervals, $U_{L,k}$ is the terminal voltage at the kth sampling intervals, $I_{L,k}$ is load current at the kth sampling intervals respectively. The time constant of polarization τ_P required to be set in advance for regression operation, and for different time constants, different correlation coefficients would be calculated; then an optimum time constant value will be achieved through finding the best correlation coefficient by genetic algorithm[13, 14].

4   Battery Experiments and Data Sampling

To identify the model parameters, a battery test bench is designed. The purpose of recognition is based on a criterion and the measurement information of the known systems to estimate the model structure and unknown parameters[11].

4.1   Experiment Setup

The schematic of the built test bench is shown in Fig. 2. It consists of (1) LiFePO$_4$ cells (the key specifications are shown in Table (1);(2) thermal chamber for environment control (The cells were placed in cell holders in the chamber); (3) a battery testing system (Arbin BT2000);(4) a PC with Arbins' Mits Pro Software for battery charging/discharging control;(5) Matlab R2013b for data analysis. The battery testing system is responsible for loading the battery cells/module based on the customer's program with a highest voltage of 60 V. The host computer is used to control the Arbin BT2000 and thermal chamber, and record many quantities, such as load current, terminal voltage, Amp-hours (Ah) and Watthours (Wh). The cell voltage can also be measured by the auxiliary channel, and its measuring range is 0~5 V. The measurement inaccuracy of the current transducer inside the Arbin BT2000 system is within 0.1%. The measured data is transmitted to the host computer through TCP/IP ports. The Arbin BT2000 is connected to the battery cells/module, which is placed inside the thermal chamber to maintain the temperature. The temperature operation range of the thermal chamber is between-55℃ and 85℃.

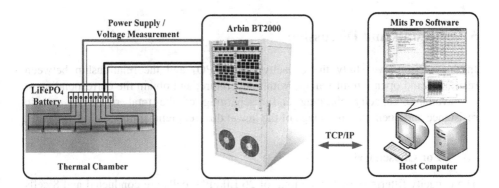

Fig. 2. Schematic of the battery test bench

Table 1. The key specifications of the test samples

Type	Nominal voltage	Nominal capacity	Upper/lower cut-off voltage
LiFePO$_4$	3.2V	5Ah	3.65V/2.5V.

4.2   Battery Test Schedule

The test schedule for our research is shown in Fig. 3, which is designed to collect the cells/pack test data under our designed program. The datasheets and the test method for characteristic test are described in details in Ref.[14]. A total of 26 LiFePO$_4$ cells are used for our research. In the capacity filtering test, we select 8 cells with similar capacities and then ready for packing. Second, characteristic test which is designed to obtain the key parameters is carried out from –10 °C to 30 °C at 10 °C interval. Noted that the characteristic test includes static capacity test which is to analyze the difference of battery pack capacities and HPPC test is executed to get the characteristics of the battery at various ambient temperatures.

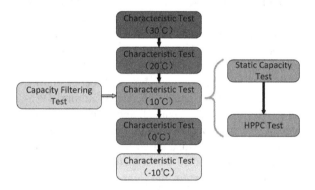

Fig. 3. The battery test schedule

5   Results and Discussion

In this section, we study the capacity of LiFePO$_4$ and the relationship between resistance and open-circuit voltage with temperature, and obtain the corresponding to the available capacity charging and discharging of the total resistance, ohmic resistance and open-circuit voltage of the law at different temperatures.

5.1   For Chosen Cells

The capacity filtering test for the total of 26 LiFePO$_4$ cells are conducted and 8 cells which has the minimum standard deviation are chosen. The capacity values are shown in Table 2.

Table 2. Cell capacities of 8 LiFePO$_4$ cells

Number	Capacity(Ah)	Number	Capacity(Ah)
Cell1	4.9337	Cell5	4.9627
Cell2	4.9517	Cell6	4.9709
Cell3	4.9545	Cell7	4.9719
Cell4	4.9620	Cell8	4.9851

5.2   The Influence of Temperature on Cells OCV Inconsistency

Open circuit voltage (OCV) and discharge capacity diagram is an important basic performance curve reaction characteristic of different types of batteries which curve is also different. At the same temperature the same test rules, the Discharge Capacity-OCV curve repeatability is very good, so the curve is also a method for correcting the SOC estimation error. It is important to study Discharge Capacity-OCV curve at different ambient temperatures. The effect of temperature on the OCV is shown in Fig. 4.

With the available battery capacity reducing, the OCV decreased gradually; At the initial discharging time, the OCV is substantially the same at different ambient temperature. At the end of discharging, the OCV vary with temperature and shows rendering bifurcation, the lower the temperature is, the lower OCV values is under the same available capacity. At the same time, the dispersion enhanced with the drop of the temperature.

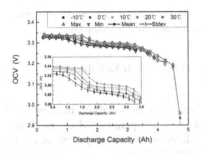 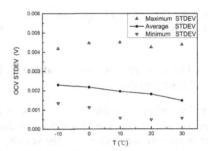

Fig. 4. Effect of different temperature on the OCV

Fig. 5. The standard deviation of the OCV at different temperature

5.3   The Influence of Temperature on Cells Ohmic Resistance Inconsistency

Internal resistance is one of the most important characteristics of accumulators because it limits their specific power and determines their thermal losses during charging and discharging. The internal resistance depends on the depth of charge (DoC) and depth of discharge (DoD) of the accumulators because the chemical composition and electro-physical properties of active electrode materials change during their charging and discharging[15].

In the whole process of battery discharging, the battery ohmic internal resistance remain substantially constant at the same temperature. At the end of the discharge, ohmic resistance increases with the loss of the available capacity.

Ohmic resistance increases as the temperature decreases while the dispersion increases for the ohmic internal resistance is mainly composed of battery plate, a column of metal fittings and ohmic internal resistance of the electrolyte. This electrolyte used in the test for the lithium is salt electrolytes and organic solvent, the electrolyte is mainly rely on electrolyte ion conductive, therefore, to a certain temperature range, temperature is reduced, ion migration velocity is reduced, the electrolyte of the ohmic internal resistance increases, as a result of the electrolyte is the main source of battery ohmic internal resistance, therefore, the ohmic resistance increased with the fall of temperature.

The lower the temperature is, the larger the standard deviation of ohmic resistance is, and the greater dispersion is. It means that ohmic resistance is more sensitive to low temperature.

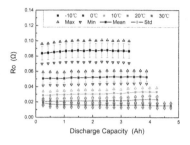

Fig. 6. Effect of different temperature on Ohmic Resistance

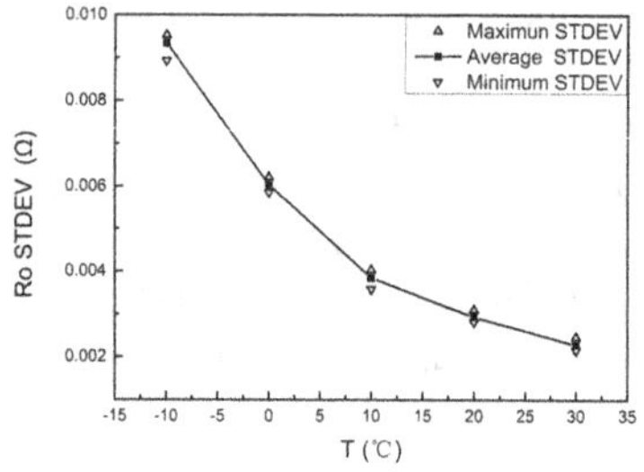

Fig. 7. The standard deviation of the Ohmic Resistance at different temperature

5.4   The Influence of Temperature on Cells Polarization Resistance Inconsistency

The polarization resistance increases with the decline of temperature, meanwhile the dispersion of polarization resistance is enhanced as the temperatures decreases. Due to falling of temperature, It decreases the velocity of the lithium ion, reduces the rate of reaction, and increases the concentration polarization and electrochemical polarization, which makes the polarization resistance increased. The difference of polarization resistance is smaller than ohmic resistance under different temperature. The polarization resistance rise earlier than that at high temperature with the going-on of discharge. The lower the temperature is, the earlier polarization resistance increases. Low temperature slows down the phenomenon of concentration on the diffusion rate.

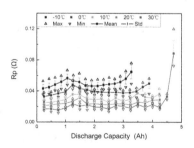
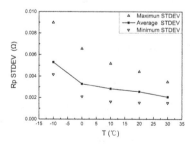

Fig. 8. Effect of different temperature on Polarization Resistance

Fig. 9. The standard deviation of the Polarization Resistance at different temperature

5.5   The Influence of Temperature on Cells Polarization Capacitance Inconsistency

As the battery discharge, polarization capacitance drops at the start time of the discharging, after that it shows a slight rise and then fall last for the end. The value increases as the ambient temperature rises. Polarized capacitor performs significant differences in each cell. As the temperature increases, the inconsistency and dispersion are enhanced.

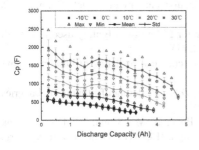

Fig. 10 Effect of different temperature on Polarization Capacitance

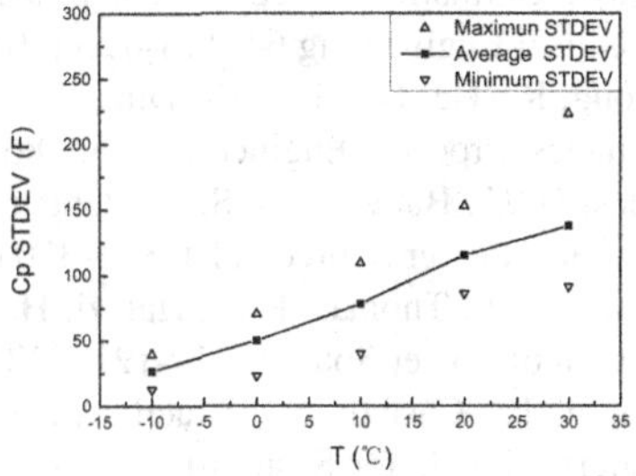

Fig. 11 The standard deviation of the Polarization Capacitance at different temperature

6   Conclusions

Based on the above analysis, the main concluding remarks can be made below.

(1) The Thevenin equivalent circuit model is selected for modeling the LiFePO$_4$ battery, and the offline identification method of the model parameters is improved by a genetic algorithm to get a more accurate relaxation time constant.

(2) The battery test bench is built and the experiment schedule is designed, which includes static capacity test and HPPC test at various ambient temperatures, and the offline model parameters are identified with the HPPC test.

(3) Ambient temperature impact the ohmic resistance of the battery pack, polarization resistance, polarized capacitors obviously. In general, the lower the temperature the greater the ohmic resistance and polarization resistance, the smaller the polarized capacitors. The dispersions are increased with decreasing temperature.

Acknowledgements. This research was supported by the Fundamental Research Funds for the Central Universities (Grant No.HIT.IBRSEM.201306).

References

1. Howlader, A.M., Izumi, Y., Uehara, A., Urasaki, N., Senjyu, T., Yona, A., Saber, A.Y.: A minimal order observer based frequency control strategy for an integrated wind-battery-diesel power system. Energy 46, 168–178 (2012)

2. Moghaddam, A.A., Seifi, A., Niknam, T., Alizadeh Pahlavani, M.R.: Multi-objective operation management of a renewable MG (micro-grid) with back-up micro-turbine/fuel cell/battery hybrid power source. Energy 36, 6490–6507 (2011)

3. Niknam, T., Kavousi Fard, A., Baziar, A.: Multi-objective stochastic distribution feeder reconfiguration problem considering hydrogen and thermal energy production by fuel cell power plants. Energy 42, 563–573 (2012)

4. Dubarry, M., Vuillaume, N., Liaw, B.Y.: From single cell model to battery pack simulation for Li-ion batteries. Journal of Power Sources 186, 500–507 (2009)

5. Chen, X., Shen, W., Cao, Z., Kapoor, A.: Adaptive gain sliding mode observer for state of charge estimation based on combined battery equivalent circuit model. Computers & Chemical Engineering 64, 114–123 (2014)

6. Xiong, R., He, H., Guo, H., Ding, Y.: Modeling for Lithium-Ion Battery used in Electric Vehicles. Procedia Engineering 15, 2869–2874 (2011)

7. Dees, D.W., Battaglia, V.S., Belanger, A.: Electrochemical modeling of lithium polymer batteries. Power Sources 110, 310–320 (2002)

8. Newman, J., Thomas, K.E., Hafezi, H., Wheeler, D.R.: Modeling of lithium-ion batteries. Journal of Power Sources 119-121, 838–843 (2003)

9. Benini, L., Castelli, G., Macii, A., Macii, E., Poncino, M., Scarsi, R.: Discrete-Time Battery Models for System-Level Low-Power Design. IEEE Transactions On Evry Large Scale Integration 9, 630–640 (2001)

10. Gao, L., Liu, S., Douga, R.A.: Dynamic lithium-ion battery model for system simulation. IEEE Transaction On Components And Packaging Technologies 25, 495–505 (2002)

11. He, H., Xiong, R., Fan, J.: Evaluation of Lithium-Ion Battery Equivalent Circuit Models for State of Charge Estimation by an Experimental Approach. Energies 4, 582–598 (2011)

12. He, H., Xiong, R., Guo, H., Li, S.: Comparison study on the battery models used for the energy management of batteries in electric vehicles. Energy Conversion and Management 64, 113–121 (2012)

13. Xiong, R., He, H., Sun, F., Liu, X., Liu, Z.: Model-based state of charge and peak power capability joint estimation of lithium-ion battery in plug-in hybrid electric vehicles. Journal of Power Sources 229, 159–169 (2013)

14. Xiong, R., He, H., Sun, F., Zhao, K.: Evaluation on State of Charge Estimation of Batteries With Adaptive Extended Kalman Filter by Experiment Approach. IEEE Transctions On Vehicular Technology 62, 108–117 (2013)

15. Kolosnitsyn, V.S., Kuzmina, E.V., Mochalov, S.E.: Determination of lithium sulphur batteries internal resistance by the pulsed method during galvanostatic cycling. Journal of Power Sources 252, 28–34 (2014)

Data Scheduling Based on Pricing Strategy and Priority over Smart Grid

Dongfeng Fang[1], Zhou Su[1,2,*], Qichao Xu[1], and Zejun Xu[1]

[1]School of Mechatronic Engineering and Automation
Shanghai University
No. 99, Shangda Road, Baoshan Dist., Shanghai 200444, P. R. China
[2]Faculty of Science and Technology
Waseda University, Tokyo 169-8555, Japan
Shine.fangdongfeng@gmail.com, zhousu@ieee.org

Abstract. Smart Grid has emerged as the next generation power systems. However, the classical scheduling methods are still not suitable for describing the exact feature of data transmission over smart grid because of its different properties from other conventional power systems. Hence, in this paper, by considering both consumers' and power supplies' perspective, different pricing strategies are presented based on user priority and load rate. And the corresponding novel scheduling algorithms are also proposed. The simulation experiments are carried out by comparing the proposed algorithms with other existing scheduling algorithms.

Keywords: Scheduling, Priority, Load Rate, Pricing, Smart Grid.

1   Introduction

Over smart grid system, the demand response of pricing is influencing the power demand time and level through the price made by market [1] [2]. There are a lot information technologies, communication technologies and control technologies used over smart grid which make the real-time pricing possible [3]. The most popular pricing strategies are time-of-use pricing (TOU), critical peak pricing (CPP) and also real-time pricing [4]. One important part of smart grid is smart meter which is an interface for both power consumers and power suppliers to exchange their power information [5][8][9].

About the real-time pricing model, a lot researches have been done before. In paper [6] there is a real-time pricing based on the model of demand side and supply terminals to maximize the consumers and power suppliers' benefits. Also, in paper [7] the authors present a pricing algorithm by setting a different pricing for interruptible and non-interruptible tasks, the simulation results prove that the algorithm can reduce the peak-to-average ratio and coefficient of variation of the aggregated profile well.

---

* Corresponding author.

K. Li et al. (Eds.): LSMS/ICSEE 2014, Part III, CCIS 463, pp. 469–475, 2014.
© Springer-Verlag Berlin Heidelberg 2014

In this paper, we analyze the data transmission over power grid based on scheduling theory. Then we propose pricing strategies by considering different factors. By accounting the pricing strategies and priority of data, the corresponding scheduling algorithms are presented. Simulation results are given by comparing the proposal with other scheduling algorithm. The proposed algorithm based on the average load rate can obtain better performances than others.

2   Data Transmission Analysis

We consider a micro-grid system with one power supplier and I power consumers. Assume that all the data are transmitted in a single channel. Also, all the data transmissions are non-interruptible. And we divide the consumers into three different priority classes. For each use, all the data are divided into different priority classes. Parameters are defined as table 1:

Table 1. Parameters Definition

$D_{i,j}$	The j-th data from user i
$LA_{i,j}$	The latest allowable end time for $D_{i,j}$
$t_{i,j}^a$	The arriving time of $D_{i,j}$
$t_{i,j}^s$	The starting time of $D_{i,j}$ transmission
$t_{i,j}^e$	The end time of $D_{i,j}$ transmission
$t_{i,j}^l$	The last time of $D_{i,j}$ transmission
PU_i	The priority of user i
$PR_{i,j}$	The priority of $D_{i,j}$
$S_{i,j}$	The size of $D_{i,j}$
$BR_{i,j}$	The limit buffer time of $D_{i,j}$
B	The bandwidth of channel

From table 1, then it must follow:

$$t_{i,j}^e > t_{i,j}^s \geq t_{i,j}^a \tag{1}$$

$$t_{i,j}^l = \frac{S_{i,j}}{B} = t_{i,j}^e - t_{i,j}^s \tag{2}$$

Based on the scheduling theories, the delay time of each data is:

$$\max\left(0, t_{i,j}^e - LA_{i,j}\right) \tag{3}$$

Assume there is a buffer for storage data before they are transmitted. For each data, they have their own limit buffer time, so the excess buffer time of each data should be:

$$\max\left(0, t_{i,j}^s - t_{i,j}^a - BR_{i,j}\right) \tag{4}$$

All users are divided into three classes, where the priority is defined as follow:

$$PU_i = \begin{cases} High \\ Middle \\ Low \end{cases} \qquad (5)$$

The priority of data is as follow:

$$PR_{i,j} = k = \begin{cases} 1 \\ 2 \\ 3 \end{cases} \qquad (6)$$

where the larger value of k means the higher priority.

Over the smart grid, the delay time of higher priority data should have a higher hierarchy than that of lower priority data. Therefore, we define the weight function of delay time of user as following:

$$\mu(PU_i) = \begin{cases} \mu_1, PU_i = High \\ \mu_2, PU_i = Middle \\ \mu_3, PU_i = Low \end{cases} \qquad (7)$$

Also, we can get the weigh function of exceed buffer time by

$$v(PU_i) = \begin{cases} v_1, PU_i = High \\ v_2, PU_i = Middle \\ v_3, PU_i = Low \end{cases} \qquad (8)$$

3   Pricing Strategies

Two pricing strategies are going to be introduced as following:
    Strategy 1: Plat pricing based on the priority of user only

$$pricing(PU_i) = C_{PU_i} \qquad (9)$$

where PU_i presents the priority of user i. This pricing strategy gives a price irrelevant with time, but only related with priority of user. This strategy can strictly control the quantities of power demand smaller the power supply. However, the pricing strategy here is not flexible where the price is not relevant with time or the performance of power grid. It is easy to cause the serious traffic of data transmission over power grid.
    Strategy 2: Based on the priority of user and average load rate.

Define DQ_{PU_i} as the data quantities on channel of user with priority PU_i. And define $\overline{MB_{PU_i}}$ as the average maximum available bandwidth of user with priority PU_i. Then the load rate function can be expressed as following:

$$LC\left(t_{i,j}^s, PU_i\right) = \frac{DQ_{PU_i}\left(t_{i,j}^s\right)}{\overline{MB_{PU_i}}} \quad (10)$$

The larger value of load rate means more serious traffic on the transmission channel. Then the pricing function should be:

$$pricing(t_{i,j}^s, PU_i, LC) = C * \left(\frac{DQ_{PU_i}\left(t_{i,j}^s\right)}{\overline{MB_{PU_i}}} - \varphi\right) + C_{PU_i} \quad (11)$$

where parameters C, C_{PU_i} and φ are pre-determined. Parameter φ presents the expected load rate. The value of C presents the factor of influence about load rate for pricing. If the value of load rate is smaller than the expected load rate, then the price of power should be lower than C_{PU_i}, which can motivate the user to use power at that time. If the value of load rate is bigger than the expected load rate, then the price of power should be higher than C_{PU_i}, which can motivate the user to shift their power consumption at that time.

4   Scheduling Algorithms

Based on influences of the power price on consumers, a data transmission scheduling algorithm is proposed, which can balance the consumer's bill and power consumption service quality. There is a weight function based on priority of data and pricing strategy. The algorithm can schedule the data by comparing the value of weight function. The larger value of weigh function means higher priority to be transmitted.

The following is the weight function based on plat pricing and priority of data:

$$\alpha\left(PR_{i,j}, pricing\left(PU_i\right)\right)$$
$$= \frac{PR_{i,j} * C_{PU_i}}{\sum_{j=1}^{J}\sum_{k=1}^{I}\left(PR_{k,j} * C_{PU_k}\right)} \quad (12)$$

From the formulation(12), the higher priority users' higher priority data have higher priority to be transmitted, and the lower priority users' lower priority data have lower priority to be transmitted.

Then, we can get the weight function based on average load rate pricing and priority of data:

$$\alpha\left(PR_{i,j}, pricing\left(t_{i,j}^a, PU_i, LC\right)\right)$$

$$= \frac{PR_{i,j} * \left(C * \left(\dfrac{DQ_{PU_i}\left(t_{i,j}^s\right)}{MB_{PU_i}} - \varphi\right) + C_{PU_i}\right)}{\displaystyle\sum_{j=1}^{J}\sum_{k=1}^{I}\left(PR_{k,j} * \left(C * \left(\dfrac{DQ_{PU_k}\left(t_{i,j}^s\right)}{MB_{PU_k}} - \varphi\right) + C_{PU_k}\right)\right)} \qquad (13)$$

From the formulation (13), higher priority data from the higher priority user with the larger value of average load rate have higher priority to be transmitted, and the lower priority data from the lower priority user with the smaller value of average load rate have lower priority to be transmitted.

5   Simulations

For the simulation, we use three types users: hospital of school, teaching building and also school dorm. Each user has its own priority value.

Set the objective function as follow:

$$f = \sum_{i=1}^{I}\sum_{j=1}^{J}\left(t_{\max LC^{-1},j}^s - t_{\max LC^{-1},j}^a\right) +$$

$$\alpha\sum_{i=1}^{I}\sum_{j=1}^{J}\max\left\{0, \mu(PU_i) * \left(t_{i,j}^e - LA_{i,j}\right)\right\} + \qquad (14)$$

$$\beta\sum_{i=1}^{I}\sum_{j=1}^{J}\max\left\{0, \nu(PU_i) * \left(t_{i,j}^s - t_{i,j}^a - BR_{i,j}\right)\right\}$$

In figure 1, it shows the lasting time of maximum of load rate, which can indicator the traffic situation over the power grid. Obviously, the priority pricing and load (PPL)   algorithm can make sure the lasting time of maximum of load rate minimum, compared with short job first (SJF) and priority pricing (PP) algorithm.

In figure 2, apparently, the PPL algorithm makes the objective function value minimum, compared with SJF and PP algorithm. In other words, the PPL algorithm can find a better point to balance the lasting time of maximum load rate, delay time and exceed buffer time.

From figure 3, we compare the objective function value based on different value of weight parameters. It proves that the PPL algorithm can make the value of objective function lower than SJF and PP algorithms.

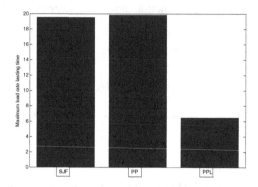

Fig. 1. Comparison of lasting time of maximum load rate from different scheduling methods

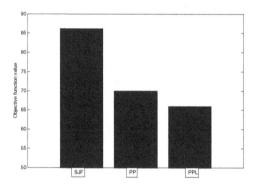

Fig. 2. Comparison of objective function value of different scheduling methods

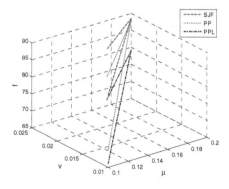

Fig. 3. Comparison of objective function value with different weight parameters

6   Conclusion

This paper has considered pricing behavior for scheduling data transmission over power grid, which is different from the traditional scheduling algorithms. A novel pricing strategy is proposed in this paper, where the pricing strategy can be regarded as a performance of load rate. Also, the priority of user is taken into account for pricing strategy. Moreover, for scheduling algorithm, the priority of date and pricing strategy are considered. As illustrated by simulations, it reduces the time of maximum load rate more efficiently; and it manages the balance of consumption bills and service qualities better.

Acknowledgments. This work was supported in part by the Ministry of Education Research Fund-China Mobile (2012) MCM20121032, Eastern Scholar Program.

References

1. Monsef, H., Wu, B.: Real-time pricing program in a smart grid environment, Simulation. Simulation 89(4), 513–523 (2013)
2. Huang, C., Sarkar, S.: Dynamic Pricing for Distributed Generation in Smart Grid. In: IEEE Green Technologies Conference, pp. 422–429 (2013)
3. Yan, Y., Qian, Y., Sharif, H.: A survey on smart grid communication infrastructures: Motivations, requirements and challenges. IEEE Communications Surveys & Tutorials 15(1), 5–20 (2013)
4. Maqbool, S.D., Ahamed, T.P.I.: Comparison of pursuit ε-Greedy algorithm for load scheduling under real time pricing. In: 2012 IEEE International Conference on Power and Energy (PECon), pp. 515–519 (2012)
5. Shahid, B., Ahmed, Z., Faroqi, A.: Implementation of smart system based on smart grid Smart Meter and smart appliances. In: 2012 2nd Iranian Conference on Smart Grids (ICSG), pp. 1–4. IEEE (2012)
6. Samadi, P., Mohsenian-Rad, A.H., Schober, R.: Optimal real-time pricing algorithm based on utility maximization for smart grid. In: 2010 First IEEE International Conference on Smart Grid Communications (SmartGridComm), pp. 415–420. IEEE,
7. Yue, S., Chen, J., Gu, Y.: Dual-pricing policy for controller-side strategies in demand side management". In: 2011 IEEE International Conference on Smart Grid Communications (SmartGridComm), pp. 357–362. IEEE (2011)
8. Fang, D., Su, Z., Xu, Q.: Analysis of Data Transmission Based on the Priority over Grid Structures. ICIC Express Letters, Part B: Applications, 751–755 (2014)
9. Xu, Q., Su, Z., Han, B., Fang, D., Xu, Z.: Analytical Model for Epidemic Information Dissemination in Mobile Social Networks with a Novel Selfishness Division. Communications in Computer and Information Science (2014)

LiFePO$_4$ Optimal Operation Temperature Range Analysis for EV/HEV

Jinlei Sun, Peng Yang , Rengui Lu, Guo Wei, and Chunbo Zhu

Harbin Institute of Technology Harbin 150001 China
{sunjinlei,lurengui}@hit.edu.cn,
{yangforyang,wg_weiguo}@sina.com,
zhuchunbo@gmail.com

Abstract. The LiFePO4 batteries are widely used in Electric Vehicle(EV)/Hybrid Electric Vehicle(HEV) because of the high energy and power density. However, high environment temperature could accelerate the aging of batteries, while low temperature could reduce output power capability. Therefore, optimal working temperature for batteries should be determined to maintain good performance in all kinds of tough conditions. In this paper, the optimal working temperature range for batteries is analyzed. The capacity loss model is applied to determine the upper limit. The lower limit is calculated taking available capacity and output power loss into consideration. Simulation and experimental results show that the working temperature range between 10°C and 40°C could ensure the performance and available capacity.

Keywords: Electric Vehicle, Hybrid Electric Vehicle, LiFePO$_4$, optimal working temperature.

1   Introduction

With the problems of energy crisis and environment becoming increasingly prominent, Electric Vehicles(EVs)/Hybrid Electric Vehicles(HEVs) have attracted more and more attention[1]. Lithium-ion batteries are becoming the best choice for solving these problems owing to the characteristics of high energy and power density[2]. But the drawbacks such as cost, safety and lifetime are the bottlenecks for EVs/HEVs taking the place of traditional vehicles. The performance of power LiFePO$_4$ tends to be greatly affected by temperature, high temperature may accelerated aging and lead to thermal run away[3]. It is reported that the slow charge transfer at the electrode/electrolyte interface leads to the poor performance at low temperature[2]. At extreme low temperature the cell capacity fades greatly comparing to the nominal capacity under room temperature [4]. Wide range working temperature has great influence on the performance and safety for EVs/HEVs. The traditional fuel vehicles have been developed over 200 years and have been able to withstand the harsh environment, while the EVs/HEVs must solve the problem of battery pack thermal management to get satisfied performance at an extreme cold or hot temperature. The optimal operation temperature range is available to provide references for TMS(Thermal Management System) and to prevent undesirable performance fade caused by environment.

K. Li et al. (Eds.): LSMS/ICSEE 2014, Part III, CCIS 463, pp. 476–485, 2014.
© Springer-Verlag Berlin Heidelberg 2014

The goal of battery thermal management is to maintain the battery within optimal temperature range. For example, the aging and resistance rise caused by high temperature and the available capacity and power fade caused by low temperature[5]. The battery thermal management methods mentioned in the literatures include: the forced air cooling[6], liquid-based thermal management system [7,8],PCM based thermal management system [9,10] and Thermo Electric Cooler(TEC) based heating/cooling. [11,12] The forced air cooling is the traditional method for cooling, the air flows across the surface of battery pack to take the heat away, this method has been used in the Toyota Prius HEV application[13]. The liquid-based thermal management takes the heat away directly or indirectly by liquid such as water, glycol, oil, acetone or even refrigerants. Thanh-Ha Tran designed a flat heat pipe cooling system, which could reduce the thermal resistance by 30% comparing with the natural air cooling[7]. Zhonghao Rao[14] developed a thermal management system whose maximum temperature could be controlled below 50°C when the heat generation rate was lower than 50 W and the maximum temperature difference is below 5°C. The phase change materials(PCM) are developed rapidly recent years, PCM absorb heat released by battery and make the temperature decrease rapidly, the heat is stored in the form of PCM. The heat releases to the battery when in extreme cold environment. The blower and pump are no longer needed in the PCM system. Selman and Al-Hallaj did some research on the PCM and take the PCM to battery thermal management system for the first time. In [15],they established 2D model for comparing four thermal methods: (1) natural convection cooling; (2) presence of aluminum foam heat transfer matrix;(3) use of phase change material (PCM); and (4) combination of aluminum foam and PCM. They came to the conclusion that the use of aluminum foam with PCM causes a significant temperature drop of about 50% compared to the first case of no thermal management. In [16] the PCM and air-based methods are compared and the advantages of heat pipe under extreme cold temperature were highlighted. Chakib Alaoui worked on the TEC heater/cooler based on Peltier effect for several years. The TEC based heater/cooler controls the temperature of cabin and battery pack and took the place of vehicle air conditioning [12]. In [11], the TEC was placed on the surface of each cell for the 24 series connected battery pack. The Coefficient of Performance (COP) under the condition of US06 was as high as 1.2 and the energy consumption is only 4% of the fully charged pack.

Although there were many methods for battery thermal management, the temperature control target is not uniform. Ref[17] argues that the highest battery operating temperature should below 40°C and the maximum temperature differences is within 5°C. The FreedomCAR Battery Test Manual [18] defines the working temperature range to be between -30°C and 52°C. The wide range of working temperature could not ensure the performance of battery pack. Thus, there should be a specific optimal working temperature range for battery pack considering the power and capacity characterizes.

In this paper, the experiments are taken first to test the temperature characterizes of battery. Then the results of Hybrid Pulse Power Characteristic(HPPC) and cold cranking tests are analyzed. The cell capacity loss model is used to analyze the aging of battery under high temperature the power fade is analyzed according to capacity loss and power capability. Finally, the optimal operation temperature range is determined.

2   Experiment Design

2.1   Measurement Equipment

The commercial $LiFePO_4$ used in the temperature characteristic experiment is 5Ah/3.2V (Voltage range 2.5V-3.65V). The test platform contains Arbin BT2000 battery tester(Output current range 0-100A, Voltage range 0-18V, Accuracy 0.02%-0.05%FSR) and Testsky temperature control box (Temperature range -40°C-200°C, Accuracy ±0.5°C). Fig.1 shows the devices for experiment.

Fig. 1. The devices for battery test and temperature control

2.2   HPPC Tests at Different Temperatures

The cell samples are placed in different temperature environments (-20°C,-10°C, 0°C, 10°C, 20°C, 30°C, 40°C, 50°C, 60°C) for 5 hours respectively. Each cell was fully charged by constant current and constant voltage (CCCV) under room temperature. The charge current pulse was 9.37A and the discharge current pulse was 12.5A. The HPPC test was taken every 10% SOC intervals with 1C rate discharge current. The HPPC test profile is shown in Fig. 2. In Fig. 2 the dotted line represents the current, the solid line represents the voltage. The experiment is stopped as soon as the call voltage reaching the cutoff voltage.

2.3   Cold Cranking Tests

According to the FreedomCAR Battery Test Manual [18], the pack should be replaced when capacity fades to 80% of the rated capacity. In order to further study the output power performance, a power pulse start test is taken according to the FreedomCAR Battery Test Manual [18]. The pulse profile is shown in Fig.3.The tests are conducted under different SOCs and temperatures, the maximum output power is measured every 10% SOC internals. The steps are as follows:
1) Charge the cell to fully charged (Constant current and then constant voltage)
2) Discharge to the target SOC
3) Rest for 5 hours under the target temperature

4) Take 3 power pulse tests at constant power, each pulse lasts 2 seconds and rest for 10 seconds. As is shown in Fig.7
5) If the discharge cutoff voltage is met, return to step1) and decrease the power value in step 4)
6) If the steps are finish, repeat step1 and increase the power value in step 4) until the maximum power is found.

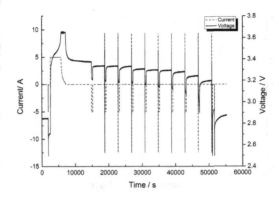

Fig. 2. The HPPC test profile

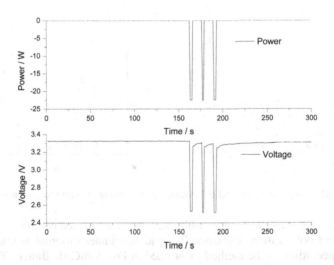

Fig. 3. The constant power start pulse profile

2.4   Results and Discussion

The results of the HPPC tests at different temperatures are shown in Fig.4-Fig.6 and that for cold cranking are shown in Fig.6.

Fig.4 shows that the discharge capacity of the same cell under different temperature conditions. It could be seen that the cell capacities are nearly the same at the temperature between 40°C and 60°C, while the cell capacity decreases obviously with the decrease of temperature, especially below 0°C. The cell capacity is 80% at 0°C and could hardly discharge at -20°C.

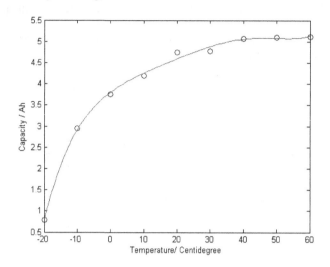

Fig. 4. The cell capacity profile at different temperatures

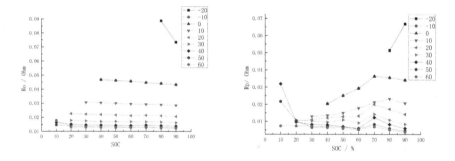

Fig. 5. The cell ohmic resistance and polarization resistance at different temperatures and SOCs

The discharge capacity fades with the decrease of temperature. The ohmic resistance and polarization resistance (Ro and Rp) under different temperatures are indentified according to the method mentioned in FreedomCAR Battery Test Manual [18]. Just as shown in Fig.4. The ohmic resistance changes a little at different SOCs at the same temperature. The ohmic resistance increases with the drop of temperature. The polarization resistance decreases with the drop of temperature, but it changes greatly at different SOC under the same temperature. Due to the discharge capacity is almost zero, the data at -20°C is not universal.

Fig.6 shows the maximum charging and discharging power at different temperatures and SOCs. The maximum charging and discharging power at target SOC is defined as the product of maximum charging/discharging voltage during pulse and the current. The charging power increases with the drop of temperature, while the discharging power fades with the decrease of temperature. At the same temperature the charging and discharging power in the full SOC range are nearly the same.

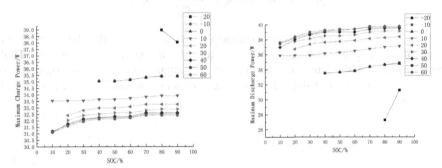

Fig. 6. The maximum charge and discharge power at different temperatures and SOCs

3   The Optimal Operation Temperature Range

As is analyzed in Section 2.3, the target cell capacities at a temperatures higher than 40°C are nearly the same, while the discharge capacities begin to fade below 0°C. Many researches claim that high temperature accelerates the aging[19] and the performance fades during low temperature [20].In this section, the operation temperature range is determined considering the current output power capability and long term lifetime.

3.1   The Determination of Operation Temperature Range Upper Limit

John Wang established the capacity loss model taking DOD, temperature, discharge rate into consideration in his research[19].

$$Q_{loss} = B \cdot \exp(\frac{-31700 + 370.3 \times C_{rate}}{RT})(A_h)^{0.55} \tag{1}$$

Where Q_{loss} is the percentage of capacity loss, B represents the pre-exponential factor, A_h the Ah-throughput, which is expressed as Ah = (cycle number)×(DOD)×(full cell capacity), and z is the power law factor, R is the gas constant. T is the absolute temperature.

Yuejiu Zheng[21] further developed the model and have confirmed the parameter B.

$$B = 10000(\frac{15}{C_{rate}})^{1/3} \tag{2}$$

The 1C rate discharge capacity loss is calculated according to equals (1) and (2)

$$Q_{loss} = 24662 \cdot \exp(\frac{-31329.7}{RT})(A_h)^{0.55}$$

$$(3)$$

The aging experiment takes considerable time and work. To explain the temperature influence on aging, we take the 1C discharge rate with 80% DOD capacity loss model to simulate and analyze. The simulations under the conditions of 10°C to 60°C (10°C internals) are taken. The results are plotted every 50 points, as is shown in Fig.7.

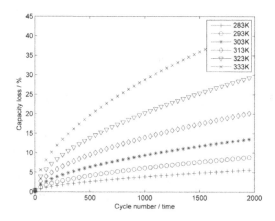

Fig. 7. The capacity loss simulation at different temperatures

When the cycle number comes up to 2000, the capacity losses below 40°C are lower than 20%. The 2000 times cycle is enough for the lifetime of both the battery and vehicle. Additionally, the maximum average temperature in summer is 40°C, the maximum capability for thermal management system is to make the temperatures in and out of the EV/HEV nearly the same. To sum up, 40°C is determined to be the upper limit of operation temperature range to maintain the performance and prevent accurate aging caused by high temperature.

3.2   The Determination of Operation Temperature Range Lower Limit

The low temperature affects the charge transfer at the electrode/electrolyte interface, which leads to the significant plating on the negative electrode during charging. It irreversibly causes the capacity loss. Low temperature affects the driving distance and output power performance for EVs/HEVs.

The maximum output power test results at different SOCs and temperatures are shown in Fig.8. It shows that at room temperature, the maximum output power is 38W with almost no change within the whole SOC range. With the decrease of temperature, the maximum output power fades gradually at the same temperature and different SOCs. For example, at -10°C and 100% SOC the maximum output power is the same as that at room temperature. However, the power differences between 10% and 100% are 10W. When the temperature comes to 10°C, the output power is similar to that of room temperature and the power differences are little with different SOC. Thus, the lower limit of operation temperature range is determined to be 10°C.

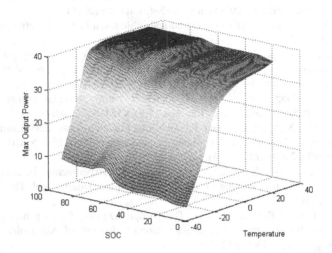

Fig. 8. The maximum output power at different temperature and SOC

4   Conclusion

In this paper, we proposed the optimal temperature operation range for batteries in EV/HEV. We first take the HPPC and cold cranking tests under different temperatures to obtain the temperature characteristics of LiFePO$_4$. And then the upper limit is determined according to the aging model. The lower limit is determined considering discharge capacity loss and output power fades. Finally, the optimal operation temperature range is proved to be 10°C to 40°C according to the experimental results. The range provides a temperature control target for pack thermal management. Working in the proposed temperature range is good for maintaining vehicle in good performance and reducing energy loss during heating or cooling.

Acknowledgments. This research was supported by the National High Technology Research and Development Program of China (2012AA111003) in part and the NSFC-EPSRC Collaborative Research Initiative in Smart Grids and the Integration of Electric Vehicles (51361130153) and Science and Technology Project of State Grid Corporation of China and the Fundamental Research Funds for the Central Universities (Grant No.HIT.IBRSEM.201306).

References

1. Lu, L., Han, X., Li, J., Hua, J., Ouyang, M.: A review on the key issues for lithium-ion battery management in electric vehicles. Power Sources 226, 272–288 (2013)
2. Bandhauer, T.M., Garimella, S., Fuller, T.F.: A Critical Review of Thermal Issues in Lithium-Ion Batteries. Electrochem Soc. 158(3), R1–25 (2011)
3. Zhang, X.: Thermal analysis of a cylindrical lithium-ion battery. Electrochim Acta 56(3), 1246–1255 (2011)
4. Tippmann, S., Walper, D., Balboa, L., Spier, B., Bessler, W.G.: Low-temperature charging of lithium-ion cells part I: Electrochemical modeling and experimental investigation of degradation behavior. Power Sources 252(0), 305–316 (2014)
5. Rao, Z.H., Wang, S.F.: A review of power battery thermal energy management. Renewable & Sustainable Energy Reviews 15(9), 4554–4571 (2011)
6. Yang, Y., Hu, X., Qing, D., Chen, F.: Arrhenius Equation-Based Cell-Health Assessment: Application to Thermal Energy Management Design of a HEV NiMH Battery Pack. Energies 6(5), 2709–2725 (2013)
7. Tran, T., Harmand, S., Desmet, B., Filangi, S.: Experimental investigation on the feasibility of heat pipe cooling for HEV/EV lithium-ion battery. Appl. Therm. Eng. 63(2), 551–558 (2014)
8. Park, Y., Jun, S., Kim, S., Lee, D.: Design optimization of a loop heat pipe to cool a lithium ion battery onboard a military aircraft. Journal of Mechanical Science and Technology 24(2), 609–618 (2010)
9. Fleming, E., Wen, S., Shi, L., Da Silva, A.K.: Thermodynamic model of a thermal storage air conditioning system with dynamic behavior. Appl Energ. 112(SI), 160–169 (2013)
10. Zhang, X., Kong, X., Li, G., Li, J.: Thermodynamic assessment of active cooling/heating methods for lithium-ion batteries of electric vehicles in extreme conditions. Energy 64, 1092–1111 (2014)
11. Alaoui, C.: Solid-State Thermal Management for Lithium-Ion EV Batteries. IEEE T. Veh. Technol. 62(1), 98–107 (2013)
12. Alaoui, C., Salameh, Z.M.: A novel thermal management for electric and hybrid vehicles. IEEE T. Veh. Technol. 54(2), 468–476 (2005)
13. Zolot, M., Pesaran, A.A., Mihalic, M.: Thermal evaluation of Toyota prius battery pack. In: 2002 Future Car Congress, Arlington, VA, United states (2002)
14. Rao, Z., Wang, S., Wu, M., Lin, Z., Li, F.: Experimental investigation on thermal management of electric vehicle battery with heat pipe. Energ. Convers Manage. 65, 92–97 (2013)
15. Khateeb, S.A., Amiruddin, S., Farid, M., Selman, J.R., Al-Hallaj, S.: Thermal management of Li-ion battery with phase change material for electric scooters: experimental validation. Power Sources 142(1-2), 345–353 (2005)
16. Kizilel, R., Sabbah, R., Selman, J.R., Al-Hallaj, S.: An alternative cooling system to enhance the safety of Li-ion battery packs. Power Sources 194(2), 1105–1112 (2009)
17. Park, C., Kaviany, M.: Evaporation-combustion affected by in-cylinder, reciprocating porous regenerator. Journal of Heat Transfer 124(1), 184–194 (2002)
18. ID, D. FreedomCAR Battery Test Manual For Power-Assist Hybrid Electric Vehicles (2003)

19. Wang, J., Liu, P., Hicks-Garner, J., Sherman, E., Soukiazian, S., Verbrugge, M., et al.: Cycle-life model for graphite-LiFePO4 cells. Power Sources 196(8), 3942–3948 (2011)
20. Yi, J., Kim, U.S., Shin, C.B., Han, T., Park, S.: Modeling the temperature dependence of the discharge behavior of a lithium-ion battery in low environmental temperature. Power Sources 244(SI), 143–148 (2013)
21. Zheng, Y., Ouyang, M., Lu, L., Li, J., Han, X., Xu, L.: On-line equalization for lithium-ion battery packs based on charging cell voltages: Part 1. Equalization based on remaining charging capacity estimation. Power Sources 247, 676–686 (2014)

Design and Simulation of a Bidirectional On-board Charger for V2G Application[*]

Weifeng Gao[1], Xiaofei Liu[1], Shumei Cui[1], and Kang Li[2]

[1] School of Electrical Egineering and Automation
Harbin Institute of Technology
[2] School of Electronics, Electrical Engineering and Computer Science
Queen's University Belfast
weifenggaofox@163.com, cuism@hit.edu.cn

Abstract. A design of on-board V2G charger was designed to meet the requirements of the V2G application. Firstly, the functions of on-board V2G charger were analyzed. Then a two-stage structure including a bidirectional AC-DC converter and a bidirectional DC-DC converter was proposed. Traditional charging mode and V2G mode were been discussed for the proposed charging system. The Dual Active Bridge topology was selected as the DC-DC converter. Moreover, the soft-switching operation over a wide output voltage range was analyzed. In the end, a 3.3 kW on-board V2G charger experimental platform was built. The experimental results demonstrated the feasibility of the proposed design and control methods.

Keywords: V2G, on-board bidirectional charger, DAB, ZVS.

1   Introduction

With the development of Electric Vehicle, the effect of EVs on electric grid should not be neglected. V2G technology gathers lots of EVs, which can not only alleviate grid load fluctuation, but also provide ancillary services like regulation and reactive power compensation, with promising prospect. Bi-directional charger is the key equipment in V2G. Compared with charging piles, under the same power density requirements, the realization of on-board V2G charger is more difficult. The reason is on-board V2G charger integrates functions of bi-directional charging and reactive power compensation. Besides, another big difficulty is efficiency improvements with wide voltage range. So far, these problems have not been well resolved. An on-board V2G designing scheme based on dual active bridge was given in this paper. Firstly requirements for V2G charger were analyzed, then system structure scheme and control strategy as well as soft switch mechanism analysis were presented. Finally 3.3kW test bench was build, and experiment results verified the feasibility of designing scheme.

---
[*] This work was supported by NSFC-RCUK_EPSRC under Project 51361130153 and funded by National High Technology Research and Development Program of China (863 Program) under Project 2012AA111003.

K. Li et al. (Eds.): LSMS/ICSEE 2014, Part III, CCIS 463, pp. 486–495, 2014.

2   Requirements of V2G Application on Bidirectional Charger

In the V2G application, electric vehicles participate in the dispatching for the power system as distributed energy sources. Thus compared to the unidirectional charger, the V2G charger should have the following functions:

1) The bidirectional power flow should be realized. The charger should provide energy for the battery in Constant Current control or Constant Voltage control. And in reverse it should be able to transmit the power to the grid, achieving the peak load shifting and frequency adjusting.

2) When operating in inverting state, AC-DC converter could be used in generating reactive power and eliminate harmonics for the grid

3) Communication function, V2G is a media that connected the control system and the EVs, the chargers charging and discharging operation can be conducted according to the instructions received from Power Grid Dispatch Center and Battery Management System.

Allow for the performance of the charger, the cost and the power density should be taken into consideration. Since the charger is designed for V2G mode, switching devices is two times as much as the traditional one. Therefore, the objective in choosing the topology and the control method is to reduce the additional circuits as much as possible for less cost and volume.

3   Bidirectional Charging System Design and Operating Mode

3.1   Bidirectional Charging System

The on-board V2G bi-directional charging system functional diagram is indicated as Fig. 1. In the proposed two-stage structure design, the charger is consisted of a bidirectional AC-DC converter and a bidirectional DC-DC converter. By controlling the bidirectional switch, the charger could be connected to the DC power supply or the AC grid.

When the charger is connected to the DC power supply system, the power flow is controlled by the bidirectional DC-DC converter independently. While integrated to the grid, the AC-DC and the DC-DC converter should cooperate with each other. If the energy is going to the battery to be restored, the DC-DC converter should be work in Constant Current charging mode (CC) or Constant Current–Constant Voltage charging mode (CC-CV). The bidirectional AC-DC is in charge for controlling the reactive power and keep the voltage of the DC link stable. As for the energy flowing to the grid, the DC-DC is used in maintaining the voltage of the DC link capacitors, and the AC-DC is used in the active and reactive power control.

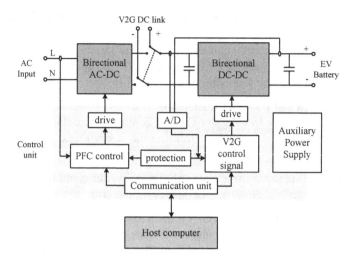

Fig. 1. On-board V2G bi-directional charging system functional diagram

3.2   Operating Mode of the Charging System

(1) Charging Mode

The charging mode is started when the energy of the electric vehicle is deficient. To prevent the overcharging and at the same time guarantee the full charging, the CC-CV charging is applied. As is shown in the Fig. 3, at the beginning of the charging process the battery voltage is low. A suitable method for this stage is CC charging during which the battery voltage rises slowly. When the voltage reaches the rated voltage, the CV charging begins. For this stage, the battery voltage is unchanged, while the output current is declining. The capacity of the battery determines the rate of the CC charging current. Taking the lithium ion battery for granted, the charging rate is generally ranging from 1/3C to 1/2C. Additionally, communication with Battery Management System (BMS) is necessary during charging mode. Once the battery is under overvoltage or overcurrent condition or even worse, the charging process should be terminated immediately.

The control diagram of the charge mode is illustrated in Fig.3. The CC charging is realized by the inner loop to control the current while the outer loop for voltage control is invalid. When the charger is operating in the CV charging mode, a dual loop is applied in which the output of the voltage loop controller is set as the reference of the current loop.

(2) V2G Mode

Unlike normal charging mode, V2G mode requires charger to frequently switch working status between charging and discharging, therefore, the charger need to control the bi-directional flow of power flow, and the power command is given by the upper V2G control center. When need to charge the battery, the charger adopts constant current control scheme in the battery side, and divides power instruction by battery voltage to obtain given current. When the power is transmitted back to grid, the charger controls the AC current by the AC-DC converter

4   Design of the Topology of Bidirectional Charger and the ZVS Analysis

The DC-DC converter applied in the bidirectional charger is Dual Active Bridge converter (DAB) as is shown in Fig.2. The DAB converter could be used in the high power level application. Isolated with high frequency transformer, it could achieve security and reliability. The switches in the two bridges could operate in the ZVS mode, increasing the efficiency of the converter with less switching losses.

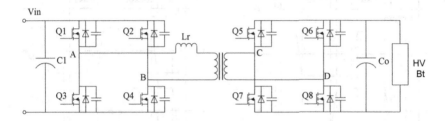

Fig. 2. Dual Active Bridge converter topology

The DC voltage from the AC-DC is the input of the DAB converter, the energy is inverted by the bridge in the primary side of the transformer, and rectified in the bridge of the second side and provide DC voltage to the battery. In the process, the direction of the power flow is controlled by the phase-shift angle φ

The most commonly used method for the control of the DAB converter is the Single Phase Shift (SPS) control as is illustrated in Fig. 7. Define φ as the angle Q_1 leading Q_5, when $\varphi > 0$, the power is transmitted to the battery. When $\varphi < 0$, it is fed back to the grid.

From Fig. 3(a) the average power in a single operating period could be derived as (1).

$$P_o = \frac{1}{T_s} \int_0^{T_s} V_{AB}(\theta) i_L(\theta) d\theta \tag{1}$$

$$P_o = \frac{V_1^2 d\varphi(\pi - |\varphi|)}{\omega L \pi} \tag{2}$$

T_s Inductor current cycle

φ phase-shifted angle in one switching period

k Ratio of the primary and secondary winding of the transformer

d The ratio between the output voltage and the input voltage

P_o Output power of the DAB converter

V_1 Input voltage of the DAB converter

V_{AB} Input voltage of the DAB converter

ω angular frequency of the DAB converter

L Power transfer inductor of the DAB converter

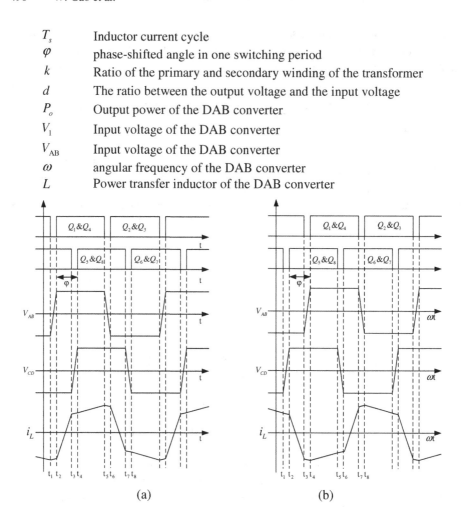

Fig. 3. Main waveforms in DAB converter with SPS method: (a) charging mode (b) discharging mode

However, in SPS method the ZVS is difficult to fulfill in light power condition, and the reactive current in the circuit causes too much pressure on devices. A PWM-SPS method showed in Fig. 7 is implemented for wide ZVS range and higher efficiency.

The power per unit is determined by the phase shift angle D_φ and the duty cycle D_{y1} of the output voltage of the full bridge on the primary side. And the expression of the power per unit is indicated as (3):

$$P_o^* = \frac{P_o}{P_{base}} = \begin{cases} d\left[4D_\varphi(1-D_\varphi)-(1-D_{y1})^2\right] & \dfrac{1-D_{y1}}{2} \leq D_\varphi \leq \dfrac{1+D_{y1}}{2} \\ 4dD_\varphi D_{y1} & \dfrac{(1-D_{y1})}{2} \leq D_\varphi \leq \dfrac{1-D_{y1}}{2} \end{cases} \tag{3}$$

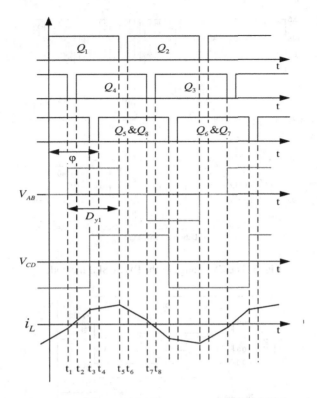

Fig. 4. Main waveforms in DAB converter with PWM-SPS method

Q_1 is turn off and Q_2 is turned on at zero voltage level when $i_L \geq 0$. Q_3 is turn off and Q_4 is turned on at zero voltage level when $i_L \leq 0$. With the constrains of the D_φ as $i_L \geq 0$ or $i_L \leq 0$.

$$\frac{8d^2 D_\varphi^2}{1-d} \leq P_o^* \leq 4d^2 D_\varphi \qquad \left(0 < D_\varphi < \frac{1-d}{2}\right) \qquad (4)$$

$$P_o^* \geq d\left[1-(1-2D_\varphi)^2 - \left(1-\frac{2d(1-D_\varphi)}{1+d}\right)^2\right] \qquad \left(D_\varphi > \frac{1-d}{2}\right) \qquad (5)$$

The blue shadow region in Fig.5 represents the ZVS range of the PWM-SPS method, which depicts according to (4) and (5). The charger operate in wide output voltage means the voltage ratio d varies in between [0, 1]. Fig.8 shows that, implementing PWM-SPS method, the charger could operate in ZVS condition when output wide range voltage by adjusting the value of D_{y1}.

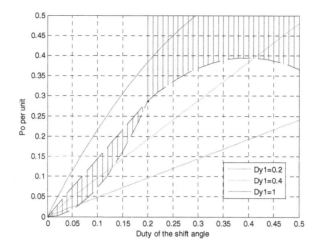

Fig. 5. The ZVS range of the PWM-SPS method

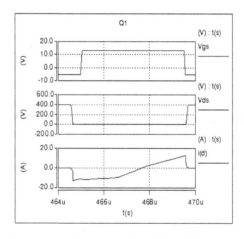

Fig. 6. The ZVS range of the PWM-SPS method

The simulation of the ZVS effect of the switch in the charger is shown in Fig. 6. It could be seen that when the drive signal is active, the switch is not immediately turned on, the inductor current is freewheeling through the anti-parallel diode. As a consequence, the U_{ds} voltage is kept in zero. The soft turning on happens at the commutation process. On the other hand, the soft turning off could be observed for the rising edge of the U_{ds} lags behind the falling edge of the I_{ds}. According to the symmetry of the operating period, the discharge process could be analyzed in similar way.

5   Experimental Results

Table 1. Structure parameter of the bidirectional charger prototype

device	value
Power ratio	3.3kW
switching frequency	100kHz
Output voltage range	220V-336V
Input voltage	400V
Output voltage ripple	1% below
Output current ripple	5% below
Efficiency	92%
Temperature range	-30°C~+85°C

Fig. 7. Test platforms of the 3.3kW on-board bidirectional charger

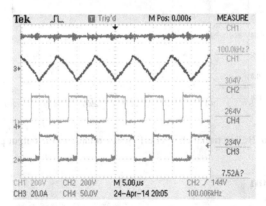

Fig. 8. DAB working condition in low power level, CH1 shows leading leg U_{ds} voltage, CH2 shows lagging leg U_{ds} voltage, CH3 shows primary current of the transformer. CH4 shows the output voltage of the DAB converter.

It is shown that in Fig. 9, for Q2 and Q6, there is no overlap between U_{gs} and U_{ds}, it could be approximately taken that I_{ds} has no common zone with U_{ds}. The switch is turned off in zero voltage. Since U_{gs} rises after U_{ds}, the switch could be turn on in ZVS condition.

Fig.10 shows how the efficiency of the DAB converter goes up with different output power increasing; the highest efficiency can reach to 92.69%.

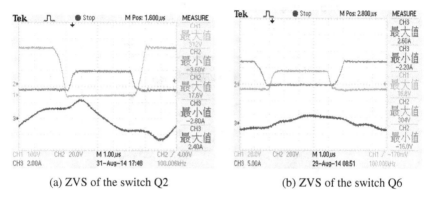

(a) ZVS of the switch Q2   (b) ZVS of the switch Q6

Fig. 9. ZVS effect of the MOSFET in DAB converter. In (a), CH1 depicts the U_{ds} of Q3, CH2 depicts the U_{gs} of Q3. In (b), CH1 depicts the U_{gs} of Q6, CH2 depicts the U_{ds} of Q6.

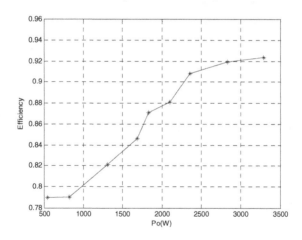

Fig. 10. Bidirectional charger efficiency related to the output power

6   Conclusion

The demands for electric vehicle V2G on-board charger were analyzed, for example, the bi-directional transmission of active power, the compensation and control of reactive power, high efficiency operation over a wide output voltage range and reliable communications, and proposes a two-stage structure of AC/DC and DC/DC converters as the design scheme for electric vehicle V2G on-board charger. In the charging mode, a constant current-constant voltage control scheme was used while in the V2G mode the constant current control scheme was adopted. A dual active bridge converter topology was employed as the bi-directional DC/DC converter unit. The analysis of soft-switching operation showed that the coordinated control of phase-shift angle and duty cycle allows the charger to achieve high efficiency over a wide output-voltage range. Finally, a 3.3kW bi-directional charger experimental prototype was designed and experiments were finished to verify the feasibility of the proposed design scheme.

References

1. Kramer, B., Chakraborty, S., Kroposki, B.: A review of plug-in vehicles and vehicle-to-grid capability. In: 34th Annual Conference of IEEE Industrial Electronics, IECON 2008, pp. 2278–2283. IEEE (2008)
2. Erb, D.C., Onar, O.C., Khaligh, A.: Bi-directional charging topologies for plug-in hybrid electric vehicles. In: 2010 Twenty-Fifth Annual IEEE Applied Power Electronics Conference and Exposition (APEC), pp. 2066–2072. IEEE (2072)
3. Alonso, A.R., Sebastian, J., Lamar, D.G., et al.: An overall study of a Dual Active Bridge for bidirectional DC/DC conversion. In: 2010 IEEE Energy Conversion Congress and Exposition (ECCE), pp. 1129–1135. IEEE (2010)
4. Vaishnav, S.N.D.: Single-stage bi-directional converter for plug-in hybrid vehicle charging and Vehicle to Grid application. The University of Texas at San Antonio (2011)
5. Alonso, A.R., Sebastian, J., Lamar, D.G., et al.: An overall study of a Dual Active Bridge for bidirectional DC/DC conversion. In: 2010 IEEE Energy Conversion Congress and Exposition (ECCE), pp. 1129–1135. IEEE (2010)
6. Oggier, G.G., Ledhold, R., Garcia, G.O., et al.: Extending the ZVS operating range of dual active bridge high-power DC-DC converters. In: 37th IEEE Power Electronics Specialists Conference, PESC 2006, pp. 1–7. IEEE (2006)
7. Oggier, G.G., Garcia, G.O., Oliva, A.R.: Modulation strategy to operate the dual active bridge DC–DC converter under soft switching in the whole operating range. IEEE Transactions on Power Electronics 26(4), 1228–1236 (2011)
8. Vaishnav, S.N.D.: Single-stage bi-directional converter for plug-in hybrid vehicle charging and Vehicle to Grid application. The University of Texas at San Antonio (2011)

Research on Simulation and Harmonics of EV Charging Stations for V2G Application[*]

Jintang Li[1], Haifang Yu[2], Shumei Cui[1], and Bingliang Xu[3]

[1]School of Electrical Engineering and Automation
Harbin Institute of Technology
[2]Changchun University of Technology
[3]Heilongjiang Electric Power Research Institute
lijt6509@163.com, cuism@hit.edu.cn

Abstract. In order to study the harmonics of EV charging stations based on V2G technology, this paper proposed a simulation model considering the structure of the bi-directional charger and the control strategy. To study the variation of THD, the bi-directional charger, the charging station and the traditional 6-pulse rectifier for a contrastive analysis has been made. The author has focused on the charging process and the number of chargers which is in the constant current and constant voltage mode, and then the two factors was combined to make a unified analysis. In the end, a comparative study for bidirectional chargers and traditional chargers has been conducted. The simulation results shows that when in constant voltage charging phase, the THD value increases significantly, with the increasing number of chargers, the THD value tends to stabilize, and if the charging power is higher, the harmonics will be smaller.

Keywords: Electrical vehicle, V2G, charging station, harmonic analysis.

1   Introduction

To solve the environment and energy problems, V2G technology has developed rapidly. V2G is the short of Vehicle to Grid, in this system electric vehicles are involved in grid scheduling as a kind of distributed device for the use of energy storage, peaking adjustment etc. the application of V2G can increase the efficiency of the grid [1]. The power flow between the grid and the charging station is bidirectional, real-time and controlled. Therefore, the EV charging station based on V2G technology is different from the traditional one. There are three types of charging stations, which are charging pile, charging station and battery replacement station [2]. No matter what type it is, the charging station will generate a wide band of harmonic components during energy interactions because of the bidirectional rectifiers and other power electronic devices. When the content of these harmonic components reaches a certain level, it will inevitably affect the normal operation of the grid. So the analysis of the

[*] This work was supported by NSFC-RCUK_EPSRC under Project 51361130153 and funded by the Science and Technology Project of State Grid Corporation of China.

K. Li et al. (Eds.): LSMS/ICSEE 2014, Part III, CCIS 463, pp. 496–504, 2014.

generation mechanism of the harmonics and it's estimation in EV charging stations will be more important.

Currently, the V2G technology-based EV charging stations is an emerging industry, so the simulation model for the charging station and harmonic analysis are still in the exploratory stage [3-7]. In 1989, Lewis proposed that a non-linear resistor can be used to approximate the high frequency power conversion circuit [8]. By this way, the single charger model can be simplified which brings convince to analyze the harmonics. PT.staat et al made a deep study on the total harmonic current distortion of the EV charger [9-11]. The article uses the central limit theorem to establish a mathematical model to analyze the harmonics, in the end the author has proposed a method to predict the current harmonics caused by a set of chargers. This model can explain the phenomenon that the harmonic currents cancel each other because of the diversity of the amplitude and phase. However, this method requires a sufficient number of chargers when analyzing, so the experiment is not applicable because of the complex implementation process. Some scholars have studied the harmonic interaction between the grid and the inverter which has the same access point, to improve the accuracy of the model; they took both the non-ideal characteristics of the switching devices and the dead band effects into consideration, by the use of the proposed output impedance model, they made a analysis of the distributed impedance network which combined several inverter and the grid [12]. Another scholar has studied the mechanism of harmonic generation on photovoltaic power plants, they took fully consideration of the non-ideal effects and modulation factor dead band part, then got the non-ideal equivalent models [13], by this way they increase the accuracy of the analysis, which has some help for the analysis of EV charging stations.

In this paper, we focused on the construction of the EV charging stations based on V2G technology, making a mathematical model of charging stations considering the internal structure of bidirectional AC-DC converter, the DC-DC convertor, the control strategy and the modulation. According to the relevant provision of the actual parameters, we made the simulation for different numbers and different charging process of chargers, in the end giving conclusions of the factors affecting harmonics.

2   The Equivalent Model and Harmonic Characteristics

2.1   Equivalent Model of the Charger

Currently, the equivalent model of the bidirectional charger for V2G application is shown in figure 1. Usually the charger consists of two parts, the pre-stage is a reversible PWM rectifier which is used for AC-DC conversion, power factor correction and also making the input current sinusoidal.   The latter stage is a high frequency DC-DC converter, it is used for adjustment of the DC voltage, constant current and constant voltage control and the isolation.

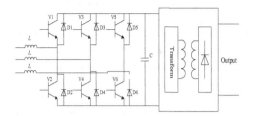 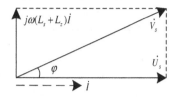

Fig. 1. Equivalent model of the charger

Fig. 2. The steady vector diagram of three-phase PWM rectifier

2.2   Harmonic Characteristics of PWM Rectifier

The steady vector diagram of three-phase PWM rectifier is shown in Figure 2. Where: U_s is the grid side voltage, I_s is the grid side current, V_s is the voltage of the AC side arm, L_z is the impedance of AC side, and L_s is the filter inductor.

We can get the formula of the fundamental and harmonic voltage in bipolar SPWM modulation by making Fourier decomposition [14]:

$$U_{ab} = \frac{\sqrt{3}}{2} m U_d \sin(\omega_r t + \phi + \frac{\pi}{3})$$ (1)

$$U_{abh} = \sum_{n=1}^{\infty} (-1)^{\frac{n+1}{2}} \frac{4}{n\pi} \sum_{k=2}^{\infty} J_k(\frac{n\pi m}{2})\{\sin[(k\omega_r + n\omega_c)t + k(\phi - \frac{\pi}{3})] -$$
$$\sin[(k\omega_r - n\omega_c)t + k(\phi - \frac{\pi}{3})]\}2\sin\frac{k\pi}{3}$$ (2)

$$U_{abh} = \sum_{n=1}^{\infty} (-1)^{\frac{n}{2}} \frac{4}{n\pi} \sum_{k=2}^{\infty} (-1)^k J_k(\frac{n\pi m}{2})\{\cos[(k\omega_r + n\omega_c)t + k(\phi - \frac{\pi}{3})] -$$
$$\cos[(k\omega_r - n\omega_c)t + k(\phi - \frac{\pi}{3})]\}2\sin\frac{k\pi}{3}$$ (3)

Where: U_{ab} is the fundamental voltage, U_{abh} is the harmonic voltage components, and in (2): n = 1, 3, 5 ..., k is an even except the integer multiple of 3, in (3): n = 2, 4, 6 ..., k is an odd except the integer multiple of 3, ω_c is the carrier frequency, ω_r is the modulation wave frequency, J_k is the Bessel function of the first kind, m is the modulation.

The above formula is the theoretically harmonics generated by the PWM rectifier. According to the harmonic voltage formula, the nth current harmonic formula is shown as follows:

$$I_{ah}(n) = \frac{U_{abh}(n)}{\omega_n(L_s + L_z)}$$ (4)

Where: ω_n is the corresponding angular frequency of nth harmonic voltage.

Usually there is very small voltage distortion in the utility grid, but the grid current can reach to a certain level because of the application of power electronic devices, therefore when comes to harmonics of charging station, we can assume that there is no voltage distortion in the grid, and the key consideration is the current distortion.

3   The Charging Station Simulation Model

3.1   The AC-DC Simulation Model

The pre-stage of the charger is a three phase reversible PWM rectifier, and the parameters of AC side are time-varying in general mathematical model, thus it is not conducive to design the control system. For this reason, we can convert the three phase stationary coordinate system (a, b, c) into a synchronous rotating coordinate system (d, q) which has the same frequency with the fundamental in the grid. By making coordinate transformation, the fundamental variable in the three phase stationary coordinate system can be converting to DC variable, thereby simplifying the design of the control system. The AC-DC simulation model is shown in figure 3.

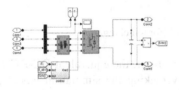 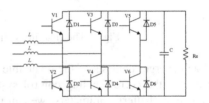

Fig. 3. The AC-DC simulation model   **Fig. 4.** The simplified model of the charger

3.2   The Simplified Simulation Model of the Charger

The major part that generated harmonics is the pre-stage PWM rectifier, and the DC-DC part can get a small impact on the AC side. Thus the simulation model can be made easily by simplifying the DC-DC part. The charging process is relatively a long time contrast with the several frequency cycles, so we can consider that the output voltage and current remain constant in several frequency cycles, by this way we can get a nonlinear resistor to characterize the DC-DC part, shown in figure 4.

The nonlinear resistor R_c can be calculated from the above figure using the following formula:

$$R_c = \frac{\eta U_b^2}{P_o} \tag{5}$$

The output voltage U_b of the PWM rectifier can be a fixed value according to the modulation mode and control strategies. So the impedance variation can be characterized when we give different charging power curve. When the Impedance curve is got, we can study the harmonic changes on the whole charging process.

3.3   The Simulation Model of the Charging Station

The power load for EV charging stations is grade 2, and the charging voltage is 380VAC (three-phase four-wire system). The distribution voltage is 220VAC, the power transformer turns ratio is 10k/380, the system short-circuit capacity is 30MVA; the charging station model is shown in Figure 5.

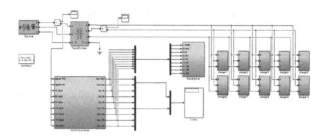

Fig. 5. The charging station model (for ten chargers)

Each charger consists of the reversible PWM rectifier, the equivalent DC-DC converter and the corresponding control system. By making the simulation model and giving some battery parameters, we can make the harmonic analysis of the charging stations.

4   Simulation Analysis of Charging Stations

4.1   Simulation for Single Charger

Currently, there are three types of batteries for electric vehicles including the lead-acid batteries, the nickel hydrogen batteries and the lithium-ion batteries. Due to the high specific energy, no memory effect and small size, Lithium-ion batteries has the most potential applications, so in this paper, the charging power curve is based on the Lithium-ion batteries, generally for EV charging stations the constant voltage and constant current charging mode has been widely used considering the battery life. Usually the battery model consists of a controlled voltage source and a constant resistor [15-16], according to a real test for a 90Ah electric vehicle battery, we can get the charging power curve shown in figure 6. For a further calculation, we can get the equivalent input resistor curve of the battery during charging shown in figure 7.

To simplify the calculation, the whole process of charging can be divided into several discrete processing, after considering various factors, the total time (270min) is divided into 18 intervals, that is for every 15 minutes we need to take a sample point as the current interval value, the equivalent input resistor is shown in figure 8.

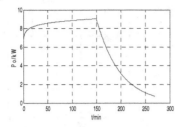 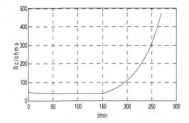

Fig. 6. The charging power curve   **Fig. 7.** The equivalent input resistor curve

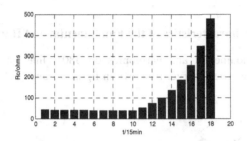

Fig. 8. The equivalent input resistor for every 15minutes

Table 1. Values of equivalent input resistance at different times

t/min	15	30	45	60	75	90	105	120	135
R_c / Ω	42.7	41.3	40.5	39.9	39.5	39.2	38.9	38.6	38.4
t/min	150	165	180	195	210	225	240	255	270
R_c / Ω	38.2	52.6	72.1	98.8	135.5	185.6	254.4	348.6	477.7

From the figure we can calculate the value of equivalent input resistance within 18 internal, shown in table 1.

According to the above table, the curve of total harmonic current distortion (THD) on 380V line is shown in figure 9. In order to make a comparative analysis, replace the original bidirectional AC-DC converter with a single 6-pulse rectifier charger of the same power (widely used in current charging stations), then also give the curve of total harmonic current distortion on 380V line shown in figure 10.

From the above, it can be seen that THD in bidirectional charging station is smaller than in traditional charging station. The variation of THD in bidirectional charging station is consistent with 6-pulse rectifier charging station on 380V line, when the constant current charging mode is converted to the constant voltage mode, the THD value increases significantly.

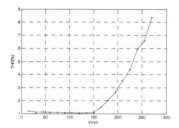

Fig. 9. THD on 380V line for bidirectional charger

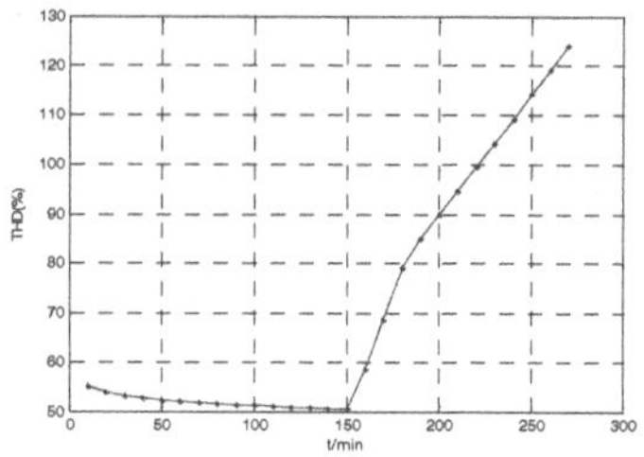

Fig. 10. THD on 380V line for 6-pulse rectifier charger

4.2   Distribution Characteristics of the Fundamental and Harmonics

The variation of fundamental and harmonics on 380V line for both bidirectional charger and 6-pulse rectifier charger is shown in figure 11 and 12.

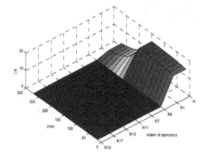
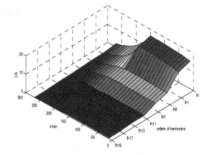

Fig. 11. Harmonics for bidirectional charging station on 380V line

Fig. 12. Harmonics for 6-pulse rectifier charging station on 380V line

From the two figures, it can be seen that 6-pulse rectifier charging station can generate higher amplitude harmonic components such as 5th, 7th, 11th harmonics and less on higher frequency domain. For bidirectional charging station the amplitude of harmonics are lower on the whole domain.

4.3   Different Numbers Impact

For analyzing the impact of THD with different number of chargers and the charging process, we mounted the charger on the charging station from 1 to 5, and the charging time is from 0 to 270min. A comparative study has also done for analyzing. The result is shown in figure 13 and 14.

From the two figures we can draw the following conclusions: the impact of THD on the grid for the two types of chargers is different. Traditional non-controlled rectifier charger can bring a lot of harmonic components to the grid while the bi-directional charger brings less. When in constant current charging phase, both the two types can cause little effect, and with the increasing number of the chargers, the THD remained stable. When in constant voltage charging phase, the THD values increases with the charging process for bidirectional chargers, the more close to the end, the higher of the THD, and the more the chargers, the higher values of the THD, but when it reaches a certain number, the THD remained stable, for 6-pulse rectifier charger, it remains constant with the increasing number of the charger.

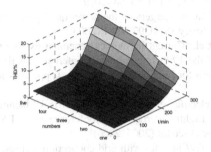   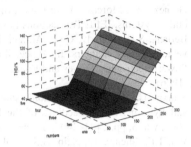

Fig. 13. The impact of THD on charging numbers and charging process for bidirectional charging station on 380V line

Fig. 14. The impact of THD on charging numbers and charging process for 6-pulse rectifier charging station on 380V line

5   Conclusion

In this paper, we focused on the electrical structure of EV charging station based on V2G technology, took fully consideration of the structural characteristics of the charging station, the modulation and the control strategy. Then made the EV charging station simulation model, for a comparative analysis, we also built a traditional 6-pulse rectifier charging station, by detailed data processing and analysis, we got the harmonic characteristics of EV charging stations as follows:

Traditional 6-pulse rectifier charger can cause large disturbance on the grid, the PWM rectifier charger cause less. But both of them need harmonic control devices when the charging station is working.

When in constant voltage charging mode, the THD value increases with the number and the charging process, but when the chargers reach to a certain numbers, the THD will remain stable which is consistent with the harmonic cancellation theory, the rule applies to both chargers.

When in constant current charging mode, the THD value changed very small with the increasing number of the chargers and the charging process. This feature is suitable for both chargers, That is to say, the higher power the charger is, the less THD it produces.

References

1. Liu, X.F., Zhang, Q.F., Cui, S.M.: Review of electric vehicle V2G technology. Transactions of China Electrotechnical Society 27(2), 121–127 (2012) (in Chinese)
2. Chen, L.L., Zhang, H., Ni, F., et al.: Present situation and development trend for construction of electric vehicle energy supply Infrastructure. Automation of Electric Power Systems 35(14), 11–17 (2011) (in Chinese)
3. Hadley, S.W., Tsvetkova, A.A.: Potential impacts of plug-in hybrid electric vehicles on regional power generation. The Electricity Journal 22(10), 56–68 (2009)
4. Gomez, J.C., Morcos, M.M.: Impact of EV battery chargers on the power quality of distribution systems. IEEE Transactions Power Delivery 18(3), 975–981 (2003)
5. Berisha, S.H., Karady, G.G., Ahmad, R., et al.: Current harmonics generated by electric vehicle battery chargers. In: Proceedings of the 1996 International Conference on IEEE Power Electronics, Drives and Energy Systems for Industrial Growth, vol. 1, pp. 584–589 (1996)
6. Yanxia, L., Jiuchun, J.: Harmonic-study of electric vehicle chargers. Electrical Machines and Systems. In: Proceedings of the IEEE Eighth International Conference on ICEMS 2005, vol. 3, pp. 2404–2407. Springer, Author Discount (2005, 2006)
7. Wu, T.F., Shen, C.L., Nein, H.S., et al.: A $1\varphi3W$ inverter with grid connection and active power filtering based on nonlinear programming and fast-zero-phase detection algorithm. IEEE Transactions Power Electronics 20(1), 218–226 (2005)
8. Lewis, L.R., Cho, B.H., Lee, F.C., et al.: Modeling, analysis and design of distributed power systems. In: Power Electronics Specialists Conference, vol. 1, pp. 152–159 (1989)
9. Staats, P.T., Grady, W.M., Arapostathis, A., et al.: A statistical analysis of the effect of elec-tric vehicle battery charging on distribution system harmonic voltages. IEEE Transactions on Power Delivery 13(2), 640–646 (1998)
10. Staats, P.T., Grady, W.M., Arapostathis, A., et al.: A statistical method for predicting the net harmonic currents generated by a concentration of electric vehicle battery chargers. IEEE Transactions on Power Delivery 12(3), 1258–1266 (1997)
11. Staats, P.T., Grady, W.M., Arapostathis, A., et al.: A procedure for derating a substation transformer in the presence of widespread electric vehicle battery charging. IEEE Transactions on Power Delivery 12(4), 1562–1568 (1997)
12. Xu, D.Z., Wang, F., Mao, H.L.: Modeling and analysis of harmonic interaction between multiple grid-connected inverters and the utility grid. Proceedings of the CSEE 33(12), 64–71 (2013) (in Chinese)
13. Xie, N., Luo, A., Chen, Y.D., Ma, F.J., Xu, X.W., Lü, Z.P., Shuai, Z.K.: Dynamic Modeling and Characteristic Analysis on Harmonics of Photovoltaic Power Stations. Proceedings of the CSEE 36, 10-17+4 (2013) (in Chinese)
14. Cao, L.W., Wu, S.H., Zhang, C.S., Wu, B.F.: A General Method of SPWM Harmonic Analysis. Power Electronics 04, 62–65 (2002) (in Chinese)
15. Shepherd, C.M.: Design of primary and secondary cells, part 2: An equation describing battery discharge. Journal of Electrochemical Society 112(7), 657–664 (1965)
16. Tremblay, O., Dessaint, L.A., Dekkiche, A.I.: A generic battery model for the dynamic simulation of hybrid electric vehicles. In: IEEE Vehicle Power and Propulsion Conference, VPPC 2007, pp. 284–289. IEEE (2007)

Procedural Modeling for Charging Load of Electric Buses Cluster Based on Battery Swapping Mode

Mingfei Ban, Jilai Yu, and Jing Ge

School of Electrical Engineering and Automation, Harbin Institute of Technology,

Harbin 150001, Heilongjiang Province, China

Abstract. To forecast and assess the impact of large-scale electric buses (EBs) to power grid, the aggregating charging load model of EBs cluster is indispensable. As EB's typical operating, including driving and parking, is a cyclical process and has obvious regularity, a procedural simulation method for aggregating charging load model of EBs cluster based on battery swapping mode is proposed in this paper. With the data come from specific buses lines and other information readily available, the behavior process of each individual in EBs cluster is continuously simulated. Then time and SOC information of battery packs emerges and is recorded. Combined with specific charging control method, corresponding aggregating charging load model becomes available. The proposed method has been verified by simulation on an actual buses line with charging/swapping station. The results show that the proposed method can grasp characteristics of EBs cluster's charging load under multiple factors, thereby improve the practicality and reliability of modeling.

Keywords: Electric Vehicle, Electric Bus, Battery Swapping Mode, Charging load model, Procedural Simulation.

1 Introduction

Aggregating charging load model of EVs cluster is essential in studying the impacts of large-scale EVs on power grid [1].In [2, 3], after simplifying the traveling distance, SOC and charging time of EVs by independent probability distributions, the charging model of EVs was built. Referring to the GPS information, one kind of more accurate result was approached in [4], and it was improved by iterative process based on conditional probability. It is simple to implement such methods [2]-[4] by adopting Monte Carlo Simulation. However, the reliability of such simplifying is dubious, as the driving and charging processes of EVs are so complicated [5]. Charging load during one day of typical EV was achieved in [6], assuming that its charging process started immediately after parking and stopped till fully charged or next travel. Moreover, based on the indexes of potential EV users, which are filtered from the existing vehicle users by using empirical data, the charging load of EVs cluster was analyzed in [7, 8]. In [6]-[8], the characteristics of EVs were predicted and modeled effectively, but the

K. Li et al. (Eds.): LSMS/ICSEE 2014, Part III, CCIS 463, pp. 505–516, 2014.

plenty of historical data which these simulations relied on are scarce or even unavailable at this stage. Meantime, considering that activities of EVs are simplified in these researches, accuracy of results is undoubtedly affected.

Furthermore, battery swapping mode has been developed into a critical energy supplying form for EVs [9]. Battery-swapping mode makes it possible to charge the batteries used in EVs under centralized management, then negative impacts caused by large-scale random charging can be effectively reduced [10]. Battery charging/swapping stations for EVs was modeled in [11] and it claimed that battery swapping mode suited public transportation better with current technical conditions. A further idea was pointed out in [12] claiming that battery swapping mode can offer better services and user experience. [13, 14] studied charging/swapping stations focusing on economic dispatch and made important progress, yet the result can hardly reflect the actual cases as no characteristics of specific charging load were involved. The optimal operation of EBs battery charging/swapping station was discussed in [15] and an effective optimal charging control method was built. However, though the research in [15] mentioned the procedure of EBs' operation, instead of providing specific research method of EBs' actual operating process, it just simplified the complex processes only by a single analyzed formula.

Overall, the current researches have done a lot of active exploration in modeling EVs' charging load and optimizing operation of charging/swapping station. Nevertheless, most of them are limited by simplifying EVs' charging operation with probability distributions or sidestepping discussion of the actual processes. And it is difficult to test them. Therefore, the reliability of these results remains to be confirmed. Besides, both of increasing number of EVs and bidirectional energy transfer modes like V2B and V2G[16, 17] call for accurate information from each EV and corresponding ancillary batteries at every moment , yet existing methods can hardly satisfy this requirement. Hence, dynamic monitoring of each individual's critical information at each link in the simulation of EVs cluster is essential. In this respect, simulating the continuous cycle process composed of EVs cluster's traveling and charging/swapping behavior, and taking a variety of influence factors into consideration while simulating is a more practical approach.

This paper takes EBs as an example and proposes a procedural simulation method for aggregating charging load model of EBs cluster based on battery swapping mode. Based on readily accessible information, such as departure schedule, driving routes and EB parameters, proposed method may condense the characteristic of EBs cluster and information of batteries' SOC accurately by simulating each EB's continuous cycle operation process.

2   Process Analyses and Modeling Approach

2.1   Process Analysis for EBs Cluster

Like traditional buses, EBs travel on a relatively fixed line. Meantime, the conditions of EBs travelling on the same line are closer to each other. Thus the behavior of each individual in EBs cluster can be regarded as an analogous continuous cycle constrained by a given process.

Then, for specific bus line, procedural cycle simulation framework of its operation can be built, using the information of departure schedule, driving routes, EB parameters and ancillary battery packs, etc. To this end, the procedural characteristics of EBs' operation are analyzed.

EBs' typical driving routes and battery charging/swapping modes are shown in Figure 1. There are two kinds of typical routes. One is single-cycle route, whose departure station and terminal are located in the same place, such as Line 1. In contrast, the other has different departure station and terminal, such as Line 2. Bus station A and B are essentially similar stations. The only difference between them is that Station A, as a battery swapping station (BSS), has battery swapping systems and spare battery packs (SBPs), so it can provide EBs with battery swapping serves. While Station C is a charging station, and it can provide both of battery charging and delivery service for other stations. In addition, Station C can also be built at original terminal to act as both a charging station and a bus station.

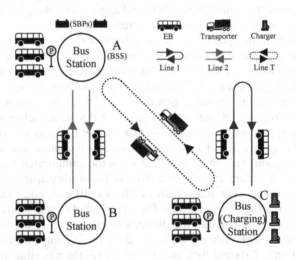

Fig. 1. EBs' typical driving routes and battery swapping/charging mode

The obvious periodicity and regularity of EB's typical operating makes it possible to simulate the whole process at this stage. Then the method of modeling the aggregating charging load of EBs clusters is feasible.

Meanwhile, due to the influences of various kinds of interferences, the differences between each cycle process within one day, even for the same EB, are inevitable. In addition, for battery swapping mode, the cycle of EB's charging-discharging process is much longer than its driving-parking one, and this process interacts with the alternation of SPBs. Thus it is too complicated to get the results or even analytic expression from historical data of conventional buses.

Taken together, based on existing data, it is easy-realizable to exploit procedural characteristics in EBs cluster's behavior to simulate the whole process of each individual in it. Besides it has the actual effect which is hard for other methods to achieve.

2.2   Modeling Method

To simplify the analysis, without considering dispatching of batteries, only Station C is taken as example in subsequent analyzes. And Station C is set to be both a bus station and a charging/swapping station. Then the total modeling method is given as follows.

Firstly, according to scheduling information and driving route of a specific bus line, establish the general framework for total simulation. After that, using this framework and taking EBs' parameters into consideration, continuously track and simulate each EB's behavior. The tracking and simulating involve EB's departure, driving, returning, swapping batteries and parking, namely every segments of EB's behavior. Then, combined with corresponding traffic conditions and other information, the SOC and arriving time of each EB are available.

When an EB arrives at Station C, whether it needs battery swapping service is determined by both of its subsequent driving demand and batteries' SOC. The information of batteries replaced, such as SOC and charging begin time, can be obtained in the process of simulation. Then total aggregating charging load of EBs can be obtained from the calculation of such information.

The whole model is designed to fit the actual operating conditions of EBs cluster. It is composed mainly of two processes. One is simulation of EBs' behavior. Its main task is to describe the states of each EB in the cluster dynamically, and get time information of each EB, including its departure time, driving time and batteries replacing time. The other is the alternating and charging processes of BPs. In this part, it focuses on the usage of batteries to give SOC information of each BP SOC at each moment.

For battery swapping mode, owing to the matching of EBs and batteries is not fixed, these two simulating processes has direct spatiotemporal contact, and their interaction is influenced by various random factors. Thus it is too complicated to externalize this series of processes by either analytical methods or probability distributions, especially when the differences between individuals of EBs and BPs are taken into considering. But, with the increasing penetration of EBs, particularly after bidirectional energy exchange mode becomes popular, the differences must be paid of enough attention. While, the proposed method may well reproduce those complex processes and get critical information of EBs and BPs, and then improves the modeling accuracy.

3   Procedural Simulation

Generally, a complete operating process of EB can be described by several time indicators, namely departing moment, driving time, arriving moment and waiting time in station. Through grasping these time indicators, the simulation and reconstruction of EBs cluster's behavior are all easy to implement.

As shown in Fig. 2, for Line 2, t_{ij-a}, the moment when EB_i arrives Station A for j-th times, is determined by the following factors: departing moment t_{ij-b}, driving time for one-way T_{ij-t1}, staying time in Station B T_{ij-s}, the other driving time for one-way T_{ij-t2}. And for Line 1, as T_{ij-s} does not exist, t_{ij-a} just depends on t_{ij-b} and T_{ij-t}.

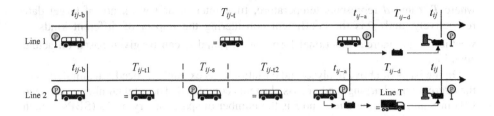

Fig. 2. EBs' typical driving and charging processes

When EB_i arrives, if it has requirement of swapping batteries, then the charging starting time t_{kh} of BP_k, the batteries replaced from EB_i, is given by

$$t_{kh} = t_{ij-a} + T_{ij-d} = t_{ij-b} + T_{ij-t} + T_{ij-s} + T_{ij-d} \tag{1}$$

where T_{ij-d} is the delay time caused by actual swapping batteries operation and waiting, and T_{ij-t} is determined by driving distance, speed and total number of passengers. For buses, the driving distance is substantially fixed, but the speed v_{ij} of each operation is hard to give a definite value at every instant. Worse still, the total number of passengers is also random. For simplicity, v_{ij} are assumed here to be different constants at different times of day, and its value depends on specific buses lines and traffic conditions.

The variation of SOC is a continuous process. Assuming all EBs are in a consistent state, then SOC_{kh}, the SOC of BP_k when it arrives at the swapping station for h-th time, is described as

$$SOC_{kh} = SOC_{k0} - \frac{d_{kh}}{s_k} + \sum_{p=1}^{h-1}(\Delta SOC_{kp} - \frac{d_{kp}}{s_k}) \tag{2}$$

where SOC_{k0} and s_k are initial value and nominal mileage(km) of BP_k, s_k is and subscripts h and p indicate the number of operations, d_{kh} is the driving distance of BP_k, as well as ΔSOC_{kp} is the SOC increment of BP_k, if after the p-th operation, BP_k does not charge, its value is 0.

Though driving route is fixed, EB's driving environment differs each time even in the same day. Moreover, the performance degradation of batteries is inevitable [18, 19]. Then formula (1) is amended as

$$SOC_{kh} = SOC_{k0} - \frac{\gamma_{kh}d_{kh}}{s_k(1-\eta_k)} + \sum_{p=1}^{h-1}[\Delta SOC_{kp} - \frac{\gamma_{kp}d_{kp}}{s_k(1-\eta_k)}] \tag{3}$$

where η_k is the aging coefficient represents the capacity decreases of BP_k, and it is decided by total charging and discharging energy $E_{k\text{-total}}$ and total usage time $T_{k\text{-total}}$, and γ_{kh} is conversion factor which convert a variety of influencing factors into driving distance. η_k and γ_{kh} can be described as

$$\begin{cases} \eta_k = f_1(E_{k\text{-total}}, T_{k\text{-total}}) \\ \gamma_{kh} = f_2(T, t, d, \rho) \end{cases} \tag{4}$$

where T, t and d represents temperature, time interval and week no. of target date respectively, and ρ is the coefficient considering the impact of different kinds of weather. The concrete functional formulas of f_1 and f_2 can be given based on actual situation.

On the basis of above analyses, take Station C as example, procedural simulation of the aggregating charging load of EBs cluster is conducted. The total number of BPs and EBs in Station C n_{bp} and n_{eb}, and s is the number of spare battery packs (SBPs) which equals to the difference between n_{bp} and n_{eb}. Fig. 3 shows the modeling process of proposed method.

Shown in Fig. 3, those in dashed line are cycle processes of EBs, and others in dash-dotted frame are alternation processes of BPs. n represents the total number of already departure trips, and n_{max} is its maximum. As after each operation of swapping, the BP that EB_i are equipped with will alternate, instead of giving the actual number, here BP_x is used. Yet in the procedural simulation, this part can be explicitly reproduced by tracking the specific process. And function Sq is defined as

$$Sq[a/b] = \begin{cases} \mod[a/b] & , \mod[a/b] \neq 0 \\ b & , \mod[a/b] = 0 \end{cases} \tag{5}$$

where $\mod[a/b]$ represents the remainder when a is divided by b.

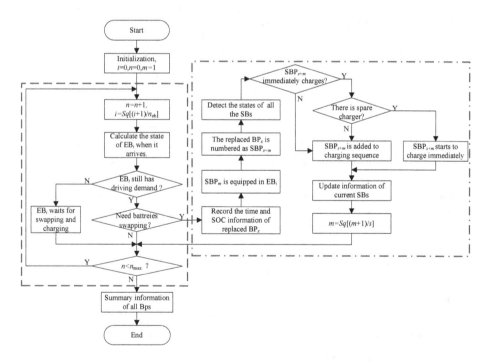

Fig. 3. Flowchart of procedural simulation

To begin with, all of the BPs are assumed to be fully charged and the maximum of SOC is 0.95. Besides, the EBs and the BPs equipped with them are numbered sequentially from 1 to n_{eb}, when the first round starts. Correspondingly the SBPs are numbered from $n_{eb}+1$ to n_{bp}.

According to the scheduling timetable, EBs depart successively. Combined with the specific situation of each period, both of the arrival time and SOC of BPs can be obtained from formula (1) and (3). Then the BP equipped in EB must be replaced when

$$\frac{\gamma_{kh}(d_{(kh+1)}+l_k)}{s_k(1-\eta_k)}>SOC_{kh} \tag{6}$$

where $d_{k(h+1)}$ is the next trip distance, and l_k is the margin of judgment.

If a BP is judged to be replaced, its information, such as arriving time and SOC, will be saved, and it will be added to the end of the SBPs sequence to wait for subsequent charging process. Meanwhile, the BP at the forefront end of the SBPs sequence will be equipped to EB. The SBPs sequence is arranged in ascending order, and the minimum subscript in current sequence is next to the total number of batteries swapping operation. The total number of SBPs is a constant equals to s.

Afterwards, this EB rejoin the departure sequence to wait for the next trip. After gathering the SOC and time information of all of the SBPs, and giving a specific charging control, the whole charging load is obtained.

When there is no more driving demand and all of the EBs return to Station C, the simulating process of the EBs ends. Then bring together information of all BPs inside the station to get the total overall charging load. And the entire simulation process ends.

With the proposed method, the whole operating process to access key information of all the EBs and BPs can be reproduced. And using these key and detailed information, optimal operation of battery swapping and charging station can be performed to reduce costs.

Taking time of use price (TOU price) as an example, given below the overall optimization method. When an EB arrives at Station C, if there is no more driving demand for the day, its BP will not be charged until the valley of TOU tariff. Otherwise, under the premise to meet the driving demand, the charging of BPs try to avoid the peak-hours.

Theoretically, the constraints of SBPs and the total batteries swapping demand must be taken into consideration in the whole process of optimization. But in practice, each EB's driving route is relatively fixed. Thus each time it returns, its arriving time and remaining power fluctuates in the vicinity of several typical values. That means that rather than using complex analytical solution, reproducing and simulating the processes of EBs is more effective and practical. This is exactly one of advantages of the proposed method.

4   Simulation Examples

4.1   Assumptions and Background

One operating circular buses line in Beijing, China is taken as an example here. Further information of the route and EBs are given in Appendix. Objective bus station in the simulation is set to be Station C in Fig.1 acting as both charging station and the bus terminal. And charging rate is 0.3C, meaning that theoretical fully charging time is about 3 hours. The swapping operation time is 10 minutes. Meantime, the number of battery generator just satisfies the charging need of all the SBPs.

The proposed model is simulated on following 4 cases:

1) Basic Case. In this case, the speed of EBs is just described by a certain average value, and impacts of other factors are not involved. After an EB arrives at Station C, the corresponding BP will be replaced begins charging directly if the EB is considered to need battery swapping service.

2) Case of Considering Traffic Information. In this case, traffic information is considered based on the general situation. Countering different departure time in a day, differentiated average speeds and other traffic information are set successively according to *2011 Beijing Traffic Development Report*.

3) Case of Considering both Traffic and Weather Information. In this case, influences of meteorological factors have been added. And it is simplified as, in a day of winter, average energy consumption increases by 10% due to air conditioning and average speed decreases by 10%.

4) Case of Considering TOU Price. On the basis of case 3), the charging load in Station C is optimized by responding to TOU price. Besides, to facilitate the management and maintenance of batteries, once a group of BP begins charging, it will not be cut-out until fully charged. And TOU price is set as *Peak time*: 10:00-15:00, 18:00-21:00, and price: 1.164 yuan. *Shoulder time*: 7:00-10:00, 15:00-18:00, and price: 0.754 yuan. *Off-peak time*: 23:00-7:00(next day), and price: 0.365 yuan.

4.2   Simulation Results

Simulation results in different situations are given in Fig.4. Fig.4 (a) to Fig.4 (d) correspond to the four cases in section 3.1 successively. And, in Fig.4 (d), on purpose of overhaul and maintenance, only half of all the battery chargers in station continue to operate after 23:00.

In Fig.4 (a), with the method of procedural simulation, aggregating charging load of EBs cluster can be given relative accurately. Meantime, by comparing Fig.4 (a) to Fig.4 (c), it can be seen that with the increasing of considered factors, characteristics of charging load show great differences among those three cases. This illustrates the imperative of considering multiple factors while modeling EBs charging load aggregation with method of process simulation.

Contrasts to Fig.4 (c), Fig.4 (d) illustrates the benefits of optimizing the operation of charging/swapping station. Only from the perspective of daily charging costs, it reduces from 8200 yuan to 4200 yuan after optimization, nearly 50% of the total. Obviously, the economic benefit is considerable.

Fig.4 (d) only shows one option of optimizations. The method of procedural simulation proposed may provide support for the optimal operation of EBs cluster and related BPs by offering a lot of usable data. Then more discussions and researches [20, 21] about optimization measures and intelligence charging control can be done with this approach.

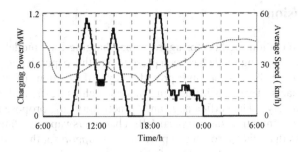

(a) Basic case

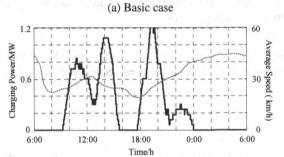

(b) Case of considering traffic information

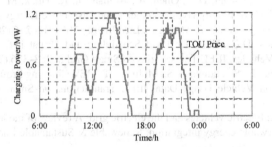

(c) Case of considering both traffic and

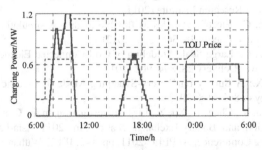

(d) Case of considering TOU price weather information

Fig. 4. Simulation results

5   Conclusion

A procedural simulation method for aggregating charging load model of EBs cluster based on battery swapping mode has been presented in this paper. It can grasp the circle characteristics of EBs cluster and provide key information of it dynamically. Meanwhile, impacts of various influencing factors such as traffic and weather can be reflected adequately. The efficiency and practicality of proposed method are demonstrated by simulation on a typical EBs charging/swapping station. Moreover, case study shows that this approach can provide data support for the optimal operation, and it may serve as an analysis tool for the future studies by considering more influencing factors in modeling process.

Acknowledgments. This work is jointly supported by NSFC-RCUK_EPSRC (No.51361130153) and the National Natural Science Foundation of China (No. 51377035).

References

1. Lopes, J.A.P., Soares, F.J., Almeida, P.M.R.: Integration of Electric Vehicles in the Electric Power System. Proceedings of the IEEE 99, 168–183 (2011)
2. Jung, J.S., Cho, Y.J., Cheng, D.L., Onen, A., Arghandeh, R., Dilek, M., Broadwater, R.: Monte Carlo analysis of Plug-in Hybrid Vehicles and Distributed Energy Resource growth with residential energy storage in Michigan. Lied Energy 108, 218–235 (2013)
3. Wu, D., Aliprantis, D.C., Ying, L.: Load Scheduling and Dispatch for Aggregators of Plug-In Electric Vehicles. IEEE Transactions on Smart Grid 3, 368–376 (2012)
4. Ashtari, A., Bibeau, E., Shahidinejad, S., Molinski, T.: PEV Charging Profile Prediction and Analysis Based on Vehicle Usage Data. IEEE Transactions on Smart Grid 3, 341–350 (2012)
5. Richardson, D.B.: Electric vehicles and the electric grid: A review of modeling approaches, Impacts, and renewable energy integration. Renewable & Sustainable Energy Reviews 19, 247–254 (2013)
6. Alexander, M.: Transportation Statistics Analysis for Electric Transportation. Technical report. Electric Power Research Institute (2011)
7. Metz, M., Doetsch, C.: Electric vehicles as flexible loads - A simulation approach using empirical mobility data. Energy 48, 369–374 (2012)
8. Rolink, J., Rehtanz, C.: Large-Scale Modeling of Grid-Connected Electric Vehicles. IEEE Transactions on Power Delivery 28, 894–902 (2013)
9. Gao, C.W., Wu, X.: A Survey on Battery-Swapping Mode of Electric Vehicles. Power System Technology 4, 891–898 (2013)
10. Liu, Y., Hui, F., Xu, R., Chen, T., Xu, X., Li, J.: Investigation on the Construction Mode of the Charging Station and Battery-Exchange Station. In: 2011 Asia-Pacific Power and Energy Engineering Conference, APPEEC 2011, pp. 1–2. IEEE, Wuhan (2011)
11. Zheng, Y., Dong, Z.Y., Xu, Y., Meng, K., Zhao, J.H., Qiu, J.: Electric Vehicle Battery Charging/Swap Stations in Distribution Systems: Comparison Study and Optimal Planning. IEEE Transactions on Power Systems 29, 221–229 (2014)
12. Nie, Y., Ghamami, M.: A corridor-centric approach to planning electric vehicle charging infrastructure. Transportation Research Part B-Methodological 57, 172–190 (2013)

13. Ge, W.J., Huang, M., Zhang, W.G.: Economic Operation Analysis of the Electric Vehicle Charging Station. Transactions of China Electrotechnical Society 2, 15–21 (2013)
14. Qian, B., Shi, D.Y., Xie, P.P., Zhu, L.: Optimal Planning of Battery Charging and Exchange Stations for Electric Vehicles. Automation of Electric Power Systems 2, 64–69 (2014)
15. Yang, Y.X., Hu, Z.H., Song, Y.H.: Research on Optimal Operation of Battery Swapping and Charging Station for Electric Buses. Proceedings of the CSEE 31, 35–42 (2012)
16. Pang, C., Dutta, P., Kezunovic, M.: BEVs/PHEVs as Dispersed Energy Storage for V2B Uses in the Smart Grid. IEEE Transactions on Smart Grid 3, 473–482 (2012)
17. Kiviluoma, J., Meibom, P.: Methodology for modelling plug-in electric vehicles in the power system and cost estimates for a system with either smart or dumb electric vehicles. Energy 36, 1758–1767 (2011)
18. Barre, A., Suard, F., Gerard, M., Montaru, M., Riu, D.: Statistical analysis for understanding and predicting battery degradations in real-life electric vehicle use. Journal of Power Sources 245, 846–856 (2014)
19. Lu, X.N., Sun, K., Guerrero, J.M., Vasquez, J.C., Huang, L.P.: State-of-Charge Balance Using Adaptive Droop Control for Distributed Energy Storage Systems in DC Microgrid Applications. IEEE Transactions on Industrial Electronics 61, 2804–2815 (2014)
20. Sheikhi, A., Bahrami, S., Ranjbar, A.M., Oraee, H.: Strategic charging method for plugged in hybrid electric vehicles in smart grids; a game theoretic approach. International Journal of Electrical Power & Energy Systems 53, 499–506 (2013)
21. Druitt, J., Fruh, W.G.: Simulation of demand management and grid balancing with electric vehicles. Journal of Power Sources 216, 104–116 (2012)

Appendix: Related Data in Cases

Related data in cases refer to that applied in [15], and Tab.A-1 is the buses departure time-table. The total number of EBs in studied station is 50. And their departure time distribute within the given period of time depending on the departure frequency. Besides, there are 20 groups of SBPs.

Tab. A-1. Buses departure time-able

Time	Departure frequency	Time	Departure frequency
06:00-07:00	20	15:00-16:00	6
07:00-08:00	30	16:00-17:00	15
08:00-09:00	12	17:00-18:00	20
09:00-10:00	10	18:00-19:00	6
10:00-11:00	10	19:00-20:00	4
11:00-12:00	10	20:00-21:00	3
12:00-13:00	10	21:00-22:00	2
13:00-14:00	8	22:00-23:00	1
14:00-15:00	6	Others	0

Tab.A-2 is relevant parameters of specific buses line and EBs.

Tab. A-2. Relevant parameters

Capacity of BP kW • h	Length of routes km	Power consumption per mile (kW • h)/km
220	38	1.3

Modeling and Voltage Stability Computation of Power Systems with High Penetration of Wind Power, Electric Vehicles and Air Conditioners

Aina Tian, Weixing Li, Jilai Yu, Ruiye Liu, and Junda Qu

Department of Electrical Engineering, Harbin Institute of Technology, Harbin 150001, China
tianaina2199@163.com

Abstract. The large-scale integration of wind power, electric vehicles and air conditioning loads would perhaps produce an adverse combined affect on the voltage stability of power system. It is of great significance to investigate the combined effects of such three factors on power system voltage stability. Firstly, the daily power curve models of the wind farms, electric vehicles and air conditioning loads are studied, and a number of typical daily power curves are given for them. Secondly, the paper analyzes the combined effects of the three factors on power system voltage stability by using different practical daily wind power models. Finally, various electric vehicle charging strategies are simulated to explore their effects in improving the system voltage stability.

Keywords: Electric vehicles, wind power, air conditioners, combined effects, power systems, voltage stability.

1 Introduction

The large-scale integration of electric vehicles (EVs) and renewable energy resources is an effective way to deal with the gradual depletion of primary energy, global warming and environmental pollution issues [1]. However, the large-scale integration of electric vehicles and renewable energy inevitably brings new challenges to power system security, reliability and economy [2].

Literature [3] studied the load characteristics of electric vehicles and the impact of different charging schemes on power system, and then a multi-time scale cooperative dispatching mathematical model was for electric vehicles and wind power established, finally the feasibility of optimizing the electric vehicle charging schedule to smooth the load fluctuations in the grid and to accommodate excessive wind power at night are analyzed, by using the measured data of North China Power Grid and Northwest Power Grid. Literature [4] conducted a preliminary study on the impact of the large integration of wind power and electric vehicles on the power system, and the corresponding security assessment and control strategy was proposed. Literature [5] constructed a random collaborative optimization scheduling model to consider the output stochastic uncertainty of the electric vehicles, wind power and solar power generation system.

K. Li et al. (Eds.): LSMS/ICSEE 2014, Part III, CCIS 463, pp. 517–528, 2014.

Since the wind farms usually absorb a certain amount of reactive power when they send active power to the system, the voltage stability issue perhaps occurs when the large-scale wind power is penetrated into the grid [6]. For a traditional air conditioner, it could be usually seen as a motor load that keeps approximately constant power, and this is very detrimental to the voltage stability of the grid because its reactive power is not dependent on the voltage. For the inverter-based air conditioner, its reactive power will decrease as the voltage reduces. Although the reactive characteristics of such air conditioner become better, the large integration of air conditioning loads is still one of the incentive reasons of the system voltage collapse [7].

However, the current research mainly focused on the combined effect of electric vehicles and wind energy resources, and did not further consider the impact of large-scale air conditioning loads on power systems, particularly on the system voltage stability. In future, the air conditioning load will be an essential part to be considered for the system voltage stability. So, exploring the combined effects of electric vehicles, wind power and air conditioning loads will be becoming increasingly important for voltage stability analysis and control, and it is to be of significance to propose a comprehensive control strategy for such three factors to improve the voltage stability of power systems.

This paper presents the daily power curve models for the wind farms, electric vehicles and air conditioning loads, and investigates the combined effects of different penetrations of wind power, electric vehicles and air conditioning loads on the system voltage profile and stability. Also, the voltage stability of the grid under different electric vehicle charging strategies is analyzed. The simulation results show that the implementation of an effective charging strategy of electric vehicles could improve the system voltage profile and stability.

2   Modeling of Wind Power, EVs and Air Conditioning Loads

2.1   Modeling of Daily EV Charging Loads

The charging load of an EV is dependent on its battery type, capacity, state of charge (SOC) and the charging mode. The main factors affecting the charging load of EVs in a region includes the number and types of electric vehicles, travel characteristics, the charging strategies, etc. [8]. In early studies, various battery models were emphasized [9]. The current research usually used the statistical method [10] to obtain some essential parameters, including the total number and mileage distributions of EVs, the initial SOC stage and its probability distribution and the probability distribution of various charging behaviors, possible charging time and duration, to model the large-scale EV charging load.

According to the "China Automotive Industry Development Report (2008)," the proportion of different types of electric vehicles can be obtained, which is shown in Table 1.

Table 1. Chinese electric car ownership statistics (2008)

number	Private car	Official car	taxi	bus
*10000	4229-5169	612-748	143-175	60-73

Based on Chinese electric vehicle charging load forecasting parameters [8], the Monte Carlo method is used to simulate the starting SOC and the initial charging time of individual EVs, and further to obtain the daily EV charging power curves under the two charging modes, i.e., uncontrolled charging mode and intelligent charging mode. Figure 1 show the results when the EV penetration level is 20%.

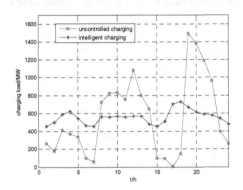

Fig. 1. The daily charging power curves of EVs

2.2   Modeling of Daily Wind Power Curves

The statistical and simulating methods are two main ways to model wind farms' characteristics [11]. The former analyses the typical wind farm output power features based on the measured wind data [12]. The latter establishes the simulation model to analyze the output characteristics of a wind farm [13]. The doubly-fed induction generators and synchronous generators were usually employed [14]. In this paper, four seasons' typical wind farms data in different parts of China – Jiangsu [15], Liaoning [16] and Jiuquan [17] are obtained based on statistical methods, and the typical daily output curves are shown in Figures 2-4, respectively.

It can be seen that the wind power output characteristics in Liaoning is similar to that of Jiuquan, but it is different to Jiangsu's wind power feature. The Jiangsu's wind power is large in the morning in both summer and winter. However, the wind power output in Liaoning and Jiuquan is large at night in both summer and winter.

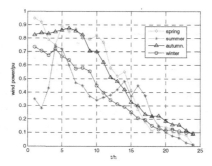

Fig. 2. The typical daily power curves of JiangSu wind farms

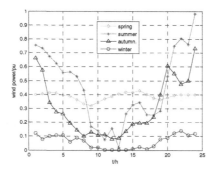

Fig. 3. The typical daily power curves of Liaoning wind farms

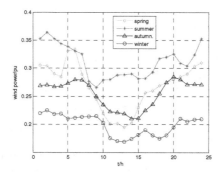

Fig. 4. The typical daily power curves of Jiuquan wind farms

2.3   Modeling of Daily Air Conditioning Loads

In spring and autumn, the air conditioning load is not heavy, so it does not have a serious impact on power systems. Therefore, both summer and winter are the main seasons to be considered for modeling of air conditioning loads. Existing models of air conditioning loads are mainly the static load model, the physical dynamic load model and the non-physical dynamic load model [18].

The air conditioning load in the grid is usually simulated by using the air conditioning load curve [19]. A typical daily air conditioning load curve is shown in Figure 5. The air conditioning load characteristic is not obviously dependent on geographical location in China. The load characteristics of each region in the particular season are basically the same.

In summer, air conditioning load is small at late night and in the early morning, and the load is large at other times. In winter, the daily air conditioning loads have relatively small change.

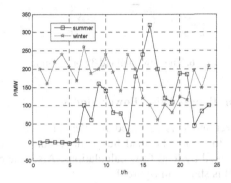

Fig. 5. The typical daily power curves of air conditioning loads

3   Simulation of Combined Effects of the Three Factors on Power System Voltage Stability

3.1   The Test System

The New England 39-bus system is used for case studies. Assuming that the bus 39 is the slack bus, the system topology is shown in Figure 6. Three wind farms are added to the grid at buses 30, 34 and 35. The system is divided into three zones according to the function, i.e., residential zone, commercial zone and industrial zone. In the simulation, the uncontrolled charging/discharging mode of electric vehicles is employed, and the EV penetration level is set to 20% and 50%. This paper only considers the Jiangsu's wind power mode and Liaoning's wind power mode since the wind power mode in Liaoning is similar to that in Jiuquan. A static load model is employed for air conditioning loads, and the air conditioning loads are added to the grid according to the proportion of the base loads in the three regions.

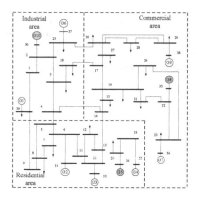

Fig. 6. The new England 39 bus system

3.2   Case Studies for Jiangsu Wind Power Modes

The Jiangsu wind power mode is assumed for the three wind farms and the total penetration level is set to 20%. The load bus voltage curve in each area shown in figures 7 and 8 are obtained by load flow calculations. Also, the curve of the voltage stability indicator L [20] is shown in figure 9.

It can be observed that, there is no voltage stability problem with residential area in the morning. On the day time, the electric vehicle charging makes the voltage stability worse. Yet, there is no big voltage stability problem with the three areas in this period. In this case, some appropriate control methods may be needed to reduce the electric vehicle charging power. At night, the voltage stability problems may occur in all the three areas, especially in the residential area. At that time, the system voltage stability may decline sharply even system voltage instability or collapse occurs, since the air conditioning load is relatively large and the electric vehicle charging is centralized. At this point, effective control strategies must be taken to shift the electric vehicle charging to other time.

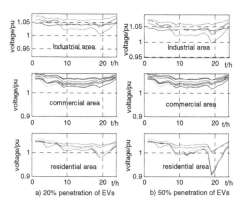

Fig. 7. The load bus voltage curves in each area in summer for Jiangsu wind power mode

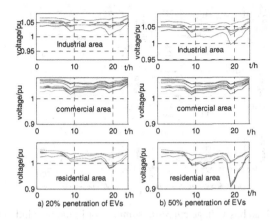

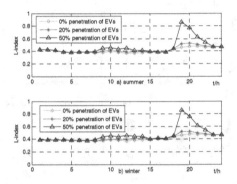

Fig. 8. The load bus voltage curves in each area in winter for Jiangsu wind power mode

Fig. 9. The L-indicator curves for Jiangsu wind power mode

3.3   Case Studies for Liaoning Wind Power Modes

The Liaoning wind power mode is assumed for the three wind farms and the total penetration level is set to 20%. The load bus voltage curves in each area shown in figures 10 and 11 are obtained by load flow calculations. The curve of the voltage stability indicator L is shown in figure 12.

It can be observed the similar phenomena with Jiangsu case. The electric vehicles charging are concentrated in the evening, at that time the air conditioning loads are heavy, so the combined effects of such three factors result in the system voltage rather worse during 19:00-21: 00 with 50% penetration level of EVs.

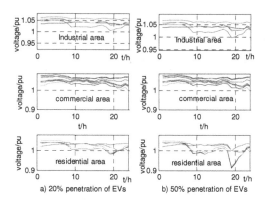

Fig. 10. The load bus voltage curves in summer for Liaoning wind power mode

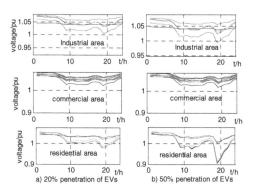

Fig. 11. The load bus voltage curves in winter for Liaoning wind power mode

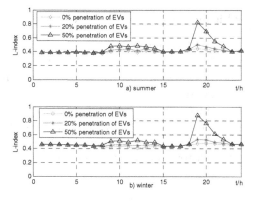

Fig. 12. The L-indicator curves for Liaoning wind power mode

4   Effect Analysis of EV Charging Modes on the System Voltage

Since the wind power output is difficult to control, and load shedding is usually the latest emergency control measure, it may be the best option to develop an effective electric vehicle charging strategy to alleviate the combined effects of the three factors. This section will analyze the impact of electric vehicle charging/discharging strategies in different seasons on the system voltage stability.

4.1   Scenarios and Parameters

The test system shown in figure 6 is used here. The Jiuquan wind power mode for the three wind farms and the total penetration level is set to 20%. The air conditioning load take the same consideration with above. Both 20% and 50% of the penetration levels of EVs are considered, and both the uncontrolled and intelligent charging modes are studied.

4.2   The Effect of Electric Vehicle Charging Modes in Summer

Figures 13 and 14 show the load bus voltage profile and stability curves, respectively. It can be seen that the intelligent charging strategies can effectively improve the system voltage profile and stability. But in an emergency situation, regulating electric vehicles may be not sufficient to ensure the system stability. In this case, a comprehensive control method including other control measures may be needed.

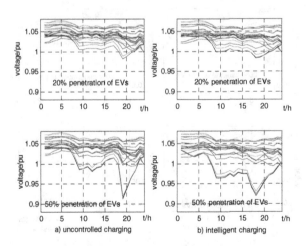

Fig. 13. The load bus voltage curves in summer

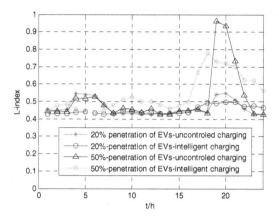

Fig. 14. The L-indicator curves in summer

4.3   The Effect of Electric Vehicle Charging Modes in Winter

Figures 15 and 16 show the load bus voltage profile and stability curves, respectively. Similarly, it can be observed that smart charging strategies can effectively improve the system voltage profile and stability. In order to ensure the voltage stability of the system, it is essential to propose effective combined regulating strategies of the electric vehicles, wind power and air conditioning loads.

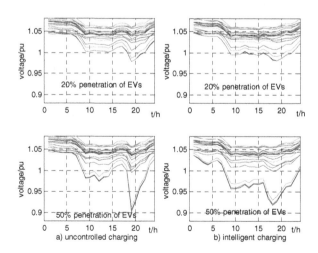

Fig. 15. The load bus voltage curves in winter

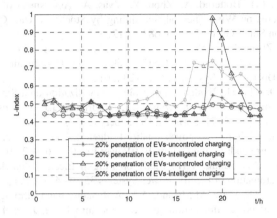

Fig. 16. The L-indicator curves in winter

5   Conclusion

This paper presents the daily power curve models of the wind farms, electric vehicles and air conditioning loads, and simulates the combined effects of such three factors on power system voltage profile and stability, with different penetration levels of electric vehicles and wind farms. Different electric vehicle charging modes are investigated to improve the system voltage profile and stability. The simulation results show that developing an effective charging strategy of electric vehicles could be of great significance to improve the system voltage. It can be concluded that, in an emergency situation, regulating electric vehicles is not sufficient to ensure the system stability, so it is essential to propose effective combined regulating strategies of the electric vehicles, wind power and air conditioning loads.

Acknowledgments. This paper is supported by the National Natural Science Foundation Council of China (51361130153).

References

1. Lopes, J.A.P., Soares, F.J., Almeida, P.M.R.: Integration of electric vehicles in the electric power system. Proc. IEEE 99(1), 168–183 (2011)
2. Ota, Y., Taniguchi, H., Nakajima, T., Liyanage, K.M., Baba, J., Yokoyama, A.: Autonomous Distributed V2G (Vehicle-to-Grid) Satisfying Scheduled Charging. IEEE Transactions on Smart Grid 3(1), 559–564 (2012)
3. Yu, D.Y., Song, S.G., Zhang, B., Han, X.S.: Synergistic Dispatch of PEVs Charging and Wind Power in Chinese Regional Power Grids. Automation of Electric Power Systems 35, 24–27 (2011)

4. Liu, C., Wang, J.H., Botterud, A., Zhou, Y., Vyas, A.: Assessment of Impacts of PHEV Charging Patterns on Wind-Thermal Scheduling by Stochastic Unit Commitment. IEEE Transactions on Smart Grid 3(2), 675–683 (2012)

5. Wang, G.B., Zhao, J.H., Wen, F.S., Xue, Y.S., Xin, J.B.: Stochastic Optimization Dispatching of Plug-in Hybrid Electric Vehicles in Coordination with Renewable Generation in Distribution Systems, 36, 22–29 (2012)

6. Karaishi, K., Oguchi, M.: Evaluation of Voltage Stabilization on a SmartGrid Simulation System for Introduction of EV. In: Mauri, J.L., Rodrigues, J.J.P.C. (eds.) GreeNets 2012. LNICST, vol. 113, pp. 60–71. Springer, Heidelberg (2013)

7. Singh, B., Bist, V.: Improved power quality bridgeless Cuk converter fed brushless DC motor drive for air conditioning system. IET Power Electron. 6, 902–913 (2013)

8. Luo, Z.W., Hu, Z.H., Song, Y.H.: Study on Plug-in Electric Vehicles Charging Load Calculating. Automation of Electric Power Systems 35, 36–42 (2011)

9. Tommasia, L.D., Gibescub, M., Brand, A.J.: A dynamic wind farm aggregate model for the simulation of power fluctuations due to wind turbulence. Journal of Computational Science 1, 75–81 (2010)

10. Yen, W.L., Littlefield, R.G., McLean, D.R., Tuchman, A., Broseghini, T.A., Page, B.J.: Battery model for the extreme ultraviolet explorer spacecraft. Journal of Spacecraft and Rockets 32, 360–363 (1995)

11. Beaude, O., He, Y.J., Hennebel, M.: Introducing Decentralized EV Charging Coordination for the Voltage Regulation. In: 2013 4th IEEE PES Innovative Smart Grid Technologies Europe (ISGT Europe), pp. 1–4 (2013)

12. Gan, D., Ke, D.P.: Wind power ramp forecasting based on least-square support vector machine. Applied Mechanics and Materials 535, 162–166 (2014)

13. Moharana, A., Varma, R.K., Seethapathy, R.: Modal analysis of induction generator based wind farm connected to series-compensated transmission line and line commutated converter high-voltage DC transmission line. Electric Power Components and Systems 42(6), 612–628 (2014)

14. Akhmatov, V., Nielsen, J.N., Jensen, K.H., Goldenbaum, N., Thisted, J., Frydensbjerg, M., Andresen, B.: Siemens Wind Power Variable-speed Full Scale Frequency Converter Wind Turbine Model for Balanced and Unbalanced Short-circuit Faults. Wind Engineering 34(2), 139–156 (2010)

15. Yang, Z.L., Zhu, Z.L., Li, R.Y., Lu, Y.Y.: Analysis on Output Power Characteristics of Jiangsu Coastal Typical Wind Farm. East China Electric Power 38, 388–390 (2010)

16. He, C.J., Wang, Y.Y., Wu, S.N.: The Application of Wind and PV Hybrid Technology to Liaoning Power System. Northeast Electric Power Technology 12, 27–31 (2009)

17. Xin, S.X., Bai, J.H., Guo, Y.H.: Study on Wind Power Characteristics of Jiuquan Wind Power Base. Energy Technology and Economics 22, 16–20 (2010)

18. Irminger, P., Rizy, D.T., Li, H.J., Smith, T., et al.: Air Conditioning Stall Phenomenon – Testing, Model Development, and Simulation. In: 2012 IEEE PES Transmission and Distribution Conference and Exposition (2012)

19. Lei, Z., Zhang, Z., Zhou, J., Chen, Y.: Analysis and Study on the Air-conditioning Load of Tianjin Grid. Journal of North China Electric Power 2, 12–15 (2013)

20. Kessel, P., Glavitsch, H.: Estimating the voltage stability of a power system. IEEE Transactions on Power Delivery 3, 346–354 (1985)

Design of Power Factor Correction System for On-board Charger[*]

Fuhong Xie[1], Xiaofei Liu[1], Shumei Cui[1], and Kang Li[2],

[1] School of Electrical Engineering and Automation,
Harbin Institute of Technology, China
[2] School of Electronics, Electrical Engineering and Computer Science
Queen's University Belfast, United Kingdom
xiefuhong2@126.com, cuism@hit.edu.cn

Abstract. As auxiliary equipment for electric vehicle, on-board charger should have high performance on harmonic, power factor and efficiency. But in high-power application, the traditional PFC technology has many restrictions and defects. Therefore, based on the parallel interleaved technique, we designed a power factor correction system suitable for on-board charger of electric vehicle. Experiment results show that the design can effectively reduce the input current ripple and improve the output voltage ripple frequency and the power factor.

Keywords: On-Board Charger, APFC, Interleaved.

1 Introduction

Nowadays, with environmental problems and energy crisis getting increasingly serious, electric vehicle has become a hot topic because of its features of low noise, zero emissions and significant advantages in energy efficient. As supporting charging equipment for electric vehicles, the research for on-board charger has important significance. Moreover, as a kind of environmentally friendly transportation, the on-board charger for electric vehicles should be environmentally friendly too. Especially when a large number of energy storage batteries connected to the grid as capacitive load, it will inevitably reduce the power factor of power grid, causing harmonic pollution and low efficiency. So, with the increasingly widespread use of electric vehicles in the future, the development of power factor correction system for on-board charger is very important.

At the early stage, people used passive network that was composed of capacitors and inductors to achieve power factor correction (PFC). However, this kind of circuit has the disadvantages of large volume and low performance in harmonic suppression of input current. Later, with the development of power electronics technology, Active Power Factor Correction (APFC) came into being. APFC technology add a power

---

[*] This research is jointly supported by the National High Technology Research and Development Program of China (863 Program) under Project 2012AA111003 and the NSFC-RCUK_EPSRC under Project 51361130153.

K. Li et al. (Eds.): LSMS/ICSEE 2014, Part III, CCIS 463, pp. 529–538, 2014.
© Springer-Verlag Berlin Heidelberg 2014

converter with a power factor correction function between rectifier circuit and output capacitance in order to ensure the stability of output voltage and control the input current harmonic and phase difference between input current and voltage. Moreover, because APFC operates at high switching frequency, APFC technology has advantage of smaller volume and power factor and high performance in efficiency. These advantages make the APFC technology have gradually become the mainstream of power factor correction development.

However, with the improvement of the converter power level, the traditional single-phase Boost PFC converter is facing increasing restrictions [1]. To solve these problems, interleaved technology is introduced into Boost PFC converter. Interleaved Boost PFC circuit can effectively reduce the input and output current ripple and simplify the design of input EMI filter.

Therefore, based on interleaved technology, we designed a 3.3kW power factor correction system for on-board charger. This paper shows the overall design of the project and gives system parameters of the circuit. And finally we tested our design and gave relevant explanation for experimental results.

2   Classification of Electric Vehicle Charging Machine

Currently, according to the differences in operational principle, there are mainly three kinds of electric vehicle charging machine [1]. The first kind is composed of uncontrolled rectifier, chopper and industrial frequency transformer, which is belong to the early product. It is characterized by low voltage ripple in DC side, good dynamic performance and frequency isolation. But its low conversion efficiency, high current harmonics in grid side and large volume prevent this kind of charger from more widespread use. Moreover, the harmonic current produced by the on-board charger contain the fifth harmonic of 60%~69%, the seventh harmonic of 40%~49% and the current total harmonic distortion (THD) is 86.2%. Therefore, generally, such charger has excessive harmonic current, so it is not suitable for access to the power grid.

The second category is made of industrial frequency transformer, three-phase uncontrolled rectifier, high-frequency transformer and isolated DC/DC converter. Though its merits are regarded as low output DC voltage ripple, good dynamic performance, high-frequency isolation and small size, it has disadvantage of high current harmonics (about 30%) in grid side and low conversion efficiency. Currently, this type of charger that its internal rectification device is three phase bridge uncontrolled rectifier is the most used charger on the market, the majority of existing research results also focus on this kind of charger. Although this type of charger produces plenty of harmonic, low-cost makes it become the mainstream of on-board charger on the market.

The third type of charger is composed of three-phase PWM rectifier and isolated DC/DC converter. It is characteristic of using PWM technology in rectifier. Although this technology increases the cost of charger, its advantages are that high power factor and low current harmonic in the grid side, also the THD can be less than 5%. Moreover, the overall volume of the device is reduced, and the output voltage ripple is decreased so that the dynamic performance is improved and the system has better conversion efficiency.

3   Comparative Study of Interleaved and Traditional Boost PFC

APFC technology can significantly reduce the current harmonics and improve power factor of the system so it become one of most practical technologies for the switching power supply design. However, with the increase of the converter power level, single-phase Boost PFC switching device is bound to withstand higher instantaneous voltage and current stress so that the device component selection will be more difficult, which not only increases the cost of the system, but also intensify the du/dt and di/dt in critical points in the circuit, causing serious radiated interference and conducted EMI. Interleaved Boost PFC circuit can effectively reduce the input current ripple, simplifying the design of input EMI filter.

Theoretically, any DC / DC converter topology can be used as the main circuit of APFC, such as Boost circuit, Buck circuit and Cuk circuit. Compared to other circuits, Boost circuit has advantages of wider range of input voltage, better effects for power factor control and easier design for the drive circuit. These characteristics make Boost circuit has been widely used in APFC. Now, research about the Boost PFC circuit is thriving, and optional integrated control chip is numerous, such as the classic control chip UC3854 and L6561.

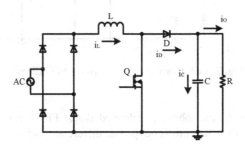

Fig. 1. Traditional Boost PFC Circuit

Fig. 1 shows a conventional Boost PFC circuit. With the shutoff and conduction of the switch device Q, the Boost inductor L releases and stores energy periodically, so that the inductor current i_L maintains a correspondingly continuous up and down state. In this circuit, only when the switching device Q shut off does the freewheeling diode D conduct, so the diode current i_D is discontinued. Meanwhile, in order to ensure a constant output power, the output terminal usually connects to a large-capacity capacitor C. The current i_C that flows through the capacitor is the difference between the diode current i_D and load current i_o. Under the conditions of constant output power, the capacitor current i_C is pulsating. Fig. 2 shows the main waveforms of the conventional Boost PFC circuit when the inductor L is working in Continuous Conduction Mode (CCM).

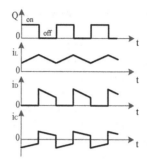

Fig. 2. The Main Waveform of Traditional Boost PFC at CCM State

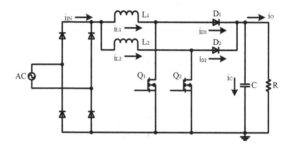

Fig. 3. Two-Phase Interleaved Boost PFC Circuit

Fig. 3 shows a typical two-phase interleaved Boost PFC circuit which consists of two identical Boost PFC circuit. The phase difference between drive signals for switching device Q_1 and Q_2 is $180°$ which makes these two Boost PFC circuits work in an interleaved state. When the duty cycle of drive signal is 50% and inductor work in the CCM state, the main waveforms of two-phase interleaved Boost PFC circuit are shown in Fig. 4.

From Fig. 4 we can see that, compared to traditional Boost PFC circuit, because interleaved Boost PFC circuit has complementary switching device driver signals, the two-phase inductor current has opposite trend in rise and decline at the corresponding time. Then, the superposition of two-phase current significantly decreases the total input and output current ripple amplitude [3-4], which can simplify the design of input EMI filter and reduce the value of output capacitor. Under the conditions of the same power level, the interleaved Boost PFC circuit can reduce current stress in switching devices and inductors by half, so that the inductor's volume is only a quarter of the volume of inductor that in the traditional Boost PFC circuit [5-7].

Therefore, although the main circuit and control circuit topology are more complex, and the power density decrease in somehow, interleaved Boost PFC circuit can effectively alleviate the EMI problem, and the system can improve its dynamic response characteristics by reducing the value of output filter capacitor. Therefore, interleaved Boost PFC technology has good development prospects.

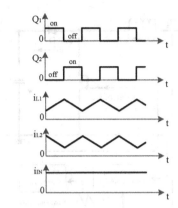

Fig. 4. The Main Waveform of Two-Phase Interleaved Boost PFC at CCM State

4   Control Method of APFC

According to the states of inductor current, the control strategies of the power factor correction are divided into Continuous Current Mode (CCM), Discontinuous Current Mode (DCM) as well as Boundary Conduction Mode (BCM). Based on the output power level, some circuits can convert between DCM and CCM, so that this control strategy is called the Hybrid Current Model (MCM). So, when the inductor works in different conduction mode, the control method of active power factor correction is completely different.

The average-current mode with CCM is the most widely used control strategy in APFC design. The converter uses the product of input voltage signal and the output voltage error signal as the reference current signal to regulate and control the input current. Based on a given reference signal, the current controller controls the input current in order to reduce the phase displacement between the input current and the input voltage. The diagram of Boost PFC circuit control by average-current mode is shown in Fig. 5.

In the Fig. 5, the control circuit uses the product of the error signal of rectifier output voltage U_d and the amplified signal of output voltage U_o as a reference current signal I_{ref}. When the input current (inductor current) signal i_L compares with the reference current signal, the high frequency component of the input current signal is averaging processed by the current error amplifier. The result of comparison between the average current error signal and the sawtooth signal provides PWM drive signal for switching device Q. Moreover, according to the output power level, the system determines the duty cycle of drive signal to ensure that inductor current is approaching the average inductor current.

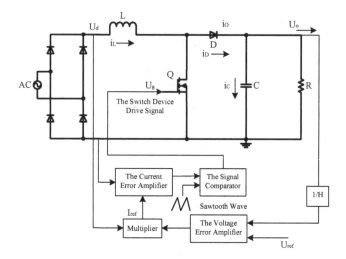

Fig. 5. The Diagram of Boost PFC Circuit Control by Average Current Mode

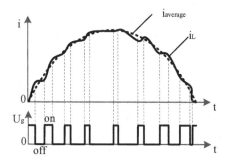

Fig. 6. The Inductor Current Waveform under the Average Current Control

In the APFC circuit under average current mode control, the system uses the current loop and voltage loop to ensure input current is approaching to a sine wave and maintain a stable output voltage. When the inductor current increases, PWM duty cycle of the comparator declines, making the inductor current reduce; otherwise increase the duty cycle and the inductor current. When the output voltage decreases, the output of the voltage error amplifier will increase, so that the reference output current from the multiplier increases, thereby increasing the inductor current and the output voltage; and vice versa, so that the output voltage drops. The inductor current waveform under average current mode control is shown in Fig. 6.

The current loop under average current mode control has a wide gain bandwidth, which makes the distortion generated by the tracking error low, so it's easy to achieve high performance in power factor correction. Meanwhile, this control mode is not sensitive to noise and has good stability, which makes it widely used.

5   Hardware Design and Experimental Results

The APFC circuit is composed of the main power circuit and the control circuit. According to the foregoing research, the main circuit adopts interleaved Boost circuit structure while the control circuit chooses a dedicated monolithic control chip UCC28070 to generate control waveform. Principle diagram is shown in Fig. 7.

UCC28070 is an interleaving continuous conduction mode (CCM) PFC controller. This integrated control chip has advantages of stable performance and numerous innovative technologies and protection measures, which can maximize the performance of the system [7].

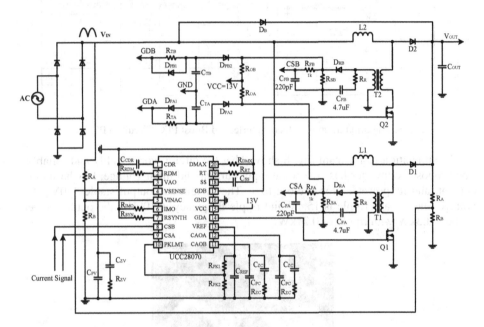

Fig. 7. The Design of Interleaved Boost PFC Circuit Based on UCC28070

The input voltage range of this converter is $U_{in} = 220\text{VAC} \pm 10\%$; the frequency of drive signal is 63kHz; the output DC voltage is $U_o = 390\text{V}$. Considering the limitation of inductor current minimum ripple and the influence of DC bias, we select the value of inductor for per PFC circuit unit is $L = 350\mu\text{H}$. As to the output capacitor, we choose $C_{out} = 2350\mu\text{F}$ after considering about the maintaining time.

Considering the output power level, each PFC circuit unit uses two identical MOSFET in parallel as switch device and actually we select IXFK64N60P. And we choose the SiC diode C3D10060A as freewheeling diode. In the actual experiment, we repeatedly adjust the parameters in voltage loop and current loop and ultimately get an excellent correction effect.

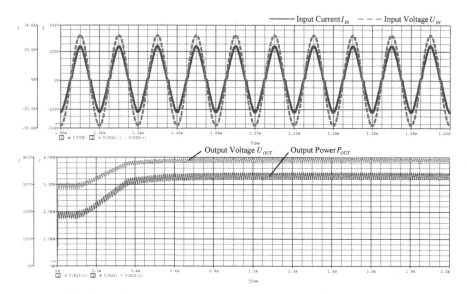

Fig. 8. The Simulation Results of Interleaved Boost PFC Circuit in PSpice

The simulation result from Fig. 8 shows that the circuit works stably and reliably. The proposed converter has sine wave input current and input voltage, low harmonic content, and its power factor is close to 1. The system can output stable 390V DC-voltage; the voltage ripple is 15.4V and ripple frequency is 100Hz; the output power is about 3.3kW with ripple of 0.27kW.

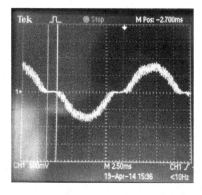

Fig. 9. The Input Current of Traditional Boost PFC Circuit

The experimental results are shown in Figs. 9 to 12. Compared to traditional Boost PFC, Fig. 9 and Fig. 11 show that the interleaved Boost PFC circuit can reduce ripple current in input current, but it is found that the input current still produces distortion. Fig. 10 shows the phase difference between the driving signals of interleaved Boost PFC circuit is 180°, indicating the system has achieved an interleaved work. Fig. 12 shows the maximum output voltage is 412V with ripple voltage of 28V, and ripple frequency is 100Hz, proving that the parallel interleaved circuit can improve the output voltage ripple frequency.

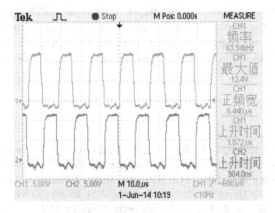

Fig. 10. Two Channel Driving Signals of Interleaved Boost PFC Circuit

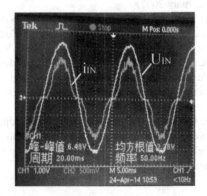

Fig. 11. The Input Voltage and Current Waveforms of Interleaved Boost PFC at Full Load

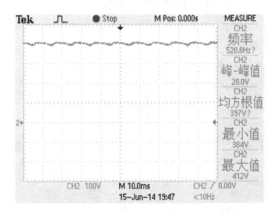

Fig. 12. The Output Voltage Waveforms at Full Load

6   Conclusion

Based on interleaved technique, this paper designs an APFC system for on-board charger and gives the topology of main power circuit and the design of control circuit. It can be known from the experimental results, the interleaved Boost PFC system can effectively reduce the input current ripple, and the power factor can be increased to more than 0.99.

References

1. Zhang, X.Q., Li, P., Hu, W.T., Xu, J.L., Zhu, J., Zhang, P.F.: Analysis on the Influence of the Harmonics for Electric Vehicle Charging Station. China Power 41(9), 31–36 (2008)
2. Huang, M., Huang, S.F., Jiang, J.C.: Harmonic Study of Electric Vehicle Chargers. Journal of Beijing Jiaotong University: Natural Science 32(5), 85–88 (2008)
3. Wang, C., Xu, M., Lu, B., Lee, F.C.: New architecture for MHz switching frequency PFC. In . In: Twenty Second Annual IEEE Applied Power Electronics Conference, APEC 2007, pp. 179–185. IEEE (February 2007)
4. Miwa, B.A., Otten, D.M., Schlecht, M.F.: High efficiency power factor correction using interleaving techniques. In: Proceedings of Seventh Annual Conference on Applied Power Electronics Conference and Exposition, APEC 1992, pp. 557–568. IEEE (February 1992)
5. Balogh, L., Redl, R.: Power-factor correction with interleaved boost converters in continuous-inductor-current mode. In: Proceedings of Eighth Annual Conference on Applied Power Electronics Conference and Exposition, APEC 1993, pp. 168–174. IEEE (March 1993)
6. Kolar, J.W., Kamath, G.R., Mohan, N., Zach, F.C.: Self-adjusting input current ripple cancellation of coupled parallel connected hysteresis-controlled boost power factor correctors. In . In: 26th Annual IEEE Power Electronics Specialists Conference, PESC 1995 Record, vol. 1, pp. 164–173. IEEE (June 1995)
7. UCC28070 Data Sheet, http://www.ti.com.cn/product/cn/ucc28070

Evolutionary Parameter Optimization of Electro-Injector Models for Common Rail Diesel Engines

Paolo Lino[1], Guido Maione[1], Fabrizio Saponaro[1], and Kang Li[2]

[1] Department of Electrical and Information Engineering, Politecnico di Bari, Bari, Italy
[2] School of Electronics, Electrical Engineering and Computer Science,
Queen's University Belfast, Belfast, BT9 5AH, UK
{paolo.lino,guido.maione}@poliba.it,
saponaro.fab@gmail.com, k.li@qub.ac.uk

Abstract. One of the major issues in innovative automotive engines is to reduce the energy consumption and pollutant emissions, at the same time, to guarantee a high level of performance indices. To this aim, common rail diesel engines can satisfy strict regulations by enhancing the model-based control of the injection process to increase the combustion efficiency. This paper presents a more accurate model for the electro-injector in common rail diesel engines. The model takes into account the mechanical deformation of relevant parts of the electro-injector and the non-linearity of the fuel flow. Model parameters are then optimized by an evolutionary strategy. Simulation shows that the optimized model can be helpful in predicting the real trend of the injected fuel flow rate when assisted with the experimental data, and in controlling the injection.

Keywords: Common rail diesel engines, electro-injection, optimization, differential evolution.

1 Introduction

The continuous growth of technology in automotive industry and the associated research have led to the development of innovative engines, electric or hybrid vehicles, and engines based on fuels different from diesel or gasoline, for example gas engines in the compressed natural gas (CNG) powered vehicles. The main motivation is to reduce both fuel and energy consumption and the pollutant emissions (gases like CO, NOx, HC, particulate matter, etc.), while guaranteeing high level of performance and robustness. Even conventional diesel engines have been improved in this sense, on the basis of the common rail technology, by developing new electro-injectors that are able to increase the combustion efficiency and, in general, the engine efficiency. However, the current regulations in Europe and other countries and regions impose severe restrictions on pollutant emissions and the requirements continuously grow for even lower emission level. In this context, this paper focuses on the parameter optimization of the injection system that would remarkably affect the performance in terms of energy consumption and pollutant emissions [1, 2, 3].

K. Li et al. (Eds.): LSMS/ICSEE 2014, Part III, CCIS 463, pp. 539–551, 2014.

In the field of modelling and control of injection systems, a well known approach considers a continuous flow through the system volumes [4, 5], and models are developed by using partial differential equations to describe the fuel compressibility and pipeline flexibility [6].

The more accurate the models are, the more powerful they can be used to diagnose faults, to explore different configurations, and to test different geometrical and functional designs. However, model complexity must be lowered to simplify the control algorithm. Then, the number of parameters has to be minimized. In synthesis, a trade-off must be reached between model accuracy and control simplicity and efficiency. To this aim, the relevant phenomena and most significant parameters are considered. In particular, the mechanical deformation of relevant components in the electro-injectors and the non-linear fuel flow to the cylinder are considered, because they determine the variations of the output injected flow rate by the electro-injectors.

Other recent works deal with pressure control in the peculiar engines that employ CNG (e.g. methane) as fuel because of its wider availability and spread than oil reserves [2, 3, 7]. To achieve reduced pollutant emissions, an accurate fuel metering is required to obtain a stoichiometric air/fuel mixture. Then, a precise electronic control of both the injection timing as specified by the opening time intervals of electro-injectors and the gas injection pressure. If the first problem can be solved with high precision, the second problem is difficult since the system operation is nonlinear due to gas compressibility. The problem of injection in CNG engines is that large parametric variations may occur and different working points must be considered according to the conditions set by manufacturers to react for different power and speed requirements. Moreover, disturbance compensation is not adequate and, above all, the dynamics is highly nonlinear. The aim is typically to regulate the pressure in the common rail diesel engine to achieve the levels required by the changing working points. Usually, standard gain-scheduled PI controllers are employed to this aim, but robustness of the control loop is not high so that new control strategies are required [8, 9, 10].

To design the control of the aforementioned innovative injection systems, it is important to build the model of the system. The model parameters only depend on well defined geometrical data of the injector and on fuel properties. The model can be optimized by using intelligent techniques based on evolutionary strategies, e.g. the Differential Evolution. Optimization is based on a performance index that takes into account the prediction error of the injected flow rate because it is one of the most important variables affecting the correct metering of fuel that is injected in the cylinder. The results of the optimization are compared with experimental results available from a real test bench.

The paper is organized as follows. Section 2 describes the considered dynamic system and introduces the model. Section 3 formulates the optimization problem and the differential evolution is employed. Section 4 presents the simulation results. Finally, Section 5 concludes the paper.

2   Common Rail Electro-Injector

In a Common Rail (CR) diesel engine the fuel flows through a low-pressure circuit, including a fuel tank and a low-pressure pump (LPP), before being delivered by a high-pressure pump (HPP) into a common rail. If the pressure of the pumped fuel overcomes a threshold, a delivery valve opens to feed the rail. The CR volume includes a sensor and an electronically controlled electro-hydraulic valve to regulate the fuel so that the rail pressure is set to a reference value. In this way, fuel from the electro-injectors in the CR can be correctly supplied, and the CR can damp the pressure oscillations due to the operation of pumps and injectors.

The main point is that the injection pressure is kept high to reduce particulate matter emissions and fuel consumption, even this leads to higher emissions of nitrous oxides, higher peak cylinder pressure, and higher pump power consumption. To synthesize, controlling the CR pressure and the injection timings allows an accurate metering of the injected fuel.

However, the injected fuel has a complex fluid-dynamics as a non-uniform distribution of the pressure in the CR arises due to the injection process. Therefore, an accurate model-based prediction of the pressure dynamics at different sections of the rail and an accurate model of the electro-injectors are most useful for fuel metering [11, 12]. Moreover, the fast operations determined by close and multiple injections establish a water hammer effect that affects the propagation of wave pressure along the pipes. Therefore, a sufficiently accurate mathematical model is necessary to design an effective controller.

In particular, the electro-injectors and their operation are of paramount importance. Therefore, modelling and control of them is important for achieving the combustion efficiency and engine performance in terms of fuel consumption and pollution reduction, and an optimized energy utilization.

2.1   The Electro-Injector

The CR electro-injector includes a control circuit which is a control chamber equipped with an electro-hydraulic valve, and a feeding circuit made by an accumulation volume (also known as the nozzle delivery volume) and a SAC volume (see Fig. 1). The SAC is a small volume at the end of the fuel flow path, directly connected to the nozzles of the injector. The CR delivers fuel to the control chamber and accumulation volume by means of high pressure pipes. A plunger-needle element is placed between the control circuit and the feeding circuit and is operated to regulate the injection flow. In particular, injection starts when the needle moves up so that the SAC and the accumulation volume are connected. Thus, acting on the upper surface of the plunger-needle element allows to properly reduce (or increase) the control chamber pressure, then to open (or close) the injector.

The control chamber has a circular symmetry and variable size due to the position of the plunger. It is connected to the high-pressure CR through the Z-hole and to a low-pressure circuit (upper part of Fig. 1) through the A-hole.

The flow section between the control chamber and low-pressure circuit is regulated by the electro-hydraulic valve that is driven by a PWM modulated voltage signal. Then, if the valve is closed, the pressure on the top face of the plunger is almost equal to the rail pressure, and the needle-plunger element is pushed down; if the valve opens, the pressure in the control chamber diminishes and the needle-plunger element is pushed up.

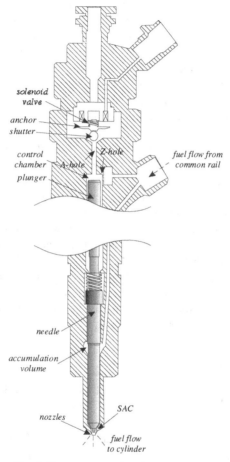

Fig. 1. The common rail electro-injector

3   The Model

The electro-injector is basically an interconnection of different volumes in which the fuel flows, as previously described. The pressure in these different subsystems is different and changes according to different dynamics. A mathematical model can be obtained by applying the continuity equation, the momentum equation, and the Newton's second law of motion for the mechanical parts. Peculiar phenomena are neglected like cavitation, effect of pressure propagation delay in the rail, variation of fuel density for different operating conditions, and variations with temperature.

Other assumptions are that the fuel has constant kinematic viscosity $v = \mu/\rho = 8$ mm^2/s, with density $\rho = 813$ Kg/m^3 and dynamic viscosity $\mu = 6.93 \cdot 10^{-3}$ N s/m^2. Moreover, the fuel is compressible and characterized by a Bulk modulus $K_f = 1.2 \cdot 10^4 \cdot [1 + p \cdot 10^{-3}]$, depending on the fuel pressure p. In standard conditions $K_f = 12000$ bar [14]. K_f represents the pressure increment for a decrease of a unitary volume, namely it holds:

$$\frac{dp}{dt} = -\frac{K_f}{V}\frac{dV}{dt} \tag{1}$$

where the fluid volume V changes by the motion of the mechanical parts, i.e. plunger and needle, and its time variation depends on the intake and outtake flow rates. The flow is given by

$$Q = sign(\Delta p)\,c_d\,A_0\,N\,\sqrt{\frac{2\Delta p}{\rho}} \tag{2}$$

where A_0 is the flow section through the holes, N is the number of holes, and Δp is the pressure gradient across A_0. c_d is a discharge coefficient that is function of the difference between the actual and ideal flows, because the flow rate is reduced by large pressure gaps and narrow orifice sections.

Moreover, the leakage is considered between coupled mechanical elements that are in relative motion and lubricated by the fluid. The leakage flow rate Q_l is proportional to the pressure drop

$$Q_l = \frac{\pi\,d\,g^3}{12\,\rho\,l\,v}\Delta p \tag{3}$$

where d is the mean diameter of the cross-section flow area, g is the radial gap, and l is the length of mechanical coupling. Then, (1) can be written as

$$\frac{dp}{dt} = -\frac{K_f}{V}\left(\frac{dV}{dt} - Q_l + \sum_i Q_i\right) \tag{4}$$

The amount of injected flow rate strictly depends on the axial position of the needle tip. The position is changed by the deformation of the plunger-needle mechanical coupling, because of the high values and variation of the pressure acting on the plunger and the needle surfaces. Then, the reduction of the axial length of the coupling increases the injection flow section below the needle tip. The section is a nonlinear function of the pressure acting on the mechanical parts, whose deformation depends on peculiar properties of the composing steel.

The dynamical behaviour of the plunger-needle complex under high pressure can be represented by several interconnected mass-spring-damper systems. Namely, both the plunger and needle are basically a series of cylinders (i.e. 5 and 2 cylinders, respectively), each characterized by a different section and mass.

Then, every intermediate element of the mass-spring-damper model is assumed to include halves of the masses of two adjacent cylinders, while the first and last elements in the serial chain include only one half of the mass of the first and last cylinders, respectively.

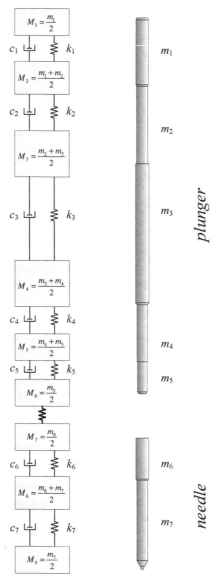

Fig. 2. Complete mechanical plunger-needle model

The link between different masses is achieved by spring and damping elements, that model elastic and viscous damping forces between the considered mechanical parts. More in details, the model of the plunger and the model of the needle are connected only by a spring representing the contact force. Figure 2 represents the complete mechanical model of the plunger-needle combination that is also shown.

Then, the main problem is to represent a continuous mechanical system by the mentioned connections and to determine the optimal parameters that allow the reproduction of an injected flow rate corresponding to the experimental trend that is typically obtained.

A free body equation can be derived by applying the Newton's second law to each mass of the mass-spring-damper model.

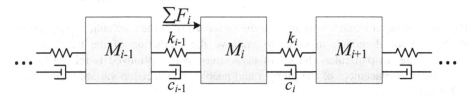

Fig. 3. Representation of adjacent masses

With reference to Fig. 3, for each mass of the chain it holds:

$$M_i\,\ddot{x}_i + c_{i-1}\,(\dot{x}_i - \dot{x}_{i-1}) + c_i\,(\dot{x}_i - \dot{x}_{i+1}) + k_{i-1}\,(x_i - x_{i-1}) + k_i\,(x_i - x_{i+1}) = \sum_i F_i \qquad (5)$$

where the symbol c denotes a viscous damping coefficient, k denotes the spring elastic constant of the ideal elements located between each mass, and $\sum_i F_i$ is the resultant of pressure forces acting on the half modeled cylinder.

Table 1. Theoretical values of the parameters to optimize

Element	Elastic constants [N/m]		Damping coefficients [Ns/m]	
Plunger	$k_1 = x_1$	$1.2517 \cdot 10^8$	$c_1 = x_8$	6.6188
	$k_2 = x_2$	$8.3272 \cdot 10^7$	$c_2 = x_9$	3.5343
	$k_3 = x_3$	$8.4554 \cdot 10^7$	$c_3 = x_{10}$	5.8156
	$k_4 = x_4$	$9.7191 \cdot 10^7$	$c_4 = x_{11}$	2.7884
	$k_5 = x_5$	$9.7116 \cdot 10^7$	$c_5 = x_{12}$	1.3937
Needle	$k_6 = x_6$	$1.9175 \cdot 10^8$	$c_6 = x_{13}$	5.3631
	$k_7 = x_7$	$5.9897 \cdot 10^7$	$c_7 = x_{14}$	2.9974

From a theoretical point of view, the spring constant is given by $k = E\,A\,/\,l_0$, where E is the Young's modulus, A is the cross section area of the considered cylinder, and l_0 is the initial length of the element. Moreover, the damping coefficient is given by

$c = 2\xi\sqrt{k\,m}$, with $\xi \in [0.001, 0.01]$. In particular, $\xi = 0.005$ is used. The forces applied by a pressure depend on the area of the section perpendicular to the motion direction. The following table provides the elastic constants and the damping coefficients that are determined based on theoretical formulas that consider geometric dimensions and Young's modulus. These values are used as a basis for the optimization.

To sum up, the injected flow rate is influenced by the deformation of the needle-plunger coupling. Thus, in this paper, the attention is focused on the optimization of parameters of the mechanical model of the plunger-needle coupling (Table 1).

4   Parameters Optimization by Evolutionary Strategies

The aim is now to identify and analyze the influence of some mechanical parameters of the dynamical model, in order to improve the prediction of the trend of the injection rate. In particular, the set of parameters to be optimized is related to the mechanical displacement of the plunger and needle. The decision variables are listed in Table 1.

The optimization problem can be expressed in the following form:

$$\min_{x \in X}\left\{ J(x) \in R \mid g(x) = 0, h(x) \geq 0 \right\} \tag{6}$$

where $J(x)$ is an objective function to quantify the model prediction accuracy, and $g(x)$ and $h(x)$ are the vector of constraints, whose components are defined in the following.

Starting from a nominal set of design values, denoted by $x_{i,ref}$, $i = 1, \ldots, 2n_p$, with $2n_p$ being the number of parameters, it is imposed that each decision variable must stay within the percentage range $[-\alpha_x\%, +\alpha_x\%]$, where α_x is a parameter that is properly set at the beginning of the optimization process. Thus, the elements of the vector of constraints can be expressed as:

$$\begin{cases} g_i(x) = x_{i+n_p} - 2 \cdot 0.005 \cdot \sqrt{x_i\,m_i} & i = 1, \ldots, n_p \\[2mm] h_i(x) = x_i - \left(1 - \dfrac{\alpha_x}{100}\right) x_{i,ref} & i = 1, \ldots, 2n_p \\[2mm] h_{i+2n_p}(x) = \left(1 + \dfrac{\alpha_x}{100}\right) x_{i,ref} - x_i & i = 1, \ldots, 2n_p \end{cases} \tag{7}$$

In the considered injection system, $n_p = 7$ (see Table 1). The optimization procedure is based on experimental data obtained from a test bench. More in details, the so-called EVI (an injection rate indicator) profile is assumed as a reference. This indicator represents the injection flow rate as a function of time, when exciting the electro-injector by a pre-defined driving current. The CR pressure is set to 1600 bar during the tests. The actual output is sampled with a frequency of 1 kHz, which is a suitable value for an accurate representation of the injection. The experimental dataset consists of 520 points for a time interval of 1.3 ms. Then, the objective function can be expressed as:

$$J = \sum_k \left| y_{\text{mod}}(k) - y_{\text{exp}}(k) \right| \tag{8}$$

where $y_{\text{exp}}(k)$ is the experimental value at sampling time k, and $y_{\text{mod}}(k)$ is the corresponding model predicted value.

To define the optimization technique, a background is preliminary recalled on the basics of the intelligent parameter optimization techniques. If $\mathbf{x} = [x_1, ..., x_n] \in R_n$ is the vector of real parameters, S is the search space including the optimized solution $\mathbf{x}^*$ and constraining the optimization procedure, and $f(\cdot)$ is an objective function to minimize, then the problem is to find $\mathbf{x}^*$ such that $f(\mathbf{x}^*) \le f(\mathbf{x})$ for every $\mathbf{x} \in S$. Since analytical solutions are difficult to find for several reasons (nonlinearity, number of parameters, lack of continuity and differentiability of f), heuristic and stochastic search are often convenient to use.

Differential Evolution (DE) improves a population of candidate solutions in order to get a solution as close as possible to the optimum. Each candidate solution is a vector of unknowns, i.e. $\mathbf{x} = [x_{1,i,g}, ..., x_{n,i,g}] \in R_n$, that represent the n parameters to be optimized. The method builds a population of solutions ($i = 1, ..., N_{\text{pop}}$) that is iteratively evolved through successive generations ($g = 0, 1, ..., G_{\text{max}}$). Evolution starts from an initial guess ($g = 0$) and goes on until the optimum is obtained or a stop criterion is met. To simplify, the procedure is stopped after a pre-fixed number G_{max} of generations. To synthesize, DE is characterized by the population size, N_{pop}, and the number of generations, G_{max}, that are usually determined by trial-and-error.

More in details, each generation undergoes the phases of mutation, recombination, and selection. Firstly, solutions are mutated to explore the search space. Secondly, solutions are recombined by crossover to increase the potential diversity in the population while including the best solutions from the previous generation. Finally, the best obtained solutions are selected. The population size is maintained constant and the search space is defined by proper limits: $\mathbf{x}_{\text{min}} \le \mathbf{x}_{i,g} \le \mathbf{x}_{\text{max}}$, where $\mathbf{x}_{\text{min}} = [x_{1,\text{min}}, ..., x_{n,\text{min}}]$ and $\mathbf{x}_{\text{max}} = [x_{1,\text{max}}, ..., x_{n,\text{max}}]$ depend on the specific optimization problem.

Evolution is initialized by taking a random generation from a uniform distribution

$$x_{j,i,0} = x_{j,\text{min}} + rnd_j \left(x_{j,\text{max}} - x_{j,\text{min}} \right) \tag{9}$$

for $j = 1, ..., n$ and $i = 1, ..., N_{\text{pop}}$, with $rnd_j \sim U(0,1)$ being a random number between 0 and 1 taken from a uniform distribution, so that the search space is covered at best.

The mutation creates a new solution as follows:

$$v_{i,g} = x_{r_{1i},g} + F \left(x_{r_{2i},g} - x_{r_{3i},g} \right) \tag{10}$$

for $i = 1, ..., N_{\text{pop}}$, with the scaling fractional factor $0 < F < 2$, and mutually exclusive integers $r_{1i} \ne i$, $r_{2i} \ne i$, $r_{3i} \ne i$, with $1 \le r_{ki} \le N_{\text{pop}}$ ($k = 1, 2, 3$).

Binomial crossover is used to recombine solutions [13] and obtain a new vector $\mathbf{u}_{i,g}$ = $[u_{1,i,g}, \ldots, u_{n,i,g}]$ by applying the following rule:

$$u_{j,i,g} = \begin{cases} v_{j,i,g} & \text{if } rnd_{j,i} \leq CR \quad \text{or} \quad j = j_{rand} \\ x_{j,i,g} & \text{if } rnd_{j,i} > CR \quad \text{and} \quad j \neq j_{rand} \end{cases} \quad (11)$$

where the crossover ratio $0 < CR < 1$ and $rnd_{j,i} \sim U(0,1)$, $j_{rand} \in \{1, \ldots, n\}$ is taken randomly from the uniform distribution so that $\mathbf{u}_{i,g}$ inherits at least one component from $\mathbf{v}_{i,g}$ and is different from $\mathbf{x}_{i,g}$.

The final operation is a selection of the solutions based on the minimization of the objective function f

$$\mathbf{x}_{i,g+1} = \begin{cases} \mathbf{u}_{i,g} & \text{if } f(\mathbf{u}_{i,g}) \leq f(\mathbf{x}_{i,g}) \\ \mathbf{x}_{i,g} & \text{if } f(\mathbf{x}_{i,g}) < f(\mathbf{u}_{i,g}) \end{cases} \quad (12)$$

To conclude, it is remarked that the speed of convergence of DE depends also on the value of F (usually selected between 0.4 and 1) and of CR (greater than 0.6 in many cases).

5   Simulation Results

The model prediction performance can be evaluated by comparing the model results and the experimental output in terms of the injected flow rate. Moreover, the simulation model is compared to: a) a non-optimized model in which the parameters are set to the constant values given by theoretical formulas (see Table 1); b) another simulation model that was recently obtained for the injector system by considering the plunger-needle complex as a unique rigid body [12].

However, the last model considered several factors: the connection between the control chamber and the low-pressure circuit by using the method of characteristics; the variation of density, Bulk's modulus, and kinematic viscosity with pressure; the deformation of the plunger-needle complex as a result of the compression forces.

Table 2 shows the optimized values of the parameters.

Table 2. Optimized values of the parameters

Element	Elastic constants [N/m]		Damping coefficients [Ns/m]	
Plunger	$k_1 = x_1$	$1.0868 \cdot 10^8$	$c_1 = x_8$	7.8334
	$k_2 = x_2$	$7.5901 \cdot 10^7$	$c_2 = x_9$	6.9801
	$k_3 = x_3$	$1.2568 \cdot 10^8$	$c_3 = x_{10}$	6.3442
	$k_4 = x_4$	$2.7072 \cdot 10^7$	$c_4 = x_{11}$	5.4278
	$k_5 = x_5$	$2.6332 \cdot 10^7$	$c_5 = x_{12}$	2.5198
Needle	$k_6 = x_6$	$2.5417 \cdot 10^8$	$c_6 = x_{13}$	2.7257
	$k_7 = x_7$	$2.5716 \cdot 10^7$	$c_7 = x_{14}$	1.3293

Figure 4 shows the optimization results for the testing dataset. The plot obtained from experimental data is compared to that obtained from the three mentioned simulation models. Note that, to improve optimization results, the data associated to the phase (for $t > 1.5$ ms) in which the injected rate is dramatically reduced are filtered out, because a very complex unmodeled fluid dynamics occurs and it is difficult to adapt the model predictions to this phase.

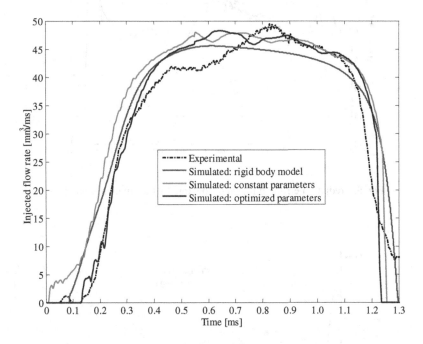

Fig. 4. Prediction of injected flow rate by a simulation model and experimental trend

A first remark is that the parameter-based models greatly outperform the rigid body model, that produces less oscillations and a smoother response but with much higher errors in the particular rising phase (roughly for $0.2 \leq t \leq 0.6$), in the intermediate period ($0.9 \leq t \leq 1.2$), and in the descending phase ($1.2 \leq t \leq 1.4$). Moreover, if the initial rising front of the flow rate ($0.3 \leq t \leq 0.5$) or the descent ($t \geq 1.1$) are considered, it is clear that the optimized model improves the non-optimized one with constant theoretical parameter values. Finally, the increase in error for $0.6 \leq t \leq 0.9$ is due to some phenomena related to pressure wave propagation, that are not represented by lumped parameter equations used herein.

To check these results, Figure 5 shows the errors between the simulated responses and the experimental one.

Finally, Figure 6 depicts the time evolution of the objective function. Both the best values and the mean values in each generation are indicated. It can be verified that the evolutionary optimization converges after few generations.

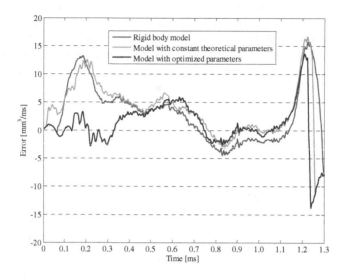

Fig. 5. Prediction errors of injected flow rate by the simulation models

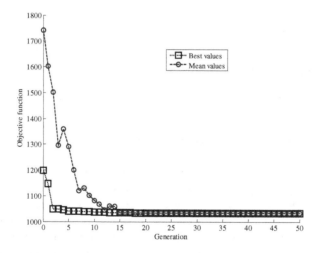

Fig. 6. Objective function

6   Conclusion

In this paper, an accurate model is introduced for an electro-injector in a common rail Diesel engine. The model takes into account both the non-linear phenomena associated to fuel flow and pressure variation in the volumes in which the injector is divided and the mechanical deformation of the main part affecting the flow section in the injection process, i.e. the plunger-needle coupling.

An evolutionary strategy was used to optimize the parameters affecting the key variable, the injected flow rate. The optimized model was compared to a non-optimized model based on constant values of the parameters obtained by closed formulas. Moreover, another model taken from recent studies was considered which assumes that the plunger and needle behave as a whole rigid body. Results indicate that optimization helps to improve the prediction performance from experimental data. Future work will further investigate and improve the optimization method, explore more performance indicators, and design innovative control strategies.

References

1. Guzzella, L.: The Automobile of the Future – Options for Efficient Individual Mobility. In: Plenary Talk, 3rd IEEE Multi-Conference on Systems and Control (MSC 2009), Saint Petersburg, Russia, July 8-10 (2009)
2. Dyntar, D., Onder, C., Guzzella, L.: Modeling and control of CNG engines. SAE Paper 2002-01-1295 (2002)
3. Lino, P., Maione, B., Amorese, C.: Modelling and predictive control of a new injection system for compressed natural gas engines. Control Engineering Practice 16, 1216–1230 (2008)
4. Heywood, J.: Internal combustion engine fundamentals. McGraw-Hill, New York (1988)
5. Streeter, V., Wylie, K., Bedford, E.: Fluid mechanics, 9th edn. McGraw-Hill, New York (1998)
6. Kouremenos, D.A., Hountalas, D.T., Kouremenos, A.D.: Development and validation of a detailed fuel injection system simulation model for diesel engines. SAE Paper 1999-01-0527 (1999)
7. International Gas Union, Global opportunities for natural gas as a transportation fuel for today and tomorrow. Report on study group 5.3 "Natural gas vehicles (NGV) (2005)
8. Lino, P., Maione, G.: Switching fractional-order controllers of common rail pressure in compressed natural gas engines. In: 19th IFAC World Congress 2014, Cape Town, South Africa, August 24-29 (to appear, 2014)
9. Lino, P., Maione, G.: Design and simulation of fractional-order controllers of injection in CNG engines. In: Kawabe, T. (ed.) 7th IFAC Symposium on Advances in Automotive Control (AAC 2013), Tokyo, Japan, September 4-7, vol. 1(pt. 1), pp. 582–587 (2013)
10. Lino, P., Maione, G.: Fractional order control of the injection system in a CNG engine. In: Proc. European Control Conference (ECC 2013), Zürich, Switzerland, July 17-19, pp. 3997–4002 (2013)
11. Saponaro, F., Lino, P., Maione, G.: Fractional modeling the wave pressure propagation in Diesel injection systems. In: International Conference on Fractional Differentiation and Its Applications (ICFDA 2014), Catania, Italy, June 23-25 (to appear, 2014)
12. Saponaro, F., Lino, P., Maione, G.: A Dynamical Model of Electro-Injectors for Common Rail Diesel Engines. In: 22nd Mediterranean Conference on Control & Automation (MED 2014), Palermo, Italy, June 16-19 (to appear, 2014)
13. Price, K.V., Storn, R.M., Lampinen, J.A.: Differential Evolution – A Practical Approach to Global Optimization. Springer, Berlin (2005)
14. Lino, P., Maione, B., Rizzo, A.: A control-oriented model of a Common Rail injection system for diesel engines. In: Proc. of 10th IEEE Conference on Emerging Technologies and Factory Automation, Catania, Italy, September 19-22, pp. 557–563 (2005)

Author Index